KARST
IMPORTANT KARST REGIONS OF THE NORTHERN HEMISPHERE

KARST

Important Karst Regions of the Northern Hemisphere

EDITED BY **M. HERAK**
Geological and Palaeontological Institute,
Faculty of Natural Sciences, Zagreb (Yugoslavia)

AND **V.T. STRINGFIELD**
Department of the Interior, U.S. Geological Survey,
Washington, D.C. (U.S.A.)

ELSEVIER PUBLISHING COMPANY Amsterdam – London – New York 1972

ELSEVIER PUBLISHING COMPANY
335 JAN VAN GALENSTRAAT, P.O. BOX 211, AMSTERDAM, THE NETHERLANDS

AMERICAN ELSEVIER PUBLISHING COMPANY, INC.
52 VANDERBILT AVENUE, NEW YORK 10017

Library of Congress Card Number: 74-151736

ISBN 0-444-40849-5

With 173 illustrations and 8 tables.

Copyright © 1972 by Elsevier Publishing Company, Amsterdam

All rights reserved. No part of this publication may be reproduced, stored in a retrieval system, or transmitted in any form or by any means, electronic, mechanical, photocopying, recording, or otherwise, without the prior written permission of the publisher, Elsevier Publishing Company, Jan van Galenstraat 335, Amsterdam

Printed in The Netherlands

Preface

Karst areas with their specific morphologic and hydrogeologic features are widespread throughout the world. Since ancient times there has been a lively interest in deciphering the relationships of strange superficial erosional relics, ponors, dolines, vauclusian springs, caves, caverns, and so forth. The discovery of saw dust, leaves and some fresh-water animals suggested an entangled connection of underground water courses which at the beginning led to rather strange explanations. The first attempts to explain karst phenomena were recorded by ancient civilisations. But not until the nineteenth century was there any clear evidence of the origin of karst features. Until recently a descriptive method of presenting karst was dominant. With the improvement of detailed geologic investigations, karst relations have gradually come to be understood as the logical consequences of various geologic forces working under specific conditions.

Practical interest, especially in the areas of water supply and the positioning of power plants, has hastened the introduction of more exact methods in karst investigations. Every day new data are collected and new theories are put forth. However, new developments often do not reach the professional journals or are published in various reviews in different languages; therefore, they are not readily available to all who are interested in them. This fact led to the publishing of this book in which are presented karst relations and the problems of various countries having advanced theoretical and practical experience.

The number of countries was limited by the size of the book as well as by the willingness of the invited experts to cooperate. We endeavoured to present contributions covering all the main types of karst.

We are conscious that we have taken only the first step in preparing a common base for wider comparison and exchange of ideas for future theoretical and practical advances in the area of karst investigation and exploration. A proposed second edition may be the result of still greater cooperation in order to complete karst notions concerning the space, ideas, methods of investigation, and application. For that purpose any further suggestions would be very much appreciated.

MILAN HERAK (Zagreb)

Contents

Preface V

Chapter 1. Historical Review of Morphologic Concepts
J. ROGLIĆ

Introduction 1
Early investigators 1
Scientific approach 3
    First phase of karst morphology investigation 3
    Second phase of karst morphology investigation 9
Main characteristics of surface karst 15
References 17

Chapter 2. Historical Review of Hydrogeologic Concepts
M. HERAK and V. T. STRINGFIELD

Introduction 19
Early investigators 20
Modern theories 21
References 24

Chapter 3. Karst of Yugoslavia
M. HERAK

Introduction 25
Lithostratigraphic and sedimentologic aspects 26
Tectonic setting of Yugoslavia 29
Phases of karstification 30
Morphologic features 31
Hydrogeologic conditions 41
    The Adriatic insular and coastal region 42
    The high karst region 46
    The inner region 52
Practical problems and their approach 53
    Practical aspects of geology 54
    Application of geophysical methods 56

Practical aspects of speleology 58
Practical aspects of hydrology 59
Surface water storage basins in karst. 62
Geologic basis for groundwater development 67
Practical problems in water supply 68
Practical aspects of forestry in the Dinaric karst 74
Practical aspects of agriculture in the Dinaric karst 75
References 76

Chapter 4. Karst of Italy
S. BELLONI, B. MARTINIS and G. OROMBELLI

Introduction 85
Lithostratigraphic and sedimentologic aspects. 85
Tectonic setting. 88
Phases of karstification 90
Morphologic features 93
    Principal karst areas. 93
        The Italian Karst 95
        Plateaus of the Friulian, Venetian and Lombardian pre-Alps 98
        The region of Apulia 101
    Lesser karst areas 104
        Italian Alps 105
        Apennines 108
        Sicily and Sardinia 112
Karst phenomena in conglomerate 113
Karst phenomena in gypsum 115
Hydrogeologic conditions 116
Practical problems and their approach 124
References 125

Chapter 5. Karst of France
J. AVIAS

Introduction 129
    Lithostratigraphy 130
    Structure 131
    Climate 131
Karst phenomena in Palaeozoic, mainly Hercynian massifs (Ardennes, Vosges, Armorican Massif, Massif Central, Montagne Noire, Palaeozoic massifs in the Alps and the Pyrenees) 133
Karst phenomena in Mesozoic and Tertiary formations 135
    Introduction 135
    Karst phenomena in Mesozoic and Tertiary formations in the Paris Basin 136
        Mesozoic formations 136
        Tertiary formations 143
    Karst phenomena in Mesozoic and Tertiary formations in the Aquitanian Basin . . 144
        Quercy 144
        Périgord and Pays des Charentes 145
    Karst phenomena of the Grands Causses 146
        Lithostratigraphy 148
        Structure 148

Climate	148
Karstology and hydrogeology	149
The Languedocian karst	153
Setting	153
Lithostratigraphy	153
Structure	153
Climate	154
Main characteristics	156
Karst in Minervois, Corbières and the Pyrenees	160
Lithostratigraphy	160
Structure	160
Karstification phases	160
Main natural units	161
Karst in Burgundy and the Jura Mountains	164
Lithostratigraphy	164
Structure	164
Phases of karstification	165
Main characteristics	165
Alpine karst and Provence karst	167
Lithostratigraphy	167
Structure	168
Karstification phases	168
Main characteristics	169
Conclusions: scientific and practical interest, resources and exploitation of French karst	180
Scientific interest	180
Geologic and hydrogeologic problems	180
Erosion and concretion mechanisms	181
Biologic and prehistoric questions	181
Practical interest	182
Tourism	182
Water supply	182
Further aspects	183
References	183

Chapter 6. Karst of Germany
D. Pfeiffer and J. Hahn

Introduction	189
Stratigraphic and sedimentologic aspects	192
Geologic development of Germany	192
Precambrian	192
Palaeozoic	192
Mesozoic	194
Tertiary	195
Rocks subject to karstification	195
Carbonate sediments	196
Saline sediments (sulphate and chloride)	197
Geologic and tectonic setting of the karst areas	198
Variscan basement	198
Thuringian Basin	200
The cuesta landscape of Swabia and Franconia	205
Westfalian Cretaceous embayment (Münsterland Basin)	208
The area of the Saxonian orogeny	209

Phases of karstification 210
    Karst types 210
        Limestone karst. 211
        Gypsum karst 212
        Saline karst 213
    The age of karstification 215
        Limestone karst. 215
        Gypsum karst 215
    Soils in karst areas 216
Hydrologic consequences 216
    Shallow karst. 218
    Deep karst 219
Problems of engineering geology in karst and leaching areas. 220
References 221

Chapter 7. Karst of Austria
F. BAUER and J. ZÖTL

Introduction 225
The Northern Limestone Alps 226
    General description 226
        Stratigraphy and lithology 226
        Tectonics 228
        The morphologic development of the Limestone Alps 228
        Main types of the North Alpine karst 229
        The development of the North Alpine karst until the end of the Glacial Epoch. . 230
        Principles of hydrology 232
        Caves 234
        Surface karst phenomena 236
    Regional examples of North Alpine karst 239
        Schneebergalpen 239
        Totes Gebirge 241
        Dachstein 241
        Hochkönig. 244
        Steinernes Meer. 244
        Karwendel Gebirge 247
        Pre-Alps. 251
The Central Styrian Karst 252
Karst of the Southern Alps. 257
Karst of the Helvetian nappe system 258
Karst of the Waschberg zone 260
Karst of the Central Alps. 260
Economic karst problems. 260
    Forestry and agriculture 260
    Water supply and hydro-electric power plants 261
Conclusions 263
References 263

Chapter 8. Karst of Hungary
F. Darányi

Introduction	267
Lithostratigraphic and sedimentologic aspects	267
Transdanubian Central Mountains	267
Northeast Range	272
Mecsek Mountains	274
Tectonic setting of the area	276
Transdanubian Central Mountains	276
Northeast Range	279
Mecsek Mountains	280
Phases of karstification	281
Transdanubian Central Mountains	281
Northeast Range	282
Mecsek Mountains	282
Morphologic features	282
Hydrologic conditions	284
Transdanubian Central Mountains	288
Northeast Range	290
Mecsek Mountains	291
Practical problems and their approach	292
References	294

Chapter 9. Karst of Czechoslovakia
J. Bystrický, E. Mazúr and J. Jakál

Introduction	297
Conditions of development of the karst phenomena	298
Geologic conditions	298
Relief and its reflection in the differentiation of the karst phenomena	307
The other physical-geographical conditions	309
Typological division of the karst in the West Carpathians	312
Central European karst of the temperate zone	314
Plateau karst – complete karst	314
Articulated karst – imperfect karst	318
Sporadic karst of the Klippen structure	322
Alpine karst	323
References	323

Chapter 10. Karst of Poland
J. Glazek, T. Dabrowski and R. Gradziński

Lithostratigraphic and sedimentologic aspects	327
Tectonic setting of the area	328
Phases of karstification	330
Morphologic consequences	333
Hydrogeologic consequences	336
Practical problems and their approach	337
Acknowledgements	338
References	339

Chapter 11. Karst of Rumania
M. D. BLEAHU

Introduction	341
Lithostratigraphic and sedimentologic aspects	341
Open-sea carbonate associations	342
The pelagic limestones	342
The mixed deposits of terrigenous and carbonate rocks	342
Shallow-water marine and marginal marine associations	342
Massive limestones or dolomites	344
Bedded limestones	344
Tectonic setting	344
Zones of pre-Alpine shields	345
Zones of Alpine folding in the Carpathians	345
Limy plateaus	345
Limy ridges	346
Isolated calcareous massifs	347
Post-tectonic Neozoic depressions	347
Phases of karstification and evolution of the karst areas	347
Morphologic features	349
Karst developed on salt and gypsum	351
Volcanic rock karst	351
Hydrogeologic conditions	352
Practical problems and their approach	352
References	353

Chapter 12. Karst of the U.S.S.R.
I. V. POPOV, N. A. GVOZDETSKIY, A. G. CHIKISHEV and B. I. KUDELIN

Introduction	355
The Russian Platform	358
Karst of the crystalline basement of the Russian Platform	359
Karst in rocks of the southern buried slope of the Baltic Shield	360
Karst of the Latvian saddle, and karst of the Belorusskaya Arch	360
Karst of the southern limb of the Moscow tectonic basin	361
Karst of the western and northwestern limb of the Moscow tectonic basin	362
Karst of the southeastern part of the Moscow tectonic basin	364
Karst of the Timan Ridge	364
Karst of the Volga-Ural Arch and the Vyatsk swell	365
Karst of the southwestern borderland of the Russian Platform and the outer zone of the Ciscarpathian Trough	366
Chalk karst in rocks of the Russian Platform	367
The Alpine folded zone in the southern and southwestern framework of the Russian Platform	367
The Crimea	367
The Caucasus	368
Karst of the eastern Carpathians	373
The Ural orogenic belt of the Hercynian syncline	376
The karst province of the west slope of the Urals	378
The karst province of the central Urals	381
The karst province of the east slope of the Urals	382
The trans-Ural karst province	384
The Turansk Platform	385
The Siberian Platform	387

CONTENTS

The orogenic belt of the eastern part of Central Asia, with a folded Palaeozoic basement . . . 395
Karst of the Trans-Baikal region 399
Karst of Palaeozoic fold structures of southern Siberia 401
    The Altay karst province 401
    The Tomsk-Kolyvansk karst province 403
    The Kuznetsk-Alatausk karst province 403
    The karst province of western and eastern Sayan 404
Underground flow in karst regions of the U.S.S.R. 405
References 411

Chapter 13. Karst of Great Britain
M. M. Sweeting

Introduction 417
The karst areas of the Carboniferous limestone 419
    Stratigraphic, sedimentologic and tectonic aspects 419
        Northwest Yorkshire 420
        Morecambe Bay district 420
        Derbyshire 420
        The Mendips 423
        South Wales 424
        North Wales 424
    Phases of karstification 424
    Morphologic features 425
        Northwest Yorkshire 425
        Morecambe Bay district 431
        Derbyshire 432
        The Mendips 433
        South Wales 435
        North Wales 436
The karst areas of older Palaeozoic limestones 436
    The Durness Limestone area 437
    The Devonian limestone area 438
Hydrogeologic conditions with special reference to the Carboniferous limestone area 438
Practical problems and their approach 439
    Problems of water supply 439
    Problems of land use 440
References 442

Chapter 14. Karst of Jamaica
H. R. Versey

General geology 445
    The Montpelier Chalks 447
    The recrystallized limestones 448
    The rubbly limestones 448
    Tectonic setting 449
Phases of karstification 450
Hydrology 451
Geomorphology 458
    The karst uplands 459

Interior valleys (poljes). 462
        The Nassau Valley 462
        The interior valleys of northern Trelawny 463
Practical aspects. 464
References 466

Chapter 15. Karst of the United States
W. E. Davies and H. E. LeGrand

Foreword 467
General areas. 469
Atlantic and Gulf coastal plain 473
    Abstract 473
    Geologic setting. 474
    Karst features 476
    Hydrogeology 477
    Practical considerations 480
Appalachians 481
    Introduction 481
    Great Valley 484
    Valley and Ridge section. 487
    Appalachian plateaus 490
    Other karst areas 494
Central Kentucky and southern Indiana 495
The Pecos Valley of New Mexico 501
Other karst regions 502
References 504

Chapter 16. Conclusions
M. Herak and V. T. Stringfield

Terminology, genesis and classification 507
Hydrology and hydrogeology 512
    Carbonate aquifers 514
    Hydrogeology and palaeohydrology 515
    Geologic structure. 516
    Methods of investigation. 517

References Index 519

Subject Index 531

Chapter 1

Historical Review of Morphologic Concepts

J. ROGLIĆ

Geographical Institute, University of Zagreb, Zagreb (Yugoslavia)

INTRODUCTION

Karst includes the circulation of water in fractures, fissures, joints and other cavities, and the development of corresponding forms in soluble, mostly carbonate rocks. The process and the produced forms reach to great depths in the rock complexes, but, in this review, as in geomorphology in general, it is mainly the surface, i.e., the visible forms, which will be treated. Since the surface features cannot be understood without knowledge of the process as a whole, the connection between them will be adequately respected. Analogous superficial forms also occur in some other deposits, e.g., in loess; these are called "pseudokarst", and will not be a subject of this article.

Limited space, the scope of the subject, and clearness required suggest a concentration on the essential problems, and on those which are interconnected in the historical development. Several papers are omitted, and those cited are used only for documentation of the explanation. This review is only an introduction to the up-to-date scientific endeavours, which are presented only by means of the basic data.

EARLY INVESTIGATORS

Like geomorphology in general, the scientific investigation of karst relief is very young, in fact the term "geomorphology" was first mentioned by NAUMANN [1858]; the first textbook on the subject was written by PENCK [1894]. Only by means of the first results of the mapping and on the basis of more reliable investigations at the end of the 19th century, were the global relief features established. At that time the basic data concerning the composition and structure of the earth were known. In connection with practical works, and especially with the construction of railways, features of relief became important elements in the natural inventory. In the course of acquiring new ideas and of solving practical problems the Alps occupied a central position and were of predominant significance.

It is normal that perialpine Vienna, at that time one of the leading scientific centers in the world, also became the center of geomorphologic investigations, especially those concerning the karst. A second important fact was that the Dinaric karst was the main source of ideas and that it thus influenced the formation of concepts. Admittedly there are in the surroundings of Vienna the Moravic karst and some calcareous areas in the Alps, but with respect to the extension and variety of their karst features, they can hardly be compared with the Dinaric karst. However, the comparative study was highly important for defining and improving scientific thought.

The classical connection of karstology with the Dinaric karst was not accidental but was dependent on the proximity of the scientific center in Vienna. Even in the present-day global endeavours, the Dinaric karst is still the most complex and attractive, and is still regarded as a classical karst area both in its content and in historical significance. Its importance is also emphasized by the many scientific terms which, regrettably, are not always correctly used.

Dinaric karst has remained almost until recent times in the shadow of cultural events, its existence being "discovered" at the beginning of this century. However, a small, narrow and marginal area between the northern part of the Adriatic Sea and the Pannonian Basin, was an important passage and a stage for interesting events and movements leading to the prehistoric legend of the Argonauts who might have come through this area from the Ister (Danube) to the Adriatic Sea. This karst threshold is exceptionally rich in specific features which attracted early attention and which required explanation. It is in this most peculiar part of the classical karst that the term "karst" has originated, based on the Indo-European word "kar" = rock. The Italian term is "carso", the Slovenian word is "kras", while "karst" is a Germanized form of the original expression. With the development of Trieste, the classical karst area got the name "karst of Trieste".

The impressive karst forms, the facts based on them, and the richness of the folk legend inspired VALVASOR [1689], a master of drawing and descriptions, to publish the legends and beliefs concerning karst which are still alive. Local people proceeded even further, in trying to explain the origin of some features. GRUBER [1781] was especially interested in long-lasting inundations of Cerkniško polje in Slovenia, considering these impressive "jamas" to be a consequence of collapses. His contemporary HACQUET [1778] gives useful data and observations on karst features. He considers dolines a consequence of the weathering of limestone. These observations were too early for the general scientific climate, and the country was outside the main flow of cultural aspirations and decisive events. Therefore, these problems had to wait during seven more decades before again becoming the subject of investigations. The French scientist VIRLET [1834] regarded the dolines in the southern Jura Mountains as the effects of collapse of the roofs of subsurface cavities, i.e., he followed the former idea of Gruber. Hacquet's idea was a forerunner of LYELL's [1839] explanation of the genesis of the "geologic organs" in the chalk of southern England by means of the solution process by atmospheric water. In France, FOURNET [1852] also explained the dolines in connection with collapses.

SCIENTIFIC APPROACH

Since the middle of the 19th century (1850–1857), when a railway was constructed over the karst threshold from middle Europe to Trieste, the karst area has become a feature of modern life. Civil engineers of the Southern Railway were confronted with karst reality which required explanation and practical measures. Karst then became an object of scientific interest.

The Viennese speleologist SCHMIDL [1854] investigated caves in the karst of Trieste with good support and for practical purposes. He maintained, in accordance with predominant views, that dolines were formed as a result of collapses. In addition to speleologic explanations, other opinions were very numerous. The English geologist PRESTWICH [1854] agreed with Ch. Lyell's explanation that dolines were formed by the infiltration of water.

Shortly after, analogous explanations were published by American geologists. OWEN [1856] studied the karst of Kentucky (U.S.A.), whilst Cox [1874] expressed the opinion that the dolines in the karst of Indiana were produced by the solution of limestones. He recorded joints through which water infiltrated. Of special importance are the investigations of the English geologist SAWKINS [1869] in Jamaica; according to his opinion conspicuous dolines (cockpits) originated due to the solution of limestones.

First phase of karst morphology investigation

A new and more complex approach to karst problems coincided with the first geologic investigations of the middle and most complicated part of the Dinaric karst, following the occupation of Bosnia and Hercegovina (1878). The leading geologists in Vienna, E. Mojsisovics, E. Tietze and E. Bittner wrote about the geologic relations of that area and gave a scientific explanation for the genesis of the karst forms [MOJSISOVICS et al., 1880]. E. Tietze supported the collapse theory. MOJSISOVICS et al. [1880] pointed out that folding and erosional forces were factors in the development of karst depressions. They compared the Dinaric dolines with the Alpine karren. Thus, Hacquet's forecast and the practice of several English and American geologists were introduced into the explanation of the Dinaric karst. DIENER [1886] explains the dolines of the karst of the Lebanon on the basis of solution. Carbon dioxide was already commonly taken into account. In a lively discussion among the Viennese geologists, the advocates of the collapse theory were in the majority, constantly introducing new contributions into the speleologic investigations, especially those of the karst of Trieste. Most prominent among them was KRAUS [1887, 1894].

At the same time geomorphology was evolving as a separate discipline concerning karst relief. In Vienna the most prominent protagonist of the new discipline was A. Penck, respected in the whole world thanks to his observations and investigations of Alpine glaciation. The master suggested to his talented student J. Cvijić, that he should prepare a doctor's dissertation on the subject of karst.

At first Cvijić was occupied with the karst of eastern Serbia. But his main ideas were based on the observations on the karst of Trieste.

CVIJIĆ's work "*Das Karstphänomen*" [1893] was a turning-point and at the same time the beginning of an intensified study of karst morphology. He also paid the most attention to dolines, "which give to the karst a special landscape character". In a cutting of the railway for Trieste, he recorded the connection between the formation of dolines and joints as Cox [1874] had done. "Normal dolines" originated under the influence of water which percolated from the surface into the interior, enlarging joints and fissures into "funnels". He argued against the generally accepted importance of collapse of the roofs of caverns. He probably considered karst as primary relief on soluble rocks. This work of Cvijić was not only a pioneering monograph on karst, but also on geomorphology in general. Its special significance lies in the fact that as a result attention was paid to the problems of karst in Penck's world-renowned school of geomorphology.

In 1899, Penck, together with his students, and in company with the American geomorphologist W. M. Davis, gave instruction on field trips in Bosnia and Hercegovina. This meeting in the karst area of two outstanding masters of geomorphology and their discussions were of decisive importance for the further development of geomorphologic concepts.

PENCK [1900] and DAVIS [1901] pointed out conspicuous plains in the karst of Hercegovina and Dalmatia. According to general views and the logical deductions of DAVIS [1884], the formation of the plains was ascribed to denudation and fluvial erosion. Thus, two more schematic views entered into the explanation of karst relief: (*a*) fluvial erosion occurred first, followed by "karstification"; and (*b*) stages of morphologic evolution or of cyclic evolution became obtrusive. Karstification and cyclic evolution dominated the explanations of karst relief during the next four decades.

The reputations of Penck and Davis, as well as their convincing concepts, had also influenced Cvijić, already a leading researcher in karst morphology. CVIJIĆ [1901] accepted the priority of fluvial erosion and used the term "karstification". Among karst features he recorded cyclic relation of evolution, whereby the coalescence of dolines creates uvalas and by their coalescence poljes are formed as the final and largest karst features.

Penck's school was active in developing theoretical treatments of new concepts. GRUND [1903], a disciple of Penck, published a work which meant a new milestone in the development of karst morphology. He accepted the existence of "karst groundwater" rising progressively from sea level towards the hinterland. Above this stagnant level, during the wet seasons, joints and fissures are periodically filled up by water called "karst water" which, being unstable, discharges again very soon. Morphologic evolution occurs toward "karst groundwater" with the final results which represent plains, in the sense of the cyclic theory of Davis. Such treatments were resolutely supported by PENCK [1904], whose own ideas were thus elaborated.

The concepts of Penck and Grund, deduced with "juridical logic", were followed by numerous disciples of the Viennese morphologic school (E. Richter, N. Krebs, etc.). Grund himself was especially active. Objections and antagonistic concepts were not strong and logical enough to counter-balance the Davis–Penck–Grund concept. Speleologists like KNEBEL [1906], MARTEL [1910] and several others pointed out that the investigations did not confirm the existence of a groundwater level in karst; moreover, individual water streams flowing along fractures and fissures were established. An especially resolute opponent of the Penck–Grund concept was geologist KATZER [1909], who knew the geology of the Dinaric karst very well and was skilled in practical knowledge. According to him water circulates in karst in separate systems of channels instead of interconnecting channels of groundwater with a water table. However, his knowledge of geomorphology was limited.

Although its opponents could not push the Penck–Grund concept aside, they did call attention to its weak points. It was obvious that the evolution of surficial karst features can not be explained without taking into account the circulation of waters through fissures and fractures. This inspired Cvijić to look for a specific explanation of the genesis of the plains which form the basis of the Penck–Grund concept.

According to CVIJIĆ [1909], the plains are formed by means of specific "karst erosion". Surficial erosion is a consequence of sinking water, while in the underground, karst rivers cut in, wash out, and enlarge the fissures. This double action characterized the erosion and planation of karst. His considerations were generalized, and without concrete and illustrative examples. He did not discuss the question of whether the plains were formed independently, or by dislocation of an originally uniform level, although the last may be concluded from the headline of the study. Afterwards he did not return to this specific explanation, but he accepted more schematic ideas of fluvial formation of the plains in karst, which were indicated by Penck and Grund. Such a procedure is curious, because in the same year SAWICKI [1909] published an analogous explanation of plain formation in calcareous terrains.

Sawicki maintained that "terra rossa", which remains after the solution of limestones, closes joints and fissures in karst making a continuous cover, i.e., "Evolutionsniveau", which prevents the sinking of the water and dominates its horizontal flow. His conclusions and deductions are based on the conditions which were recorded in the Slovakian karst. It is noteworthy that not so long ago BIROT [1954a] returned to this explanation of plains, as proposed by Cvijić and Sawicki.

In 1910 GRUND [1910] published his work on the karst relief of the Dinaric Mountains. Very skilled in constructions and deductions he argued against objections to his theory concerning the groundwater level in the karst. On the basis of his observations in the karst of Bosnia and Hercegovina, he delivered his concepts on the evolution of the main karst features and especially of poljes and plains on limestones. He ascribes the formation of poljes to the tectonic movements distinguishing different types. With this work, and the arguments connected with it, the concept of Penck and

Grund concerning the evolution of karst relief gained new support.

In 1914 GRUND [1914] completed his concept on the evolution of karst relief, and herewith ended the activity of this ambitious and skillful scientist, who found his tragic end at the beginning of World War I. Following Cvijić, he considered the doline a basic feature ("Leitform") of karst, like the valley in fluvial relief. Deepening and evolution of karst forms occur towards the level of the groundwater which represents the "basis of corrosion". In the early stage, the dolines are separated by ridges. In the mature stage, the ridges are reduced to their most resistant parts, protruding in the form of isolated hills among which there is present a labyrinth of karst depressions. He considered "Cockpit Country" (Jamaica), described by DANEŠ [1908], to be a good example of such a mature relief. In the later stage, plains formed by corrosion and accumulation with isolated hums which are lowered by solution processes predominate. In this way the "corrosional peneplain" is finally formed. In fact, GRUND [1914] accepted Cvijić's concept (then 14 years old), but he brought the evolution to a causal connection with the uniform groundwater, which does not end within the poljes, but continues to the open plains.

Grund's explanation of cyclical evolution led to the end of the period of the creation of general concepts and schemes supported by relatively few observations. This coincided with the beginning of the World War I, as a result of which systematic field investigations and the fruitful exchange of views ceased.

Before the war a significant work was published by a Hungarian geologist, TERZAGHI [1913], but its echo had to wait two decades before being realized. Terzaghi carried out geologic investigations in the Gacko polje in west Croatia in connection with the planning of a storage basin for the water of the river Gacka, to be utilized for a power plant at Senj. Being the result of detailed and complex observations, his work differs from previous work in that he recognized and recorded some features which were overlooked earlier. He established that the bottom of the Gacko polje is a calcareous plain which could not have been formed by fluvial erosion, because the depression is closed without trace of younger tectonic dislocations. According to his opinion, the plain had been modelled in karst depressions, reaching to the groundwater level. At this level lakes were formed, and lacustrine clays which hindered the sinking of the water were deposited. At the margin of the level of inundation, there occurred denudation and enlargement of the plain. Though Terzaghi did not determine the very nature of the denudation in question, it is important that he pointed out the necessity of a specific explanation of the formation of isolated plains, and that he introduced the idea of "marginal denudation", or corrosion. The plains in limestones are not only horizontally isolated, but they can also be formed at various elevations depending on local levels of inundation. Terzaghi also observed the intensive solution of limestones in wooded areas, explaining it as being caused by biotic processes.

During the World War I Cvijić lived in France, teaching karst. There, he prepared his most complex work [CVIJIĆ, 1918] a synthesis of his views of the evolution of karst relief which was published in 1918. He treated water circulation and the development of karst features in common, together with emendations of earlier views. Cvijić accepted the idea of groundwater and abandoned his earlier idea of the priority of karst process. A short fluvial phase was the forerunner of karstification, causing the formation of large plains! The karst process occurs in connection with sinking water, vertically through the whole calcareous mass, regardless of sea level. The karst process goes on to the impervious base and until the complete solution of the calcareous rocks (Fig. 1). Karst waters discharge into the sea along the "line of hydrostatic balance", which is determined by the relation of the pressure of the sea water and that of fresh water coming from the karst interior. Recharge of atmospheric water causes the fluctuations of the groundwater. As a result of this, Cvijić distinguished three "hydrographic zones" of karst, i.e.: (a) a zone of permanent inundation; (b) a zone of periodical inundation; and (c) a dry zone. This reminds one of some of the views of PENCK [1904], and can be considered a logical consequence of Grund's concept. Cvijić, also, pointed out the significance of climate, distinguishing between "northern" and "Mediterranean" types of karst. This had already been described in his "*Karstphänomen*".

World War I caused not only the end of a phase of morphologic investigation of karst, but also a turning-point with essential changes in the general circumstances influencing the evolution of geomorphologic concepts. The defeat of the Austro-

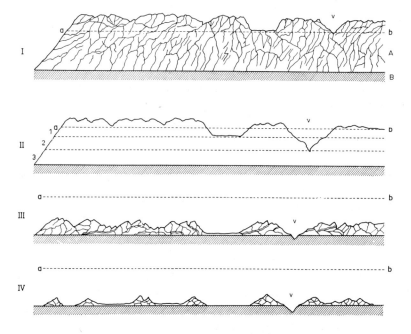

Fig.1. Evolution of water circulation and development of karst relief. *I, II, III* and *IV* are phases of karst development. A = Calcareous cover; B = impervious base; a–b = lower limit of the first "hydrographic zone"; v = allogene river; 1 = dry zone; 2 = periodically inundated zone; 3 = permanently inundated zone. (After CVIJIĆ, 1918.)

Hungarian monarchy resulted in almost complete lethargy in the Viennese center of karst investigations, which had already lost very much by Penck's move to Berlin.

The leading karstologist Cvijić was exhausted by the problems of war; relatively alone and engrossed with numerous problems, he intended to complete his observations and assemble his ideas, already partially emendated during the war. However, he later returned to his earlier views, probably because GRUND [1914] also had turned to them. The karren, the smallest features of karst relief, according to CVIJIĆ [1924a], are developed in their own cycle, disappearing at the end of the cycle. Thus, they are not the basis for the development of larger forms, as may be concluded from his earlier works. Speaking of morphologic types of karst, he acceped Grund's idea of "Halbkarst", calling it in the Greek form "merokarst", and also introduced "holokarst". Like Grund, he ascribed combined characteristics derived from karst and from fluvial relief to "merokarst". "Holokarst" develops with all its features only in pure limestones; "true poljes" are limited only to holokarst. Here is not a word about cyclic evolution, as in 1901, but of differences based on lithology.

Cvijić prepared a new monography on karst. It was reportedly finished, delivered to the publisher, and then lost. The work was eventually found and published [CVIJIĆ, 1960], 33 years after the death of the author. It consisted of views on various periods which partially exclude one another, and which are impossible to bring into harmony. This posthumous work of the most authoritative karstologist of the first phase is a good indication of how far this science had reached in that period. The work of PENCK [1924] on "subsurface karst" marks the end of the first phase of the development of karst morphology. He modified his concept, which was enthusiastically accepted and elaborated by his disciples. He returned to the first concept of CVIJIĆ [1893] concerning the priority of karst processes on soluble rocks.

During the first phase of karst morphology important karst features were recorded which were not in accordance with the predominant theories. However, they were of great importance for the further development of karst morphology. The following are the main points:

(*1*) The relationships between soluble and other rocks reflect tectonic displacements. Moreover, tectonic movements may effect the compactness and jointing of soluble rocks. The tectonic influences were pointed out in general terms, but some researchers ascribed predominant significance to them (Katzer, Grund, etc.).

(*2*) The relation between soluble and insoluble rocks may modify the circulation of waters in joints, fissures and fractures and influence the development of karst features. Confined water circulation was discussed by GRUND [1910], TERZAGHI [1913] and JOVANOVIĆ [1924], but the interpretations were not identical. Distribution and relations between soluble and impermeable rocks may result in differential erosional processes and reflect a variety in landscape of karst areas.

(*3*) Karst relief is formed on soluble rocks. Therefore, the lithology is decisive. However, the formation of karst features depends on whether the soluble rock is exposed or covered by allogene deposits. This leads to distinctions between "bare" and "covered" karst [RICHTER, 1907]. KATZER [1909] pointed out the different evolution of denuded and "wooded" karst. But, he considered the thickness of soluble rocks even more important, and distinguished "shallow" and "deep" karst. Taking into account the transitional character of dolomites, GRUND [1914] introduced the term "Halbkarst".

(4) Water is the medium for solution process. The solution depends on climate but includes other ecologic factors also. Climatic-ecologic factors were pointed out in numerous papers [CVIJIĆ, 1893; SAWICKI, 1909; TERZAGHI, 1913; GRUND, 1914; etc.], but there was no method of defining the corresponding influences. Distinguishing between so-called "Mediterranean" and "northern" or "middle-European" karst is not sufficient. More reliable observations and conclusions regarding the various climates are needed to recognize and define the climatic influences satisfactorily.

(5) Partisans of the Penck–Grund concept on succession of the fluvial erosion-karst process, were not able to explain the circulation of water through fractures and the results of speleologic investigations. Following logical deductions from the concept, the opinion prevailed that the depth of circulation in fractures, and the development of subsurface cavities, depends upon underground water level or fluvial erosion. However, this supposition opposes the natural facts. The main difficulty of the Penck–Grund concept was caused by the consideration that the fluvial or "normal" erosion is predominant. Another problem was connected with the inclusion in Penck–Grund's explanation of the predominant scheme of cyclic evolution of relief proposed by Davis. Basing his thoughts on insufficient data concerning characteristics of karst relief in humid and warm climates, GRUND [1914] in fact accepted the earlier ideas of CVIJIĆ [1901], which were later abandoned by the latter [CVIJIĆ, 1918]. Thus the Penck–Grund concept was stronger in the confrontation with its opponents than in its own inner logic.

After World War I, the investigations of karst relief stagnated. Followers of the respected masters were not numerous, and did not produce worthwhile contributions. Thus, the characteristics of the first phase remained as they were, rich in general ideas and poor in real analyses.

Second phase of karst morphology investigation

The beginning of the second phase of the morphologic investigation of karst is connected with the work of the well-known morphologist of the Viennese group LEHMANN [1932]. He tried to solve the problem over which the Penck–Grund concept had stumbled, i.e., the nature of the circulation of water in fractures, fissures and joints, and its relation to the evolution of karst relief. According to his own experiences and those of the others, and with the aid of experiments, Lehmann came to the conclusion, that subsurface rivers promoted by speleologists, and groundwater as defined by Penck and Grund, may be considered as two extremes which do not reflect the real natural relations. Katzer's concept of entangled systems of flows through subsurface fractures corresponds the best with reality, though strongly attacked by followers of the Penck–Grund theory. Lehmann's work showed convincingly that the surficial evolution of karst and the circulation of water in fractures have to be explained in mutual connection.

KAYSER [1934] established that step-like elevated plains on the margin of the depression of Scuttari Lake (Skadarsko jezero) were shaped by marginal corrosion around the marshes near former springs, i.e., by means of a process which was applied by TERZAGHI [1913] to explain the plain in Gacko polje.

LEHMANN [1936] published a detailed study of the karst of Java. He had at hand more reliable data (detailed maps and geologic documentation) and more time than the former researchers. He established that isolated hills were specific characteristics of karst in humid and warm climates. He accepted the term "Kegelkarst", used by LEHMANN [1926], elaborating the results of the Handel–Mazetti expedition along southern China. H. Lehmann modified the explanation of Grund, according to which "Kegelkarst" represented a mature stage of evolution of karst relief. On the contrary, he considered it a karst relief specific for a warm and humid climate, where the isolated hills are characteristic similar to the dolines in moderate latitudes. This point of view is in accordance with the geomorphologic results of that time. Beside the traces of Pleistocene glaciation, specific features of earlier climatic phases were also recorded. The study of relief in the areas with different climates supplied reliable indicators of estimates and comparison. The role of climate became of one the essential elements in geomorphologic investigations, especially concerning karst, since the climate considerably modifies solutional processes. The formation of isolated hills in warm and humid climates was, according to H. Lehmann, due to marginal corrosion. The terrain among the hills, was covered by soil which hindered the sinking in of water. Soon after, it was recorded [ROGLIĆ, 1938, 1940] that the terraces on the calcareous margins of the Dinaric poljes are cut by corrosion along the discharge side of flooded plains (Fig. 2). Records of investigated poljes show that they are bounded on impervious rocks and surficial discharge, i.e., that they are shaped by processes of differential erosion. The distribution of limestones and impervious deposits is a consequence of tectonics.

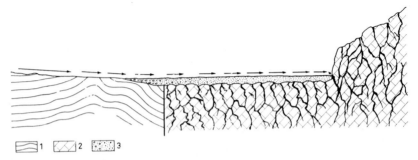

Fig.2. Enlargement of a corrosional plain. 1 = Impervious beds; 2 = limestones; 3 = alluvial cover; → = flow direction. (After ROGLIĆ, 1938.)

These investigations were the beginning of the second phase of karst morphology, free of former concepts and based on more reliable facts and more comprehensive studies, but they were tragically interrupted by World War II. Soon after the war, investigations of karst morphology were revived and increased in volume and intensity in an essentially different way than those in the past.

In the post-war period there are spontaneously maintained global scientific contacts and exchanges of opinions. The different countries are more easily accessible. Observations and data are multiplying, concepts are being modified and combined. International and regional meetings of scientists are frequent. Especially meritorious was the Commission of Karst Phenomena of the International Geographical Union.

Speleology is very greatly intensified, arousing the interest in close cooperation in both national and international fields (already four international congresses have been held). With more exact observations and data, the corresponding conclusions have been improved. In the karst areas important practical work in the prospection and exploitation of ore deposits has been carried out, as well as the construction of hydroelectric plants, tunnels, etc. Scientific concepts have necessarily become adapted to new ideas and facts.

Investigations have been made in all the karst areas in warm and humid climates. In addition to Java and Jamaica where already in the past investigations had been made, karst areas of southern China, Indo-China, Celebes, Puerto Rico, Cuba, Yucatan, Mexico, etc. have been studied. The karst of all these areas is characterized by isolated steep hills ("mogotes" in Cuba). The genesis of these isolated elevations is a consequence of the enlarging of corrosional plains (Fig. 3).

The fact that solution is intensive in warm and humid climates is not in accordance with the fact that colder waters adsorb more carbon dioxide, i.e., that they are more aggressive. It has been recorded that the water in warm climates is enriched in carbon dioxide as a result of its passing through the crowns of trees and through the cover of humus [LEHMANN et al., 1963]. This "younger" carbon dioxide is especially aggressive, and the intensive solubility of limestones in a humid and warm climate can be explained in this manner [BIROT, 1954b]. The vegetation and organic components of the soil retain insoluble particles of rock that form an impermeable cover which

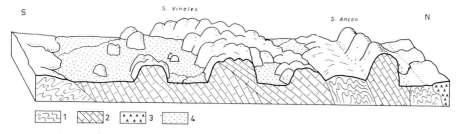

Fig.3. Block diagram in Sierra de los Organos (Cuba). *1* = Impervious rocks of the Pizzaras Formation; *2* = Jurassic and Lower Cretaceous limestones; *3* = serpentine (structural relations simplified); *4* = flood plains. (After LEHMANN et al., 1963.)

prevents the sinking in of water and concentrates the solution at the contact with the soluble rocks. Thus, climate and the corresponding ecologic conditions modify the karst process to a high degree.

Investigations have been made also in other countries, especially in cold and humid ones, where intensive solution may be logically expected. The hardness of flowing water is a consequence of material in solution. Measurements of hardness in southern Alaska [CORBEL, 1960] have shown that the solution of carbonate rocks in cold and humid climates is especially intensive. The barren surfaces of cold belts make the rapid sinking of water possible, with the consequence that the solution process occurs mainly in the deeper parts, if it is not prevented by permafrost. On the contrary, in dry climates solution is slight, limestone being a very resistant rock. Thus the climate modifies the karst process.

Sinking waters enlarge joints and fissures and develop complicated underground cavities, which are, as a result of speleologic investigations, known even down to depths of more than 1,000 m. The exploitation of ore deposits reaches still deeper, whilst bore holes drilled for the prospection of oil and groundwater have reached several thousand metres in depth. Karst cavities are especially significant in formations that yield fluids. Many cavernous formations yield oil and some of the most productive aquifers are cavernous limestone.

The water circulation in fractures is recorded even at several thousand metres below sea level [CORBEL, 1957]. Deep circulation is also established in young Tertiary limestones of the Caribbean area [CORBEL, 1959] with the surface slightly above sea level. Cvijić's first concept of karst as a primary process in soluble rocks is confirmed. There is no reason to support the Davis–Penck concept of primary fluvial erosion, which turned the investigation of karst morphology towards schematic combinations. A compromise proposed by CVIJIĆ [1918], who assumed the fluvial phase to be a short starting period, cannot be retained even theoretically, because the plains which ought to be formed as a result, are conspicuous and large.

The supposition that the soluble rocks were at the beginning consolidated without any sinking of water and that the circulation of water occurs through the fractures contradicts the fluvial erosion itself [ROGLIĆ, 1960. In consolidated rocks the presence of groundwater is not possible and consequently there cannot be any sources which may feed the rivers. At the surface of a consolidated rock there are no weathering products, and consequently no outwash as an essential element of fluvial erosion. More detailed investigations have shown that the plains (from which the idea of a fluvial phase of development of karst relief first arose) cannot be explained by means of the action of flowing water. The plains are isolated in relief, usually flat, without fluvial gravel. They are developed proportionally to the purity of the limestones and at different elevations. They were formed by marginal corrosion on the contact of impervious rocks and karst, in the periods of convenient tectonic and ecologic conditions. Thus, it is obvious that they can be formed at various elevations [ROGLIĆ, 1952, 1957].

Collected facts and logical deductions qualify the karst process as primary and normal in soluble rocks. It starts when the "aggressive" water, i.e., water capable of causing solution, comes in contact with soluble rocks, as finally stated by PENCK [1924], initiator of the concept according to which fluvial erosion should be the primary process and karstification the subsequent one. Circulation in fissures and the consequent development of karst features, occurs down to great depths, regardless of the sea level, as reported by CVIJIĆ [1918]. The karst process is specific, and normal for soluble rocks. It depends on climate, which modifies solution in space and time and further, on the purity and texture of the rocks, and on the relation of soluble to insoluble rocks.

Dinaric Mesozoic and Palaeogene limestones are characterized by over 99.5% $CaCO_3$. All of it is dissolved without leaving insoluble remains which could close joints and fissures and modify the circulation of water and the development of karst features.

The development of karst is also influenced by the texture of the rocks, which varies from thick beds to thinly bedded carbonate deposits. Thick layers disintegrate into smaller blocks, among which are formed joints of various dimensions. The solution process is predominant, while further mechanical disintegration is unimportant. On the contrary, the thinly bedded limestones are subjected to mechanical changes. On the surface there is broken material which moves on the surface.

Impure limestones and dolomites are of special significance. Insoluble remains of weathered marly limestones are washed out on the surface.

Pure dolomites are transformed into dolomitic sand, on the surface as well as in fractures. They are washed out and at the same time they prevent the sinking of water. Thus exchange in composition causes a combination of water sinking and solution with weathering and wash-out; the two processes occur in common causing "fluviokarst" relief [ROGLIĆ, 1960]. Combining two morphogenetic factors, the term "fluviokarst" is more adequate than "Halbkarst". Covered karst modifies solution processes in such a degree that marginal corrosion can prevail.

Exchanges of soluble and impervious rocks to a greater extent cause differential evolution, reflecting an entangled and dynamic picture of relief. In the case of impervious rocks, surface streams modify the surface forms, while in karst they sink or, as allogene rivers, flow through narrow canyons. The discharge of water and removal of waste material through the karst underground indicate the depth and dimensions of cavities in karst. Differential evolution depends upon the position of impervious rocks: (*1*) whether they are the base of the soluble rocks; or (*2*) they are only a shallow blanket covering the pervious ones; or (*3*) the soluble rocks partially sink into the impervious base. In the first mentioned case, permanent flow discharges on the impervious base continuing through the karst underground or through canyons. In the second example, water can flow below the impervious cover, as is the case in many of the poljes in the Dinaric karst. In the third case, the impervious base governs the directions of flow and the karst development.

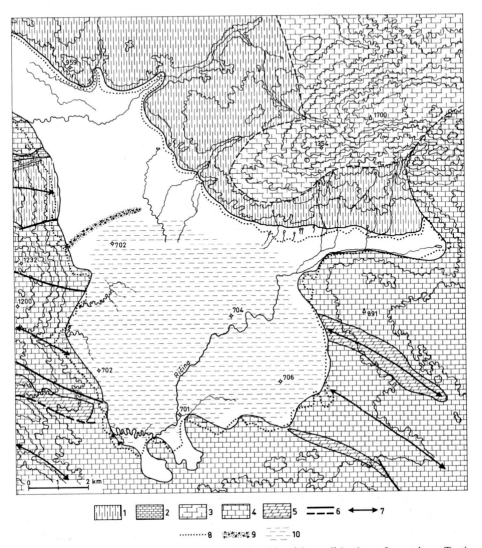

Fig.4. Buško Blato (southeastern part of the Livanjsko polje). *1* = Impervious Tertiary lacustrine deposits; *2* = Palaeogene limestones; *3* = Mesozoic limestones; *4* = limestones with deep karst features; *5* = weathered dolomites and impure limestones; *6* = faults; *7* = strike of the folds; *8* = margin of the polje; *9* = littoral sandbank of the Pleistocene lake; *10* = periodically inundated area. (Southern part of the figure after RALJEVIĆ, 1967.)

Differential erosion changes the relationship between impervious and pervious rocks. In various climates soluble and impervious rocks are unequally subjected to erosional processes and specific features are formed. In the karst relief, elements are preserved that originated in earlier space and climatic conditions. Such a relief, inherited from different lithologic and climatic relationships, is well conserved in karst, because there is no surface dissection, or it is exceptional.

The colder Pleistocene climate and the erosional processes connected with it are reflected in a more intensive disintegration of rocks and in the transport of loose materials during the period of rapid melting of ice. Swallow holes ("ponors") could not accept the increased quantity of river waste which is deposited in enlargements. Inundations of plains in poljes and hydrologic relationships are inherited from Pleistocene climatic conditions [ROGLIĆ, 1964].

Isolated and conspicuous plains, developed on the pure carbonate rocks of the Dinaric karst, encircle the canyons in which Pleistocene glaciofluvial deposits are present. The plains indicate that before the Pleistocene, conditions pertained which led to marginal corrosion (Fig. 4).

New investigations have defined the development of surface karst and its allogene and ecologic modifications fairly well. But satisfying explanations are still lacking for the solution processes and the formation of cavities in the depths which also have consequences at the surface. Application of experiments in corrosion by means of mixing waters of different properties i.e., "mixed corrosion" [BOEGLI, 1964], is an important step towards solving this problem and explaining the complex karst morphology.

Thus the actual concept of karst morphology pays close attention to the various components which influence the solution process. Simple explanations proved to be unjustifiable and it is not realistic to deduce rules according to the features developed in one area only, as was the custom in the earlier investigations of karst morphology. The first scientific ideas about karst relief were acquired in the Dinaric karst, together with recorded features and their mutual relationships. But this was insufficient for a general morphologic concept of the variable karst on earth. Moreover, it was established that some features were erroneously defined and related to each other.

MAIN CHARACTERISTICS OF SURFACE KARST

Finally, the main characteristics of the surface karst forms are summarized:

Karren (lapies) are the most characteristic features of the barren calcareous surfaces. They attracted the attention of the early researchers of the Alpine karst. In his "*Karstphänomen*", Cvijić did not pay them adequate attention. He linked the genesis of a "normal doline" with vertical joints. This was afterwards misinterpreted and enlarged upon the evolutional connection between karren and dolines. CVIJIĆ himself [1924b] considered as karren those karst forms which develop in an isolated cyclic process and which do not pass over into any larger form. Karren, as characteristic and variable karst features, continue to arouse interest [BOEGLI, 1960]. Vegetation and shallow cover modify the evolution of karren.

Deep karst, such as the Dinaric, is especially characterized by karren which are formed due to joints and fissures. They are bound to change into deep open pits, *jamas* (sing. jama), which are most important features of the deep karst. They connect the surface and the underground cavities, having pecularities similar to *ponors* (swallow holes). So far relatively little is known about jamas, because they are only accessible with great difficulty, or sometimes even not accessible at all.

Small, shallow dish-like depressions (*kamenice*, sing. "kamenica") on consolidated calcareous blocks are a clear expression of marginal corrosion due to a lack of vertical joints which might enable the sinking of water and thus corrosion in a vertical direction.

Dolines are forms which determine the "landscape character" of the classical karst and thus they were in the center of interest as fundamental karst features from the very beginning. Dolines are also developed horizontally, in spite of a general sinking of waters and a normally vertical deepening of the karst forms. Cvijić was also convinced that the "landscape characteristics" of dolines, the so-called "pockmarked karst", was formed due to earlier soil cover on the calcareous base. This cover retarded the sinking of water and enabled corrosion of the base and enlargement of forms. The continuity and composition of the soil cover can modify and direct the discharge and solution process so much that there occurs only marginal corrosion, and isolation of hills which determines the "landscape character". Among the hills corrosional plains are developed. The "Kegelkarst" of the humid and warm climates is of this kind.

Pitted dolines, formed due to collapses, are reflections of subsurface cavities. The fact that they are not numerous and that their form is peculiar indicates the only limited importance of this process.

Uvala is a general term in Yugoslavia and is not limited only to karst. It was put into the terminology of karst in order to have a transient form between dolines and poljes which was needed in the cyclic explanation of karst evolution. Without the cyclic concept, there is no further need for a transient form. Anyhow the term remained, characterizing larger depressions in karst. They were formed owing to the differences in lithology and tectonic predisposition, and were modelled by means of processes of differential erosion [POLJAK, 1951].

Polje became a "karstic" attribute together with the introduction of the concept of cyclic evolution in the karst morphology. The name itself indicates something strange in a rocky, karst landscape. Poljes include impervious deposits and surface fluvial modelling. The contact of different rocks is most often a consequence of tectonic dislocations. Alluvial cover can modify the karst process in an adjacent calcareous area, enabling marginal corrosion and the enlargement of depressions. Poljes are best developed in the areas where impervious beds are in connection with thick and pure limestones. In the adjoining karst underground, the cavities are spacious. The discharge of water and removal of waste materials have taken place through them. Actual conditions in the poljes represent an inheritance of the climatic-erosional modifications due to Pleistocene glaciation [ROGLIĆ, 1964]. Being a strange element in a karst landscape, poljes are not specific karst features.

Classical concepts of karst phenomena were born of insufficient knowledge of the Dinaric karst. The investigations of different karst countries which have followed, and a general improvement in karst knowledge have widened our outlook on the subject which makes it necessary to revise old concepts and to inspire the creation of more complex ones.

REFERENCES

Birot, P., 1954a. Problèmes de morphologie karstique. *Ann. Géograph.*, 63(337):161–192.
Birot, P., 1954b. Esquisse d'une étude zonale de l'érosion en pays calcaires. *Erdkunde*, 8(2):121–122.
Boegli, A., 1960. Kalklösung und Karrenbildung. Internationale Beiträge zu Karstmorphologie. *Z. Geomorphol., Suppl.*, 2:4–21.
Boegli, A., 1964. Mischungkorrosion—ein Beitrag zum Verkarstungsproblem. *Erdkunde*, 18(2):83–92.
Corbel, J., 1957. Les karsts du Nord-ouest de l'Europe et de quelques régions de comparaison. Etude sur le rôle du climat dans l'érosion des calcaires. *Inst. Études Rhodaniennes, Mém. Doc.*, 12:541 pp.
Corbel, J., 1959. Les karsts du Yucatan et de Florida. *Bull. Assoc. Géograph. Franç.*, 282/283:2–14.
Corbel, J., 1960. Erosion en terrain calcaire (vitesse d'érosion et morphologie). *Ann. Géograph.*, 66(366):97–120.
Cox, E. T., 1874. *Fifth Annual Report of the Geological Survey of Indiana.* Geol. Surv. Ind., Indianapolis, Ind., pp.280–305.
Cvijić, J., 1893. Das Karstphänomen. Versuch einer morphologischen Monographie. *Geograph. Abhandl.*, 5(3):218–329.
Cvijić, J., 1901. Morphologische und glaziale Studien aus Bosnien, der Hercegovina und Montenegro, 2. Die Karstpoljen. *Abhandl. Geograph. Ges.*, 3(2):1–85.
Cvijić, J., 1909. Bildung und Dislozierung der dinarischen Rumpffläche. *Petermanns Geograph. Mitt.*, 6:121–127; 7:156–163; 8:177–181.
Cvijić, J., 1918. Hydrographie souterraine et évolution morphologique du karst. *Rec. Trav. Inst. Géograph. Alpine*, 6(4):376–420.
Cvijić, J., 1924a. Types morphologiques de terrains calcaires. *Glasnik Geograph. Društva*, 10: 1–7.
Cvijić, J., 1924b. The evolution of lapiés. *Geograph. Rev.*, 1924: 26–49.
Cvijić, J., 1960. La géographie des terrains calcaires. *Monograph. Acad, Serbe, Sci. Arts*, 341(26): 1–212.
Daneš, J., 1908. Geomorphologische Studien im Karst-Gebiete Jamaikas. *Intern. Geograph. Congr., 9th, Geneva, 1908*, 2:178–182.
Davis, W.M., 1884. Geographic classification by a study of plains, plateaux and their derivatives. *Proc. Am. Assoc.*, 33:428–432.
Davis, W.M., 1901. An excursion in Bosnia, Hercegovina and Dalmatia. *Bull. Geograph. Soc.*, 3(2):47–50.
Diener, C., 1886. *Libanon.* Hölder, Wien, 412 pp.
Fournet, J., 1852. Note sur les effondrements. *Mem. Acad. Lyon*, 176–186.
Gruber, Th., 1781. *Briefe Hydrographischen und Physikalischen Inhalts aus Krain.* Krauss, Wien, 162 pp.
Grund, A., 1903. Die Karsthydrographie. Studien aus Westbosnien. *Geograph. Abhandl. (Penck)*, 7(3):1–200.
Grund, A., 1910. Beiträge zur Morphologie des dinarischen Gebirges. *Geograph. Abhandl. (Penck)*, 9(3):1–236.
Grund, A., 1914. Der geographische Zyklus im Karst. *Z. Ges. Erdkunde*, 1914: 621–640.
Hacquet, B., 1778. *Oryctographia Carniolica oder Physikalische Erd-beschreibung des Herzogthums Krain, Istrien und zum Teil der Benachbarten Länder.* Gottlob und Breitkopf, Leipzig, 1:162 pp.; 2:186 pp.; 3:184 pp.
Jovanović, P., 1924. Zagaceni karst (L'eau dans le karst barré). In: *Recueil des Travaux Offert à M.J. Cvijić.* Beograd, pp.397–409.
Katzer, F., 1909. *Karst und Karsthydrographie. Zur Kunde der Balkanhalbinsel.* Kajon, Sarajevo, 94 pp.
Kayser, K., 1934. Morphologische Studien in Westmontenegro, 2. *Z. Ges. Erdkunde*, 1/2:26–49.

KNEBEL, W., 1906. *Höhlenkunde mit Besonderer Berücksichtigung des Karstphänomens—Die Wissenschaft*. Braunschweig, 15:198 pp.
KRAUS, F., 1887. Über Dolinen. *Verhandl. Geol. Reichsanstalt*, 2:54–62.
KRAUS, F., 1894. *Höhlenkunde*. Gerolde Sohn, Wien, 308 pp.
LEHMANN, H., 1936. *Morphologische Studien auf Java*. Engelhorn, Stuttgart, 114 pp.
LEHMANN, H., KROMBELIN, K. and LOTSCHERRT, W., 1963. Karstmorphologische, geologische und botanische Studien in der Sierra de Los Organos auf Cuba. *Erdkunde*, 10(3):185–204.
LEHMANN, O., 1926. Die geographische Ergebnisse der Reise (Handel-Mazetti) durch Guidschou (Kweitschou). *Denkschr. Akad. Wiss.*, 100:77–99.
LEHMANN, O., 1932. *Die Hydrographie des Karstes*. Enzyklopädie der Erdkunde. Deuticke, Leipzig, 212 pp.
LYELL, CH., 1839. On the tubular cavities filled with gravel and sand called "Sand-pipes" in the chalk near Norwich. *London Edinburg Phil. Mag. J. Sci.*, 15:257–266.
MARTEL, E. A., 1910. La théorie de "Grundwasser" et les eaux souterraines du karst. *Géographie*, 21:126–130.
MARTEL, E.A., 1921. *Nouveau Traité des Eaux Souterraines*. Doin, Paris, 840 pp.
MOJSISOVICS, E., 1880. Zur Geologie der Karsterscheinungen. *Z. Deut. Österreich. Alpenver.*, 11:111–116.
MOJSISOVICS, E., TIETZE, E. and BITTNER, E., 1880. Geologie von Bosnien. *Jahrb. Geol. Reichsanstalt*, 1880:2–272.
NAUMANN, C. FR., 1858. *Lehrbuch der Geognosie*. Engelmann, Leipzig, 576 pp.
OWEN, D.D., 1856. *Annual Report of the Geological Survey in Kentucky*. Geol. Surv. Ky., Frankfort, Ky., pp.169–172.
PENCK, A., 1894. *Morphologie der Erdoberfläche*. Engelhorn, Stuttgart, 1:471 pp.; 2:696 pp.
PENCK, A., 1900. Geomorphologische Studien aus der Hercegovina. *Z. Deut. Österreich. Alpenver.*, 31:25–41.
PENCK, A., 1904. Über Karstphänomen. *Vorträge Ver. Naturwiss. Kenntnisse*, 44(1):1–38.
PENCK, A., 1924. Das unterirdische Karstphänomen. *Recueil de Travaux Offert à M. J. Cvijić*. Beograd, pp.175–197.
POLJAK, J., 1951. Je li krška uvala prijelazan oblik između ponikve i krškog polja. *Geograf. Glasnik*, 13:25–48.
PRESTWICH, J., 1854. On some swallow holes on the chalk hills near Canterbury. *Quarterly J.*, pp.222–224; 241.
RALJEVIĆ, B., 1967. Geološki i hidrološki odnosi područja Buškog Blata (Geological and hydrological relations of the wide area of Buško Blato). *Geol. Vjesnik*, 20:273–283.
RICHTER, E., 1907. Beiträge zur Landeskunde Bosniens und der Herzegowina. *Wiss. Mitt. Bosnien Herzegowina*, 10:383–545.
ROGLIĆ, J., 1938. Morphologie der Poljen von Kupres und Vukovsko. *Z. Ges. Erdkunde*, 7/8:291–316.
ROGLIĆ, J., 1940. Geomorphologische Studien über das Duvanjsko polje in Bosnien. *Mitt. Geograph. Ges.*, 83:1–26.
ROGLIĆ, J., 1952. Les surfaces de corrosion dans le karst dinarique. *Proc. Gen. Assembly, 8th, Intern. Congr., 17th, Washington, D.C.*, pp.366–369.
ROGLIĆ, J., 1957. Zaravni na vapnencima. *Geograf. Glasnik*, 19:103–134.
ROGLIĆ, J., 1960. Das Verhältnis der Flusserosion zum Karstprozess. *Z. Geomorphol.*, 4(2):116–128.
ROGLIĆ, J., 1964. Les poljes du karst dinarique et les modifications climatiques du Quaternaire. *Rev. Belge Géograph.*, 88(1/2):105–125.
SAWICKI, L., 1909. Ein Beitrag zum geographischen Zyklus im Karst. *Geograph. Z.*, 15:185–204.
SAWKINS, J. Q., 1869. Reports on the geology of Jamaica. *Mem. Geol. Surv. London*, 340 pp.
SCHMIDL, A., 1854. *Die Grotten und Höhlen von Adelsberg, Lueg, Planina und Lass*. Braumüller, Wien, 316 pp.
TERZAGHI, K., 1913. Beiträge zur Hydrographie und Morphologie des kroatischen Karstes. *Mitt. Jahrb. Kgl. Ungar. Reichsanstalt*, 8(6):256–369.
VALVASOR, J.W., 1689. *Die Ehre des Herzogsthums Krain*. Endter, Nürenberg, 1:696 pp.; 2:835 pp.; 3:730 pp.; 4:610 pp.
VIRLET, J., 1834. Observations faites en Franche-Comté sur les cavernes et la théorie de leurs formation. *Bull. Soc. Géol. France*, 6:148–163.
WHITE, CH. A., 1870. Report of the geological survey of the state Iowa. *Geol. Surv. Iowa*, 1:169–172.

Chapter 2

Historical Review of Hydrogeologic Concepts

M. HERAK and V.T. STRINGFIELD

Geological and Palaeontological Institute, Faculty of Natural Sciences, Zagreb (Yugoslavia)
Department of the Interior, U.S. Geological Survey, Washington, D.C. (U.S.A.)

INTRODUCTION

The sinking rivers and brooks, the deep swallow holes, the water-bearing caves, the vauclusian and submarine springs, as well as the katavothres are the karst phenomena which enriched the human imagination from its very beginning, becoming the subject of tales and of supernatural explanations of the karst interior. The concept of underground concentrated flows (rivers and brooks) connecting the caves was the first natural interpretation of the underground hydrological pattern. The primitive observations of these mutual connections relating to the delay of the appearance of turbidity in the lower lying springs and on the appearance of leaves and other objects in the upwelling water, led to primitive tests in order to provide proofs for such concepts, and to exaggerated reports of dead animals, such as horses, coming through the karst channels from the swallow holes to the lower lying springs.

The coincidence of the distribution of the typical karst phenomena and of the early development of the ancient civilizations in the Mediterranean areas explains why the first clear statements of the hydrology of karst phenomena can be found in the old Greek and Roman written documents. The groundwater and its connection with the sea were noted by Homer, Thales of Milet, Anaxagoras, Plato, Aristotle, Lucretius, and Seneca. The most interesting opinions are the following.

Thales (640–547 B.C.) believed that the sea water is forced by the wind from the sea into the earth, and due to the weight of the rocks it is driven upwards. The springs and rivers are fed by this water.

Aristotle (384–322 B.C.) considered the groundwater a consequence of the gathering of drops of water and of the transforming of the "air" into water (condensation) within the underground cavities due to the low temperature. This water is stored in the large underground lakes, and it feeds the springs by means of which the water returns to the sea. The largest springs are situated at the foot of the mountains.

Lucretius (98–55 B.C.) described the circulation of water between the sea and the earth and discussed its loss of salt. He mentioned, also, the submarine spring in the sea at Aradus.

The origin of ground water and springs and the circulation of water within the earth were the problems which received much attention later on, but for a long time there was little improvement of the early concepts of groundwater, especially within the karst. Not until the 19th century did karst become a subject of scientific studies; the scientific studies mostly with descriptions of caves, dolines, swallow holes, and other karst forms. The hydrogeologic problems were also treated as fragments of the karst curiosities.

EARLY INVESTIGATORS

The first attempts to analyse the hydrogeologic relations of a large karst area on a geological basis are described in little-known papers of BEYER et al. published in German and Croatian in Zagreb [1874] under the common title *"Lack of Water in the Karst of the Military Zone of Croatia"*. E.Tietze and G. Pilar were geologists, and A. Beyer was a civil engineer. Thus, it was possible that the concept of the hydrogeological relations in the karst was intimately connected with water supply, the problem remaining unsolved to some degree until recent time.

Beyer spoke only of isolated flows supposed to occur in the numerous underground caves and caverns, and possibly also in subsurface lakes. He noted the coastal springs and considered the possibility of the flow of fresh water below the Adriatic Sea towards the isles.

Tietze pointed out the significance of the sinking rivers and brooks and considered them peculiarities of the karst. He assumed the existence of the subsurface brooks which never appear on the surface, flowing underground into the sea. In addition to the horizontal cavities (occupied by isolated flows) there must be more or less vertical systems of fissures, too, which may be connected with the horizontal ones. The storage of water occurs in the labyrinth of channels or in the larger accumulations, which are situated sporadically below the level of the valleys and surface lakes. According to Tietze, this is the reason the water sinks into the underground. Both the rate of precipitation and the temporary distribution of the water influence the subsurface water. After heavy rains the channels are periodically replenished, causing even rising of water in the background and possibly issuing at the places were during the low level it sank into the rocks. He believed that by means of the simple hydraulic laws apparently all irregular changes of water levels and even the appearance of the artesian water during the high water level may be explained. The systems of joints and fissures are supposed to reach to the basal impervious beds (shales and sandstones).

Pilar tried to define the tectonic frame of the distribution of the subsurface water in the Croatian karst, supposing major folds as general regulators of the hydrogeologic pattern. He stated that precipitation rapidly disappears through the joints and fissures into the underground, forming deeply lying and moving "water layers" or "water networks". During heavy rains the inflow is greater than the discharge and consequently the water rises in the subsurface troughs, and under pressure it rises up through the fissures to the surface again. So, it happens that some areas, which are generally dry, are periodically inundated. He saw the reason for the karstification in the endodynamical processes, pointing out that the time of the beginning is not easy to determine. However, he assumed that it happened after the Miocene. The continuous intensification of jointing and fissuring of the rocks caused a gradual disappearing of the surface water accumulations, especially of younger lakes.

The considerations of Beyer, Tietze and Pilar are in fact the forerunners of the later ideas which are based on two opposing concepts: one is based on a supposed continuous flow of the groundwater in karst; the other is concerned with the isolated flows.

MODERN THEORIES

GRUND [1903] assumed groundwater in the karst interior to be stagnant. Its base level coincides with the sea level, but it rises towards the karst background. Only above this level the water moving towards the sea is considered the karst water. It occurs at various levels and where the hydrostatic pressure of the karst water exceeds the hydrostatic pressure of the sea water, the karst water issues even at the sea bottom. The precipitation raises the level of the groundwater to the bottom of the poljes periodically causing their flooding. The fluctuations may explain all hydrogeologic relations connected with the poljes, including the estavelas.

KATZER [1909] distinguished "shallow" and "deep" karst, the latter being connected with an irregular hydrogeologic pattern. Erosion in the shallow karst has reached to the impervious basement, while in the deep karst it has remained within the carbonate rocks. The water in the deep karst may be divided into several more or less separate bodies with different water levels. He based his concept on several observations; some springs on the higher level may be permanent while the lower ones are periodic; among the springs with intermittent flow some react to localized precipitation whereas others do not; at nearly the same level there may be springs, estavelas and swallow holes. Therefore, he considered that the karst is jointed and fissured in several directions. Also he noted tubes, caverns and irregular cavities connected sporadically by siphons. He thought that the inundation of poljes cannot be explained by regular rising of the continuous groundwater because even during its rise the water sinks through the swallow holes.

MARTEL [1911, 1921] described numerous investigations of caves and caverns, partially by help of tracers, which indicate the existence of subsurface streams and torrents.

BEEDE [1911] assumed three stages in the development of the karst hydrology. In the first, water is diverted into the karst interior. The second stage is characterized by major entrenched surface streams. In the third one, surface drainage predominates again.

CVIJIĆ [1918] denied that groundwater in the karst is stagnant but he was not against a sinking and rising groundwater which, according to him, cannot be explained by underground channels alone. He pointed out especially the dynamics in the development of the hydrologic pattern in the deep Dinaric karst. Normal valleys were formed at the beginning of evolution of the relief. The water filled all joints and narrow fissures, forming a kind of phreatic zone. But this phase did not last for a long time for the cavities in the lower part of the carbonate rocks were continuously widening and consequently the water sank, leaving the upper part of the rocks almost dry. The valleys turned to dead ones with newly formed swallow holes. This stage was not stable. The down cutting process continued towards the impervious base. The consequences are reflected in the formation of three different zones: (*1*) the dry zone limited to the upper part of karst with almost dry caves and caverns; (*2*) the second zone which is a transition from the upper to the lower zones; its level is at the bottom of the valleys and poljes and along the coastal cliffs; in this zone with few caverns the water normally percolates towards the lower zone; sporadically there are underground streams; and (*3*) the lowest zone in which the water circulates mostly by slowly descending towards the depth through the cavities which may locally have the characteristics of siphons. Over the impervious base the phreatic zone may be formed. Because of the limited discharge, the increase of recharge causes rising of water towards the upper zones, or even up to the surface feeding periodical springs, causing inundations. During the process of karstification the upper zones become enlarged more rapidly than the lowest one. Simultaneously the karst relief is continuously decreasing. By mutual action of the factors mentioned the carbonate masses become reduced to fragmental karst relief above the water lying on the impervious bed.

LEHMANN [1932] contributed to the further development of the concept of the groundwater in karst by applying the laws of hydrodynamics in his attempts to explain the water circulation in the karst interior. He supported the idea of occurrence of various and complicated systems of cavities (among them tubes, siphons, etc.), and of several water levels. The hydrogeologic pattern is not uniform but extensively variable with turbulent or laminar water flow depending on the type, form and size of water-bearing cavities.

Although modern investigators have made considerable progress in understanding the hydrology of karst, some differences of opinion—including those regarding the occurrence and movement of water in carbonate rocks—remain. One group, including some speleologists, who are familiar with streams in caves, think that the circulation of water is restricted to underground streams flowing more or less similar to surface streams; the other group, including many hydrogeologists think of groundwater as occurring in a zone of saturation in carbonate aquifers at some base level. They believe streams in the zone of aeration in caves occupy solution caverns and channels which were formed chiefly when the upper part of the zone of saturation occupied that part of the limestone.

There is general agreement that solution of soluble rocks accounts for the large permeability of some productive carbonate aquifers, but there is still some lack of agreement on the question of the conditions under which the solution occurs.

Prior to 1930, ideas of the conditions under which solution of carbonate rocks formed caves, caverns and other secondary permeability features were rather vague and some of the principal investigators believed that caverns, as well as other solution channels were formed chiefly in the zone of aeration, above the water table. MEINZER [1923, p.132] argued that limestone is dissolved most rapidly above the water table where there is abundant and rapid percolation and where percolating water contains carbon dioxide, which is necessary for dissolving limestone. He recognized that crevices and solution cavities above the water table are not available as reservoirs of groundwater because only the cavities below the water table yield permanent supplies to wells. He cited others who also thought that solution cavities were formed in the zone of aeration. In 1930, DAVIS [1930, pp. 475–628] challenged the idea that caverns are developed above the water table and that solution below the water table is insignificant. The conclusion that caverns are formed chiefly in the phreatic zone, as reached by Davis, probably was influenced by his visit in 1899 to some karst areas of Europe and his association with A. Penck and some other principal investigators of these areas, as discussed in the chapter "Historical Review of Morphologic Concepts". He accepted GRUND's theory [1903] that caverns are formed by solution below the water tables. However, Davis recognized deep circulation in the phreatic zone. The conclusions reached by Davis were accepted by BRETZ [1942] who gave much additional evidence in support of the theory that caverns are formed chiefly by solution below the water table.

In studies of the general aspects of movement of water in limestone some investigators [SWINNERTON, 1932, p. 665; RHOADES and SINACORI, 1941, p. 794] recognized that theoretically, groundwater moves in arcuate paths, following lines of flow that have their origin at the top of the zone of saturation in recharge areas and then rise to an outlet or point of discharge. Diagramatically these lines of flow can be represented by a family of curves, the spacing of which is closest near the area of outlet and widest along the top of the zone of saturation as the distance from the outlet increases. In uniformly permeable material where geologic structure does not control the direction of movement, the waters may be expected to have an arcuate pattern of flow. However, even in the initial stages of groundwater circulation in limestone and other carbonate rocks, the lines of flow are modified. The concentration of flow lines with the greater velocities in that area will result in an enlargement of the outlet and a consequent shallowing of the more remote arcuate paths. The more direct paths will become larger than the less direct paths and will permit progressively large flows at the expense of other passages.

In recent years hydrogeologists have focused special attention on karst because the principal recharge areas of carbonate aquifers, both artesian and non-artesian, are karstified and also because an understanding of the development of karst is essential in studies of buried karst, palaeohydrology, and secondary permeability due to solution of soluble rocks. At the present time, investigators of karst are making progress with the use of modern methods, concepts and data which are discussed in numerous papers and symposia. Some of the results of studies in karst in different parts of the world are given in the following chapters and are summarized at the end of this report.

REFERENCES

BEEDE, J. W., 1911. The cycle of subterranean drainage as illustrated in the Bloomington quadrangle (Indiana). *Proc. Indiana Acad. Sci.*, 20:81–111.
BEYER, A., TIETZE, E. and PILAR, GJ., 1874. *Die Wassernot im Karste der Kroatischen Militärgrenze.* Albrecht und Fiedler, Zagreb, 159 pp.
BRETZ, J. H., 1942. Vadose and phreatic features of limestone caverns. *J. Geol.*, 50(6):675–811.
BURDON, J.D. and PAPAKIS, N., 1963. *Handbook of Karst Hydrogeology.* Karst Groundwater Investigations and Institute for Geology and Subsurface Research, Athens, 276 pp.
CVIJIĆ, J., 1918. Hydrographie souterraine et évolution morphologique du karst. *Rec. Trav. Inst. Géograph. Alpine*, 6(4):376–420
DAVIS, W.M., 1930. Origin of limestone caverns. *Bull. Geol. Soc. Am.*, 41(3):475–628.
GRUND, A., 1903. Die Karsthydrographie. Studien aus Westbosnien. *Geograph. Abhandl. (Penck)*, 7(3):1–200.
KATZER, F., 1909. *Karst und Karsthydrographie. Zur Kunde der Balkanhalbinsel.* Kajon, Sarajevo, 94 pp.
KEILHACK, K., 1917. *Lehrbuch der Grundwasser und Quellenkunde.* (2nd Edition.) Borntraeger, Berlin, 640 pp.
LEHMANN, O., 1932. *Die Hydrographie des Karstes. Enzyklopedie der Erdkunde.* Deuticke, Leipzig, 212 pp.
MARTEL, E.A., 1911. *L'Évolution Souterraine.* Flammarion, Paris, 301 pp.
MARTEL, E.A., 1921. *Nouveau Traité des Eaux Souterraines.* Doin, Paris, 838 pp.
MEINZER, O.E., 1923. The occurrence of groundwater in the United States (with a discussion of principles). *U.S., Geol. Surv., Water Supply Papers*, 489:321 pp.
RHOADES, R.F. and SINACORI, M.N., 1941. Pattern of groundwater flow and solution. *J. Geol.*, 49(8):785–794.
SWINNERTON, A.C., 1932. Origin of limestone caverns. *Bull. Geol. Soc. Am.*, 43(3):663–693.

Chapter 3

Karst of Yugoslavia

M. HERAK

Geological and Palaeontological Institute, Faculty of Natural Sciences, Zagreb (Yugoslavia)

INTRODUCTION

The karst features are widespread in Yugoslavia, covering one third of the whole country (Fig. 1). The main belt begins in the northwest on the frontier towards Italy and strikes to the southeast along the Dinaric Mountains to the frontier of Albania. This belt is called the Dinaric karst (local names krš and kras). It is the classical karst area not only as regards the mode of development of the karst features but also as

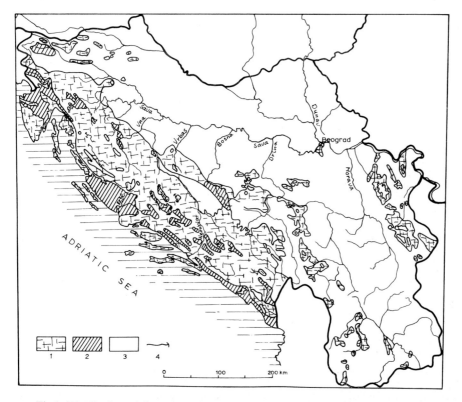

Fig.1. Distribution of the main karst areas of Yugoslavia. *1* = Surface karst; *2* = karst covered by Tertiary and Palaeozoic (allochthon) clastics; *3* = non-karstic areas; *4* = sinking streams.

an area of study. It inspired the first theoretical concepts and continuously awakens the scientific and practical interest of Yugoslavian and foreign investigators. For this reason some local terms were accepted, and are still used, in the international terminology (ponor, doline, uvala, polje, etc.).

Though the differences among the various karst areas in Yugoslavia are of great interest, the features concerning lithostratigraphy, tectonics, and phases of karstification are common bases for correlation. They will therefore be analysed first.

LITHOSTRATIGRAPHIC AND SEDIMENTOLOGIC ASPECTS

Throughout its early geologic history Yugoslavia was part of the Mediterranean geosyncline (Tethys). It was not until the Late Palaeozoic that carbonate sediments were deposited in quantities favourable for karstification. We shall therefore present the lithostratigraphic sequence starting with the Carboniferous Period. The schematic geological column (Fig.2) shows that the Carboniferous System consists chiefly of various clastics (sandstones, conglomerates and shales). The limestones and partly dolomites are distributed lenses in the clastic deposits. In the Permian, the relations remain the same in most of Yugoslavia. But in some areas, especially in Lika (middle Croatia), the Upper Permian consists almost completely of dolomites with small lenses of limestones and in other areas of evaporites (chiefly gypsum and anhydrite). The Lower Triassic is composed mainly of thinly bedded calcareous sandstones and schistose siltstones, with intercalations of allochtonous carbonate rocks. The upper

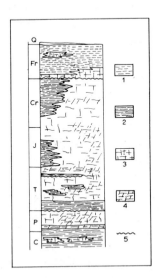

Fig.2. Schematic lithostratigraphic column. *1* = Tertiary clastic deposits; *2* = Palaeozoic and Mesozoic clastics; *3* = limestones; *4* = dolomites; *5* = main unconformities; *C* = Carboniferous; *P* = Permian; *T* = Triassic; *J* = Jurassic; *Cr* = Cretaceous; *Tr* = Tertiary; *Q* = Quaternary.

part of the Lower Triassic is characterized by a gradual diminishing of the clastic deposits and by an overwhelming abundance of carbonate sediments. The Middle Triassic is represented by various facies, with shallow water limestones with dasyclad algae (Dasycladaceae) dominating. The clastic facies are only locally distributed and are represented by lenses in the carbonate complexes. In the Anisian they consist predominantly of shales and subordinately of conglomerates with carbonate clastic components. They are locally associated with some igneous rocks. In the Ladinian the clastic elements are represented by siliceous, calcareous, argillaceous and pyroclastic elements, also sporadically associated with igneous rocks.

Basal sediments of the Carnian Stage of the Upper Triassic in the Dinaric Mountains usually consist of clastics (calcareous conglomerates and alternating series of sandstones and pelites with sporadical admixtures of tuffaceous sediments and tuffs), and only exceptionally of bauxitic lenses and dolomites. The upper part of this stage, as well as the Norian and Rhaetian, consists of more or less uniform dolomites often grading vertically and laterally into reef limestones. In the other parts of Yugoslavia the carbonate transition series from the Middle to the Upper Triassic is almost regular, with local predominance of limestones.

Jurassic carbonate sediments in the Dinaric Mountains are very well developed. Mostly, there is a continuous sequence of limestones and dolomites, bedded or massive. There are some bauxites also present and in the Upper Jurassic there are local chert intercalations in the well-bedded limestones (Lemeš Beds). The incomplete sequence of the carbonate Jurassic is recorded in the Julian Alps and Karawanken, and that of the clastic Jurassic in west Serbia. A hiatus during the Jurassic is supposed to exist in north Croatia. Bosnia is characterized by a variability of facies, consisting of carbonate and clastic deposits with some igneous components. In eastern Serbia the Lower Jurassic (Lias) is predominantly clastic, the Middle Jurassic (Dogger) equally clastic and carbonate, and the Upper Jurassic (Malm) has predominantly a carbonate facies with cephalopods and reef limestones.

The Cretaceous of the Dinaric Mountains is almost entirely carbonate (limestones and dolomites) with the exception of the Durmitor Mountain, where Upper Cretaceous flysch is also developed. The rest of Yugoslavia is characterized by a more or less clastic (flysch-like) Cretaceous, with limestone lenses in the Lower Cretaceous of the eastern part (Urgonian) and sporadically in the Upper Cretaceous along the whole belt (reefs by rudistids). Locally there are areas of igneous activity.

The Lower Palaeogene of the Dinaric Mountains consists mainly of limestones (with some coal in the basal Liburnian or Kosina Beds). Most important are the limestones with large Foraminifera (*Alveolina* and *Nummulites*). The Upper Palaeogene, on the contrary, consists mostly of flysch and other clastic sediments, accompanied locally by coal (Promina Beds). In the other districts of Yugoslavia, the Palaeogene is only represented by coal-bearing clastic freshwater or paralic deposits (mostly Upper Oligocene).

The Neogene is composed mainly of marls, clays, sandstones, sands, coal and locally (in the northern part of Yugoslavia) of Leitha Limestones with coralline algae in the Middle Miocene.

The Quaternary includes the different weathering products, calcareous tufa in the valleys of the karst rivers and in the lakes, further loess, glacial deposits, etc.

The main lithostratigraphic complexes favourable for the development of karst features are limited to limestones and dolomites, and only partially to marbles, gypsum and anhydrite.

Notwithstanding the variations in appearance and texture of limestones and dolomites, their composition is almost uniform with the exception of some lateral and vertical transition zones or thin intercalations. The data from numerous analyses in the Dinaric Mountains (for which I am grateful to Mrs. B. Šćavničar), clearly show the uniform composition. The limestones vary in $CaCO_3$ content as follows: Middle Carboniferous 90–95%, Upper Carboniferous 85–98%, Lower Permian over 93%, Lower Triassic 80–95%, Middle Triassic 98–99%, Lower Jurassic 92–97%, Middle Jurassic over 97%, Lower Cretaceous 95–98%, Upper Cretaceous 98–100%, Palaeogene 98%. The main horizons of the dolomites contain $CaMg(CO_3)_2$ in the following percentages: Upper Permian 90%, Middle Triassic over 90%, Upper Triassic up to 95%, Lower Jurassic 30–80%, Middle Jurassic about 95%, Cretaceous 95–100%. The higher percentages are more frequent than the lower ones. Between the Middle Triassic and Palaeogene (containing the main carbonate rocks) the largest percentage of $CaCO_3$ in the limestones is 97–100%. The maximum range of $CaMg(CO_3)_2$ in the dolomites is 90–100%. These differences (especially in the limestones) are too small to be considered a factor of great selective influence on the karst development of the different areas. But probably they cannot be neglected as regulators of karst development in some details, especially if they are combined with variability of texture. The thickness of bedding seems also to be important. Non-bedded sediments are often characterized by numerous more or less scattered dolines ("pock-marked" karst). In the well-bedded rocks the dolines are not so frequent, and they are more or less oriented. But, uvalas are frequently present. In the Palaeogene carbonate conglomerates (not always completely cemented) relics of erosion are often present.

The argillaceous and siliceous clastic deposits, as well as the igneous rocks, can be considered the factors hindering karst development. They dominate the Late Palaeozoic and the Lower Triassic in quantity and distribution. The carbonate rocks in these deposits are mostly present only as lenses. The Middle and Upper Triassic are characterized by quite opposite relations. The clastic impervious sediments are included in the carbonate complex as smaller or larger lenses, and can only locally control the water circulation and the karstification which is connected with it. The Jurassic, Cretaceous and Lower Palaeogene in the Dinaric Mountains are completely favourable to karstification, because they have almost no clastic impervious elements. On the contrary, in the other districts of Yugoslavia, only the Upper Jurassic and Cretaceous carbonate lenses are karstified. Likewise, in the younger sediments, only Miocene limestone and Pleistocene tufa are characterized by some karst phenomena.

TECTONIC SETTING OF YUGOSLAVIA

Though the orogenetic and epirogenetic movements were active through the whole geologic history of Yugoslavia, the Alpine orogeny formed the latest structural and morphological features. Yugoslavia can be divided into several major structural units (Fig.3). They have been the subject of different concepts, but so far none of them can be considered as finally accepted. The differences concern not only the number and limitation of the various units, but also their character, especially the rate of transport. This is especially obvious within the Dinaric Mountains proper ("high karst zone"), which are principally interpreted in three different ways, i.e., as: (*1*) a large overthrust; (*2*) a part of a mega-anticlinorium; and (*3*) a sum of structural units, comprising all kinds of disturbances from normal faults and folds over the imbricate structures to the overthrusts with various but not too extensive rates of transport. The explorable karst features correspond the best with the third concept, but this does not affect the deep relations where the main overthrust contact zone may be placed. In that case, the surface structural features should be classified only as secondary manifestations.

Elementary structural units have had more influence on the development of the karst features. Different systems of joints, fissures and fractures including faults, transformed the homogeneous carbonate rocks into secondary pervious complexes with more or less directed subsurface water courses. The folded systems influence the underground water by forming watersheds, accumulations or barriers due to which the water is impounded, periodically rising or even overflowing. Tangential structures can intensify the disturbances with various effects on water circulation. Finally, unconformities and disconformities should also be mentioned as weak zones stimulating the flow of subsurface water.

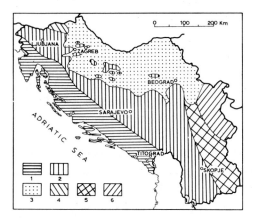

Fig.3. Major tectonic areas of Yugoslavia. *1* = Outer Dinaric units (Adriatic and high karst); *2* = inner Dinaric and south Alpine units; *3* = Pannonian Basin; *4* = eastern Alps; *5* = Serbo-Macedonian Belt; *6* = Carpatho-Balkanian Belt.

PHASES OF KARSTIFICATION

The development of the Yugoslavian karst areas was gradual, but with increasing volume and intensity. There exists some parallelism between the karstification and diastrophic movements within the Hercynian and Alpine orogenies.

Near the beginning of the Middle Permian, some Carboniferous limestones were exposed for a short time to outside influences, but without far-reaching consequences.

Between the Permian and the Triassic the same thing happened to the Permian carbonate rocks, in connection with the Pfalzian phase. Afterwards the Palaeozoic sediments were covered by Lower Triassic impervious deposits.

Oscillations, fracturing and local emersions during the Montenegrinean phase (in the Anisian) and the old Labinian phase (between the Anisian and the Ladinian), did not especially stimulate the process of karstification, because the main complexes of carbonate rocks were not yet deposited.

The new Labinian phase (between the Ladinian and the Carnian) caused a widespread emergence of jointed and fissured Middle Triassic carbonate rocks in the recent Dinaric karst area. These new relations stimulated the development of some solution dolines. But this process was soon interrupted by a new general transgression.

In the northern parts of Yugoslavia, an emergence at the end of the Triassic or during the Lower Jurassic (old Cimmerian phase?), accompanied by structural changes, introduced a longer period (until the Middle Cretaceous), during which the Triassic carbonate rocks were submitted to the influences of circulating water. Then, a widespread transgression provoked a prevailingly clastic sedimentation until the end of the Cretaceous. During the same time the central and southern region were continuously governed by typical marine conditions with only sporadic and local oscillations.

In the eastern part of Yugoslavia, a break in the sedimentation during the old Cimmerian phase was very short, and already in the Lower Jurassic the marine sedimentation continued.

The Laramian phase, between the Cretaceous and the Palaeogene, is characterized by the rising of large land masses accompanied locally by intensive structural changes. Manifold disturbances caused relative block movements, with the consequence that the carbonate sediments of different stratigraphic horizons were more or less connected. Thus, for the first time in the geologic history of the Yugoslavian region, the potential for more intensified subsurface water circulation and widespread karstification was favourable. In the northern and eastern part of Yugoslavia this process lasted until the end of the Palaeogene. But in the recent Dinaric Mountains it was already interrupted during the Laramian phase. The subsidence of larger areas provoked new transgression and marine sedimentation first of limestones, then of impervious flysch-like sediments.

The Middle Eocene movements caused differentiation of relief by the forming of some heights and depressions (mainly connected with one another). This permitted weathering processes of the elevated areas and accumulation of the products in the depressions, with the final consequence of some peneplanation. The subsurface water circulation was very small. The Pyrenean phase (Eocene–Oligocene), and that of the Savian at the end of the Oligocene, caused far-reaching structural changes, i.e., the rising of land masses on a wide scale accompanied by folding, faulting and even the forming of thrusts and overthrusts. Old sedimentation basins disappeared; new basins were created. Since than and up to the Recent time, smaller or larger surfaces of the mountains have been continuously exposed to the weathering forces which provided favourable conditions for intensified subsurface water circulation and development of karst features.

In northern Yugoslavia, the Pannonian Basin was formed. It was inundated during almost the whole Neogene leaving only the tops of the mountains exposed like islands. Thus, the possibility of the evolution of the karst in that area was diminished, first because of the water cover and later due to a thick cover of mainly impermeable Neogene sediments. Only the Miocene limestones locally enlarged the potential karst areas.

The phases during the Neogene (Styrian, Attian and Rhodanian) caused further structural changes. New basins of sedimentation were formed.

Large areas of lowland probably also existed, with a very high subsurface water level and mainly diffuse discharge into the basins, causing flooding and dissolution on a large scale. The same might happen in connection with periodic and final emptying of the basins.

The Pleistocene was introduced not only by the climatic changes but also by new structural and morphologic evolution, especially in the Dinaric Mountains. The disturbed Pliocene deposits indicate young faulting. The erosional relics of the Late Neogene at high elevations (at several hundreds or even over a thousand metres) prove an intensive general rising of land masses, accompanied by differentiated block movements on new fractures or on rejuvenated older ones. The most distinctive effects are reflected by uplift and subsidence of different areas. The areas of subsidence include the poljes where water was active before and after the diastrophic movements. The effects of the previous water action have been inherited as some morphologic features on the polje bottom, e.g., hums which may represent previous mosors, etc. The later influences are more or less reflected in their recent shape.

The changes of the climate and the rate of the diastrophic movements regulated the periods of accumulation and of removal of younger (Pleistocene) deposits from the poljes forming Recent flat bottoms. To all the described processes, an increasing cover of vegetation should be added which completed the Recent active factors in the karst areas.

MORPHOLOGIC FEATURES

In the Yugoslavian karst areas karren (lapies), dolines, pits (jamas), ponors (swallow holes), dry valleys, caves and caverns as single forms, and uvalas, poljes and karst

Fig.4. Two sets of karren on Palaeogene carbonate breccias of north Velebit. (Photograph by V. Matz.)

Fig.5. Dolines in Mesozoic carbonate rocks in the Biokovo Mountain. (Photograph by I. Bralić.)

plains as larger complex forms are developed. They all vary in shape and dimensions.

The karren are manifold corrosional ornamentations on calcareous beds of different age from the Triassic to the Tertiary. They are spread more or less all over the karst areas (Fig. 4).

The predominantly funnel-shaped dolines range in diameter from a few to over a hundred metres (Fig. 5). They are mainly the products of mechanical and chemical action of the water percolating through joints and fissures (Fig. 6) and they are the only exceptionally effects of subsurface collapses. Their distribution is highly influenced by the tectonic fracture zones (linear arrangement) or squeezed areas (scattered). They may indicate the strike of some tectonic zones.

The pits (jamas) are among the most frequent forms in the Dinaric karst. Most of them are as much as a few metres in diameter and from a few to several hundreds of metres deep. Their vertical walls suggest connections with the collapses in the karst

Fig.6. Pools in the karst of Čvrsnica Mountain (Veliki Vilinac) indicating the development of solution dolines. (Photograph by V. Matz.)

Fig.7. Terminal swallow hole (Djulin ponor) of the Dobra River near Ogulin. (Photograph by M. Malez.)

interior. The distribution shows no obvious regularity. According to the data obtained from S. Božičević pits have been recorded and measured with a total depth of about 45 km (in Croatia over 25 km, in Slovenia over 12 km, in Serbia ca. 2.5 km, in Bosnia and Hercegovina over 2 km, in Montenegro over 1 km, and in Macedonia ca. 0.7 km). Actual depths are estimated as several times greater. Some of the deepest measured pits are "Jama" at Raspor (Istria, Croatia) 450 m, "Jazben" (Slovenia) 365 m, "Duboki do" (Montenegro) 350 m, "Jama Dubinka" (Lika, Croatia) 328 m, etc.

The ponors (swallow holes) are variable in form. Very often they are funnel-shaped like dolines. Some of them coincide with the entrances of caves (Fig. 7). The rest correspond to the local concentrations of fissures and fractures without special morphologic distinctions. All the main ponors are in the tectonically disturbed zones in the steep or step-like karst areas. The most important are those on the margins of the poljes.

Dry valleys are numerous. In karst regions they formerly were occupied by surface streams that were later diverted through ponors. In these areas surface-water flow is replaced predominantly by underground flow.

The subsurface cavities are developed in all horizons of the carbonate rocks. But the caves and caverns are mostly in the Jurassic and Cretaceous limestones, the Palaeogene calcareous conglomerates and locally in the Middle Miocene limestones. There is no obvious regularity of their distribution (Fig. 8). In general, however, they are in fractured and fissured rock masses which made possible an increasing water circulation parallel with the tectonic disturbances. S. Božičević collected data of the measured systems of caves and caverns in the Yugoslavian karst. They amount to 430 km in length (in Slovenia 150 km, in Croatia 110 km, in Serbia 70 km, in Bosnia and Hercegovina 55 km, in Montenegro 30 km, and in Macedonia 15 km). The actual total length of all the caves and caverns is estimated to be several times greater. "Postojnska jama" in Slovenia is the longest one (Fig. 9). The whole system measures 16,424 m. The next longest is "Vjetrenica" (Popovo polje in Hercegovina) with 7,503 m. These are followed in Slovenia by "Križna jama" 6,949 m, "Velika Karlovica" and "Mala Karlovica" 6,500 m, "Predjama" system 5,782 m, "Planinska jama" 5,350 m, "Pološka jama" 5,200 m, "Škocjanska jama" 5,088 m, "Cerovačke pećine" (Croatia) 3,650 m, "Bogovina pećina" (Serbia) 3,517 m, etc. The dripstones are variable (Fig. 10, 11).

Many of the caves contain very precious palaeontological remains preserved in Pleistocene and Holocene deposits. M. Malez has collected the most important data. In Slovenia the remains of Palaeolithic culture and of fossil animals are discovered in "Potočka Zijalka", "Mokrička jama", "Betalov Spodmol", "Parska Golobina", "Črni Kal", "Mornova Zijalka", "Njivice", "Špehovka", etc. [BRODAR, 1938; OSOLÉ, 1965]. In Croatia analogous discoveries have been made in the following caves: "Vindija" [VUKOVIĆ, 1950; MALEZ, 1961], "Vilenica", caves at Varaždinske Toplice, "Šandalja", cave of Romualdo, "Bukovac pećina", "Cerovačke pećine",

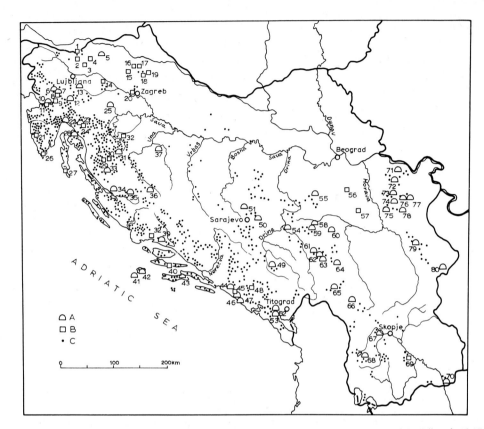

Fig.8. Distributions of various speleologic objects in Yugoslavia (Malez and Božičević, 1965). A = Larger caves and caverns in general; B = caves with palaeontologic, palaeoanthropologic and archeologic remains; C = smaller speleologic objects and undefined karst features; 1 = Potočka zijalka; 2 = Mokriška jama; 3 = Mornovijalka; 4 = Špehovka; 5 = Huda luknja; 6 = Predjamski grad; 7 = Postojnska jama; 8 = Betalov spodmol near Postojna; 9 = Parska golobina; 10 = Škocijanska jama; 11 = Črni kal; 12 = Križna jama; 13 = Taborska jama; 14 = Njivice; 15 = Hušnjakovo (Krapina); 16 = Velika pećina (Ravna Gora); 17 = Vindija near Voća; 18 = Vilenica near Novi Marof; 19 = Caves of Varaždinske Toplice; 20 = Veternica (Medvednica); 21 = Romualdova pećina (Limski kanal); 22 = Sporožna; 23 = Lokvarska pećina; 24 = Bukovac; 25 = Vrlovka; 26 = Šandalja near Pula; 27 = Medvjedja pećina; 28 = Vitezić pećina; 29 = Pećina near Ličko Lešće; 30 = Bezdanjača (Vatinovec); 31 = Caves of Plitvice Lakes; 32 = Baračeve pećine; 33 = Samogradska pećina; 34 = Pećine Velike Paklenice; 35 = Cerovačke pećine; 36 = Titova pećina (Drvar); 37 = Hruštovača; 38 = Pisana Stina (Opor Mountain); 39 = Vrnjača; 40 = Grapčeva pećina (Hvar); 41 = Biševo (Vis); 42 = Titova pećina (Vis); 43 = Pišurka (Korčula); 44 = Vjetrenica; 45 = Djurkovina; 46 = Močiljska pećina; 47 = Šipun (Cavtat); 48 = Crvena stijena near Petrovići; 49 = Ledenica (Durmitor); 50 = Novakova pećina; 51 = Biambarska pećina; 52 = Lipska pećina; 53 = Obodska pećina; 54 = Propastva pećina; 55 = Petnička pećina; 56 = Risovača near Arandjelovac; 57 = Pećina below Jerino brdo; 58 = Podpećka pećina; 59 = Stopića pećina; 60 = Prodanova pećina; 61 = Ledena pećina; 62 = Ušačka pećina; 63 = Tubića pećina; 64 = Raška pećina; 65 = Radavačka pećina; 66 = Kraljica pećina; 67 = Peštera Dona Duha; 68 = Golema peštera; 69 = Makarovec in the walley of Babuna; 70 = Peštera Bela voda; 71 = Dubačka pećina; 72 = Ceremošnja; 73 = Beljanička pećina; 74 = Divljakovačka pećina; 75 = Ravanička pećina; 76 = Verkinjica; 77 = Zlotska pećina; 78 = Bogovinska pećina; 79 = Prekonoška pećina; 80 = Vetrena Dupka.

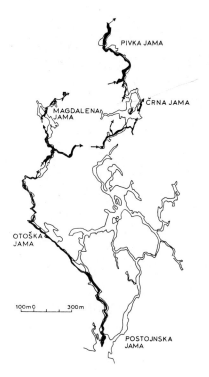

Fig.9. Simplified cave system of Postojna with underground streams. (After GOSPODARIČ, 1969.)

caves at Brina etc. In Montenegro: "Crvena stijena" [BASLER et al., 1966]. In Serbia: "Risovača" and the cave under the Jerino brdo [GAVELA, 1963]. In Macedonia: "Makarovec" [GAREVSKI, 1956]. Of special importance among them are the Croatian caves "Hušnjakovo", "Velika pećina" and "Veternica", where the remains of culture and animals were found in addition to the bones of prehistoric man.

In "Hušnjakovo" there were discovered over 600 bones (among them also carbonized ones) of *Homo neanderthalensis* var. *krapiniensis*, also fire-places, Mousterian tools etc. [GORJANOVIĆ-KRAMBERGER, 1906]. This is the richest palaeoanthropologic locality in Europe.

In "Velika pećina" deposits with Palaeolithic and Neolithic remains were preserved. In the transitory horizon between the Mousterian and the Aurignacian a fragment of a skull of the Neanderthal type was found [MALEZ, 1967]. This is the second discovery of this type in Yugoslavia (after that of "Hušnjakovo").

"Veternica" is also a locality for finding traces of prehistoric man. In the Upper Pleistocene and Holocene deposits there was found a skull fragment of *Homo sapiens fossilis* with numerous remains of fossil animals and culture [MALEZ, 1965].

Among the numerous Neolithic and younger discoveries, "Grapčeva pećina" on the island Hvar is the most important site [NOVAK, 1955].

Uvalas (a few hundred metres long) and poljes (several kilometres long) are the

Fig.10. Dripstones in Postojna Cave. (Photograph by F. Habe.)

Fig.11. Dripstones in Vrelo Cave near Fužine in Gorski kotar. (Photograph by V. Redenšek.)

Fig.12. Polje Staševica in Hercegovina with several hums. (Photograph by S. Božičević.)

larger closed depressions (Fig. 12). Though different in dimensions they are similar in shape. Uvalas are commonly dry, and the poljes are characterized by very complicated hydrogeological relations. ŠERKO [1947] lists 221 larger closed depressions in the areas of the Yugoslavian karst. Some of them are lakes or swamps, the others periodically inundated or even dry. In general, the poljes are heterogeneous in respect to geological relations, but a great abundance of impervious rocks show that their depressions were formed with the aid of tectonic forces. Generally, the shape of all poljes is similar: elongated, closed, with steep slopes and with very complicated hydrogeologic patterns. The lithostratigraphic elements are very different, especially below the bottoms of the poljes. They may be more or less covered by the impervious Tertiary sediments, or only by Pleistocene deposits, or by both of these on an old karst relief. In some poljes the impervious and older (Triassic or Palaeozoic) sediments are exposed. In some of them the Mesozoic carbonate rocks dominate. These differences show clearly that such similar depressions in lithostratigraphic mediums so variable originated mainly as a result of diastrophic movements (Neogene–Pleistocene).

The large karst plains are in a geologic sense also heterogeneous. The base of the Istrian Plain is an anticline of Jurassic and Cretaceous limestones. The plain of Gacka lies on a Cretaceous syncline. The plain of Lika is situated in the area which is affected by Palaeozoic and Triassic impervious sediments. The plain of northern Dalmatia is relatively closed against the sea by a fold system, and that of Zadvarje by Eocene flysch. The plain of Una and Korana was influenced by Neogene deposits, etc. Thus, their general pattern is connected with the tectonic setting of the areas and with the erosional forces. The final shape is due to the solution processes. Therefore they are also called corrosional plains.

Since the solution processes, though not uniquely responsible for karst development, are the most typical and final factors of karstification, attempts have been made to determine the corrosion intensity in different karst areas. According to GAMS [1965] the differences may be established even within a single drainage area. They are due to the different petrographic composition, soil cover, vegetative cover, land use, kind of vertical water circulation, number of caverns (where the hardness of dripping water diminishes because of dripstone deposition), etc. Nevertheless, he believes that the differences among the single drainage areas as closed units are comparable, and he attempted to calculate the amounts of dissolved rocks in different river systems. The results are given in m^3/km^2 per year: Ljubljanica (73), Krka in Dolenjsko (70), Soča (82), Kupa (113), Dobra (74), Mrežnica (67.2), Korana (56), Lika (46), Krka in Dalmatia (75), Zrmanja (56), Neretva with the Trebišnjica (61) etc. As can be seen, even systems close to each other can vary to a great extent (e.g., Kupa and Dobra) without any regional regularity.

The hard karst waters aid the tufa deposition due to biological and hydrochemical causes, making beautiful falls across the river beds and the lakes (Fig. 13).

Fig.13. Terminal falls of the Plitvice Lakes (left) and of the Plitvica River (right), head of the Korana River. (Photograph by M. Herak.)

HYDROGEOLOGIC CONDITIONS

It seems convenient to distinguish three main regions characterized by specific hydrogeologic phenomena: (*1*) the Adriatic insular and coastal region; (*2*) the high karst region; and (*3*) the inner region.

The Adriatic insular and coastal region

The Adriatic region comprises the area of the Adriatic Sea and the coastal zone. The karst features are present on the sea bottom, on the islands, and along the coast. The northwestern part of the sea bottom emerged from the Oligocene towards the end of the Pleistocene, i.e., in the time characterized by increasing karstification. The melting of ice and snow caused the rising of the sea level, and due to the diastrophic movements some subsidence of the present northwest sea bottom occurred—its northwest part being flooded from the southeast. This is the reason why submarine channels and local depressions of continental origin are present at the bottom. Some of the depressions are outlets of fresh water as submarine springs or vruljas (Fig. 14).

The islands are isolated morphologic and hydrogeologic units. They are mostly elongated and composed of Jurassic, Cretaceous and Palaeogene dolomites and limestones. Their elevated parts most frequently represent anticlines or parts thereof. The depressions correspond partially with the synclines. But they can also be the results of water action on the crests of the anticlines or due to longitudinal fault systems. The primary permeability is confined to the bedding planes. It is very moderate but allows some movement of water in the islands, and a predominantly diffuse mixing with the sea water. The secondary permeability, due to the tectonics, is of greater extent, and depends of the fault zones, fissures and joints. These trend diagonally, along and across the islands. The cross and diagonal fractures and fissures are very important because the main subsurface fresh water courses flow along them. The basic regulators of the subsurface water network are the dolomites in the cores of

Fig.14. Submarine spring in the Bay of Kaštela near Split. (Photograph by S. Alfirević.)

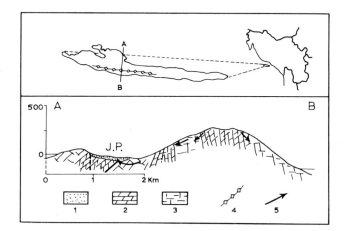

Fig.15. Simplified hydrogeologic relations of the island Hvar. *J.P.* = Jelšansko polje; *1* = Quaternary; *2* = Cretaceous dolomites; *3* = Cretaceous limestones; *4* = watershed; *5* = directions of subsurface water flow.

the folds. They act as subsurface watersheds. Since the folds are most often inclined to the south, the northern flanks of the anticlines are larger and, consequently, greater amounts of water are directed to the north coast of the islands characterized by an anticlinal structure (Fig.15). If the structural basis is a syncline, the water flows partially in the opposite direction, i.e., towards the centre of the syncline into the Quaternary aquifers (Fig.16). Exceptionally there is also discharge into freshwater lakes (Fig.17). If the fault-zones are intensive and have different trends, the mixing with sea water occurs within the mass of the islands.

Similar relations are also present along the coastline of the mainland.

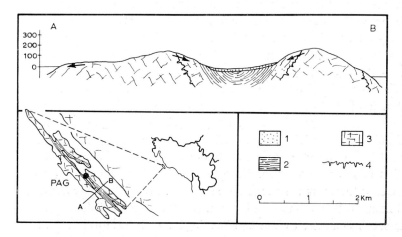

Fig.16. Simplified hydrogeologic relations of the island Pag (MAGDALENIĆ, 1965). *1* = Quaternary; *2* = Palaeogene flysch; *3* = Cretaceous and Palaeogene limestones; *4* = unconformity.

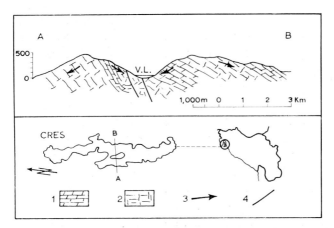

Fig.17. Simplified hydrogeologic relations of the island Cres (MAGAŠ, 1965). *1* = Cretaceous dolomites; *2* = Cretaceous limestones; *3* = directions of water flow; *4* = faults.

The relative regulators of the water courses in the islands and along the coastal zone, are the thicker intercalations of the dolomites and the flysch zones. Their function depends on their karstified base, through which the fresh water percolates and mixes with the sea water. Therefore, behind such relative barriers, brackish water can very often be recorded. Such conditions also make possible submarine springs (vruljas). These are upwellings of fresh or brackish water within the sea. The strongest submarine springs are along the coast, e.g., "Vrulja" at Donja Brela (eleven localities of upwelling), "Vrutak" and "Žrnovnica" on the littoral at Makarska, and vruljas in the surroundings of Split, Šibenik, in the Podvelebitski Canal, along the coasts of the islands Brač, Vis, Korčula, etc.

Another hydrogeologic phenomenon very characteristic of the coastal areas, is the salt-water encroachment into the inland and its mingling with fresh water. Due to the type of secondary permeability, the rate and mode of penetration is variable and the mixing does not occur uniformly. As a result of these conditions brackish water is found far inland from the coastline in some places and brackish springs occur at different levels.

The problems connected with these phenomena were analysed by several authors in order to determine how to prevent salt-water encroachment in the aquifers. V. Cukor collected the main data which will be cited in the following text. LEHMANN [1932] analysed old concepts of Fouqué and Wiebel, based on the sucking in of the salt water. He contributed to further development of the concept and gave a hydrodynamic explanation of the phenomenon. He operated with two connected fissures. The main one extends towards the spring above sea level. From its locally narrowest part the second fissure extends to the sea bottom. Increased speed of the water flow would cause the sucking-in of salt water and its mingling with fresh water. But such examples are only exceptionally to be found in nature, while in most cases the relations are quite opposite. Therefore GJURAŠIN [1942, 1943] tried to explain the encroachment of the salt water into the land and the origin of brackish springs, vruljas, sea estavelas and sea ponors on the basis of the differences in the specific gravity between the salt and fresh water. Though this concept was based only on single tubes of known properties, whilst in nature the systems have mostly more or less undefined cavities, it has remained as a useful theoretical basis for investigations in the field. KUŠČER [1950] accepted his ideas and explained several examples in the field, but also taking partially into account the sucking-in phenomenon. The general functioning of a fissure system can be imagined as follows: We start with the summer, when a coastal spring yields brackish water (Fig. 18). In the autumn, due to the recharge of new fresh water, the pressure increases. The spring water becomes less and less saline, then fresh, and finally the fresh water even issues from the second fissure leading to the sea bottom (vrulja). By decreasing the recharge of new fresh water and of the pressure, the process is reversible with the extreme consequence of sinking of the salt water into the previous vrulja, becoming now a ponor or, generally speaking an estavela. Important factors that affect the process include not only the recharge of fresh water from precipitation, but also the tides changing the sea level [KUŠČER, 1950; PETRIK, 1961; JEVREMOVIĆ, 1966, etc.]. In recent times several authors have revived the ideas of Berthollet and Fick concerning diffusion as the cause of the mingling of the fresh and salt water, adding new ideas to the interpretation [KNEŽEVIĆ, 1962, etc.]. The general conclusions of these investigators are based either on merely theoretical assumptions or solely on a few natural phenomena. But it is difficult to believe that a generally valid concept can be established for all or for the best known phenomena. The variations in nature are so extensive that every new example should be separately analysed.

Even some continuous surface water streams are connected with the Adriatic hydrogeologic region, e.g., the Lika, Mirna, Raša, Rječina, Zrmanja and Bojana, and the lower courses of the Krka, and Neretva which flow into the sea, but their springs are within the high karst region. The springs of the Mirna, Raša and Rječina

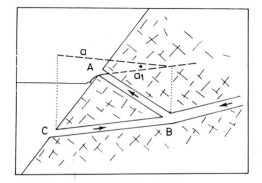

Fig.18. Simplified sketch of a brackish spring by M. Petrik. a = Piezometric level for C–B (due to greater specific gravity of the sea water); a_1 = piezometric level for B–A.

are due to the complete or relative barrier of Palaeogene flysch which causes the overflow of ground water, completely (Mirna) or partially (Rječina). In the last-mentioned case, some water flows under the flysch. Also, there is some loss of water in the river bed within the carbonate rocks. The springs of Zrmanja are due to the true barrier consisting of Lower Triassic sediments. Within the carbonate river bed, swallow holes also occur. Krka ends in a bay with brackish water, Cetina through a gorge, Neretva in a plain, and Bojana issues from Skadarsko jezero (Scuttari Lake) and flows to the sea over a covered flat surface.

Also along the coast, near the sea, some accumulations of fresh or brackish water are to be found as lakes, and "oka" ("eyes").

The annual rainfall in the Adriatic insular zone varies from 600 mm (southern isles) to 1,200 mm (northern isles), but due to the Mediterranean climate, its distribution and consequently the water regime, is irregular in time. This is the reason for the lack of water during the dry seasons.

The high karst region

The high karst region occupies the most elevated parts of the Dinaric karst, comprising numerous heights (up to over 2,000 m), poljes and large plains. The limit of the Adriatic insular and coastal zone is not sharply defined. The northeast border is marked by springs of the main southern tributaries of the river Sava. Surface streams are rare. Two of them flow continuously to the sea (the Cetina and the Neretva), but some water disappears into the ponors on the way. The others represent true sinking rivers and brooks (ponornice). Some of them have underground connections. Their surface courses are bound partly by the impervious base and partly by the structures, faults, synclines and anticlines, with impervious seaward flanks which cause the rise of the groundwater and make possible the existence of surface brooks and rivers. Along this zone trends the watershed between the Adriatic and the Black seas. It partly corresponds to the elevations (mountains and ridges), but mostly it is bound by deeper lithostratigraphic and tectonic relations. It so happens that it strikes along the depressions or plains.

The unconformity between the subsurface watersheds and land forms is frequent in the areas with the relatively impervious ("hanging") barriers which retain surface streams (notwithstanding underground flow in the opposite direction). In Slovenia very characteristic examples are the Planinsko polje and the Cerkniško polje (Fig. 19). The geological data [after PLENIČAR, 1962] and the results of dye tests [GAMS, 1965; GOSPODARIČ, 1969] show that the surface waters of the Pivka Basin are bounded by the Palaeogene flysch. The Pivka sinking river flows into the cave "Postojnska jama" and continues underground in the same direction towards the Planinsko polje through Cretaceous rocks. On the contrary, the water discharged into the cave "Predjama" turns underground towards the west, i.e., towards the Vipava River along a very intensive fault zone while the flow parallel to that of the Pivka and the Planinsko

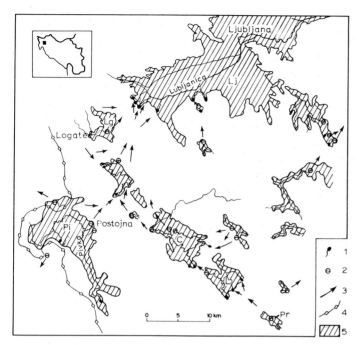

Fig.19. Hydrogeologic system in high karst of Slovenia (GAMS, 1965; GOSPODARIČ, 1969). Pi = Pivka Basin; Pr = Prezid; L = Loško polje; C = Cerkniško polje; Lg = Logaško polje; LJ = Ljubljansko barje; 1 = springs; 2 = swallow holes; 3 = directions of water flow; 4 = watershed; 5 = depressions (including the poljes).

polje is at least partly impounded by Upper Triassic dolomite. Owing to these conditions the watershed between the Adriatic and the Black seas is complicated and cannot be completely defined by land forms or the topography.

The relations in the Cerkniško polje cause only local irregularities, but typical ones which are due, in part, to the attitude of the Triassic sediments. According to the geologic data received from PLENIČAR [1962], on the eastern side of the polje between the Bloška Planota and Borovnica, the impervious Triassic sediments form the

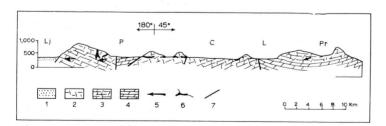

Fig.20. Cross-section along the poljes in Slovenia. Pr = Prezid; L = Loško polje; C = Cerkniško pole; P = Planinsko polje; Lj = Ljubljansko barje; 1 = Quaternary; 2 = limestones; 3 = limestones and dolomites; 4 = dolomites; 5 = direction of water flow; 6 = underground cavities; 7 = faults. (After GOSPODARIČ, 1969.)

normal foot-wall of the Jurassic. Though the whole complex might have been moved, the impervious Triassic sediments function as a barrier on the surface and below the Jurassic (Fig. 20). This is the reason for the surface streams on the northeastern margin of the Cerkniško polje. The same explanation is valid for the main spring Obrh. In addition to the margins of the polje, as well as below the Quaternary on the polje bottom, is another belt of the Triassic, consisting of Upper Triassic dolomite bordered by Jurassic and Cretaceous carbonate rocks. Under normal conditions (as an inlier) it would be a very good barrier. But, in this case it is overthrust on the younger carbonate rocks and secondary faulted. This is the reason for groundwater below the dolomite and locally across it. Hence, the differences among the land forms, surface streams, and underground drainage characterized by bifurcation can be understood. One direction is parallel to the long axis of the polje and the other is across it (towards the north) more or less opposite to the surface stream Cerkniščica.

The Glamočko polje (Fig. 21, 22) and the Gatačko polje (Fig. 23, 24) are additional examples of the trend of the watershed across depressions between the Adriatic and

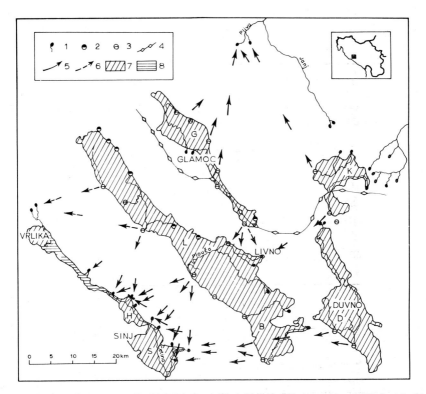

Fig.21. Hydrogeologic system of the poljes in Bosnia, Hercegovina and Dalmatia (data obtained from the Elektroprivreda Dalmacije). S = Sinjsko polje; G = Glamočko polje; K = Kupreško polje; H = Hercegovina; B = Bosnia; D = Dalmatia; 1 = springs; 2 = estavelas; 3 = swallow holes; 4 = watershed; 5 = directions of subsurface water flow (proved); 6 = directions of subsurface water flow (assumed); 7 = poljes; 8 = storage basin of Peruča.

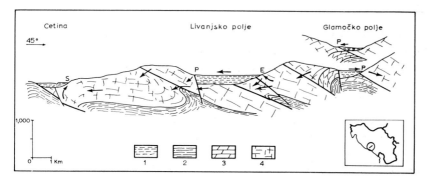

Fig.22. Simplified cross-section of the poljes area in Bosnia and Dalmatia. S = Springs; E = estavelas; P = swallow holes; 1 = Neogene clastic deposits; 2 = Triassic clastics; 3 = Mesozoic dolomites; 4 = Mesozoic limestones. (After J. Papeš.)

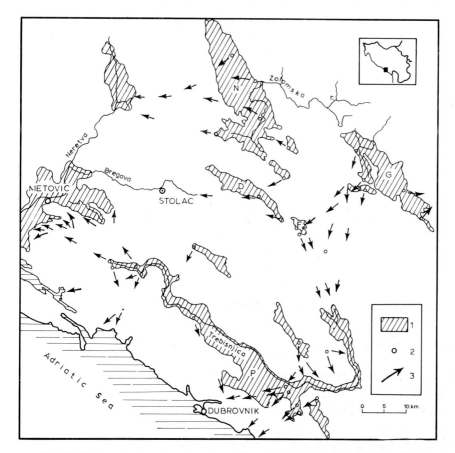

Fig.23. Hydrogeologic system in eastern Hercegovina (data obtained from the Energo-invest Sarajevo). P = Popovo polje; D = Dabarsko polje; N = Nevesinjsko polje; F = Fatničko polje; G = Gatačko polje; 1 = depressions; 2 = localities of dye testing; 3 = directions of subsurface water flow.

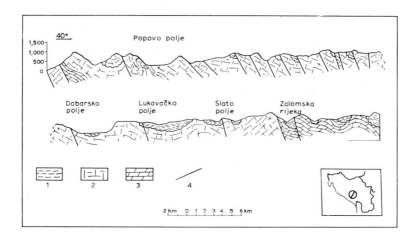

Fig.24. Simplified cross-section of the poljes area in eastern Hercegovina by K. Torbarov. *1* = Mesozoic dolomites; *2* = Mesozoic and Palaeogene limestones; *3* = Tertiary clastic deposits; *4* = faults.

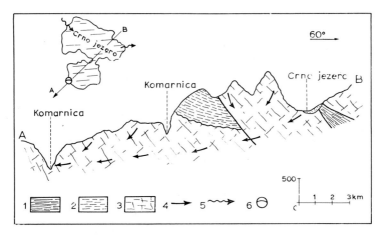

Fig.25. Simplified hydrogeologic cross-section in Montenegro in the area of Crno jezero (Black Lake) and the river Komarnica (BEŠIĆ, 1965). *1* = Lower Triassic clastics; *2* = Cretaceous flysch; *3* = Mesozoic carbonate rocks; *4* = directions of subsurface water flow; *5* = directions of surface water flow; *6* = swallow holes.

the Black seas. There are ponors oriented to the north and to the south. Local bifurcations exist in the Fatničko polje and Popovo polje.

Also, Crno jezero in the Durmitor area lies on a local watershed, loosing its water northwards and southwards (Fig. 25). The numerous local watersheds are very often incongruent with the topography. Many of the distinctive mountains, consisting predominantly of limestones, are pierced by subsurface streams and do not function as watersheds at all.

Among the most conspicuous features of the high karst region are estavelas

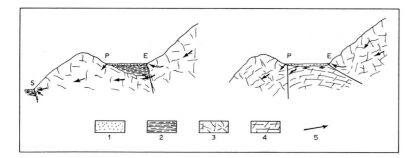

Fig.26. General sketch of two types of estavelas and swallow holes. S = Spring; P = ponor (swallow hole); E = estavela; 1 = Quaternary; 2 = Tertiary clastic deposits; 3 = Mesozoic dolomites; 4 = Mesozoic and Palaeogene limestones; 5 = directions of water flow.

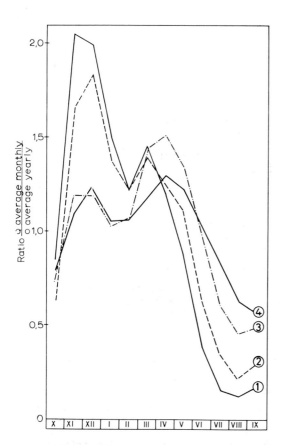

Fig.27. Variability in discharge for the period 1925–1955 for the following rivers. 1 = Trebišnjica (station: the Arslanagić bridge); 2 = Cetina (station: Trilj); 3 = Bosna (station: Zenica); 4 = Pliva (station: Volari); I–XII = January–December. (After MIKULEC and TRUMIĆ, 1969.)

(Fig. 26) caused by relative barriers and oscillations of water flow, which reach locally deep below sea level, varying from local water tables to the more or less isolated water courses. The variability occurs laterally as well as vertically. It is intensified by the rate of precipitation (1,000 to over 5,000 mm or about 1,500 mm average), irregularly distributed in time due to the influences of the Mediterranean climate. The normal maximum occurs in the second half of the autumn, and a secondary one in the spring with temporarily unbalanced discharge (Fig. 27). According to MIKULEC and TRUMIĆ [1969] the average annual discharge in the Dinaric Mountains is ca. 15 m^3/m^2. The run-off is about 70%.

The inner region

The inner region is widespread on the north and east of Yugoslavia, but it comprises mostly isolated and often small karst areas, which are frequently shallow and incomplete. The poljes are absent. The karst areas are often dammed by the impervious rock complexes. The tectonic forms are very complicated, but only moderately relevant to the general karst relations, The whole region is characterized by a well-developed continuous surface water network. The loss of water is only local, within the single-catchment areas. The local watersheds are frequently well defined, but they are not always congruent with the land forms. The water from precipitation percolates more or less vertically downward to the water table and then moves laterally toward a surface river. The deep subsurface streams are rare. The groundwater below the river beds influences the appearance of springs. The small "oka" (eyes) are direct offsprings of the groundwater. The annual precipitation ranges from 650 to 1,200 mm. Its distribution is regular, not being affected by the Mediterranean regime. The run-off is smaller then in the previous zone.

Two areas deserve short separate comments, i.e., the Alpine karst and that of the eastern Serbia.

The Alpine karst is developed especially in the Julian Alps and Kamnik Alps. According to GAMS [1965], this karst is predominantly of a high-mountain character developed in very thick Triassic limestones (over 2,000 m). The annual precipitation is abundant (1,600–3,000 mm). The karren, grooves, small sinks with vertical walls (kotliči = small kettles) and gently sloping dolines are frequent above the forest line (between 1,500 and 1,900 m). The potholes are especially numerous as a consequence of the predominant vertical water flow owing to the considerable thickness of the limestones (from the highest parts to the bottom of the valleys about 2,000 m). The caves are rare. The connection between the non-karstic areas and the streams which once might have flowed through the caves has not been established.

The following comments on the karst areas of the Carpatho-Balkanian Belt (eastern Serbia) are based on data received from MILOJEVIĆ [1962]. The karst phenomena are developed in the carbonate rocks of the Triassic, Jurassic, Cretaceous and Miocene, with predominance of the Jurassic and Lower Cretaceous. The limestones

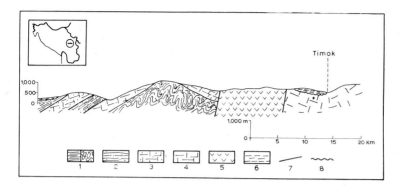

Fig.28. Simplified cross-section through the Carpatho-Balkanian karst areas (MILOJEVIĆ, 1962). *1* = Palaeozoic schists; *2* = Permian sandstones; *3* = Triassic limestones; *4* = Jurassic and Cretaceous limestones; *5* = igneous rocks; *6* = Tertiary deposits; *7* = faults; *8* = unconformity.

and dolomites are distributed in several narrow, elongated zones which are tectonically disturbed (folded, jointed, fractures and overthrust). Owing to the allochthon tectonics, they are mostly overlain and underlain by impermeable Palaeozoic rocks (Fig.28). In this karst all morphologic features but poljes are developed. As to the hydrogeology they represent open structures in which the recharge area and the area of underground-water distribution are interconnected. The recharge depends on the precipitation and surface streams coming from the non-karstic areas. The water of such streams in the karst belt of Kučaj–Beljanica disappears, leaving numerous karstified dry valleys. Impervious Tertiary deposits locally cover some karst areas, but without influencing the water flow below them. The precipitation varies from 650 mm (northern part of Golubac) to 1,000 mm (Stara Planina). It recharges the karst groundwater, sinking to the non-karstic base. The discharge occurs at numerous springs with great variations (Q_{min}/Q_{max} varies from 1/2 to 1/20).

PRACTICAL PROBLEMS AND THEIR APPROACH

The specific characteristics of the karst areas, especially unbalanced water regimes, i.e., isolated surface flows, general lack of water in the dry seasons and surplus during the wet months, and also large areas without a cultivable soil, scarce vegetation etc., long ago imposed many practical problems concerning living conditions for the population. The first task was to enlarge the cultivable soil by preventing inundations or by irrigation, and to find out how to provide more potable water. At first the improvements were only local and on a small scale. Not until recent times, when the karst water became recognized as a potential source for producing electrical energy, were the problems analysed in mutual connection. The widening or tightening of swallow holes, the surficial or underground accumulations, the derivations of water from one catchment area to another etc., can have far-reaching effects on the lower-lying areas. Therefore there were investigations on a large scale including geologic mapping, hydrogeologic and speleologic explorations, dye tests, analyses of run-off, testing of the degree of permeability of various carbonate areas and horizons, tightening by grout curtains, etc. Some of the important achievements and experiences will now follow.

Practical aspects of geology

Plans for hydroelectric power plants, and water supply stimulated intensive geologic explorations in order to reconstruct structural, morphologic and hydrogeologic relations relevant to the practical applications in karst areas. We shall point out only the most important ones. In many cases it was possible to show that the tectonic position of clastic impervious series affected the hydrogeologic conditions giving different roles to some homogeneous units, and a similar role to some heterogeneous units in relation to facies and age.

Concerning the Palaeozoic sediments, it has been established that at most localities in the Dinaric Mountains, they are inliers, and consequently that they function as true barriers, cutting off the underground karst streams and affecting local watersheds. But examples were also found where Palaeozoic complexes formed only hanging parts of smaller or larger overthrusts and therefore having only the function of a relative barrier, as for example in the area of the upper flow of the Kupa River (Fig. 29). This required a more careful approach to the hydrogeologic interpretation based merely on the lithologic properties of the rocks.

The Triassic clastic sediments most frequently form the outliers of the Palaeozoic rocks, and define their hydrogeologic function. The Lower Triassic is the most widespread of these sediments with only moderate lateral variations. Its function as an aquiclude is similar to that of the Palaeozoic inliers. Middle and Upper Triassic clastics appear generally as lenses within the carbonate complexes. Thus, their hydro-

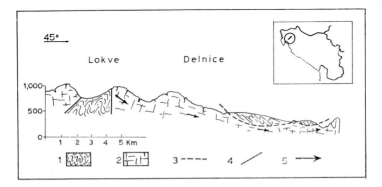

Fig.29. Simplified hydrogeologic cross-section through Lokve–Delnice Kupica (tributary of Kupa River) in Gorski kotar. *1* = Upper Palaeozoic clastics; *2* = Mesozoic limestones and dolomites; *3* = overthrust; *4* = faults; *5* = directions of subsurface water flow.

geologic function is localized. The Cretaceous and Palaeogene flysch and flysch-like sediments as a whole also represent aquicludes. But their stratigraphic position (within or in the hanging wall of the Mesozoic and Palaeogene carbonate rocks) lessens their hydrogeologic role. Only in the thrust faults and overthrusts as well as in the folds, they can achieve quite a significant function as relative or exceptionally true barriers (Fig. 30). The Neogene clastics, due to their position, only have a function of relative barriers. In connection with this some storage basins are confined within the clastics, others flood the calcareous border areas without loss of water.

Faults are also established with uplifted blocks consisting of impervious sediments. The role of such faults is hydrogeologically positive. They give rise to numerous overflow springs. In many cases the anticlines also have the same effect. A most striking example is the anticline on the north side of Velika and Mala Kapela (Fig. 31), which cuts off the waters of those mountains flowing underground to the north, although the core of the anticline consists mostly of dolomites (Triassic and Jurassic). In this way the springs of Vitunj, Zagorska Mrežnica (Fig. 32A), Dretulja, and the estavelas of Begovac Lake (Fig. 32B), Lička Jesenica etc., originate. The same can be said for the anticlinal structures in the polje of Štikada in Lika (Fig. 33).

Even the larger synclines in the carbonate rocks have a positive effect on hydrogeologic relations, the Gacka River issues and flows in a syncline. These and other similar examples show that the major tectonic structures define the fundamental hydrogeologic relations even in the Dinaric karst. This fact stimulated detailed investigations also in the monotonous carbonate complexes. The results required further exploratory work and influenced the projects.

The major tectonic structures also helped to understand some important morphologic features in karst. As for the large karst plains, it may be mentioned once more that they are geologically heterogeneous and belong to larger tectonic units which may be taken into account in all practical considerations concerning these plains.

The relations concerning the poljes are quite different. The main problems are

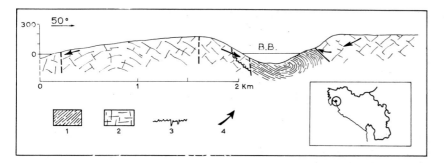

Fig.30. Cross-section of the area of the Bay of Bakar in the Croatian littoral by Ž. Vulić. *1* = Palaeogene flysch; *2* = Cretaceous and Palaeogene carbonate rocks (mostly limestones); *3* = unconformity; *4* = directions of subsurface water flow; *B.B.* = Bay of Bakar.

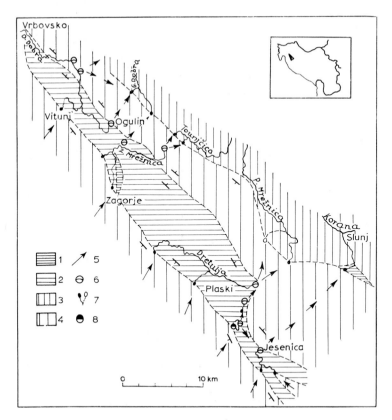

Fig.31. Sketch of the anticline Vrbovsko–Jesenica with its hydrogeologic pattern (BAHUN, 1967). 1 = Triassic clastics; 2 = Triassic and Jurassic dolomites (inlier); 3,4 = Jurassic and Cretaceous limestones with some dolomites; 5 = directions of subsurface water flow; 6 = ponors (swallow holes); 7 = permanent and periodical springs; 8 = estavelas.

concentrated along their margins, where springs and estavelas (inland margin) and ponors (seaward margin) occur.

The microtectonic structures mainly regulate the local hydrogeologic pattern. Systems of joints and fissures have to be established in order to localize the technical measures connected with the projection of accumulation basins etc.

Application of geophysical methods

The supplementary methods in hydrogeologic investigations are chiefly the methods of electric sounding, electric mapping, "mise à la masse", and electric well-logging (separately or in combination). They are used to determine the depth of karstification and the boundaries among the zones of differing numbers of cavities, with or without water; the prevailing directions of karstification within a chosen horizon combined with the prevailing subsurface water flows; the thickness, distribution and rate of

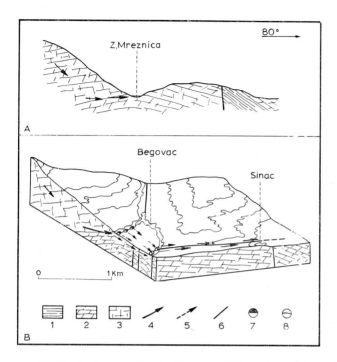

Fig.32. A. Cross-section in the area of the spring of Zagorska Mrežnica (BAHUN, 1967). B. Block diagram of the periodical lake Begovac in Mala Kapela (concept by M. Herak; reconstruction by S. Bahun). *1* = Triassic clastics; *2* = Mesozoic dolomites; *3* = Jurassic and Cretaceous limestones; *4* = directions of subsurface water flow; *5* = directions of surface water flow in the lake; *6* = faults; *7* = estavelas; *8* = swallow holes.

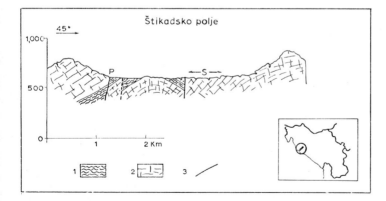

Fig.33. Cross-section in the area of Štikadsko polje near Gračac in Lika. *S* = Main zone of springs; *P* = main zone of swallow holes; *1* = Middle Triassic clastics; *2* = Triassic and Jurassic carbonate rocks; *3* = faults.

permeability of clastics covering a karst area; the deep-lying aquicludes; the boundaries between the fresh water and brackish (or salt) water; and the direction and rate of underground water flow by a single well, etc. The data obtained are an important contribution to the explanation and practical evaluation of different hydrogeologic phenomena in the Dinaric karst areas, especially in the Adriatic zone [KRULC, 1967; ARANDJELOVIĆ and MILOŠEVIĆ, 1967, etc.].

Practical aspects of speleology

Speleology was among the oldest exploratory investigations of the Dinaric karst; it is concerned not only with the karst peculiarities, but also with various practical problems. Many of the caves were inhabited by prehistoric man during Pleistocene time. Afterwards they were only sporadic shelters from storms, and from enemies during war times, etc. Some of them were used as lairs by various animals, especially cave bears, hyenas, etc. Therefore the cave deposits may be collecting places for very precious fossil remains, while the caves with rich and attractive dripstones are very useful for the development of tourism.

Water from caves has been used as temporary or permanent sources for primitive water supply. The first exploration works were mostly performed by amateurs. Increased practical and theoretical interest, however, was arousing more and more

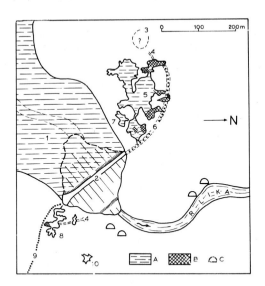

Fig.34. Storage basin in the valley of the Lika River (Croatia), accomplished by technical measures in the Palaeogene carbonate breccias. A = Accumulated water; B = injection mass; C = entrances of caves; 1 = storage basin "Kruščica"; 2 = dam "Sklope"; 3 = subsurface cavity registered by bore hole; 4 = investigation gallery; 5 = Horvat's Cavern; 6 = grouting gallery and grout curtain; 7 = Golubarnik Cave; 8 = Poljak's Cave; 9 = grout curtain; 10 = Ivo's jama. (After BOŽIČEVIĆ, 1969).

professional interest. But not until recent times has this activity been organized and supported adequately in proportions to its significance.

Planning of hydroelectric projects, especially requires a clear understanding of the speleologic phenomena in order to help to solve the hydrogeologic relations and to establish the conditions for artificial storage basins and the construction of grout curtains, etc. (Fig. 34). According to personal information from S. Božičević, about 12,000 speleologic phenomena (caves, caverns, pits and ponors) have been explored in Yugoslavia. But this is probably only about a fifth of those existing.

Practical aspects of hydrology

Since the Dinaric karst has many sinking rivers and large areas without surface streams, the main interest was paid to the subsurface connections among various catchment areas. The first ideas were more empirical and qualitative and often based on non-verified observations. But for application, reliable and quantitative data are needed. Therefore different tests and measurements were undertaken in order to establish underground connections, and to define the catchment areas as precisely as possible. Also efforts were made to define the run-off in the Dinaric karst.

Dye tests were performed by different organizations, but were only partly controlled. Therefore some of the results obtained are not reliable. Fluorescent materials, uranine "0", eosine, salts, spores, and isotopes were used as tracers. The main results (Fig. 19, 21, 23, 35) show how intricate the connections can be. In some cases it was possible to determine quantitative relations from dye-tests. At Ogulinska Dobra–Gojak 83% of dye was recorded and at Plouča–Rumin 93.9%, but at Lička Jesenica only 38.93%. The deficit of dye can be due to the absorption or dissipation in the underground, to solar radiation, or to loss in unknown directions. It is also necessary to take into account the limitations of the method which is based on visual estimation. The test of Plouča–Rumin offers a very instructive example, because the dyed water issued only from one spring called "Veliki Rumin". The test was organized and interpreted by A. Stepinac. The following formula was used to compute the run-off of the dyed water:

$$B = 10^{-x} \cdot Q$$

where B is the run-off of dye in g/sec; Q is the run-off of water in cm^3/sec; and x is the rate of dilution of dye.

Although the underground connection between the "Čaića ponor" and the spring "Veliki Rumin" is a direct one, the subsurface water flow is not simple (Fig. 36). In addition to main stream system, there exist at least two more streams causing secondary maxima of the discharge of the dyed water. Furthermore, the rise of the water level may result in the connections not only with "Veliki Rumin", but also with the other springs.

The mean velocity of the water was observed by different dye tests to see the effects

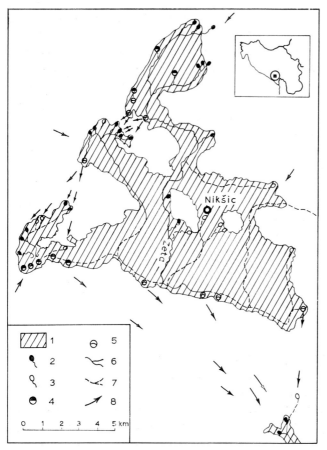

Fig.35. Nikšićko polje in Montenegro. *1* = Depression; *2* = permanent springs; *3* = periodical springs; *4* = estavelas; *5* = swallow holes; *6* = permanent surface streams; *7* = periodical surface streams; *8* = directions of underground flow. (After VLAHOVIĆ, 1958.)

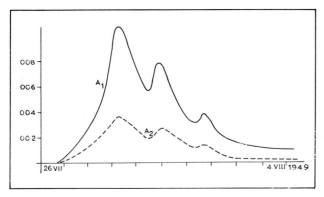

Fig.36. Discharge of the dyed water at the spring Veliki Rumin, the tributary of the Cetina River (STEPINAC, 1969). A_1 = Quantity of the dye; A_2 = intensity of the dye.

of the gradient. The results confirmed the negative assumption based on the fact that only apparent gradients estimated on the shortest distances can be determined, and further that the average velocity depends also on the widths of the cavities and other factors. We checked 107 examples at random in the northern and the middle parts of the Dinaric Mountains (Slovenia, Dalmatia, Bosnia and Hercegovina). The apparent gradient ranges from 0.001 to 0.076 (average 0.0216). The mean velocity varies from 0.06 cm/sec to 12 cm/sec (average 3.6 cm/sec).

The problem of the rate of percolation and evapotranspiration has so far been a subject of merely qualitative estimates. Therefore it was impossible to determine the capacity of the underground aquifers on the basis of water balance and on the changes of piezometric level. Anyhow, there is no doubt that the variability of recharge is largely depending on evapotranspiration and distribution of precipitation in space and in time. The correlation between the precipitation and run-off is analysed as follows from two examples by STEPINAC [1969]:

(1) The catchment area of Ličanka in the Gorski kotar (western Croatia) is mostly covered by forests. The surface is 38 km^2; average precipitation in a period of 10 years is 2,609 mm; and the medium run-off in the same period is 1.92 m^3. The annual variability of precipitations is 1,630–4,060 mm and of the run-off 0.83–3.16 m^3/sec.

(2) The catchment area of Radoblja, near Mostar (Hercegovina) comprises 83 km^2, mostly of uncovered karst. The average precipitation for a period of 15 years is 1,575 mm; medium run-off 2.47 m^3/sec. The annual range of precipitation is 884–2,312 mm, and the run-off is 1.13–3.50 m^3/sec.

In both cases, the annual run-off is in a tolerable correlation with the precipitation, it is better at Radoblja than at Ličanka. But the monthly variations (and discrepancies) are greater at Radoblja, because its catchment area is larger and therefore under various influences (rainfall in the hinterland, retardation of run-off in the karst interior of melting snow, evaporation, etc.).

In general, as has already been stated, the precipitation in the Adriatic and Dinaric high karst regions has an irregular intensity and distribution. Some 60–70% of the precipitation in the Dinaric Mountains falls between October and March. In the dry months almost all of the precipitation returns to the atmosphere due to evapotranspiration. In addition there is a time lag in the recharge of the aquifers. Therefore, specific recharge and run-off coefficients must be analysed only for longer periods.

Annual run-off coefficients at Ličanka are 0.48–0.72, and at Radoblja 0.48–0.80. The coefficients are in reverse proportion to the amount of precipitation.

In the Dinaric karst the catchment areas are very frequently difficult to define precisely. Therefore hydrogeologic analyses have to be applied to the groups of springs, which are wide enough to attain the required accuracy. In general, high waters in karst are prolonged, low waters shortened.

Surface water storage basins in karst

The abundance of water in the Dinaric karst, extensive divergencies in its distribution in time, and also water courses at high elevations aroused the idea of constructing storage basins for the utilization of water for producing electric energy, for irrigation, etc. According to REŠTAROVIĆ [1969], practically every litre of water falling 80–90 m represents the power of 1 kW. The steepness of the slopes and of the step-like elevations of the karst depressions makes it possible, with relatively short derivation tunnels, to obtain concentration of considerable heads of water which turn the Dinaric karst into a very economic potential source of energy and of water supply. This, however, is only on the condition that a part of the high water is retained for the dry season. The variability of discharge, as already stated, is large. According to REŠTAROVIĆ [1969] the approximate ratio of the minimum, avarage and maximum values is:

$$Q_{min} : Q_{mean} : Q_{max} = 1 : 10 : 100$$

Sometimes the differences are still larger, but rarely smaller due to underground retentions. The artificial accumulations may secure daily, monthly or yearly balance in power production. If balance for even longer periods were necessary, then the differences in mean discharge in dry, average and wet years would be in the ratio of 0.6 : 1 : 1.5 [REŠTAROVIC, 1969].

The accumulations in the Dinaric karst which are planned, in course of construction, or already accomplished are numerous, and they are distributed within the catchment areas of Ljubljanica, Ličanka, Rječina, Lika, Gacka, Zrmanja, Cetina, Neretva, Trebišnjica, Zeta, Morača, Kupa, Una, Vrbas, Drina, etc. The amounts of water are large, e.g., in the catchment area of Cetina 35%, Neretva 20%, Trebišnjica 36% of the average run-off.

The accumulated water can be used for power plants, agriculture, fishing and water supply for population and industry. The biological minimum water level needed in order to preserve the governing relations in karst should also be taken into consideration. New accumulations have to take this into account.

In respect to agriculture the problem is twofold. In the construction of storage basins, cultivable soils are covered by water and are lost for cultivation. On the other hand, the reservoirs make possible irrigation and the intensification of agriculture. According to the calculations of the prices of energy and agricultural products [REŠTAROVIĆ, 1969] the limit of priority between storage basins and agriculture lies in the Dinaric karst at approximately 700 m above sea level.

The geologic problems in planning hydroelectric power plants are manifold. The first problem concerns the basin, in respect to volume and imperviousness. According to the natural conditions there may be only limited possibilities for constructing storage basins, e.g., Jablanica in Hercegovina, which is underlain by the Lower Triassic. The same can be said for Ličanka and Lokvarka in Gorski kotar, which are confined within the Late Palaeozoic, and Rječina in the Croatian littoral within the

Palaeogene flysch. But in the last-mentioned cases the accumulation is limited, because of the relatively low junction of the impervious clastics and karstified carbonate rocks. Especially interesting are the conditions in the Rječina Valley, where water courses exist below the flysch on which a "hanging" storage basin is under construction.

In all other cases smaller or greater inundations of the carbonate margins will occur and therefore it is necessary to secure the storage basins from loss of water by means of technical measures (tightening of swallow holes and fractures, grout curtains, etc.). Some characteristic examples will be cited.

The convenient morphology and abundance of water which caused periodical inundations of the Cerkniško polje and the Planinsko polje in Slovenia, as well as their elevation above sea level (Planinsko polje: 447 m, Cerkniško polje: 550 m), made them attractive as potential storage basins (Fig. 19). The whole area is underlain by carbonate rocks. The oldest rocks are Upper Triassic dolomite followed by Jurassic dolomite and limestone and Cretaceous limestone. Most of the depression of the Planinsko polje is underlain by Upper Triassic, mostly watertight dolomite. Only the northeastern side consists of Jurassic and Cretaceous limestones. On the same side the Idrian thrust fault strikes; along this fault the Jurassic and Cretaceous pervious rocks are overthrust on the Triassic dolomite. Along this tectonic zone secondary, variously directed disturbances are also present. This is the reason that the main underground discharge of the polje occurs on this margin through numerous ponors. Some ponors within the polje (in the northern part) are also to be explained by fractures below the Quaternary which guide water directly to the main zone of ponors. The whole body of Triassic dolomite is more or less compact and as a whole impervious, forming a barrier to the waters coming from the southwestern side. The consequence is visible in the springs and surface water course of the sinking river Unica. The accumulation of water is possible due to tightening of the ponors within the storage basin preventing the water from coming in direct contact with the Jurassic and Cretaceous rocks along the fault line.

The Cerkniško polje is situated in the southern prolongation of the same tectonic belt. But the differences in detail, concerning the tectonic conditions, make the problem of water storage more complicated. Along the Cerkniško polje, both the previously mentioned main fault zones can be followed. Upper Triassic dolomites do not underlie the Quaternary below the whole polje bottom. They are tectonically restricted to the central and northeastern part. In connection with this, however, the underground water also comes from the southwestern side (through Javorniki), the main ponors are situated on that side, which enables the water to flow in two directions, i.e., to the Planinsko polje and also under the bottom of the Cerkniško polje and along the secondary faults towards the springs of Ljubljanica (Bistra and Lubija). Secondary tectonic disturbances are present on both margins of the polje. Those at Cerknica seem to be especially important. The relations described show that the loss of subsurface water from the accumulation could be prevented only by large and deep grout curtains, in order to cut off not only the localized water courses through Upper Triassic dolomite, but also those within the Cretaceous (or possibly Jurassic) below the Triassic. The local tightening of the ponors could not secure the accumulation during the low-water stand in the surrounding area.

In the catchment area of Cetina River several storage basins and power plants are planned. Two of them are already finished, i.e., Peruča (Fig. 21) and Pranjčevići. The storage basin of Peruča has the main role of remote control of the discharge for the power plant at Split, which is the last plant of the whole system.

Peruča was one of the first storage reservoirs planned to be constructed almost completely in a karst area consisting mainly of Jurassic and Cretaceous carbonate rocks. In that area the Cetina River connects widened depressions similar to poljes by means of short canyons, but without zones of swallow holes. It was first necessary to overcome inherited prejudices concerning the governing concepts of karst. To solve the existing hydrogeologic problems and to foresee the changes which would occur after the accumulation with a dam 60 m high has been constructed, it was necessary to take into consideration a large area including the Troglav and Svilaja Mountains, the first on the northern and the second on the southern side of the Cetina River. The problems were centred around three main questions: (*1*) how would the karst springs on the margin of the small poljes and in the valley of the Cetina River react after being flooded; (*2*) was it possible for the accumulated water to disappear from the poljes on the karstified slopes of the Cetina Valley; and (*3*) how to prevent the accumulated water from passing around and below the dam.

The general geologic conditions of the larger area showed, in spite of an extensive karstification on the slopes of the valley and of the poljes along the Cetina River, that the storage of water and the flooding of some springs are possible due to the structural and hydrogeologic conditions in the Troglav and Svilaja Mountains. In the body of the Troglav, the gradient of the water rises from Cetina to the Livanjsko polje. Besides this, a local anticline prevents the water from flowing along the river side. Svilaja represents a generally complex form with a compressed dolomitic core and a southern slope partly dammed by impervious Lower Triassic clastics. Under these conditions, the karstified banks and slopes can only increase the storage of water by underground accumulation. There only remained the prevention of discharge of water from the basin below and on the sides of the dam along joints and fissures, as well as through several springs on the left, i.e., northeast side which, under the pressure of the accumulated water, may be diverted into the swallow holes or estavelas.

The detailed study [PAVLIN, 1961] showed that this was possible by means of the completion of a thixotropic grout curtain composed of 70–75% clay and 25–30% cement below and on both sides of the dam. After completion of the grout curtain, a study showed that the total seepage, with practically maximum elevation of water level, amounted to approx. 0.7% of the average discharge on the same profile of the Cetina River. Thus, the results were better than had been foreseen. No traces of underground erosion have been noticed.

In the lower course of the Cetina, the storage basin has been established in the pure karst consisting of Cretaceous and Palaeogene carbonate rocks. An explanation of the hydrogeologic conditions on the basis of tectonics was not possible. Therefore, bore holes ranging in depth from 10–250 m have been drilled, in which the permeability has been measured and water level checked during a period of several years. The data obtained showed that the permeability decreases with depth. Besides this, there is evidence of interconnection of the underground water network with irregular changes in closely spaced bore holes. To prevent loss of water from the storage basin, grout curtains and local concrete walls were constructed [PAVLIN and MLADINEO, 1957; MAGDALENIĆ, 1965].

The important role of technical measures in transforming a karst depression into a storage basin may be shown by the example of Buško Blato in the eastern prolongation of the Livanjsko polje (Fig. 21). The surroundings and the bedrock of the polje are Cretaceous limestones and dolomites and partly Palaeogene calcareous conglomerates and breccias; the bottom is covered partly by Quaternary deposits and partly by Neogene marls. The springs are situated on the eastern margin of the polje, the ponors on the western and southwestern side. The springs are intermittent, and during the summer there is only occasional inflow into the polje. In this season the water only percolates below the Neogene and Quaternary bottom cover towards the ponors. The only controlling factors on the subsurface flow are major and minor folds in the Cretaceous deposits. In order to construct a storage basin it is necessary not only to tighten the ponors, but also to cut off the water in the bedrock below the Quaternary and Neogene deposits. For this purpose, a grout curtain has been planned by E. Nonweiller and his collaborators. The bottom of the Buško Blato is 700–710 m above sea level. Checking the water table in numerous bore holes shows that the stabilized water levels never sink below 660 m. Similar data were obtained by means of tests of the water permeability. According to these tests the zones of high permeability do not extend below 640–650 m above sea level. At the same time, this depth was estimated as being the maximum depth which the grout curtain must reach from the surface in order to cut off the shallower streams and to prevent loss of water from the storage basin by converting the springs into estavelas.

In eastern Hercegovina it is planned to build three large storage dams. Among them Grančarevo on the Trebišnjica sinking river (Fig. 23) is the largest (123 m high). The spring of the Trebišnjica will be submerged 70 m below the level of the accumulated water. The loss of water will be diminished by cement grouting [MIKULEC, 1965].

Another example of a storage basin can be found in the periodically inundated Nikšićko polje (Fig. 35) in Crna Gora (Montenegro) which really consists of a group of single depressions. According to BEŠIĆ [1959], all of the larger area consists predominantly of Mesozoic dolomites and limestones. Sporadically there are Lower Triassic clastics and Middle Triassic igneous rocks in the upper part of the Gračanica River as the inlier of the whole area. Smaller belts of Cretaceous and Palaeogene flysch are found in the northwest and southeast margin of the polje, as well as in the polje bottom below Quaternary fluvio-glacial deposits of various thickness (up to above 20 m). In general the structural relations are characterized by folds disturbed by thrust faults, resulting in a locally disturbed imbricate structure with partially disturbed flanks of anticlines and synclines (Fig. 37). The anticlines cut off the subsurface water courses and cause permanent springs. Synclines with impervious cores most frequently have only the role of relative barriers, making conditions favourable for periodical springs and estavelas. The same hydrogeologic phenomena may occur as a consequence of local faulting. The swallow holes (ponors) are peculiar only to distinctive fissures and fault zones, without subsurface barriers. The distribution of springs (permanent and periodical), estavelas and swallow holes (ponors) reflect the structural conditions described. The main permanent springs lie on the northeastern margin of the polje and are impounded by Middle Triassic dolomites and limestones (with a deep core of Lower Triassic clastics). The other permanent springs, periodical springs and estavelas are caused by local barriers connected with the imbricate structure trending from the northwest to southeast. The swallow holes indicate fault zones with open subsurface flow. In the Nikšićko polje they are mainly concentrated on the southeastern margins which have subsurface connections with the springs of the lower Zeta, confirming that there are no real barriers which could prevent all the waters of the upper Zeta from disappearing underground in that direction. The dye test confirmed the conditions described and enabled a reconstruction of the underground connections along a network of channels, fissures and joints between the upper and lower Zeta [VLAHOVIĆ, 1958]. Thus it was possible to define the area in which technical measures were necessary for the construction of the storage basin.

There are other very interesting examples but in general the principles described are applicable.

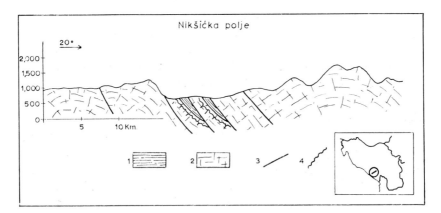

Fig.37. Simplified cross-section in the area of the Nikšićko polje (BEŠIĆ, 1965). 1 = Jurassic, Cretaceous and Palaeogene carbonate rocks; 2 = Palaeogene flysch; 3 = faults; 4 = unconformity.

Geologic basis for groundwater development

The lack of potable water in large areas of the Dinaric karst is still present, which has since ancient times stimulated attempts to detect and develop underground waters in the karst. Special attention was paid to the islands and to the coastal region. But not until recent decades was there a scientific approach to the problem on a clear geological basis. The complex structural relations and sporadic alternation of thick aquifers and relatively thin aquicludes exclude the possibility of one single approach to attempts in development. The possible areas of accumulated water are:

(1) More or less flat areas with a kind of connected groundwater. Little-disturbed flat areas containing groundwater are rare in the Dinaric karst and cannot be considered as sources of large quantities of water.

(2) Synclines and synclinoria. The underground water in synforms is less interesting because of existence of the surface streams in the area.

(3) Partially dammed limbs of antiforms (anticlines and anticlinoria). The antiforms, without aquicludes in their flanks, may be abundant in water which is difficult to recover. If the barriers are present, the water overflows them and there is no point in looking for it in the underground.

(4) Especially fractured zones with a good intake area.

Thus the only favourable conditions to catch additional quantities of potable water in the Dinaric karst area are in the backgrounds of relative barriers where water periodically issues, but does not flow out of the area, and in evident fracture zones if they are not directly connected with sea water. The usefulness for development depends on the amounts of water. Further, it is necessary to pay close attention to biological properties, because of the easy contamination along waterways in karst cavities.

To illustrate the foregoing considerations we shall mention some characteristic examples.

Owing to its especially agreeable climate, the island Hvar is a favourite resort. The lack of water necessitated explorations in order to provide new amounts of fresh water. The geological conditions are favourable with respect to both the lithostratigraphical sequence and the tectonic setting of the island. There are alternating Cretaceous dolomites and limestones. Only locally, along the southern side of the island, do Palaeogene limestone and flysch strike in a small belt. The main body of the island represents an anticline with dolomite in its core. On the flanks, the limestones and dolomites are alternating (Fig. 15). The dolomitic core is compressed and forms an aquiclude. The second dolomitic belt is an aquiclude only in general. But they allow transmission of water along the fractures caused by the unequal thrust of the area towards the south. Nevertheless, these zones function at least as relative barriers. The limestones are the main aquifers. Owing to the tectonic conditions for water supply, the limestones which are confined by the dolomitic belts are the most significant. Their whole surface is the recharge area, but the discharge is localized at their lowest point in the Jelšansko polje, which is covered by Quaternary pervious deposits.

The dolomite hinders the water flow under the Quaternary directly towards the sea and causes its overflow into the Quaternary, where it issues partly as a small spring, but the majority continues its underground flow towards the sea through the Quaternary. Since the Quaternary overlying the dolomite is relatively shallow, it was possible to capture the water by ditching, and to collect it in the reservoir. The limestone aquifers which are in direct contact with the sea or behind the Palaeogene flysch, are not satisfactory for water supply because of the mingling of fresh water with salt water within a large subsurface area.

According to MAGDALENIĆ [1965]—see Fig. 16—a syncline trends along the middle of the island Pag, having Palaeogene flysch in its core and covered by pervious Quaternary deposits. Its flanks consist of Cretaceous and Palaeogene limestones. They form ridges with the watershed, causing the water to flow to the sea and to the centre of the island, where it overflows the contact of the limestones and flysch as small springs or is collected in the Quaternary deposits where it can be developed by means of wells.

Practical problems in water supply[1]

The velocities of subsurface water flow in fissured rocks are much higher than in fine-grained unconsolidated aquifers. The drainage of such rocks is rapid and, consequently, the underground storage of water is relatively small and of shorter duration. These facts make such regions deficient in water even if they receive enough precipitation for an adequate water supply under favourable conditions. Only areas which have been found to yield sufficient quantities of water at all times should be chosen for the purposes of developing subsurface supplies.

In karst regions preliminary investigations are necessary, including pumping tests in critical conditions, and continuous observations and records over at least a full year, covering the extreme meteorologic and hydrologic conditions. The quality of the water should also be studied, whenever it is of importance in the development, especially if the water is to be used for public, domestic and industrial supplies.

In many cases and over wide areas of carbonate rocks, it has been found that water from a single ponor may appear on the surface— in streams, lakes or springs— in a multitude of places which may be very far from each other; equally a single spring may receive water from several ponors or sink holes between which no other connection may exist. Such connections are not necessarily permanent, and may function only under specific conditions, such as a certain discharge or at a specific piezometric level in the channels. Conditions of such connections can be disclosed by systematic and continuous observations, especially by the use of tracers, and sometimes by physical, chemical and biological analyses.

Water can be tapped either on the surface or underground. On the surface, it can

---

[1]Written by M. Petrik.

be taken from streams, lakes and springs. The use of lakes and streams in karst regions does not differ from such use in other regions, but the use of spring water does.

In carbonate rocks the springs most commonly flow from fissures. Very often, there is not a single outlet (Fig. 38) but rather several, disposed over a larger area. Mostly they are not at exactly the same elevation. Usually not all of them flow continuously. At low discharge, it is only the lowest one that functions and the rest comes successively into function as the discharge increases.

In the development of water supplies the springs are first uncovered by removal of the weathered rock. In the case of a single spring the water-collecting chamber is constructed at the point of its emergence, with the foundations of the chamber itself and of its lateral wings, bedded down in the impervious substratum. In the case of a group of springs the chamber is situated at an elevation below the lowest spring and connected with the individual springs by a common infiltration gallery or a system of drainage channels, all of them equally seated in the impervious layer. The water surface in the chamber should never be raised above the elevation of the lowest member of the group. In the construction, only mild explosives should be used.

A special kind of spring is the type found on the sea coast, because it usually shows a higher salinity, which is in some cases always present and in others only periodically manifesting itself, only at high discharge or at low flow.

The intrusion of saline water into the fresh water of the spring may occur in several ways. One of them is through a closed water channel discharging fresh water at a small elevation above the sea and containing a narrow sector in which the piezometric level at higher discharges drops beneath the level of the sea. If that narrow place is connected with the sea by another fissure, sea water will be sucked through it into the channel in periods of high discharge, and the water in the channel will become brackish [LEHMANN, 1932]. Since such an arrangement cannot be expected to occur often, the case must be considered as exceptional. In addition, the inflow of salt water occurs at times of high discharge only, and these occur in wet weather; but coastal springs usually show increased salinity in dry weather, at low discharges. These cases were explained by GJURAŠIN [1942, 1943] and later by KUŠČER [1950].

GJURAŠIN [1942, 1943] supposed a closed channel which—at a point of discharge and a little lower than the surface of the sea—splits into two branches, one having its mouth above and the other one below sea level. Under certain conditions, sea water enters the lower branch, mixes with fresh water at the branching point and emerges as brackish water from the mouth of the upper branch. This case is to be considered as common. KUŠČER [1950] defined the conditions of coastal springs which are: (*1*) never brackish; (*2*) always brackish; (*3*) brackish only at low discharge; and (*4*) brackish only at high discharge.

Some mixing of fresh and saline water can also result from the entrance of sea water into the fissures at high tides or in tempests, from high waves. Mixing by diffusion seems to be negligible, being a very slow process both in quiescent and in moving water. It is by far overshadowed by the other processes of mixing.

Fig.38. Vauclusian spring Buna, the tributary of the Neretva River, near Mostar. (Photograph by S. Božičević.)

The coastal springs of any greater significance always appear in zones of pronounced tectonic disturbance with wide zones of more or less disintegrated rocks. For this reason, fresh water very seldom appears from a single orifice, but rather from a number of them, dispersed over a larger area of the shore and of the bottom of the sea, especially where spring complexes are covered by the sea. On the eastern Adriatic coast there is a number of such zones, where underground streams discharge waters from the hinterland in clusters of coastal and submarine springs. Development of such springs is an almost hopeless task on account of the high degree of disintegration of the rocks, and the presence of the sea. Endeavours to utilize subterranean members of such groups of springs seem equally hopeless.

Another possibility of development of water in carbonate rock regions is to draw water from the underground. In this case, water can be drawn either from a point where it is visible and accessible, such as from a subterranean stream in a cave, a sink-hole or a pit, or from a bore-hole driven into a body of ground water. If an underground stream is to be used, the limit of development is set by the minimum flow of the stream which is measurable. On the contrary, the volume of water filling a system of fissures or other voids is not measurable. In single well-defined basins it might be calculated on the basis of prolonged observations and measurements of the quantities of incoming and outgoing water, if conditions permit such observations.

Finally there is underground impounding of water. The greatest difficulty with this method seems to be concerned with the calculation of the volume of the voids. It might be calculated in the most favourable case only in that part which is actually filled with water, but not in that which is to be added for the impounding by a grouting technique.

The water in fissured rocks is more exposed to rapid infiltration from the surface without the retarding and straining effect of the passage through fine-grained material. Therefore it is more likely to be changed both in quality and in quantity by such infiltration than in the case of water in fine-grained unconsolidated strata.

During stable conditions of long rainless periods, its quality usually has the following characteristics.

The temperature of waters which flow in closed underground channels at greater depth below the surface of the soil is near the average annual temperature of the region, with fluctuations within 1°C. Consequently, it decreases inversely with the elevation. However, abundant springs at low elevation may show appreciably colder water if it comes from regions of a high elevation. In the Dinaric region of carbonate rocks the temperature of water varies between 5 and 18°C.

Dissolved oxygen is always found in water at the point of emergence. The saturation with oxygen is usually between 50 and 100% which is an argument for good aeration in underground systems of circulation. Occasionally, supersaturated water may appear in a spring, which probably indicates a warming up of the water in the upper part of the soil, before emergence.

Free carbon dioxide is usually present in concentrations lower than 10 mg/l.

Higher concentrations may be found in waters coming from areas under cultivation.

The alkalinity is usually bicarbonate only. In most cases it is equal to carbonate hardness. In marshy areas and regions under cultivation, however, it may be greater than the carbonate, and even the total hardness.

The usual hardness varies between 100 and 270 mg/l, if it is expressed as $CaCO_3$. In mountains it is usually lower than in the plains, in non-cultivated regions it is also lower than in cultivated ones or in marshy areas. In pure limestone, all hardness —or nearly all—is calcium hardness; in waters issuing from pure dolomite, 50 % or a little more is calcium hardness and the rest is magnesium hardness, if both are expressed as $CaCO_3$.

The chloride concentration is usually below 10 mg/l. Close to the sea coast this concentration is usually higher, and may even be much higher, due to the fact that the spray of sea water transported inland by the wind, is deposited on the soil and vegetation, and then washed by rains into the soil. Excessively high salinity, up to 1,000 mg Cl/l or even more, is sometimes found in coastal springs or systems, due to the mixing of salty and fresh water.

Nitrogen, phosphorus and silica are usually present in very small concentrations or in traces only.

In dry periods, the water as a rule shows little or no turbidity, a small biochemical oxygen demand (expressed by $B.O.D._5$), or the consumption of potassium-permanganate or biochromate. It also shows small M.P.N. (most probable number) of coliform organisms.

In rainy periods, especially in their initial phases, the picture may be essentially different. There is usually a high turbidity, an increase of nitrogen compounds, chloride and sulphate concentrations, a great demand for oxygen, and the bacteriological picture becomes radically worse; in fact, the most marked changes occur in the density of bacteria, especially of coliforms, and in turbidity. Such changes appear suddenly and quickly and usually render the water unsuitable for public and domestic supplies without treatment. The character and degree of such pollution depend greatly upon the nature of the drained area. It may be slight and more or less innocuous in hilly, uninhabited and uncultivated areas, but very much heavier and dangerous in populated areas.

For such reasons, water from carbonate rocks should not be used for human consumption nor for the food industry without treatment. The minimum treatment, only for the best kind of water, should be disinfection. It should be practiced not only when necessary, but constantly, as a precautionary measure. The introduction of obligatory disinfection in all major water-supply systems in the Dinaric region in the post-war period eliminated water-borne diseases in communities in which they occurred periodically or were endemic.

Additional methods of treatment should be used if water shows signs of permanent or periodic pollution. Increased turbidity is always a sign of actual or potential pollution, but water may be polluted without announcing the fact by increased turbidity. Settling alone is of little avail in the removal of turbidity, the matter causing turbidity being present in very fine particles which do not settle readily. It is, therefore, to be induced by coagulation. Since the alkalinity of waters from carbonate rocks is always sufficiently great, the usual coagulants will give satisfactory flocs which can be easily removed by settling and subsequent rapid sand filtration.

The water from carbonate rocks can be, therefore, adequately purified in all normal cases by conventional methods of treatment.

The peculiar conditions in carbonate rock regions require methods of protection of their water which differ from those which are used in other regions.

Of special importance are the following facts:

(*1*) In many areas the soil is thin and may even be absent. This results in poor filtering and rapid entrance of water from the surface into the ground.

(*2*) There are practically no fine-grained aquifers, and consequently there is no filtration of the water during its subsurface circulation. A natural purification of water by mechanical processes is therefore impossible.

(*3*) The size of the voids in the aquifers is great and allows water to flow with comparatively high velocities. Polluted water may, therefore, require only a short time from its entry into the ground to the point of its emergence and use—perhaps a few days only. That time is usually not sufficient for the achievement of the biological processes of natural self-purification. Thus, natural purifications by biological processes can be regarded as only partially successful and, therefore, unreliable.

(*4*) Due to the frequency, multiplicity and ubiquity of the fissures and of their interconnections, potential pollution must be suspected from the surface even at small distances from the point of water emergence.

These circumstances demand the policy of water protection to be essentially stricter, much less of a local and much more of a regional and general character than is the case in other regions. That policy should not only be a matter of interest to health authorities and authorities of water management, but should also be an item of general planning within regional boundaries.

Industrial plants and even large towns that discharge noxious wastes into the environment should be located in the lowest parts of carbonate rock regions, i.e., on the sea coast or in the lowest reaches of streams, in places from which no pollution can be transported by water into other bodies of surface- or ground water.

Within the entire carbonate rock region, uncontrolled storage of both liquid and solid wastes should be abolished, and no discharge of such wastes should be permitted unless they have been submitted to appropriate treatment.

Regional water-supply systems and inter-municipal sewerage systems should be prefered and promoted wherever they are feasible.

In all public water-supply systems, water should be adeqately treated and always disinfected.

Practical aspects of forestry in the Dinaric karst[1]

The forests in the Dinaric karst are influenced by various factors especially by variable climates and soils. The maritime forests are exposed to strong sun's radiation; the low karst forests to high temperatures, long summer droughts, considerable warming of the soil, and strong winter winds. In addition to the forests of the high karst area higher precipitation is influenced by fog, dew and atmospheric humidity, etc. The soils vary from terra rossa (most abundant) to brown carbonate soils and various rendzinas, etc. Further, the geomorphologic factors are very pronounced. In connection with this, three types of forests are commonly distinguished in the Dinaric karst:

(*1*) The eu-Mediterranean forests, to be found in the belt which includes southern Istria, the coastline southward of Zadar and a major part of the islands. There evergreen forests prevail of *Quercus ilex* and different degradation (dwarf) forms (maquis, densely or sparsely stocked garrigues and stony fields). In addition there are forests characterized alternatively by: *(a) Quercus coccifera; (b) Pinus halepensis; (c) Pinus pinea;* and *(d) Pinus nigra* var. *dalmatica*.

(*2*) The sub-Mediterranean maritime forests which sporadically reach an altitude of over 1,000 m. Here are developed forests characterized by the following communities: *(a) Quercus pubescens / Carpinus orientalis; (b) Qu. pubescens / Ostrya carpinifolia; (c) Qu. pubescens / Qu. macedonica; (d) Qu. pubescens / Qu. frainetto; (e) Qu. pubescens / Qu. petraea; (f) Pinus nigra* var. *austriatica; (g) Castanea sativa*, etc. There, also, prevail various degraded forms (coppices, densely or sparsely stocked scrubs, and stony fields).

(*3*) In the region of the high and continental karst montane and subalpine forest communities are found. On the sea-board side the forest associations are characterized alternatively by: *(a) Fagus silvatica; (b) F. silvatica / Picea abies;* and *(c) Pinus montana* var. *mughus*. On the continental side the following communities are present: *(a) Quercus petraea / Carpinus betulus; (b) Fagus silvatica; (c) F. silvatica / Abies alba; (d) F. silvatica / A. alba / Picea abies*, and various subalpine types of forests.

The forests have been the best maintained in the high karst area while the maritime forests are more or less involved in degradation. The most poorly preserved forests are in the vicinity of the sea, where a stony desert occurs over large areas.

The soil and vegetation of the Dinaric karst have been subjected to long-lasting destruction. The forests were removed through cutting, burning, grazing, browsing and uprooting, in order to provide construction materials and fuel, for defense purposes, for expansion of pastures, new agricultural crops, new settlements, etc. All this seriously interfered with natural balance between soil and vegetation and consequently forests were devastated over large areas. Only 21 % of the Dinaric karst is covered by forests of normal growth, while 55 % is accounted for by various degraded forms which need to be rehabilitated.

---

[1] Written by M. Anić.

Concerning silvicultural interventions, the regeneration of matured stands by means of natural methods represents an important activity with respect to the future care of forests. The main meliorative interventions relate to the reforestation of denuded areas and to the conversion of poor and degraded forms into more valuable stands. In this process, auxiliary or transitory species improve the soil and prepare it for future stands. In reforestation, as well as in other improvement works, the following autochthonous tree species are used: *Pinus nigra; P. halepensis; Quercus ilex; Qu. pubescens; Fraxionus ornus; Celtis australis; Ostrya carpinifolia; Pinus silvestris; Picea abies* etc. Also suitable foreign plants are introduced: *Pinus pinaster; P. brutia; Cupressus sempervirens; Cedrus atlantica; C. libani; C. deodara; Pinus strobus; Pseudotsuga taxiofolia* etc.

In addition to the exploitation of wood the forests are useful for other purposes. The establishment of shelterbelts (i.e. windbreaks) has, in the maritime belt, a special significance for the protection of settlements and of agricultural crops against winds, desiccation and salt-water deposits, etc. Further, the forests have positively an influence on the hydrogeologic conditions. In the vast areas of the degraded karst, only forests can restore and increase soil productivity, etc.

Therefore, the cultivation of forests and reforestation are among the most significant practical activities in the Dinaric karst area.

Practical aspects of agriculture in the Dinaric karst

The development of agriculture in karst areas requires detailed knowledge of the genesis, composition and physical properties of relatively scarce karst soils in order to find measures for their improvement by changing of their composition and by application of natural phytoecology.

The karst soils were first studied by petrologists. Their interest was initially directed towards the genesis and composition of terra rossa and to its relation with the carbonate bedrock [TUĆAN, 1912]. The pedologists afterwards enlarged the field of investigation to all calcareous soils. It has been shown that they are polygenetic, highly dependent on the bedrock, and on specific hydrology.

The main productive areas in karst are the numerous poljes distributed in a step-like fashion at different elevations. According to JELAVIĆ [1967] they cover 261,000 ha or 24 % of the arable lands, and they are the only potential centres of an intensive agriculture. The genesis of their soils depended on their morphology, hydrology, elevation and the composition of their surroundings. The completely closed poljes are characterized by underground recharge and discharge. The spring waters may be the only cause of periodic flooding and deposition of sediments with high content of $CaCO_3$ (according to Jelavić it amounts 20–50 %). The open poljes are also influenced by surface flows bringing non-carbonate components also into the basins. JELAVIĆ [1967] analysed the lime content in the soils of several poljes and he found that it also depends on the elevation of the polje. The variability ratio for the poljes investigated is the following: the lowest depressions contain 80–92 % of lime; at 100–400 m elevation 17–45 % (in the surface layers); at 700–900 m elevation 6–20 %; and in the highest poljes 0.2–2 %.

Irrigation can be planned only in coordination with the utilization of waters for hydroelectric power plants, with the tendency for the depressions at lower elevations to be cultivated and for those at higher elevations to be used as water storage basins for producing power and for irrigation. As already stated, the limit of priority between storage basins and agriculture is approximately at 700 m above sea level.

Owing to the various elevations and the different climatic influences on the cultivable areas, it is possible to grow various agricultural products; in the eu-Mediterranean and sub-Mediterranean areas: rath-ripe vegetables, southern fruit trees (especially olive trees, vineyards, tobacco, etc.). At high elevations, with exception of the cultivable poljes, only the pastures are worth mentioning.

The scarceness of the cultivable karst areas imposes the application of measures not only for their improvement, but also for their protection against erosional processes in order to prevent irreversible devastation. The first step in that direction is to take into account the natural vegetation–soil relations.

REFERENCES

ALFIREVIĆ, S., 1966. Hydrogeological investigations of submarine springs in the Adriatic. *Assoc. Intern. Hydrogéol. Congr.*, *Belgrade*, *1963*, 6:255–265.

ARANDJELOVIĆ, D. and MILOŠEVIĆ, LJ., 1967. Hydrogeological investigation in Yugoslavian karst areas by the application of geophysical methods. *Assoc. Intern. Hydrogéol. Congr.*, *Hannover*, *1965*, 7:248–254.

BAHUN, S., 1967. *Geološka Osnova Krškog Područja Izmedju Slunja i Vrbovskog*. Thesis, Univ. of Zagreb, Zagreb, 166 pp.

BAKIĆ, M., 1966. Hydrogeological importance of results of prospecting for water on the Isle of Brač. *Assoc. Intern. Hydrogéol. Congr.*, *Belgrade*, *1963*, 6:299–307.

BARTEL, K., 1957. Vodoprivredni problemi kraškog područja u Bosni i Hercegovini. In: *Krš Bosne i Hercegovine*. Savezno savj. o kršu, Split, 3:197–199.

BASLER, DJ., MALEZ, M. and BRUNNACKER, K. 1966. Die rote Höhle (Crvena Stijena) bei Bileća, Jugoslawien. *Eiszeitalter Gegenwart*, 17:61–68.

BATURIĆ, J., 1961. Neki rezultati ispitivanja cirkulacije vode u obalnom području. *Yugoslav. Speleol. Congr.*, *2nd*, *Zagreb*, *1958*, pp.35–37.

BAUČIĆ, I., 1961. The importance of impermeable sediments in origin, evolution and form of underground cavities in Dinaric karst regions. *Symp. Intern. Speleol.*, *Como*, 5:1–5.

BAUČIĆ, I., 1963. The principal phases in the development of cavities in the Dinaric karst. *Intern. Congr. Speleol.*, *2nd*, *Vienna*, 2(1):207–209.

BAUČIĆ, I., 1965. Hydrological characteristics of the Dinaric karst in Croatia with a special regard to the underground water connections. *Naše Jame*, 7(1/2):61–72.

BEHLILOVIĆ, S., 1956. Geološka grada područja rijeke Trebišnjice. *Geološki Glasnik Sarajevo*, 2:21–34.

BEŠIĆ, Z., 1959. *Geološki Vodič Kroz NR Crnu Goru*. Geol. društvo Crne Gore, Titograd, 559 pp.

BEŠIĆ, Z., 1965. Hydrologic characteristics of the karst regions in the Socialist Republic of Montenegro with special reference to subterranean connections. *Naše Jame*, 7(1/2):97–106.

BEYER, A., TIETZE, E. and PILAR, GJ., 1874. *Die Wassernot im Karste der Kroatischen Militärgrenze*. Albrecht and Fiegler, Zagreb, 159 pp.

Bohinec, V., 1957. Slovenačke kraške jame i njihov privredni značaj. In: *Krš Slovenije*. Savezno savj. o kršu, Split, 1:243–249.
Boreli, M. and Pavlin, B., 1965. Approach to the problem of the underground water leakage from the storages in karst regions. Karst storages Buško Blato, Peruča and Kruščica. *Proc. Symp. on Hydrology in Fractured Rocks, Dubrovnik*. A.I.H.S.-UNESCO, 1:32–63.
Božičević, S., 1969. Pećine, jame i ponori s vodom u području Dinarskog krša. *Simp. o Općoj Vodnoj Ekonomiji na Kršu, Split, 1966—Krš Jugoslavije, Jugoslav. Akad.*, 6:105–136.
Brodar, S., 1938. Das Paläolithikum in Jugoslawien. *Quartär*, 1:140–172.
Brodar, S., 1955. Paleolitik u karstu. *Yugoslav. Speleol. Congr., 1st, Postojna, 1954*, pp.70–84.
Burlica, Č., 1967. Režim vlažnosti zemljišta na krečnjaku. *Kongr. Jugoslav. Društva za Prouč. Zemljišta, 3rd, Zadar*, pp.607–614.
Bušatlija, I., 1963. Neki problemi hidrografije i morfologije Nevesinjskog polja i njegovog oboda. *Yugoslav. Speleol. Congr., 3rd, Sarajevo, 1962*, pp.57–63.
Buturović, A., 1955. Ispitivanje bosansko-hercegovačkog krša. *Yugoslav. Speleol. Congr., 1st, Postojna, 1954*, pp.34–35.
Cholley, A. and Chabot, G., 1930. Notes de morphologie karstique du polje de Lika au Popovo. *Ann. Géograph.*, 39:270–285.
Cvijić, J., 1893. Das Karstphänomen. Versuch einer morphologischen Monographie. *Geograph. Abhandl. (Penck)*, 3:1–114.
Cvijić, J., 1895a. Pećine i podzemna hidrografija u istočnoj Srbiji. *Glas Srp. Akad. Nauka*, 46:1–101.
Cvijić, J., 1895b. *Karst—Geografska Monografija*. Drž. štamp. kralj. Srbije, Belgrade, 176 pp.
Cvijić, J, 1901. Morphologische und glaziale Studien aus Bosnien, der Hercegovina und Montenegro, 2. Die Karstpoljen. *Abhandl. Geograph. Ges.*, 3(2):1–85.
Cvijić, J., 1918. Hydrographie souterraine et évolution morphologique du karst. *Rec. Trav. Inst. Géograph. Alpine*, 6(4):376–420.
Cvijić, J., 1924. *Geomorfologija, 1*. Drž. štamp. kralj. Srba, Hrvata i Slovenaca, Belgrade, 588 pp.
Cvijić, J., 1926a. Cirkulacija vode i erozija u karstu. *Glasnik Srp. Geograf. Društva*, 12:1–16.
Cvijić, J., 1926b. Pojam karsta i pregled karsnih oblika. In: J. Cvijić (Editor), *Geomorfologija, 2*. Drž. štamp. kralj. Srba, Hrvata i Slovenaca, Beograd, pp.1–17.
Cvijić, J., 1950. Stare otoke Popova polja i hidrografske zone u karstu. *Glasnik Srp. Geograf. Društva*, 30(1): 3–9.
Ćirić, B., 1963. Le développement des Dinarides yougoslaves pendant le cycle alpin. *Mém. Soc. Géol. France*, 2:565–582.
Čubrilović, P., 1963. La résurgence de la rivière Dubrovačka reka et ses liaisons fixées avec des abîmes à Popovo polje. In: *Guide pour Voyage—Assoc. Intern. Hydrogéol. Congr., Belgrade, 1963*, pp.27–32.
Djerković, B., 1966a. Hidrogeološke karakteristike bliže okoline Nevesinja. *Geološki Glasnik Sarajevo*, 11:189–225.
Djerković, B., 1966b. Izdizanje Velež planine i evolucija karsnih oblika na njoj. *Referati VI Sav. Geol. Jugoslav., Ohrid*, 1:552–573.
Djikić, S., 1957. Historijski razvoj devastacije i degradacije krša u Bosni i Hercegovini. In: *Krš Bosne i Hercegovine*. Savezno savj. o kršu, Split, 3:125–137.
Djurić, V., 1954. Prilog proučavanju erozije tla i njezina posledica u kršu iznad Bakarskog zaliva. *Zbornik Radova Prir., Mat. Fak. Beograd*, pp.53–62.
Franić, D., 1966. Problèmes hydrologiques de l'approvisionnement d'eau de l'île de Lastovo. *Assoc. Intern. Hydrogéol. Congr. Belgrade, 1963*, 6:281–288.
Fukarek, P., 1958. Prilog poznavanju granica krša u Bosni i Hercegovini. *Geograf. Pregled*, 2:11–18.
Gams, I., 1962. Slepe doline v Sloveniji. *Geograf. Obzornik*, 7:263–301.
Gams, I., 1963. Meritve korozijske intenzitete v Sloveniji in njihov pomen za geomorfologijo. *Geograf. Vestnik Ljubljana*, 34:3–18.
Gams, I., 1965a. On the types of tufa-depositing waters and on the corrosion intensity in the northwestern Dinaric. *Proc. Symp. on Hydrology in Fractured Rocks, Dubrovnik*. A.I.H.S.-UNESCO, 63:7 pp.
Gams, I., 1965b. Aperçu sur l'hydrologie du karst slovène et ses communications souterraines. *Naše Jame*, 7:51–60.
Gams, I., 1965c. Speleological characteristics of the Slovene karst. *Naše Jame*, 7:41–50.

GAMS, I., 1967. Perspektive fizične geografije krasa. *Geograf. Obzornik*, 14(2):47–50.
GAREVSKI, R., 1956. Geologische und paläontologische Forschungen in den Höhlen von Mazedonien. *Fragmenta Balcanica*, 1:1–4.
GAVELA, B., 1963. O paleolitiku Srbije. *Arheološki Vestnik*, 13/14: 85–99.
GAVRILOVIĆ, D., 1965. Ein Beitrag zur Kenntnis des Karstes in Serbien. *Naše Jame*, 7(1/2):107–117.
GJURAŠIN, K., 1942. Prilog hidrografiji primorskog krša. *Tehnički Vjesnik*, 4:107–112.
GJURAŠIN, K., 1943. Prilog hidrografiji krša. *Tehnički Vjesnik*, 1:1–17.
GODEK, I., 1957. Bujičarstvo i problem erozije tla na kraškom području Hrvatske. In: *Krš Hrvatske*. Savezno savj. o kršu, Split, 2:221–228.
GORJANOVIĆ-KRAMBERGER, D., 1906. Der diluviale Mensch von Krapina in Kroatien. In: *Studien über die Entwicklungsmechanik des Primatenskelettes mit besonderer Berücksichtigung der Anthropologie und Descendenzlehre*. Walkhoff, Wiesbaden, pp.59–277.
GORJANOVIĆ-KRAMBERGER, D., 1912. Plitki krš okolice Generalskog stola u Hrvatskoj. Predjel oko Lešća. *Glasnik Srp. Geograf. Društva*, 1:42–47.
GORJANOVIĆ-KRAMBERGER, D., 1914. Nekadanji otvoreni tok Dobre i Krški ravnjak u Ogulinu. *Vijesti Geol. Povjerenstva*, 3(4):101–106.
GOSPODARIČ, R., 1969. Prirodne akumulacije vode v jamah porečja Ljubljanice. *Simp. o Općoj Vodnoj Ekonomiji na Kršu, Split, 1966—Krš Jugoslavije*, Jugoslav. Akad., 6:157–174.
GRUBIĆ, A., 1963. The stratigraphic position of bauxites in the Yugoslavian Dinarids. *Symp. sur les Bauxites, Oxydes et Hydroxydes d'Aluminium—Jugoslav. Akad.*, Zagreb, 1:51–79.
GRUND, A., 1903. Die Karsthydrographie. Studien aus west Bosnien. *Geograph. Abhandl. (Penck)*, 7(3):1–200.
GRUND, A., 1910. Beiträge zur Morphologie des dinarischen Gebirges. *Geograph. Abhandl. (Penck)*, 9(3):348–570.
GUŠIĆ, B., 1957. Čovjek i kras. *Krš Jugoslavije, Jugoslav. Akad.*, 1:23–61.
HABE, F., 1963. Hidrološki problemi severnega roba Pivške kotline. *Yugoslav. Speleol. Congr., 3rd, Sarajevo, 1962*, pp.77–84.
HABIČ, P., 1961. Nekaj oblik akumulacije in značaj sedimentov v kraških jamah. *Yugoslav. Speleol. Congr., 2nd, Zagreb, 1958*, pp.101–106.
HABIČ, P., 1966. Hidrografski problemi Visokega krasa med Idrijco in Vipavo. *Geograf. Obzornik*, 13(3/4):104–108.
HABIČ, P., 1967. Hidrografska rajonizacija krasa v Sloveniji. *Simp. o Općoj Vodnoj Ekonomiji na Kršu, Split, 1966—Krš Jugoslavije, Jugoslav. Akad.*, 6:79–91.
HADŽI, J., 1965. Bemerkungen zu einigen biospeläologischen Problemen des dinarischen Karstes. *Naše Jame*, 7(1/2):21–31.
HAMRLA, M., 1959. O pogojih nastanka premogišč na krasu. *Geologija*, 5:180–264.
HERAK, M., 1956. O nekim hidrogeološkim problemima Male Kapele. *Geološki Vjesnik*, 8:19–37.
HERAK, M., 1957. Geološka osnova nekih hidroloških pojava u dinarskom kršu. *Yugoslav. Geol. Congr., 2nd, Sarajevo*, pp.523–539.
HERAK, M., 1959. Prilog geologiji i hidrogeologiji otoka Hvara. *Geološki Vjesnik*, 12:135–148.
HERAK, M., 1960. Geologija Gračačkog polja u Lici. *Geološki Vjesnik*, 13:31–56.
HERAK, M., 1962. Tektonska osnova hidrogeoloških odnosa u izvorišnim područjima Kupe i Korane (s Plitvičkim jezerima). *Referati V Sav. Saveza Geol. Društava F.N.R.J.*, Beograd, 3:17–25.
HERAK, M., 1965. Geologische Übersicht des dinarischen Karstes. *Naše Jame*, 7(1/2):5–11.
HERAK, M., BAHUN, S. and MAGDALENIĆ, A., 1969. Pozitivni i negativni utjecaji na razvoj krša u Hrvatskoj. *Simp. o Općoj Vodnoj Ekonomiji na Kršu, Split, 1966—Krš Jugoslavije, Jugoslav. Akad.*, 6:45–78.
HRANILOVIĆ, H., 1901. Geomorfološki problemi hrvatskoga krasa. *Glasnik Hrv. Naravosl. Društva*, 13:93–133.
JANJIĆ, M., 1966. Cavernen im Karst. *Assoc. Intern. Hydrogéol. Congr. Belgrade, 1963*, 6:289–292.
JELAVIĆ, A., 1967. Klasifikacija krških polja s obzirom na njihov pedološki sastav. *Kongr. Jugoslav. Društva za Prouč. Zemljišta, 3rd, Zadar*, pp.549–556.
JENKO, F., 1955. Kraška hidrogeologija in geomorfologija v luči novih raziskav na Dinarskem krasu. *Geologija*, 3:266–268.
JENKO, F., 1957. Vodoopskrba na slovenačkem kršu. In: *Krš Slovenije*. Savezno savj. o kršu, Split, 1:213–215.

JENKO, F., 1965. L'hydrographie du karst. *Proc. Symp. on Hydrology in Fractured Rocks, Dubrovnik.* A.I.H.S.-UNESCO, 1:173–182.
JEVREMOVIĆ, M., 1963. Les résultats des recherches hydrogéologique au domaine de Marina-Rogoznica. In: *Guide pour Voyage— Assoc. Intern. Hydrogéol. Congr., Belgrade, 1963,* pp.56–59.
JEVREMOVIĆ, M., 1966. Hydraulic characteristics and classification of brackish springs in the Adriatic zone of the Dinaric karst. *Assoc. Intern. Hydrogéol., Congr., Belgrade, 1963,* 6:293–297.
JOVANOVIĆ, B.P., 1946. Podzemna oburvavanja u krasu. *Zbornik Radova Srp. Akad. Nauka,* 51:1–46.
JOVANOVIĆ, P.S., 1928. Karsne pojave u Poreču (Makedonija). *Glasnik Skop. Naučnog Društva,* 4:1–46.
JURAS, I. and ĆIRIĆ, M., 1967. Stanje i problemi tala krških područja. *Kongr. Jugoslav. Društva za Prouč. Zemljišta, 3rd, Zadar,* pp.13–14.
JURKOVIĆ, I. and SAKAČ, K., 1963. Stratigraphical, paragenetical and genetical characteristics of bauxite in Yugoslavia. *Symp. sur les Bauxites, Oxydes et Hydroxydes d'Aluminium—Jugoslav. Akad., Zagreb,* 1:253–263.
KAJMAKOVIĆ, R. and PETROVIĆ, B., 1961. Dubina i intenzitet karstifikacije u zavisnosti od geološko-tektonskih uslova i hidrauličkog gradijenta pada. *Yugoslav. Speleol. Congr., 2nd, Zagreb, 1958,* pp.67–70.
KATZER, F., 1909. *Karst und Karsthydrographie. Zur Kunde der Balkanhalbinsel.* Kajon, Sarajewo, 94 pp.
KERNER, F., 1912. Beitrag zur Thermik der Karstquellen. *Verhandl. Geol. Reichsanstalt,* 14:327–330.
KERNER, F., 1917. Quellengeologie von Mitteldalmatien. *Jahrb. Geol. Reichsanstalt,* 66(2):145–293.
KEVO, R., 1969. Krške vode kao posebno zaštićeni objekti prirode. *Simp. o Općoj Vodnoj Ekonomiji na Kršu, Split, 1966—Krš Jugoslavije, Jugoslav. Akad.,* 6:521–534.
KIŠPATIĆ, M., 1912. Bauxite des kroatischen Karstes und ihre Entstehung. *Neues Jahrb. Mineral., Abhandl.,* 34:513–552.
KISPATIĆ, M., 1915. Neuer Beitrag zur Kenntniss der kroatischen Karstes. *Glasnik Hrv. Naravosl. Društva,* 27:52–58.
KNEŽEVIĆ, B., 1962. Hidraulički problemi karsta. *Saopštenja,* 25:1–13.
KOBER, L., 1952. Leitlinien der Tektonik Jugoslawiens. *Posebna Izd. Srp. Akad. Nauka,* 189(3):1–64.
KOCH, F., 1928. La géologie du karst. In: A. UGRENOVIĆ (Editor), *Le Karst Yougoslave.* Direction générale de Forêt du Royaume des Serbes, Croates et Slovènes, Zagreb, pp.1–15.
KODRIČ, M., 1957. Geološka podloga i tla slovenačkog kraškog područja. In: *Krš Slovenije.* Savezno savj. o kršu, Split, 1:9–13.
KOMATINA, M., 1964/1965. Prilog rešavanju problema odredjivanja hidrogeoloških razvodja i pravaca cirkulacije podzemnih voda u karstu. *Vesn. Zavoda Geol. Geofiz. Istraživanja N.R. Srbije, (B),* 4/5:63–79.
KOMATINA, M., 1966. Prilog proučavanju hidrogeoloških odnosa centralnodinarskog karsta. *Vesn. Zavoda Geol. Geofiz. Istraživanja N.R. Srbije, (B),* 6:113–129.
KOSMAČ, J., 1957. Mineralna vrela i prirodna lječilišta na slovenačkom kršu. In: *Krš Slovenije.* Savezno savj. o kršu, Split, 1:251–252.
KOVAČEVIĆ, P., 1960. Prilog metodici klasifikacije krša. *Naša Poljoprivreda Šumarstvo,* 6(4):17–24.
KREBS, N., 1908. Neue Forschungsergebnisse zur Karsthydrographie. *Petermanns Geograph. Mitt.,* 54:166–168; 263–264.
KRULC, Z., 1966. Die Beziehungen zwischen den geologischen, hydrologischen und geoelektrischen Eigenschaften des Gebirges im jugoslawischen Karst unter besonderer Berücksichtigung der Verhältnisse beim Bau von Kraftwerkanlagen. *Boll. Geofis. Teor. Appl.,* 8:83–113.
KRULC, Z., 1967. Einige Ergebnisse geophysikalischer Untersuchungen der unterirdischen Wasserverbindungswege im Karst. *Intern. Speleol. Congr., 4th, Ljubljana, 1965—Fizična Speleol.,* 3:327–334.
KUNAVER, P., 1955. Raziskovanje našega visokoalpskega krasa. *Yugoslav. Speleol. Congr., 1st, Postojna, 1954,* pp.51–54.
KUŠČER, I., 1950. Kraški izviri ob morski obali. *Razprave Slov. Akad. Nauka,* 1:97–147.
KUŠČER, I., 1961. Metode raziskovanja obomorskih kraških izvirov. *Yugoslav. Speleol. Congr., 2nd, Zagreb, 1958,* pp.39–43.
LAHNER, G., 1917. Der westmontenegrische Karst und sein hydrologischer Zusammenhang mit der Bucht von Cattaro. *Petermanns Geograph. Mitt.,* 63:297–302.

LEHMANN, O., 1932. *Nouveau Traité des Eaux Souterraines*. Doin, Paris, 838 pp.

LUKOVIĆ, M., 1951. O izvorima i podzemnim vodama Gorskog Kotara i Hrvatskog Primorja. Prilog hidrogeologiji ovih oblasti. *Geol. Anali Balk. Poluostrva*, 19:155–176.

LUKOVIĆ, M., 1960. Neka opažanja o kretanju podzemnih voda u karstu prilikom izgradnje izvesnih velikih objekata u karstu Jugoslavije. *Vesn. Zavoda Geol. Geofiz. Istraživanja N.R. Srbije, (B)*, 1:65–77.

LUKOVIĆ, M., 1960. Kratak pregled sadanjeg poznavanja hidrogeoloških i inženjerskogeoloških karakteristika terena na teritoriji NR Srbije i zadataka koji iz toga proizlaze. *Vesn. Zavoda Geol. Geofiz. Istraživanja N.R. Srbije, (B)*, 1:9–14.

LUTOVAC, M., 1969. Privredni značaj kraških voda u severoistočnim delovima Crne Gore i na Staroraškoj visoravni. *Simp. o Općoj Vodnoj Ekonomiji na Kršu, Split, 1966—Krš Jugoslavije, Jugoslav. Akad.*, 6:347–355.

MAGAŠ, N., 1965. O depresiji Vranskog jezera na otoku Cresu i geološkim odnosima njegovog užeg područja. *Geološki Vjesnik*, 18(2):255–261.

MAGDALENIĆ, A., 1965. Hidrogeološka interpretacija bazena Prančevići na Cetini. *Geološki Vjesnik*, 18(2):385–404.

MALEZ, M., 1956. Djulin ponor u Ogulinu. *Geološki Vjesnik*, 8:153–172.

MALEZ, M., 1960. Pećine Ćićarije i Učke u Istri. *Acta Geol. Jugoslav. Akad.*, 2:163–263.

MALEZ, M., 1961. Speleološki objekti jugozapadne Like. *Acta Geol. Jugoslav. Akad.*, 3:107–241.

MALEZ, M., 1964. Prilog poznavanju speleoloških odnosa na Glamočkom i Duvanjskom polju. *Krš Jugoslavije, Jugoslav. Akad.*, 4:171–200.

MALEZ, M., 1965a. Cerovačke pećine. *Izdanja Speleol. Društva Hrv.*, 1:1–42.

MALEZ, M., 1965b. Pećina Veternica,1. Opći speleološki pregled; 2. Stratigrafija kvartarnih taložina. *Acta Geol. Jugoslav. Akad.*, 5:175–237.

MALEZ, M., 1967. Paleolit Velike pećine na Ravnoj gori u sjeverozapadnoj Hrvatskoj. *Arheološki radovi rasprave Jugoslav. Akad.*, 4(5):7–68.

MALEZ, M. and BOŽIČEVIĆ, S., 1965. The Medvjedja pećina (Bear Cave) on Lošinj Island, a rare case of a submerged cave. *Problems of the Speleological Research—Intern. Speleol. Conf., Brno, 1964*, pp.211–216.

MANAKOVIK, D., 1965. Le karst de la Macédonie. *Naše Jame*, 7(1/2):119–120.

MARIĆ, L., 1964. Terra rossa u karstu Jugoslavije. *Predavanja Održana Jugoslav. Akad. Znan. Umjet.*, pp.5–58.

MARIĆ, L., 1967. Klasifikacija i geokemijska migracija nekojih makroelemenata i mikroelemenata u JZ Dinaridima (Jugoslavija). *Kongr. Jugoslav. Društva za Prouč. Zemljišta, 3rd, Zadar*, pp. 539–547.

MARTEL, E. A., 1901. L'origine des polje du karst. La géographie. *Bull. Soc. Géograph.*, 4:190–195.

MATONIČKIN, I. and PAVLETIĆ, Z., 1965. Les formes zoogènes des tufs et leur formation dans la région des lacs de Plitvice en Yougoslavie. *Hydrobiologia*, 26:292–300.

MELIK, A., 1955. Kraška polja Slovenije v Pleistocenu. *Acta Géograph.*, 3:7-61.

MIJATOVIĆ, B., 1962a. Hidrogeološka uloga eocenskog fliša pri formiranju slatkovodne izdani u karstifikovanim krečnjacima Dinarske primorske zone. *Yugoslav. Geol. Congr., 3rd, Budva, 1959*, 2:27–39.

MIJATOVIĆ, B., 1962b. Neka zapažanja o razvitku dinarskog primorskog karsta. *Referati V Sav. Geol. Jugoslav., Beograd*, 3:1–7.

MIKULEC, S., 1965. Problèmes sur la construction des bassins de retenue dans l'Herzegovine de l'est. *Proc. Symp. on Hydrology in Fractured Rocks, Dubrovnik*. A.I.H.S.-UNESCO, 2:590–601.

MIKULEC, S. and TRUMIĆ, A., 1969. Akumuliranje vode na kršu i problem njena optimalnog iskorištavanja. *Simp. o Općoj Vodnoj Ekonomiji na Kršu, Split, 1966—Krš Jugoslavije, Jugoslav. Akad.*, 6:279–298.

MILETIĆ, P. and BABIĆ, Ž., 1963. Uloga mikrotektonike kod rješavanja lokalnog režima podzemne vode u kršu. *Geološki Vjesnik*, 16:233–242.

MILIĆ, Č. S., 1954. Prilog poznavanju morfološke raznolikosti vrtača u zagaćenom krasu. *Zbornik Radova Srp. Akad. Nauka*, 39(7):1–18.

MILOJEVIĆ, B., 1952. Les formes karstiques de la côte dinarique. *Lab. Géograph., Univ. Rennes, Volume Jubilaire*, pp.198–208.

MILOJEVIĆ, N., 1962. O uslovima pojavljivanja snažnih karsnih vrela. *Yugoslav. Geol. Congr., 3rd, Budva, 1959*, 2:5–26.

MILOJEVIĆ, S., 1937. Nekoliko napomena o morfološkoj raznolikosti vrtača u golom kršu. *Glasnik Geograf. Društva*, 23:1–16.

MILOJEVIĆ, S., 1938. Pojavi i problemi krša. Proučavanje u dinarskom kršu istočne Srbije. *Posebno Izd. Srp. Akad. Nauka, Prir. Mat. Razr.*, 123:32–160.

MILOŠEVIĆ, LJ., 1957. Odredjivanje podzemnih vodenih tokova i pećina geoelektričnim metodama u karstu u okolini Trebinja. *Vesn. Zavoda Geol. Geofiz. Istraživanja, N.R. Srbije*, 14:337–345.

MILOVANOVIĆ, B., 1964/1965. Epirogenetska i orogenetska dinamika u prostoru Spoljašnih Dinarida i problemi paleokarstifikacije i geološke evolucije holokarsta. *Vesn. Zavoda Geol. Geofiz. Istraživanja N.R. Srbije, (B)*, 4/5:5–44.

MILOVANOVIĆ, B., 1966. Les problèmes de la paléokarstification dans les Dinarides externes. *Assoc. Intern. Hydrogéol. Congr., Belgrade, 1963*, 1:57–58.

MURKO, V., 1957. Ekonomski problemi krša u N.R. Sloveniji. In: *Krš Slovenije*. Savezno savj. o kršu, Split, 1:255–275.

NOVAK, D., 1964/1965. Hidrogeologija područja Osapske reke. *Vesn. Zavoda Geol. Geofiz. Istraživanja N.R. Srbije, (B)*, 4/5:81–91.

NOVAK, G., 1955. *Prethistorijski Hvar. Grapčeva Spilja*. Jugoslav. Akad., Zagreb, 381 pp.

NOSAN, A., 1963. Notice explicative pour le domaine Opatija–Ilirska Bistrica–Kozina. Grottes de Škocjan près de Divača–Postojna. In: *Guide pour Voyage—Assoc. Intern. Hydrogéol. Congr., Belgrade, 1963*, pp.82–85.

OSOLÉ, F., 1965. Les stations paléolithique dans des grottes en Yougoslavie. *Naše Jame*, 7(1/2):33–40.

PAPEŠ, J. and SRDIĆ, R. 1969. Opći hidrogeološki odnosi na teritoriji Bosne i Hercegovine. *Simp. o Općoj Vodnoj Ekonomiji na Kršu, Split, 1966—Krš Jugoslavije*, Jugoslav. Akad., 6:93–104.

PAPEŠ, J., LUBURIĆ, P., SLIŠKOVIĆ, T. and RAIĆ, V., 1964. Geološki odnosi šire okolice Livna, Duvna i Glamoča u jugozapadnoj Bosni. *Geološki Glasnik Sarajevo*, 9:87–122.

PAVLETIĆ, Z. and MATONIČKIN, I., 1964. Studii ecologici delle biocenosi sulle deposizioni travertinose. *Arch. Oceanograph. Limnol.*, 13:197–205.

PAVLIN, B., 1961. Realisation du bassin d'accumulation de Peruča dans le karst dinarique. *Congr. grands barrages, 7me, Rome*, 25(85):1–28.

PAVLIN, B. and MLADINEO, L., 1957. Istražni radovi za akumulaciju u kršu. *Gradjevinar*, 2(3):1–12.

PAVLIN, B., BAHUN, S. and FRITZ, F., 1963. L'usine hydroélectrique Senj. *Guide pour Voyage—Assoc. Intern. Hydrogéol. Congr., Belgrade, 1963*, pp.69–76.

PAVLIN, B., MAGDALENIĆ, A. and ZLATOVIĆ, D., 1963. Le système hydro-énergétique de Split. (a) Le bassin d'accumulation Peruča—karst dinarique. In: *Guide pour Voyage—Assoc. Intern. Hydrogéol. Congr., Belgrade, 1963*, pp.45–51.

PERIĆ, J., 1963. Podzemne akumulacije kao hidrotehnički objekti u svrhu regulisanja režima proticaja karsnih i drugih izvora. *Vesn. Zavoda Geol. Geofiz. Istraživanja N.R. Srbije, (B)*, 3:67–97.

PETKOVIĆ, K., 1958. Neue Erkenntnisse über den Bau der Dinariden. *Jahrb. Geol. Bundesanstalt (Austria)*, 101:1–24.

PETRIK, M., 1957. Hidrološki režim jezera Vrana. *Krš Jugoslavije*, Jugoslav. Akad., 1:109–151.

PETRIK, M., 1960a. Hidrografska mjerenja u okolici Imotskog. *Ljetopis Jugoslav. Akad.*, 64:266–286.

PETRIK, M., 1960b. Prilozi limnologiji jezera Vrane. *Krš Jugoslavije*, Jugoslav. Akad., 2:105–192.

PETRIK, M., 1961. Temperatura i kisik Plitvičkih jezera. *Rasprave Jugoslav. Akad.*, 11(3): 31–37.

PETRIK, M., 1967. Seenschutz im Karst. *Fäderation Europäischer Gewässerschutz, Informationsblatt*, 14:30–32.

PETROVIĆ, J.B., 1955. Hidrografsko-morfološke i privredne odlike krša. *Zemlja i Ljudi*, 5:30–36.

PETROVIĆ, B. and PRELEVIĆ, B., 1965. Hydrologic characteristics of the karst area of Bosnia and Herzegovina and a part of Dalmatia with special consideration of underground water connections. *Naše Jame*, 7:79–87.

PLENIČAR, M., 1962. Razvoj krednih karbonatnih stena u slovenačkom karstu i kršne pojave u njima. *Referati V Sav. Geol. Jugoslav., Beograd*, 3:27–31.

POLJAK, J., 1913. Pećine hrvatskog krša,1. Pećine okoliša lokvarskog i karlovačkog. *Prirodosl. Istraživanja Jugoslav. Akad.*, 1:28–48.

POLJAK, J., 1914. Pećine hrvatskog krša, 2. Pećine okoliša Plitvičkih jezera, Drežnika i Rakovice. *Prirodosl. Istrazivanja Jugoslav. Akad.*, 3:1–25.

POLJAK, J., 1924. Pećine hrvatskog krša, 3. Pećine Hrvatskog Primorja od Rijeke do Senja. *Prirodosl. Istraživanja Jugoslav. Akad.*, 15:219–266.

POLJAK, J., 1952. Pojava starih krških oblika i njihova veza sa rudnim ležištima područja Debeljaka na sjevernom Velebitu. *Geološki Vjesnik*, 2(4):99–110.
POLJAK, J., 1958. Razvoj morfologije i hidrogeologije u dolomitima dinarskog krša. *Geološki Vjesnik*, 11:1–20.
POLŠAK, A., 1963. Les rapports hydrogéologique des lacs de Plitvice. In: *Guide pour Voyage—Assoc. Intern. Hydrogéol. Congr., Belgrade, 1963*, pp.77–81.
POTOČIĆ, Z. and MURKO, V., 1957. Ekonomski problemi krša Jugoslavije. In: *Krš Jugoslavije*. Savezno savj. o kršu, Split, 5:297–365.
RACZ, Z., ŠILJAK, M. and MALEZ, M., 1967. Višeslojni profili na području kontinentalnog krša Hrvatske i pitanje porijekla pojedinih horizonata. *Kongr. Jugoslav. Društva za Prouč. Zemljista, 3rd, Zadar*, pp.581–590.
RADOVANOVIĆ, M., 1929. Pećina Vjetrenica u Hercegovini. Morfološko hidrografska studija. *Spomenik Srp. Akad.*, 68:3–113.
RADOVANOVIĆ, S. and PAVLOVIĆ, P., 1894. Les phénomènes du karst dans la Serbie orientale. *Ann. Géograph.*, 4:56–61.
RAKOVEC, I., 1956. Pregled tektonske zgradbe Slovenije. *Yugoslav. Geol. Congr., 1st, Bled, 1954*, pp.71–83.
REDENŠEK, V., 1959. Popis špilja i ponora u Hrvatskoj. *Naše Planine*, 11:179–186; 229–236; 276–278.
REMY, P. A., 1953. Descriptions des grottes yougoslaves (Herzegovine, Dalmatie, Crna Gora et ancien Sandjak de Novi Pazar). *Glasnik Prirod. Muzeja Srpske Zemlje, Ser. B*, 5(6):175–233.
REŠTAROVIĆ, S., 1969. Energetsko iskorišćenje voda u slivovima vodotoka primorsko-dalmatinskog krša i Kupe. *Simp. o Općoj Vodnoj Ekonomiji na Kršu, Split, 1966—Krš Jugoslavije, Jugoslav. Akad.*, 6:237–277.
REŠTAROVIĆ, S. and ZLATOVIĆ, B., 1957. Elektroprivreda na kršu. In: *Krš Jugoslavije*. Savezno savj. o kršu, Split, 5:209–221.
RIDJANOVIĆ, J., 1966. Orjen—la montagne dinarique. *Radovi Geograf. Inst. Sveuč.*, Zagreb, 5:1–103.
ROGLIĆ, J., 1951. Unsko-koranska zaravan i Plitvička jezera. Geomorfološka promatranja. *Geograf. Glasnik*, 13:49–68.
ROGLIĆ, J., 1954. Polja zapadne Bosne i Hercegovine; prilog poznavanju prirodnih osobina i ekonomskog značenja. *Yugoslav. Geograph. Congr., 3rd, Sarajevo*, pp.45–58.
ROGLIĆ, J., 1957. Zaravni na vapnencima. *Geograf. Glasnik*, 19:103–131.
ROGLIĆ, J., 1961. Odnos morske razine i cirkulacije vode u kršu. *Yugoslav. Speleol. Congr., 2nd, Zagreb, 1958*, pp.45–48.
ROGLIĆ, J., 1963. Glaciation of the Dinaric Mountains and its effects on the karst. *Intern. Congr. Quaternary, 6th, Warszawa, 1961, Rept.*, 3:293–299.
ROGLIĆ, J., 1964. Les poljes du karst dinarique et les modifications climatiques du Quaternaire. *Rev. Belge Géograph.*, 80(1/2):105–125.
ROGLIĆ, J., 1965a. The depth of the fissure circulation of water and the evolution of subterranean cavities in the Dinaric karst. *Problems of the Speleological Research—Intern. Speleol. Conf., Brno, 1964*, pp.25–35.
ROGLIĆ, J., 1965b. The delimitations and morphological types of the Dinaric karst. *Naše Jame*, 7(1/2):12–20.
ROGLIĆ, J. and BAUČIĆ, I., 1958. Krš u dolomitima izmedju Konavoskog polja i morske obale. *Geograf. Glasnik*, 20:129–137.
RŽEHAK, V., 1965. Speleological curiosities of the Bosnian and Herzegovinian karst. *Naše Jame*, 7(1/2):12–20.
SAKAČ, K., 1966. O paleoreljefu i pseudopaleoreljefu boksitonosnih područja krša. *Geološki Vjesnik*, 19:123–129.
SAVNIK, R., 1965. Beitrag zur Kenntnis der Karsthydrographie in Slowenien. *Congr. Intern. Spéléol., 1re, Paris*, 1:241–246.
ŠERKO, A., 1947. Kraški fenomeni v Jugoslaviji. *Geograf. Vestnik*, 19:43–70.
SIKOŠEK, B. and MEDWENITSCH, W., 1965. Neue Daten zur Fazies und Tektonik der Dinariden. *Verhandl. Geol. Bundesanstalt*, 1965:86–102.
SPIROVSKI, J., 1967. Zemljišta na krečnjaku kod Demir-Kapije (S.R.M.). *Kongr. Jugoslav. Društva za Prouč. Zemljišta, 3rd, Zadar*, pp.573–579.
SREBRENOVIĆ, D., 1965. Hidrološke veličine i odnosi u jednom kraškom sistemu. *Zavod Hidrol. Geodetskog Fak. Zagreb*, 1:17.

STEMBERGER, M., 1955. Problem erozije tla u NR Hrvatskoj s osobitim obzirom na krš. In: *Naučne Osnove Borbe Protiv Erozije—Prvo Savetovanje, Beograd*, pp.43–55.

STEPANOVIĆ, B., 1956. Hidrogeološki problemi vodosnabdevanja na teritoriji F.N.R.J. *Yugoslav. Geol. Congr., 1st, Bled, 1954*, pp.209–218.

STEPINAC, A., 1951. Povezanost protoka dvaju krških vodotoka. *Elektroprivreda*, 4:17–20.

STEPINAC, A., 1969. Otjecanje u Dinarskom kršu. *Simp. o Općoj Vodnoj Ekonomiji na Kršu, Split, 1966—Krš Jugoslavije, Jugoslav. Akad.*, 6:207–235.

SWEETING, M.M., 1955. The karstlands of Yugoslavia. *Trans. Cave Res. Group Great Britain*, 4(1): 51–72.

TERZAGHI, K., 1913. Beiträge zur Hydrographie und Morphologie des kroatischen Karstes. *Mitt. Jahrb. Ungar. Reichsanstalt*, 20(6):255–369.

TREIBS, W., 1954. Im Karst von Adelsberg und Triest. *Z. Deut. Geol. Ges.*, 105:141–142.

TUĆAN, F., 1912. Terra rossa deren Natur und Enstehung. *Neues Jahrb. Mineral., Abhandl.*, 34:401–430.

TUĆAN, F., 1933. Einblicke in die Geochemie des dinarischen Karstes. *Bull. Intern. Acad. Yougoslave*, 27:37–48.

TURINA, L., 1913. Hidrografski, geološki i tektonski odnošaji jednog kraškog predjela sjeverozapadne Bosne. *Glasnik Zemlje Muzeja Bosne Hercegovine*, 25:253–306.

UGRENOVIĆ, A., 1957. Krš kao naučni problem. *Krš Jugoslavije, Jugoslav. Akad.*, 1:5–22.

VIDOVIĆ, M., 1961. O tektonici Jadranskog primorja od Boke Kotorske do Neretve. *Geol. Anali Balk. Poluostrva*, 28:143–154.

VILIMONOVIĆ, J., 1966. Prilog proučavanju prihranjivanja podzemnih voda u karstu. *Vesn. Zavoda Geol. Geofiz. Istraživanja N.R. Srbije, (B)*, 6:130–136.

VLAHOVIĆ, V., 1958. Hidrogeologija Nikšičkog polja (Crna Gora). *Geološki Glasnik Titograd*, 2:243–266.

VUKOVIĆ, S., 1935. Istraživanje prethistorijskog nalazišta u spilji Vindiji kod Voće. *Spom. Varažd. Muzeja*, pp.73–80.

VUKOVIĆ, S., 1950. Paleolitska kamena industrija spilje Vindije. *Hist. Zbornik*, 3:241–256.

WRABER, M., 1967. Genetska veza izmedju vegetacijskih i talnih jedinica na karbonatnom području visokog krša Slovenije. *Kongr. Jugoslav. Društva za Prouč. Zemljišta, 3rd, Zadar*, pp.557–564.

ZOGOVIĆ, D., 1967. Petrogenetic properties of dolomites as a factor of their different hydrogeological role in Dinaric karst. *Assoc. Intern. Hydrogéol. Congr., Hannover, 1965*, 7:267–270.

ZUBČEVIĆ, O., 1958. Dabarsko i Fatničko polje; prilog poznavanju kraške hidrografije. *Geograf. Pregled*, 2:19–33.

Chapter 4

Karst of Italy

S. BELLONI, B. MARTINIS AND G. OROMBELLI

Institute of Geology, University of Milan (Italy)

INTRODUCTION

Carbonate rocks and, in general, water-soluble rocks are widely distributed all over Italy, and therefore a more or less intensive development of karst phenomena is present in all regions. Notwithstanding this fact, the areas which present such a complete karst morphology as to constitute the predominating note in the landscape are infrequent and of modest dimensions, with the exception of the Italian Karst and the Apulia region.

Numerous studies have been made on karst and karst hydrology in Italy, most of them furnishing descriptions of single phenomena, with particular emphasis on speleology. On the other hand, regional studies are rare and often deal with areas which are not the most important in this regard.

This contribution is intended to give a synthesis of the distribution and the characters of the karst phenomena in Italy. We do not propose, nor is it possible, to give a complete description; we only wish to emphasize the most typical and representative aspects of the karst in this country. Therefore in the following pages we shall examine the various karst areas beginning with the Italian Karst, the most typical and the best known. Later, we shall describe the other regions from east to west along the Alps, and from north to south along the Italian Peninsula, paying particular attention to those areas which seem to us the most significant.

LITHOSTRATIGRAPHIC AND SEDIMENTOLOGIC ASPECTS[1]

Karst in Italy is developed upon rocks which are often very different in composition, texture and structure, and which belong to formations ranging from the Cambrian to the Pleistocene. Fig. 1 represents the areal distribution of these rocks, divided into three principal types: carbonate rocks, conglomerate and gypsum. The same figure also shows that these soluble rocks prevalently outcrop along the Alpine range, the Apennines, in the Apulia region, and are finally scattered into irregular outcrops in Sicily and Sardinia.

Along the Italian side of the Alpine range, the formations affected by karst are

---

[1] Written by G. Orombelli.

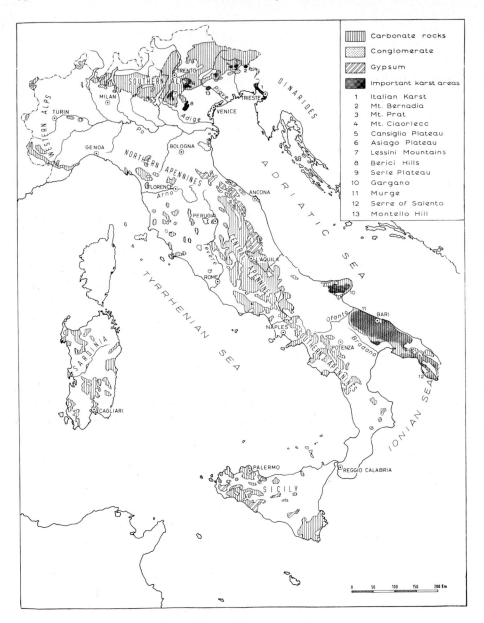

Fig.1. Soluble rocks and karst areas in Italy. (Based on DAINELLI, 1940; and NANGERONI, 1957a.)

almost exclusively localized, with different characters and extensions, in the Dinarides, the Southern Alps and the Western Alps.

In the *Italian Karst*, which is the extreme western border of the Dinarides, soluble rocks outcrop along a narrow strip trending northwest–southeast, and are represented by Upper Cretaceous–Middle Eocene, well-bedded limestones (San Michele Limestones).

In the *Southern Alps* the soluble rocks outcrop along an east–west trending belt, which extends from Lake Maggiore towards the east, becoming gradually wider. The main formations affected by karstification are: the Permo-Triassic and Carnian gypsums and anhydrites; the Middle–Upper Triassic, massive or thick-bedded limestones, dolomitic limestones and dolomites (mostly the Esino Limestone, Schlern Dolomite, Marmolada Limestone, and "Dolomia Principale" formations); the Liassic, generally well-bedded limestones, dolomitic limestones and cherty limestones (chiefly the "Corna", Moltrasio Limestone and "Calcari Grigi" formations); the Cretaceous reef limestones; the thick-bedded limestone breccias intercalated in the Eocene flysch of Friuli; and the Pontian thick-bedded polymictic conglomerates.

In the *Western Alps* the outcrops of soluble rocks appear to be extremely dispersed and of generally modest extent, with the exception of a narrow southern belt trending northwest–southeast. The lithostratigraphical units which present karst features are, according to CAPELLO [1938], the Permo-Triassic gypsum and anhydrite, the Triassic cellular dolomite, the Triassic–Jurassic dolomites and dolomitic limestones, and, to a lesser degree, the Jurassic calcareous schists and the Jurassic–Cretaceous–Eocene limestones and marly limestones.

Whereas along the northern portion of the Apennine range karst-bearing rocks outcrop in small scattered areas, they constitute an extended and almost continuous belt along the whole central–southern portion.

In the *Northern Apennines* karst phenomena are developed principally in the internal Tuscan area, particularly in the massive carbonate formations of the Upper Triassic–Liassic belonging both to the metamorphic autochthonous and parautochthonous complex of the Apuanian Alps ("Grezzoni" and "Marmi" formations), and to the "Tuscan Series" ("Calcare Cavernoso" and "Calcare Massiccio" formations).

In the external area of the northern Apennines, where the mainly marly-arenaceous rocks of the "Ligurian–Emilian Series" outcrop, the karst phenomena are developed only locally in the "Alberese" marly-calcareous flysch (Paleocene–Eocene), and in the Upper Miocene gypsum.

In the *Central–Southern Apennines*, karst-bearing rocks outcrop in a continuous belt which extends from Monte Catria (Marches) to Monte Montea (north Calabria), and which reaches its maximum width in Abruzzi. From a lithostratigraphical standpoint this belt corresponds to the "Umbrian" and "Latium–Campania" zones: in the first zone soluble rocks are essentially represented by Liassic massive limestones and cherty limestones, in the second by Upper Triassic–Eocene mainly massive limestones, dolomitic limestones and dolomites. A narrow strip of Upper Miocene gypsum, which locally presents karst phenomena, extends along the Adriatic border of the Central Apennines.

In *Apulia*, three principal outcrop areas of karst-bearing rocks can be distinguished: the Gargano headland, the Murge, and the Serre of Salento. The first area consists of Jurassic reef dolomites, dolomitic limestones and limestones, and of Cretaceous skeletal limestones merging laterally into well-bedded cherty limestones in the most easterly region. The Murge and the Serre of Salento are made up of Cretaceous well-bedded limestones.

Miocene, Pliocene and Quaternary calcarenites, which locally present modest karst phenomena, are also present in this region.

In *Sicily*, most of the karst-bearing rocks outcrop in the western half of the island and in the southeastern extremity. For the greater part they are made up of Upper Triassic–Liassic dolomites, dolomitic limestones and limestones, Cretaceous rudistid limestones, Eocene and chiefly Miocene detrital limestones, and Upper Miocene gypsum.

In *Sardinia* the outcrops of soluble rocks are of modest extent and mostly scattered in the border regions of the island. According to VARDABASSO [1955] they are made up of Cambrian and, to a lesser degree, of Silurian and Devonian limestones, of Lower–Middle Triassic dolomites, dolomitic limestones and limestones, of Jurassic–Cretaceous limestones and dolomites, and Miocene skeletal-detrital limestones.

TECTONIC SETTING[1]

Almost all the formations which present karst phenomena in Italy belong to the Mesozoic–Palaeogene and have been more or less intensely affected by the Alpine tectogenesis, with the exception of the Mesozoic formations of Sardinia which remained undeformed. The Alpine tectogenesis greatly deformed the internal sectors of the Alpine, Dinaric and Apennine geosynclinal branches, and had a much smaller effect on the more external sectors (Italian Karst, Apulia). Later on both were dissected by the post-Alpine tension faulting. We shall now outline the structural setting of the principal outcrop areas of soluble rocks described in the preceding section, paying particular attention to the most typical karst areas.

The *Italian Karst*, which includes the Karst of Monfalcone and of Trieste, belongs to the external Dinarides. Here the tectonic setting is characterized by a gentle asymmetrical fold, called the "Basso Carso" anticline, with a northwest–southeast axial trend and a shorter and steeper south–west limb, which is locally overturned near Trieste. The Karst of Trieste is developed on the southern flank of this fold while the Monfalcone Karst includes all the western extremity of the anticline whose axis turns here slightly towards the south, and passes near the lake of Doberdò. The "Basso Carso" anticline is dissected by normal faults, the most important of which have the same trend as the axis of the fold. The slight dips of the limestone beds, especially

---

[1] Written by G. Orombelli.

nearing the very flat anticline crest, notably favoured the intense karstification of this area (see Fig. 3, p. 96).

The *Southern Alps* show a structure essentially marked by folds, thrusts, overthrusts and faults, trending predominantly east–west. Axial planes of fold and thrust planes generally dip towards the north. Overthrusts have been observed in many sectors; these overthrusts, although at one time believed to be an exception, are certainly more common and important than would appear from the field survey. In the northern area of the Lombardian and Venetian pre-Alps the karst-bearing formations are folded, faulted and locally overthrust. Farther south these formations display homoclinal structures on which the greatest developments of karst phenomena can be found. Beyond these structures the strata bend abruptly southwards in a knee fold, which constitutes the border of the Po Plain.

In the Friulian pre-Alps, where karst morphology is developed to a higher degree, it is possible to recognize two areas from north to south. The first area has a tectonic setting distinguished by an imbricate structure, while the second corresponds to the characteristic ellipsoidal zone, described by various authors. Here, along the entire pre-Alpine belt and to the south of a great displacement—the peri-Adriatic overthrust —a series of large anticlines is developed, approximately trending east–west and with north-dipping axial plains. Among these are the anticlines of Bernadia, Ciaorlecc, etc., which we shall examine later. In fact the more or less flat crest of these folds offers suitable conditions for karstification.

The *Western Alps* show a particularly complex geological structure, marked by nappe tectonics. Soluble rocks in this region can be found in the sub-Briançonnais, Briançonnais and Piedmont zones, and are generally extremely deformed and therefore unsuitable for karst development. For this reason karst phenomena are localized in rather restricted areas.

The *Apennines*, particularly in the central–southern portion, where karst-bearing rocks largely outcrop, present a tectonic setting marked by generally asymmetrical folds, break thrusts and overthrusts trending northwest–southeast with a thrust direction towards the northeast. These structures are dissected by a close network of normal faults predominantly striking in parallel and transversal directions to the preceding tectonic trends. The main tectonic and physiographic features of the central–southern Apennines are essentially due to these normal faults which produced horst, graben and step-fault structures. The most typical karst features in this region, such as poljes and karst plateaus, clearly appear to be controlled by these structures.

Apulia shows a rather simple structural setting, and from the Gargano to the Serre of Salento only slight differences can be observed. The Gargano is considered to be a horst, weakly folded and dissected by numerous normal faults, mostly striking northwest–southeast, in a parallel direction to the Apennine trends. Step-fault structures often occur, giving place to secondary tectonic elements.

The Murge region presents a very gentle fold structure, dissected by normal faults which bound the limestone plateau principally on the western side. The dip of the beds is very slight everywhere, and this favoured an intense karstification.

In the Serre of Salento the fold structure is more in evidence than elsewhere. Here, long and narrow asymmetric anticlines, trending northwest–southeast, are developed. Eastern flanks are shorter and steeper and often cut by normal faults.

Faults are in fact present in the whole area, especially on the highest hills of the peninsula (the Murge of Salento), which show a horst character (see Fig. 5, p. 102).

In *Sicily* soluble rocks outcrop in two distinct areas, as we mentioned in the preceding section. The western region is characterized by folds, thrusts and overthrusts trending northwest–southeast, with a thrust direction towards the southwest. The southeastern region displays a less complex structure, marked by gentle folds dissected by horst, graben and step-faults structures.

In *Sardinia* the Palaeozoic formations have been deformed by the Caledonian and Hercynian tectogenesis, and are therefore strongly folded; on the other hand, the Mesozoic and Tertiary formations were not affected by the Alpine tectogenesis, but only by vertical movements and present subhorizontal bedding, affected only locally by tension faults.

PHASES OF KARSTIFICATION[1]

The difficulty previously mentioned in tracing a homogeneous and well-balanced picture of the Italian karst phenomena becomes more acute in this section, which will therefore be fragmentary in relation to the elements at our disposal.

In the geological evolution of Italy there have been several phases of emergence, and we may presume that during some of these, the development of karst phenomena took place locally.

In the Iglesiente area of Sardinia some karst cavities in the Cambrian limestones (as for example the Santa Barbara Cave), show calamine and barite concretions or incrustations; further, some calamine deposits seem to fill doline-shaped depressions. According to VARDABASSO [1955] these mineralizations are connected with the Hercynian magmatic phase, but, particularly in the first case, the same author does not exclude even a much later origin due to deposits from solutions enriched through contact with Hercynian ore bodies.

Bauxite deposits, distributed in various Italian regions, are clear evidence of karst activity during the Cretaceous. In the Italian Karst the effects of the Senonian emergence, causing numerous bauxite deposits in Istria, which indicate a notable karstification phase, were felt almost as far as Trieste. Further to the west, on the Karst of Monfalcone, the marine sedimentation was not interrupted.

In the peninsular portion of Italy the principal bauxite deposits are in Abruzzi (Rocca di Mezzo, Ovindoli, Lecce dei Marsi), in Campania (upper Liri Valley) and

---
[1] Written by S. Belloni.

particularly in Apulia, in the Gargano (San Giovanni Rotondo), in the Murge (Gravina, Spinazzola, Corato and Minervino) and in the Salento Peninsula. All these deposits appear to be fillings of very irregular depressions and testify to a sub-aerial phase during the first half of the Late Cretaceous [CRESCENTI and VIGHI, 1964]. In this phase an intense karstification of the Cretaceous limestones took place and gave rise to dolines, abysses and wells. Cretaceous bauxites are also known in Sardinia, near Olmedo.

The localities mentioned are grouped in scattered and rather small areas. The Late Cretaceous emergence was probably not of a general character.

The areas which today display a karst morphology had no common palaeogeographic evolution, but this took place in the various Italian regions in different ways and at different times.

The geomorphologic evolution of the *Italian Karst* began immediately after the ultimate emergence, probably in the Early Miocene. During all the Miocene this region was completely covered by the Eocene flysch deposits, and a sub-aerial drainage was developed. At the end of this period a regional uplift took place with a rejuvenation of the geomorphic cycle. The deepening of the valleys thus reached the limestone formations and gave rise to the karstification. This evolved slowly during the Pliocene and became more intense at the beginning of the Quaternary owing to the complete removal of the flysch cover. Alignments of dolines and caves, controlled by Pliocene (and Quaternary?) normal faults, are evidence of a rather young age of many karst phenomena. Further, the cave deposits are not older than the Lower Pleistocene [MAUCCI, 1961a; MARTINIS, 1962a].

Even today, there are few data relative to the evolution of the karst phenomena in the Italian Alps. This evolution must necessarily have had its beginning immediately after the complete emergence which, in the Central–Western Alps, can be attributed to the Oligocene, while in the Eastern Alps it was concluded during the Miocene. During the Pontian, all of the karst areas of the Alpine belt had certainly emerged. The majority of authors are unanimous in maintaining that the karstification began during the Pliocene, while the maximum development took place during the Pleistocene. In this region the single karst phenomena are dated essentially by the relationships with the Quaternary Glacial forms and by the age of the cave deposits, determined both by fossil faunas and prehistoric artifacts.

In the *Southern Alps*, according to SAIBENE [1951], the subterranean drainage in the Grigne Group had its beginnings already from the First Glacial Period, while the surface karstification can be attributed to the post-Glacial. According to PRACCHI [1943] the dolines of the Andossi area, near Madesimo, were partially obliterated during the Würm Glacial. In our view, the Montello Hill (see Fig. 6, p. 114), made up of Pontian conglomerate, is of particular interest; on the western slope of this hill there is a series of terraces on which a gradual diminution in the intensity of the karst phenomena, from the highest to the lowest, can be observed. According to the age attributed to those terraces by S. Venzo [UFFICIO IDROGRAFICO MAGISTRATO

Acque Venezia, 1963], it can be supposed that the karstification in this area began during the Pleistocene and that it continued during the greater part of the Pleistocene. Karst features observed in Pleistocene conglomerates (Ledro Valley, etc.) confirm a recent karstification of some areas.

In the *Western Alps* the greater part of the caves, such as those of Bossea, were attributed by Rovereto [1923] to the Bühl Stade, while according to the same author the dolines in gypsum, such as those in the upper valley of Aosta (Checrouit), are believed to be younger than the Würm Glacial.

The fossil faunas and the prehistoric findings discovered in the cave deposits are, for the greater part, attributed to the Würm (Sambughetto, Buco del Piombo, etc.), while only a minor part can be attributed to an older age, as far back as the Mindel–Riss Interglacial (Lessini Mountains).

The greater part of the limestone masses of the *Apennines* had already emerged during the Late Miocene–Early Pliocene and the emergence was completed during the Villafranchian. At this age a primitive drainage pattern was established, controlled by the principal tectonic trends and the karst phenomena probably began, without, however, giving rise to great morphological effects, owing to the small depth of the water table. The Late Villafranchian normal faulting then dissected the drainage pattern and, on the whole, caused an uplift of the region in respect to the base level of erosion and to the water table. Consequently the karst phenomena were accentuated with the formation of dry valleys and large enclosed depressions. The karstification then probably continued with varying intensity, with halts and starts during the Glacial and Interglacial phases, probably reaching its maximum during the Riss. Several large pits and collapse sinks in the Sabini Mountains can be attributed to a pre-Würm age, and in fact traces of the Würm Glacial often appear superimposed over pre-existant karst features. The Cavallone Cave in the Maiella is attributed to the same age. A successive renewal of karst activity took place after the Würm with the formation of small dolines and lapiés [Segre, 1948; Ortolani and Moretti, 1950; Demangeot, 1965].

According to Castaldi [1961], the karst phenomena in the Sorrento Peninsula reached their maximum development during the Günz, and continued until after the Mindel. Finally, according to Marcaccini [1964], in the Apuanian Alps the karstification was superimposed over a glacial morphology, the karst features at present being in a fossilization stage.

The development phases of karstification in *Apulia* are not well known. According to Colamonico [1951], the karst topography and the subterranean drainage within the Cretaceous limestones began during the Quaternary. However, the same author also puts forward the hypothesis that the present karst of Apulia can be attributed to an older activity, preceding the Pliocene. According to Lazzari [1958], an intense phase of karstification was developed in the Murge and in the Salento during the Pliocene; the karst activity was then renewed during the Sicilian and reached its maximum during the Tyrrhenian. During that time the largest caves of the region were probably formed, among them the Zinzulusa Cave near Castro, which was studied by Lazzari.

There was certainly more than one phase of karstification in Apulia. The oldest is the one connected with the bauxite deposits already mentioned and which is dated as Turonian in the Gargano and in the Murge. Even during the Tertiary there is evidence of emersion, which certainly favoured karstification, even within the calcarenites. Finally a Quaternary karst activity appears obvious if one takes into account the phenomena present in the Pliocene and Quaternary calcarenites.

Few data are available regarding the age of the karst phenomena in *Sicily* and in *Sardinia*. According to MARINELLI [1917] the karstification of the Upper Miocene gypsum in the Agrigento area (Sicily) began after the post-Pliocene uplift and is still in a youthful stage. In Sardinia, according to FUREDDU and MAXIA [1964], the Nettuno Cave was formed between the Pliocene and the Pleistocene and was already in a fossilization phase during the Mindel–Riss Interglacial.

MORPHOLOGIC FEATURES[1]

The karst morphology of the various outcrop areas of soluble rocks in Italy is controlled by the lithologic character, the tectonic setting and the geomorphic evolution briefly described in the preceding pages.

Completely favourable conditions for the development of a karst morphology can be found only in few areas along the southeastern border of the Alps and in Apulia. In these areas the solutional features are extremely well developed in a great variety of forms, and dominate the landscape. We shall indicate these regions as "principal karst areas".

Karst features are also present in all the other outcrop areas of soluble rocks, but owing to certain unfavourable conditions they are less developed and generally display only few morphologic types. Such areas in which karst features are present, but do not dominate the landscape, will be described by us as "lesser karst areas" (Fig. 1, 2).

Principal karst areas

Among the principal karst areas to be described in Italy are the Italian karst, the plateaus of the Friulian, Venetian and Lombardian pre-Alps and the Apulia region. The Montello Hill is another significant but small karst area; it is made up of conglomerate and is therefore described under that heading.

---

[1] The first part of this section *(Principal karst areas)* was written by G. Orombelli, the second part *(Lesser karst areas)* by S. Belloni.

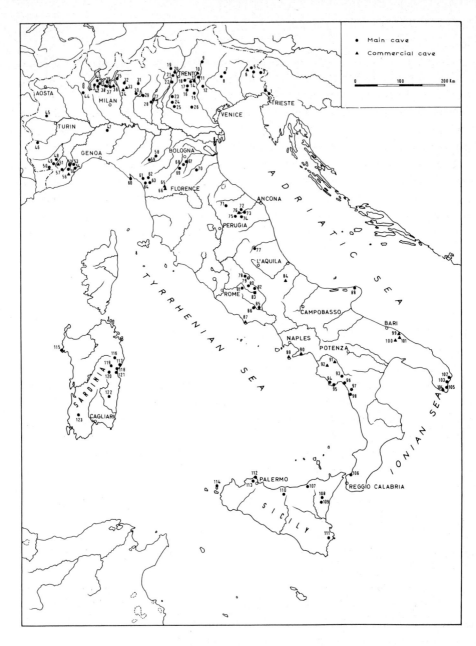

The Italian Karst

The portion of the karst plateau which is in Italy belongs to the "Basso Carso" or "Carso Proprio" (Lower Karst or True Karst).

In turn, the Italian Karst can be divided into the Karst of Monfalcone (or Karst of Gorizia) and the Karst of Trieste, separated by the "Vallone" of Doberdò (Fig. 3).

The *Karst of Monfalcone* is a plateau at about 100–150 m above sea level, made up of Upper Cretaceous–Middle Eocene well-bedded limestones (San Michele Limestones) gently folded into an anticline, trending east–west, interrupted by normal faults [MARTINIS, 1962a]. The predominant morphological feature is the great frequency of dolines and the notable development of lapies (karren). However, larger karst features are also present in this area, such as the poljes of the lakes of Doberdò, Pietra Rossa and Sablici.

The lake of Doberdò is in a depression bounded by normal faults, trending west-northwest–east-southeast. It has a depth of 6 m at low water, which can increase to 12 m at exceptionally high water; on the northern side there are three swallow holes which feed or empty the lake. The lake of Pietra Rossa, immediately to the south of the above, is situated in a depression trending west-northwest–east-southeast,

---

Fig.2. The distribution of the main caves in Italy: *1* = Grotta Gigante; *2* = Cristalli Abyss; *3* = San Giovanni d'Antro Cave; *4* = Viganti Cave; *5* = Grotta Vecchia di Villanova; *6* = Grotta Nuova di Villanova; *7* = Val Cellina caves; *8* = Bus de la Lum; *9* = Franzei Cave; *10* = Castel Tesino caves; *11* = Bus de la Bela; *12* = Pederobba Cave; *13* = G. B. Trener Cave; *14* = Bigonda Cave; *15* = Tre Cantoni Cave; *16* = Bus de la Rana; *17* = Busa del Sciason; *18* = Costalta Cave; *19* = Torrione di Vallesinella Cave; *20* = Battisti Cave; *21* = Laman Abyss; *22* = Bus del Diaol; *23* = Spluga della Preta; *24* = Ponte di Veia northern Cave; *25* = Buso del Piatte; *26* = Covolo della Guerra; *27* = Bus Coalghes; *28* = Buco del Frate; *29* = Bus del Quai; *30* = Buco del Corno; *31* = Bus di Tacoi; *32* = Lacca del Roccolino; *33* = Sogno Cave; *34* = Meraviglia Cave; *35* = Fiumelatte Cave; *36* = Buco del Piombo; *37* = Guglielmo Cave; *38* = Buco della Volpe; *39* = Zocca d'As; *40* = Cunardo caves; *41* = Monte Tre Crocette Cave; *42* = Bus de la Scondurava; *43* = Bus di Remeron; *44* = Buco della Bondaccia; *45* = Pugnetto Cave; *46* = Balma di Rio Martino; *47* = Camera Cave; *48* = Dossi upper Cave; *49* = Caudano caves; *50* = Col del Pass Abyss; *51* = Bossea caves; *52* = Buranco Rampiun; *53* = Fate Cavern; *54* = Arene Candide Cave; *55* = Arma Pollera; *56* = Strega Cave; *57* = Vene Cave; *58* = Vallestra Cave; *59* = Monterosso Rise; *60* = Colombi Cave; *61* = Tana dell'Uomo Selvatico; *62* = Revel Abyss; *63* = Tana che urla; *64* = Antro del Corchia; *65* = Maona Cave; *66* = Giusti Cave; *67* = Farneto Cave; *68* = Spipola Cave; *69* = Buca del Diavolo; *70* = Tana del Re Tiberio; *71* = Tassare Cave; *72* = Mezzogiorno Cave; *73* = Buco Cattivo; *74* = Infinito Cave; *75* = Monte Cucco Cave; *76* = Frasassi Cave; *77* = Acquasanta Cave; *78* = Pietrasecca swallow-hole; *79* = Luppa swallow hole; *80* = Inferniglio Cave; *81* = Arco Cave; *82* = Santullo Well; *83* = Collepardo Cave; *84* = Cavallone Cave; *85* = Pastena Cave; *86* = La Vettica Abyss; *87* = Guattari Cave; *88* = Manaccore Cave; *89* = Grotta Azzurra of Capri; *90* = Smeraldo Cave; *91* = Pertosa Cave; *92* = Castelcivita caves; *93* = Bussento caves; *94* = Grotta Azzurra of Palinuro; *95* = Cala delle Ossa; *96* = Dragone Cave; *97* = Maratea Cave; *98* = Madonna di Praia a Mare Cave; *99* = Palazzese di Polignano Cave; *100* = Putignano Cave; *101* = Castellana caves; *102* = Romanelli Cave; *103* = Zinzulusa Cave; *104* = Diavolo Cave; *105* = Grotta Grande di Ciolo; *106* = Trémusa Cave; *107* = San Teodoro Cave; *108* = Gelo Cave; *109* = Archi Cave; *110* = Minnonica Well; *111* = Perciata Cave; *112* = Pietra Selvaggia Abyss; *113* = Addaura caves; *114* = Genovese Cave; *115* = Nettuno Cave; *116* = Tumba e Nurai Abyss; *117* = Grottone di Biddiriscottai; *118* = Toddeitto Cave; *119* = Su Anzu Cave; *120* = Su Bentu Cave; *121* = Bue Marino Cave; *122* = Su Marmuri Cave; *123* = San Giovanni Cave. (After ALMAGIÀ, 1959; and NANGERONI, 1957a.)

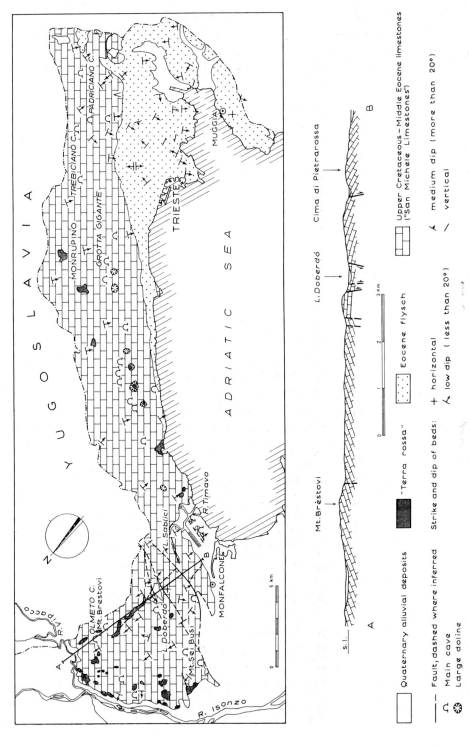

Fig.3. The Italian Karst: geologic sketch map with the main karst features and geologic cross-section. (Simplified after BERTARELLI and BOEGAN, 1926; MARTINIS, 1962a; and UFFICIO IDROGRAFICO MAGISTRATO ACQUE VENEZIA, 1963.)

controlled by a normal fault. The lake of Sablici lies within a contiguous polje, with the same trend and geological setting. There is a direct subterranean communication between the three lakes, which have no connection with the nearby rises of the Timavo, but are fed by the waters of the river Vipacco, which runs immediately to the north of the plateau.

The most common dolines are a few metres deep, with sloping sides either at a sharp angle to the flat floor or in a very gentle curve.

Funnel-shaped dolines and open pits are also present. Dolines are irregularly scattered over the entire area, locally in groups and locally aligned in northwest–southeast or northeast–southwest directions, corresponding to the principal fracture trends [VENZO and FUGANTI, 1965]. From this point of view the six dolines of Monte Sei Busi are a typical example. Compound dolines are frequent and the largest among them are in the northern part of the area; they have very gently sloping sides, flat floors and are as much as 200 m in diameter.

Lapies are evident at the southwest extremity of the Doberdò Plateau, where it slopes gently to the southwest. All the aspects of a karst surface are present, from the semi-covered to the bare and relict lapies and to a debris plain, locally named "griza".

The typical "terra rossa" is widespread in this region, particularly inside the flat-floored dolines, where recent borings have found a maximum thickness of 12 m.

The subterranean karst phenomena appear particularly developed in the northern region of the plateau, and in the southeast near the village of Iamiano. Caves extend in the direction of the dip, while shafts predominantly correspond to the vertical or subvertical fractures. On the whole all of the caves show the effects of the tension faulting which distinguishes the region. Many are aligned along the main fault trends. Both the caves and the shafts are only small; the largest cave is the Monte Olmeto Cavern, 186 m long, while the deepest explored shaft is that of Monte Sei Busi, which reaches a depth of 40 m.

The *Karst of Trieste*, to the southeast of the preceding area, is a plateau extending at an average height of about 250 m, of which only a small part is within Italian territory. This corresponds to the southern limb of the same anticline of the Karst of Monfalcone and is made up of the same formations. Contrary to the preceding area, the dolines are less frequent, but some of them attain very notable dimensions with the major axis ranging from 250 to 500 m. These are mainly funnel-shaped and show a northwest–southeast alignment.

More complex and larger karst depressions are present, such as the great uvala of Monrupino [MAUCCI, 1961a].

Semi-covered, bare and relict lapies and debris plains are common. Subterranean features are much more largely developed and imposing than in the Karst of Monfalcone. Among the caves are: the Grotta Gigante, an extremely vast cavern, 160 m high, in which instruments for the study of earth tides (two horizontal pendulums) have been installed; the Padriciano Cave, 270 m deep with an extension of 600 m, and the Trebiciano Cave, 329 m deep with a length of 200 m, along the bottom of which flows the river Timavo [BOEGAN, 1938]. This river is described in the hydrologic section.

During the last few years numerous morphologic studies have been carried out in this region [MARUSSI, 1941a,b; D'AMBROSI, 1956, 1961, 1966; MAUCCI, 1961a,b] In these studies particular attention is paid to the problems of the palaeogeographic evolution of the region and of the origin of the different karst phenomena, often from an original standpoint. Recently FORTI and TOMMASINI [1967] and FORTI [1968], in a detailed study of the local lithologic sequence, emphasized a close relationship between karst forms and lithology.

Plateaus of the Friulian, Venetian and Lombardian pre-Alps

Along the pre-Alpine belt of the Southern Alps numerous limestone plateaus are to be observed, generally of modest extent, in which a complete karst morphology is developed. From east to west the principal and best known are the plateaus of Bernadia, Monte Prat, Ciaorlecc, Cansiglio, Asiago, Lessini, Berici Hills and of Serle near Brescia (Fig. 1).

The first four are to be found in the well-known zone of the "ellipsoids", the pre-Alpine belt distinguished by a series of asymmetric brachyanticlines with axial planes dipping towards the north, and made up of a core of Mesozoic limestones surrounded predominantly by a Tertiary flysch. These brachyanticlines form heights with fairly flat summits which, together with the lithological and structural characteristics, permit the development of an intensive karst topography.

The *plateaus of Bernadia, Monte Prat and Ciaorlecc* in the Friulian region reach elevations of between 800 m and 1,000 m above sea level and show a large doline development, flat for the greater part in the first two, funnel- and pit-shaped in the third, with an average diameter of about 50 m. The dolines tend to be grouped in the flat areas, forming uvalas locally; they also tend to occur in steps in the slightly sloping areas and aligned along small blind valleys. A notable development of lapiés completes the karst landscape [MARINELLI, 1897; FERUGLIO, 1914, 1923].

Similar features, more widely extended, are present in the Venetian pre-Alps, on the *plateau of Cansiglio*, which rises to an elevation of 1,200–1,300 m. Also present in this area is a large depression (polje), in turn sub-divided into three secondary basins, the floor of which is dotted with swallow holes, shallow flat-floored or funnel-shaped dolines, sub-alluvial dolines and uvalas, often aligned from northeast to southwest, according to the most common directions of the fractures and joints [COPPADORO, 1903; DE GASPERI and FERUGLIO, 1910; LEHMANN, 1959; CASTIGLIONI, 1960, 1964].

The subterranean morphology in the four areas examined is represented by numerous caverns and shaft caves, among which the best known are the Bus de la Lum, an abyss 225 m deep, opening onto the Cansiglio and the Villanova caves, of which the largest is 3,665 m long, carved out in the limestone breccias of the flysch, which is developed around the Mesozoic core of the Bernadia brachianticline [DE GASPERI, 1916; FERUGLIO, 1954].

The *plateau of Asiago* or plateau of Sette Comuni, extends to the north of Vicenza between the Astico and Brenta valleys. This is a vast, gently undulating plateau, which from a maximum elevation of 2,341 m (Cima Dodici) slopes gently down to about 800 m. It is made up of Upper Triassic–Cretaceous, predominantly limestone formations; the structural setting is distinguished by a slightly dipping homocline sequence, very gently folded, which curves sharply in a knee fold along the borders of the plain. The major karst development is to be found in the "Dolomia Principale" (Upper Triassic) and in the "Calcari Grigi" (Liassic), while in the "Biancone" (Lower Cretaceous) there is a minor development [DE MARCHI, 1911]. The dolines are generally small and funnel-shaped and are particularly numerous in the flat areas. Flat-floored, marshy dolines of greater dimensions are also present locally. Some dolines with asymmetric flanks occur along the slopes. Characteristic and relatively numerous are open pits which are about 50 m in depth and width, with vertical walls, flat floors, partially covered by a roof with a hole to the land surface or by remains of a collapsed roof (Busa del Geson, Stoanhaus, Tanzerloch). Ice caves are also frequent.

Lapies and debris plains are very common, particularly on the "Calcari Grigi". Caves extending horizontally are rare; among these are the Bus de la Rana with a total length of 3,700 m, and the caves of the Oliero rises, at the foot of the plateau on the Canale di Brenta.

On the Asiago Plateau surface drainage is lacking, with the exception of a few water courses which have cut deep canyons, as for example, the Val d'Assa.

The *Lessini Mountains* to the north of Verona between the Adige and Leogra valleys, constitute a mountainous group in which it is possible to distinguish a higher region to the north, between 1,200 and 1,500 m above sea level, a central plateau between 900 and 1,200 m, and a hilly region to the south descending gradually to the plain. The area with the largest development of karst phenomena is the central plateau, made up of a homoclinal sequence of Triassic dolomites and Jurassic–Lower Cretaceous limestones, dissected by normal faults.

In this area the surface karst features do not appear uniformly distributed, but are generally grouped along palaeovalleys [PASA, 1954]. Here areas of dolines, as well as uvalas, small poljes, dry valleys and largely diffused lapies can be observed. Numerous dolines, mostly funnel-shaped, are found in the Monte Tomba region; and dolines aligned along small dry valleys in the area of Monte Pastello, Verona Hills, Erbezzo–Bosco Chiesanuova, etc. Large isolated dolines can be observed in the Corno d'Aquiglio and Val Fredda areas, and uvalas in the regions of Erbezzo–Bosco Chiesanuova, Tracchi–Grezzano, to the east of Monte Bellocca and in the Campo Fontana area.

Small poljes occur in the upper Fumane Valley, in the Ronconi–Ceré area, in the Pantena Valley, etc.

Dry valleys are very frequent, particularly in the Erbezzo–Bosco Chiesanuova region. To the north of Monte Tesoro there is the natural bridge of Veja, remnant of the roof of a former cavern, now transformed into a collapse sink.

The subterranean karst features are largely represented by shaft caves, generally

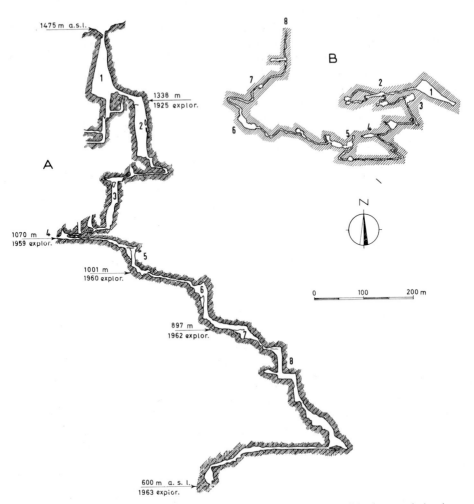

Fig.4. The Spluga della Preta Abyss. The numbers *1–8* make possible the correlation between the section and the plan view. A. Section. B. Plan view. (Simplified after PASINI et al., 1965.)

aligned along the areas with a karst surface topography. Among the most important caves is the Spluga della Preta near the Corno d'Aquiglio, the deepest abyss in Italy[1] (879 m deep, with a horizontal extension of 416 m), opening into the floor of a doline (Fig. 4).

The *Berici Hills*, to the south of Vicenza, constitute a small gently undulating plateau, which rises from the plain up to 444 m above sea level. This plateau is made up of Upper Cretaceous–Miocene limestones, marly limestones, marls and tuffs, weakly folded in a periclinal structure. Surface karst features are particularly frequent

---

[1] Recently in the Monte Ganin Group (Julian Alps), the Gortani Abyss, developed in Upper Triassic dolomites and dolomitic limestones has been explored down to a depth of 892 m.

where the Lower Oligocene limestones outcrop, and they are represented essentially by numerous circular or elliptic dolines, mostly shallow and flat-floored with a maximum diameter of 200 m and a depth of about 20 m. Dolines are usually in the highest flat areas of the southwestern Berici Hills and are generally filled with "terra rossa" and are without evidence of swallow holes.

Dolines on slopes, with the major axis following the direction of the maximum slope are rare. Further, there are numerous scattered dolines on the whole plateau, with diameters between 100 and 200 m and generally shallow.

Major features are the karst depression between Monte San Gottardo and Monte Mottolone and that of Monte Grande, with a maximum size of 500 m and a depth of about a dozen metres. The second-mentioned is transformed into a temporary lake in exceptionally rainy seasons.

The valley of Pozzolo–Bocca d'Orno, dried by subterranean capture, is connected with the karst evolution of this area.

Lapies are very frequent, particularly to the north of Lumignano. There are numerous shaft caves, particularly near Arcugnano and Lumignano; among the deepest is that at Valmarana, 93 m deep, which opens into the floor of a doline. Caverns are less numerous; among them are the San Gottardo Cave, more than 150 m long, and the Grotta della Guerra, 115 m long, through both of which runs a perennial stream [FABIANI, 1902a,b; DA SCHIO et al., 1947].

The most westerly plateau with a typical karst topography is that of *Serle* (or Cariadeghe) near Brescia, where the Liassic "Corna" Formation outcrops, with mostly horizontal or subhorizontal bedding. In this area the surface karst phenomena [LAENG, 1923; PRACCHI, 1943] are represented by numerous dolines, with a frequency of about 60 to the km^2 and with a diameter varying from a few metres to over 100 m. Funnel-shaped dolines are prevalent while open pits are more rare. The dolines are distributed all over; on the slopes they are aligned along small dry valleys, in steps, often with adjoining borders. In the larger dolines, smaller ones can sometimes be found. The shape varies in relationship to the bedding and the slope, assuming symmetric, and more often, asymmetric forms. Often at the bottom of the dolines a narrow shaft cave opens.

Well-developed lapies are to be found along the borders of the plateau.

The region of Apulia
Karst is one of the most significant geographical aspects which combine to determine the unity of the Apulia region, a direct consequence of the homogeneous geological setting. Karst features appear the most evidently in the vast areas of the Gargano and Murge plateaus and of the Serre of Salento, where Cretaceous limestones outcrop (Fig. 5). In these three areas the karst displays common characteristics although with different intensity, and it is marked by the presence of all the principal surface and subsurface morphological types, though with lesser frequency than in the Italian Karst.

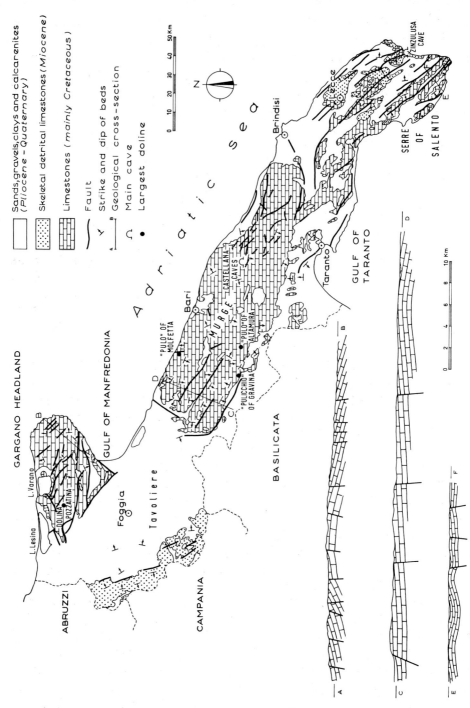

Fig.5. The Apulia region: geologic sketch map and cross-sections. (Based on MARTINIS, 1962b,c, 1965; COTECCHIA, 1963; SERVICIO GEOLOGICO D'ITALIA, 1963–1970; and CAMPOBASSO and OLIVIERI, 1967.)

Common features are the grouping of the dolines along the borders of the plateaus, the presence of great long karst depressions (poljes and blind valleys) mostly trending northwest–southeast, according to the main structural trends, the scarcity of lapies, the absence of a surface drainage pattern and the infiltration of the water through a close joint network rather than through well-developed swallow holes. Further, "terra rossa" (red mediterranean soil) is largely diffused and characteristic of this region [COLAMONICO, 1951].

The *Gargano Plateau* rises isolated from the surrounding region and protrudes into the Adriatic Sea to form a headland, with a maximum elevation of 1,056 m above sea level. This is the area with the greatest karst development in Apulia and in the whole Italian peninsular region. Karst features are extremely numerous and varied, often with different characteristics from those common to all the Apulia: lapies, dolines of the most varied shapes and dimensions, swallow holes, poljes, dry valleys and caves. The predominant aspect is the extremely intense development of the dolines on large areas, particularly above 500–600 m [MARCACCINI, 1962]. Dolines are especially frequent in the flat areas, along the mountain tops and on the terraces, mainly in the central-western region, where Jurassic–Cretaceous white limestones and dolomites outcrop. The most extensive sinkhole plains are in the eastern part of the terrace of San Giovanni Rotondo, at a height of 500–600 m, on the top plateau of the headland, between 800 and 1,000 m, and further east in the Monte Spigno–Monte Sacro area (maximum frequency 80 dolines to a square kilometre). The greater part of the dolines are subcircular with rather constant diameters of about 100 m. They are often aligned along dry valleys and coalesce locally to form uvalas. Also some dolines are of exceptional size. Among these, the most notable is the Pozzatina doline, 5 km to the south of San Nicandro, of elliptic form, 675 m long, 440 m wide and with a maximum depth of 130 m; the sides slope at about 30° and the floor is flat [BISSANTI, 1966).

The poljes, called "piani" as in the Apennines, are generally elongated and predominantly in northwest–southeast and east–west direction, parallel to the two main fault systems which dissect all of the headland. The poljes of San Giovanni Rotondo, Sant'Egidio and San Marco in Lamis, are aligned from east to west along the southern part of the headland. In the central part of the Gargano numerous small poljes and dry valleys trending northwest–southeast can be observed at a height of 600 m (Piano Canale, Piano San Vito, Piano San Martino, Piano Ceresaldi).

According to BALDACCI [1957] the caves are prevalently horizontal and occur chiefly in the northeastern region. On the other hand, shaft caves are more frequent in the Foresta Umbra area and in the southern region, and are connected to the surface by swallow holes, which are generally independent from the dolines [CAPPA, 1962].

The *Murge Plateau* extends in northwest–southeast direction from the Ofanto Valley to Francavilla, between the Adriatic Sea and the Bradano Valley. On this plateau it is possible to distinguish the higher Murge region, which constitutes the actual plateau (at about 400–650 m) separated by an escarpment from the lower

Murge region, which is a series of terraces descending in steps towards the Adriatic. The main karst aspect of this area is the almost complete absence of a surface drainage pattern. Surface karst features are scattered and particularly well developed on the higher Murge.

Among the larger forms are the enclosed or semi-enclosed elongated depressions, mostly in northwest–southeast direction, very shallow and with a flat floor covered by "terra rossa". The dolines, generally scattered, locally in groups, often aligned, have many different shapes and sizes. The most common type is circular with a flat floor and diameters between 50 and 100 m and depths ranging from 2 to 6 m; the floor is covered with "terra rossa" and the swallow hole is not often visible. The water drainage is generally very slow, so that sinkhole ponds are often formed temporarily. Contrasting features are shown by the so-called "puli", which are large cavities, very deep and with very steep sides, as for example the Pulo di Altamura (maximum diameter 725 m, depth 96 m) and the Pulicchio di Gravina di Puglia (maximum diameter 560 m, depth 110 m). The floor is flat and becomes marshy in winter.

Swallow holes, locally known as "capivento" if they are small, "grave" if larger, are also present. Open pits and collapse sinks, such as the bell-shaped one at Toritto, also occur locally. The karst landscape is completed on the Adriatic margin by the "lame", rather shallow ravines nearly always dry, cutting the borders of the terraces, and on the Ionian margin by the "gravine", larger in size and resembling canyons.

Subterranean forms in this area are relatively small with the exception of the caves of Castellana, one of the most noted subsurface complexes in Italy. In these caves two parts can be distinguished; the "grave", a very large dome open at the top, 60 m in height, on the floor of which the central part of the roof had fallen, and the actual caves extending for about 2 km prevalently in horizontal direction along a northeast–southwest fracture. These caves are extremely rich in depositional features of every form and type.

The *Serre of Salento* are long and narrow hills, trending north-northwest–south-southeast, which reach a maximum height of about 200 m. In this area the karst phenomena are similar to those of Murge, but less impressive. Here also the main karst characteristic is the absence of a surface drainage pattern. Dolines are rare, swallow holes more frequent, locally known as "vore". In the coastal region, between the Otranto and Cape Santa Maria di Leuca, the karst assumes a different character with the development of debris plains and signs of lapies. There are also numerous caves, mostly due to marine erosion; on the other hand the Zinzulusa Cave is certainly of karst origin and extends for 140 m, rich in beautiful depositional features. It was originally partly filled with deposits of guano, now completely exploited.

Lesser karst areas

The greater part of the Italian regions in which soluble rocks outcrop do not present a complete karst morphology except in very restricted areas, although the different karst features are often frequent and locally imposing.

We shall now describe separately the essential features and the distribution of the lesser karst areas in the Italian Alps, the Apennines, Sicily and Sardinia.

Italian Alps

The karst phenomena do not appear to be uniformly distributed along the Alpine range, but display an increasing intensity from the west to the east, where, as we have seen, they often assume a morphological predominance.

On the *western Italian Alps* the karst features are in general scattered and not of notable importance. A series of monographs on this subject has been written by CAPELLO [1938, 1950, 1952, 1955]. The surface forms are particularly frequent in the east Maritime Alps and pre-Alps and less diffused in the Cottian Alps. Lapies, dolines and enclosed depressions are frequent in the Middle Triassic–Cretaceous dolomitic limestones, limestones and marly limestones of the Marguareis Mountain, Mongioie and the Mondolé group (Maritime Alps). This area is the most important in the Western Alps for its extent, variety and diffusion of karst phenomena. Another very interesting region is that of the Maritime pre-Alps near Mondovì, where Triassic–Jurassic dolomites, dolomitic limestones and calcareous schists outcrop, characterized by numerous dolines, often aligned along blind valleys. The small valley of Carnino in the upper basin of Tanaro is, according to ROVERETO [1923], an example of "reculé" in this region. Small groups of dolines are present in the calcareous schists of the Monviso area (Cottian Alps) and of the Devero Valley (eastern Lepontine Alps).

The subterranean karst features are more or less developed in all the Western Alps; however, they assume particular importance in the Monte Antoroto-Mondolé, Monte Fenera, Cuneo-Mondovì, Monviso, Monginevro–Ambin, Lanzo Valley and upper Sesia–Toce areas. The largest caves generally extend horizontally. Among these we can mention the caves of Caudano in the Mondovì area, which open into the Middle Triassic limestones and have a total development of 2,190 m, and above all the Bossea Cave opening into the Triassic dolomitic limestones and extending for 1,984 m, sloping gently upwards 217 m. This latter cave follows an east–west trending fracture; it is notable for numerous halls rich in speleothems, and through it flows a stream which locally forms small lakes. In the Monviso area the Rio Martino Cave opens into the Mesozoic calcareous schists, and extends about 600 m. In the Lanzo Valley, the Grotta Maggiore of Pugnetto which is 703 m long, and opens into the same formation, is notable.

In the Strona Valley (upper Sesia–Toce area) the Caverna Grande delle Streghe is carved into the pre-Triassic marble, for about 1,000 m.

Shaft caves are also present in the Western Alps, such as the subterranean complex of Piaggia Bella in the Marguareis Massif, formed by three communicating caves with 5,800 m of lateral passages at a depth of 689 m, and the Raymond Gaché abyss, 558 m deep [BADINI and GECCHELE, 1965].

In the *central Italian Alps* the karst phenomena are principally localized in the

Lombardian pre-Alps. In this region the subsurface morphology is generally better known than the surface; studies of a general character were made by PRACCHI [1939, 1943] and by LIGASACCHI and RONDINA [1955]. According to these authors, karst in Lombardy is prevalently developed in the pre-Alpine limestone belt, particularly between 800–1,800 m above sea level, but varying in intensity from place to place. The major surface karst features are on flat or slightly sloping surfaces, especially where the bedding is also subhorizontal. The distribution and the intensity of the karst phenomena gradually diminish from east Lombardy towards the west, and do not appear to be connected with the age of the formations. Finally, in the eastern part the dolines seem to be in a more advanced stage than those in the western part.

With reference to the subsurface karst, it can be observed that from east to west the size of the caves increases both horizontally and vertically, and also that the altitude of the cave mouths becomes higher. In eastern Lombardy only the subterranean karst appears to be connected with that of the surface. Caves seem to be localized chiefly along fracture lines. In general, there is a slightly greater number of the vertically extended caves than the horizontal. The speleothems in the caves are infrequent.

The caves in the Varese district are an exception, as they are more largely extended horizontally. They are mostly bedding caves and are rich in speleothems.

The most common surface forms are the lapies, which are mainly to be found in the Triassic formations of the Esino Limestone (Monte San Martino and Piani di Bobbio in the Grigne Group), in the "Conchodon Dolomite" (Prasanto, near Monte Rai), and in the Liassic "Corna" Formation (Botticino). The greatest diffusion of lapies is to be found at altitudes between 500 and 2,100 m.

Dolines are only locally developed in areas of limited extent, where suitable topographical conditions are present (glacial cirques, flat tops, etc.).

In the Grigne Group, numerous dolines can be observed in the Moncodeno glacial cirque between 1,800 and 2,300 m, and several dolines partly filled by Recent alluvial deposits and by talus scree on the Alpe Campigne, at a height of 1,750 m, in the Esino Limestone [SAIBENE, 1951, 1955]. Lesser doline areas are found at Bazena, Val Fredda and Val Cadino in the Esino Limestone, on the Bossico Plateau, at a height of 800–900 m, in Val Cavallina above Entratico, on the western slope of Monte Alben, on Monte Cancervo, on the Piani di Bobbio and Artavaggio, in the "Dolomia Principale" Formation, on the Selvino Plateau in the Rhaetian limestones and on the Andossi Mountain near Madesimo in the Triassic formations of the Spluga syncline. Scattered dolines are found on Monte Campo dei Fiori, in the Upper Brianza between the Palanzone and Faello Mountains, above the village of Torno in the Liassic Moltrasio Limestone and on Pizzo Arera and in Val Canale in the Esino Limestone.

Subterranean karst features are mostly diffused on the Maddalena and Palosso Mountains in the "Corna" Formation, in the Brembo Valley between San Pellegrino and the southern border of the Bergamo pre-Alps in the "Dolomia Principale"

Formation and in the southern region of the Como Lake, in the Moltrasio Limestone.

Shaft caves prevail on the Maddalena and Paitone Mountains, in the area lying between the Chiese and Garza rivers, one of the deepest of which is the Buco del Roccolino (105 m deep). The Buco del Frate, 230 m long, is the largest of the horizontal caves.

In the Trompia Valley (Monte Palosso and Monte Pernice) horizontal caves are more numerous than are shaft caves; one of the horizontal caves is Buco del Fuso, about 100 m long.

In the middle Brembo Valley vertically extending caves are the most frequent. These caves are of greater dimensions than those mentioned above. On the mountains surrounding the Como Lake, mainly in the Liassic Moltrasio Limestone and in the Norian "Dolomia Principale" formations, horizontal caves, often of impressive extent are common. Among the largest are the Buco della Volpe, about 1,000 m long and 178 m deep, and the Guglielmo Cave, a succession of shafts and passages, 452 m deep with a horizontal development of 950 m, both in the Liassic limestones, and the Buco del Piombo in the Upper Jurassic–Lower Cretaceous "Maiolica" Formation, which extends for 780 m along a fracture, with a gently rising slope. Through this cavern runs a stream which has formed cave deposits in which vertebrate fossils and artifacts have been found.

Subterranean karst features are also scattered in the Iseo Lake region, mainly on the western slopes, in Val Seriana, in the upper Brembo Valley, in the Grigne Group, on Monte Campo dei Fiori and in Val Ganna. At the foot of the western side of the Grigne rises the Fiumelatte, a subterranean periodic stream, dry in the winter months and active from spring to autumn. The underground course has been explored for about 250 m.

In the *eastern Italian Alps* the karst phenomena which predominate in some restricted areas already described, also appear to be present—although to a lesser degree—on the Monte Baldo Group, in the Trento district, on the Feltre and Belluno pre-Alps and in Friuli.

Lapies are widely diffused, particularly where the Liassic "Calcari Grigi" Formation outcrops, as in the highest area of the Pasubio, on the Lavarone Plateau and on the Vette and Pelse plateaus near Belluno. Lapies are also present in the Ladinian limestones of the Marmolada.

Small doline plains can be found locally, where a suitable topography is present. Numerous dolines and small uvalas connected with subterranean deep shafts can be observed in the Monte Baldo and Altissimo group [PASA, 1954], along remnants of north–south trending palaeovalleys. A notable example of a palaeovalley is the Lumini Plain near San Zeno di Montagna. Dolines are often present where the "Calcari Grigi" Formation outcrops, in the Trento district, on the Monte Stivo–Bondone and Monte Casale–Gazza–Paganella ranges and on the Pasubio, Folgaria and Lavarone plateaus. Lesser doline areas can be found in the Ladinian and Norian dolomitic limestones of the upper Non Valley, in the "Calcari Grigi" Formation on the Dolomites, in the Badia, Gardena and Funés valleys, at the Alpe di Siusi, Alpe Flavona and Ampezzo regions [TRENER, 1933].

In the Feltre pre-Alps there are a few dolines in the Pietina glacial cirque on the Vette Plateau [DAL PIAZ, 1907] and in the Belluno pre-Alps on the Pelse Plateau, where the "Calcari Grigi" Formation outcrops.

Shallow flat-floored or funnel-shaped dolines, often in steps along small dry valleys, are to be found in the Friuli near San Pietro al Natisone, where Tertiary limestone breccias and conglomerates outcrop [FERUGLIO, 1913].

Caves are numerous in the Monte Baldo Group, particularly on the northern side. In the Trento district the cave of the Torrione di Vallesinella (Brenta Group), opening on the face of the mountain at a height of 2,350 m in the "Dolomia Principale", is marked by a predominantly horizontal extension and by frequent forms of corrosion. Mechanical erosion seems to have prevailed over chemical in the entire cavern. Also notable are the caves of Castel Tesino, 379 m long, G. B. Trener, 3,400 m long, and Bigonda, 3,020 m long in the Sugana Valley [NANGERONI, 1957a,b].

Apennines

In close connection with the geological setting, the karst phenomena are widely distributed all along the central–southern Apennine range, from Monte Catria (1,702 m) in the Marches region, to Monte Montea (1,986 m) in northern Calabria; in the Apuanian Alps and in the region between the Arno and Tiber rivers (Tuscany, part of Umbria, and Latium) they are only locally developed, in small areas (Fig. 1).

Along the central–southern Apennine range the Volturno Valley separates two distinct sectors: the northern is distinguished by large, often continuous karst areas, and the southern is marked by areas both smaller and generally more scattered.

In both sectors the major karst features seem to be essentially limited to two main morphological types, the karst basin and the karst massif [SEGRE, 1947].

Karst basins. The morphological type of a karst basin is a large depression of varying forms and sizes, at relatively high altitudes, generally between 800 and 1,200 m. The most common and characteristic types are sub-elliptical in shape, mainly elongated following the Apennine trend, that is northwest–southeast, with major axes of a few kilometres and minor axes of some hundred metres and with a central flat area filled by alluvial and eluvial deposits from which hums rise locally. Small funnel-shaped sub-alluvial dolines are frequently present in the floors of the basins. The surrounding slopes are often asymmetric and the steepest generally coincide with a fault scarp; downstream the basins are closed by rocky sills rising up to about 20 m from the floor and generally cut by small V-shaped valleys. The catchment areas of some basins are not completely enclosed, but are open on one side owing to the capture by headward erosion of one or more valleys. However, the surface drainage from basins is nearly always at a minimum or nil, and the outflow of the waters takes place through one, or occasionally more, generally eccentric swallow holes, often located near the slope corresponding to a fault scarp.

Some basins contain small lakes, or the lowermost part is marshy. Sometimes they may be temporarily flooded during the periods of heaviest rain, but never as a consequence of water rising from the swallow holes. In Italian terminology these basins are generally called "piani" or "campi" and are believed to be similar to Dinaric poljes. They differ, however, from these in some particular aspects, such as their smaller sizes and generally the presence of only one active swallow hole, which serves only as a drain for the water and not as a rise. NANGERONI [1957a, 1960], LEHMANN [1959], SESTINI [1963] and DEMANGEOT [1965] are in agreement that "piani" are indeed true poljes.

The Apennine poljes are generally in groups or in linear series. Among the most noted and characteristic are the following: the "piani" of Colfiorito, between Foligno and Nocera Umbra, at an average height of 750–800 m; the "piani" of Castelluccio, to the east of Norcia, at an average height of 1,300 m, consisting of a large enclosed basin formed by four distinct poljes with numerous dolines, perennial ponds and hums, and the "piani" of Santa Scolastica, to the southwest of the former, at a height of 700–800 m, probably an open polje; the "piani" of Cornino, Rascino Lake, Campo Lasca and Antrodoco, between Rieti and l'Aquila; the numerous "piani" on the southwestern slopes of Gran Sasso d'Italia, near Castel del Monte, Santo Stefano and Calascio, among them the "piani" Prosciuta, Chiano, Tagno, and Calascio, at a height of about 1,200 m, aligned west-northwest–east-southeast, and the "piani" of Viano and Vuto, at a height of about 950 m, greater than the preceding ones, still aligned in the same direction; the "piani" of Rocca di Cambio, Rocca di Mezzo, Rovere and Ovindoli, between the Sirente and Velino Mountains; the "piani" of the Carseolani, Simbruini and Ernici Mountains, aligned northwest–southeast, among which is the "piano" di Arcinazzo, at a height of 850 m, about 8 km long, with scattered hums and numerous dolines and uvalas, and also that of the Canterno Lake, which was formed at the beginning of the 19th century through an obstruction of the swallow hole, and which dried up several times later [SEGRE, 1947, 1948; ORTOLANI and MORETTI, 1950].

Among the karst lakes in the Apennines, in addition to the one cited, there are also some small temporary lakes which form during heavy rain near Sulmona and in Irpinia, and the permanent lake of Matese, in Campania, about 8 km long and 2 km wide, at present used for hydroelectric purposes [NANGERONI, 1957a].

Karst massifs. These are wholly limestone mountain groups, bounded by steep slopes, which support an undulating plateau with enclosed basins, blind valleys, doline areas, swallow holes and uvalas. The karst massifs, lacking in surface drainage, constitute the most deserted and uninhabited areas of the Apennines. At times, however, they appear as narrow, long ridges consisting of an alignment of generally rather high, arid peaks with few passes. Massifs and ridges follow the main Apennine trend, from northwest to southeast, in many belts separated one from the other by deep valleys, along which Tertiary marly-arenaceous formations outcrop. Among the main belts are those of the Sibillini Mountains, the Gran Sasso–Maiella Mountains, the Velino–Sirente–Meta Mountains, the Simbruini–Ernici Mountains and the Lepini–Aurunci Mountains in the central Apennines, and the Matese and Alburni Mountains in the southern Apennines.

The heights surrounding the "piani" of Castelluccio in the Sibillini Mountains, present a small development of surface karst features with lapies, scattered or grouped dolines, locally in steps, and uvalas [LIPPI BONCAMBI, 1948].

In the Gran Sasso Massif numerous dry valleys are noted, as for example the Valle Cupa and the upper part of the Fossa Paganica, which descend to a characteristic "piano" joining the Campo Imperatore Depression. In this area there are numerous doline areas, most of which are at altitudes of between 1,400 m and 1,600 m; among the most characteristic are those to the north of Castel del Monte, where shallow flat-floored dolines prevail, often joined to form small uvalas, or aligned to form small dry valleys. Funnel-shaped dolines can be observed on the hums which rise from the alluvial floor of the "piano" Moltigno. Numerous sub-alluvial dolines on till deposits are present in the area Le Coppe at Campo Imperatore. The best-developed lapies can be observed above Castel del Monte, near the Fossa Paganica, and in the "piani" of the Fogno area. A particular feature of this region is the more or less total lack of "terra rossa" [ORTOLANI and MORETTI, 1950].

In the Maiella Massif there are some dry and, to a lesser degree, blind valleys, controlled by faults, and in the highest flat surfaces there are some areas with numerous funnel-shaped, small-sized dolines [DEMANGEOT, 1965].

Numerous and characteristic areas with dolines, compound dolines, slope dolines in steps, blind valleys and small enclosed basins are to be found in the regions of the Carseolani, Simbruini, Ernici, Lepini and Aurunci Mountains in Latium. In this region collapse sinks are also frequent and typical, as for example the Santullo Well, 60 m deep and 155 m wide, formed by the breakdown of the roof of a large cavern. Remains of the roof are still evident along the borders, and from these and on the walls numerous stalactites hang. Collapse sinks in alluvial or tuffaceous deposits, formed by deep subsurface erosion and which often hold more or less temporary ponds, are still characteristic and common in the same region and are known locally as "sprofondi" [SEGRE, 1948].

In the southern Apennines the karst topography with dolines and dry valleys, occur in the Alburni Mountains and, to a lesser degree, in the Matese Group.

The subsurface morphology seems to be more developed in the southern than in the central Apennines. The principal caves known are the following: the Cavallone Cave in the Maiella Massif, which opens at a height of 1,357 m, halfway up a mountain face, and extends for 850 m; the Pàstena Cave, in the Ausoni Mountains, extending about 2 km and the Pertosa Cave, in the Alburni Mountains, extending 2,270 m, both rich in magnificent speleothems and through which flow subterranean streams; the Castelcivita Cave, near Salerno, 4,200 m long, with frequent rock mills, numerous small lakes and siphons and rich in speleothems, among which the cave pisolites are very typical; and the Bifurto Abyss, 683 m deep, in the Pollino Massif (Calabria), an occasionally active swallow hole, with characteristic corrosion features. In the southern Apennines some coastal caves are also known, such as that at Cape Palinuro near Salerno, the Grotta di Smeraldo near Amalfi and the Grotta Azzurra in Capri. These caves are at least partly due to marine erosion.

Finally in the southern Apennines there are several rivers, part of whose courses are underground, and which are mentioned in the hydrogeological section.

Several mostly small karst areas are scattered over the region lying to the north of the Arno and between the Arno and Tiber rivers; in the Apuanian Alps subsurface forms prevail, and to the south of the Arno the surface forms.

Small-sized, shallow, flat-floored doline areas are on the Montagnola Senese, near Massa Marittima and on the hilly region to the northeast of Argentario in the Upper Triassic "Calcare Cavernoso" Formation, on Monte Cetona in the Liassic "Calcare Massiccio" and on the Calvana Mountains in the Cretaceous–Eocene "Alberese" limestones. Scattered dolines can be observed on the Apuanian Alps, on the northern slope of Monte Tambura and on Monte Pisano in the Liassic "Marmi" and in the Upper Triassic "Grezzoni" formations. Slope dolines, opened downstream, are present on Monte Morello and on Montagnola Senese.

Near Siena, there is the polje of Pian del Lago in the "Calcare Cavernoso", covering an area of about 5 km^2; in the same formation near Massa Marittima several large depressions, once marshy but now drained, open on one side with a low sill and with one or more swallow holes on the floor, can be observed.

Finally, lapies do not seem to be widely diffused; some typical examples are in the "Calcare Massiccio" Formation on Monte Pisano and in the Panie Group (Apuanian Alps).

Subterranean features are chiefly encountered in the "Grezzoni" and "Marmi" formations in the Apuanian Alps. Caves are often very deep and generally controlled by the joint network. Among the best known caves are: the Antro del Corchia, the second deepest abyss in Italy, opening at 1,150 m, extending 1,800 m horizontally, with a depth of 805 m; the Tana dell'Uomo Selvatico, an active swallow hole, extending 1,330 m with a depth of 318 m; the Enrico Revel Abyss, a single well 316 m deep, the deepest of its kind in Italy [BADINI and GECCHELE, 1965]. The Tana che Urla, an irregular slightly descending passage 372 m long, developed between the "Calcare Cavernoso" and the underlying "Verrucano" Formation, is to be mentioned among the predominantly horizontal caves. Shelter caves, mostly open at the base of the "Calcare Cavernoso" at the contact with the underlying schist rocks, are frequent and characteristic in these regions, and locally known as "tecchie".

About a dozen caves generally small, (with the exception of the Fontana Buia Cave which extends horizontally for about 350 m) are found in the "Alberese" limestones on the Calvana Mountains.

On Monte Argentario in the "Calcare Cavernoso" there are several caves and the largest is that at Punta degli Stretti, which is some hundred metres long [MARCACCINI, 1961, 1964].

Sicily and Sardinia

Details about karst in Sicily are rare except as regards the gypsum outcrop areas, to be described in a later section. According to SAIBENE [1957] karst phenomena are locally developed in the Mesozoic limestones of the Madonie Mountains, the Palermo region, the area between Castellammare and Trapani, the Sicani, Nebrodi and Peloritani Mountains, in the Miocene limestones of the region lying between Ragusa and Siracusa, and in the Pliocene limestones of the region between Marsala and Sciacca.

In the summit areas of the Madonie Mountains dolines and large karst depressions have been observed. Characteristic features of this region are the intense weathering of the outcropping limestones through micro-karst phenomena and the absence of lapies.

Numerous caves are known, many of which are also due to marine erosion, such as for example, the caves of Monte Pellegrino near Palermo, at present lying some twelve metres above sea level. The Addaura Cave, at 70 m above sea level and extending 1,600 m, in which characteristic helictites are found [NANGERONI, 1957a,b], is in the same area.

In Sardinia there are numerous but generally small karst areas. Surface karst is only slightly developed [VARDABASSO, 1957; FUREDDU and MAXIA, 1964] and is chiefly marked by lapies which are present on almost all the limestone outcrops, particularly in the Baunei and Dorgali area, on Monte Sarri and Monte Albo, and in the Iglesiente and Nurra regions. Dolines have been observed on the mountains of Oliena and on Monte Albo, where a small polje (Campu e Susu), 300 m in diameter, has also been noticed.

Natural wells and swallow holes are also present, but the dominant feature of the Sardinian karst is the large number of caves, the largest of which extend horizontally.

The principal caves are in the Middle–Upper Jurassic limestones of the Tavolara Island, in the Mesozoic limestones of Monte Albo, of the Nuoro district (Su Anzu Cave, the longest in Italy, extending more than 8,000 m, Su Bentu Cave, extending about 7,000 m; Bue Marino Cave, 4,100 m long and 15 m deep), of the Ogliastra plateaus (Golgo Abyss, 270 m deep), in the Cambrian limestones of the Iglesiente region (cave of San Giovanni di Domusnovas, 1,200 m long and 35 m deep), in the Miocene limestones of the Sassari district and in the Mesozoic limestones of Alghero and Cape Caccia. In this latter region is the cave of Nettuno, 840 m long and 42 m deep, opening at 1 m above sea level. Inside there is a brackish lake about 130 m long and 15–40 m wide, from which massive columns, joined to the roof, emerge. The walls and the roof are covered with beautiful speleothems, which are also present on the bottom of the lake, bearing witness to a recent submergence of this area [SOMMARUGA, 1952].

KARST PHENOMENA IN CONGLOMERATE[1]

The most conspicuous Italian example of a complete karst morphology in conglomerate is in the Montello, a large low hill rising isolated in the Treviso Plain (Fig. 6.) It is composed of polymictic conglomerate with pebbles mainly of dolomitic limestone, subordinately of chert, sandstone, porphyry, ultrabasic rocks and granite, and with calcite cement. Pebbles are often pitted, well-rounded and ranging from a few centimetres to 10–15 cm in size. The conglomerate is thick bedded with lens-shaped marly clayey or sandy intercalations. This formation has been attributed to the Pontian [MARTINIS, 1955].

The conglomerate is generally covered by a more or less thick eluvium.

Along the western border the hill is cut by a series of terraces which were attributed by S. Venzo [UFFICIO IDROGRAFICO MAGISTRATO ACQUE VENEZIA, 1963] to the Günz, Mindel and Riss. Beds are generally horizontal or slightly dipping, except in the border regions. The structure of the Montello is marked by a gentle anticline, trending west-southwest–east-northeast parallel to the orographic axis. This anticline is very evident in the western side, and flattens towards the east [MARTINIS, 1955].

The morphology of this region was studied by TONIOLO [1907] who recognized three areas with different characteristics. The eastern area is a plateau with dolines and uvalas. Shallow flat-floored and funnel-shaped dolines, with an average diameter of 100 m are numerous, open pits are rarer. With the exception of a restricted area on the northernmost side of the hill, where the dolines (with a frequency of 74 to one square kilometre) are mainly isolated, all the remaining part of this sector is marked by the coalescence of dolines forming uvalas, with major axes of 400–500 m, with a frequency of 46 to one square kilometre. Further, several blind valleys, trending parallel to the northern and southern borders of the hill, are characteristic of this area. Finally, in the eastern side of the hill there are almost all of the subterranean forms, represented by caves generally extending horizontally, through which run small perennial streams which have produced typical corrosion forms. Among these caves there is one, called Tavaran Lungo, which extends more than 300 m.

The central area is the highest part of the hill and corresponds to the structural high of the anticline. Shallow flat-floored and funnel-shaped dolines, with an average diameter of 100 m, are still very numerous (50 to one square kilometre) and are irregularly scattered. These often coalesce to form uvalas of a lesser size than those in the preceding area and are generally aligned along dry valleys. The dry valleys, mainly trending north-northwest–south-southeast, are the main feature of this area. They generally issue from a uvala at the top of the hill and present a convex profile (average slope 6%) interrupted by a series of dolines in steps. Along the southern flank of the hill the dry valleys terminate at about 225 m above sea level, where a sharp accentua-

---

[1] Written by G. Orombelli.

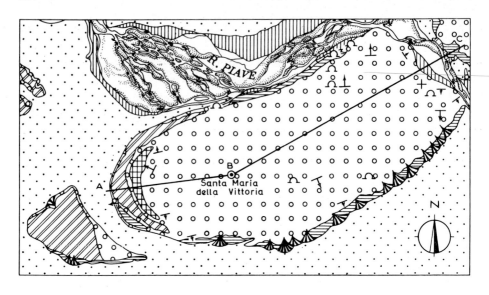

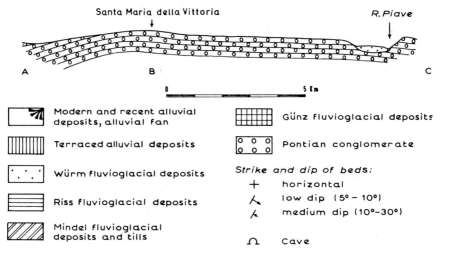

Fig.6. Geologic-morphologic sketch map and cross-section of the Montello Hill. (After Martinis, 1955; and Ufficio Idrografico Magistrato Acque Venezia, 1963.)

tion of the slope takes place. Along the northern flank the dry valleys terminate at about 200 m above sea level on a bordering terrace.

The western area is marked by a series of terraces on which the frequency of the dolines decreases downwards. Shallow flat-floored and funnel-shaped dolines are present, with the greatest frequency and variety on the highest terraces (60 to one square kilometre). The terrace underneath, attributed to the Günz, presents a far smaller number of mainly shallow flat-floored dolines, often aligned in groups parallel to the terrace. At a lower height a further large terrace can be observed, attributed to the Mindel, with rare shallow flat-floored dolines (16 to one square kilometre). Finally on the lowest terrace, attributed to the Riss, there is no evidence of karst topography[1].

In the entire belt of Pontian conglomerate of the Venetian and Friulian pre-Alps, karst phenomena are frequent, although not to the same degree as on the Montello Hill. Mention can be made of the numerous and beautiful dolines near Polcenigo [DE GASPERI, 1916] and those near Ragogna and Susans in Friuli [LORENZI, 1903]. Finally, several dolines are known in the Pleistocene conglomerate (Riss–Würm Interglacial) of the Ledro Valley near Trento [TRENER, 1933] and also several caves in the Pleistocene conglomerate, locally named "Ceppo", near Paderno d'Adda (Lombardy).

KARST PHENOMENA IN GYPSUM[2]

Karst phenomena are not present in all of the Italian regions in which gypsum outcrops, but when present all of the features of a typical karst area are developed (Fig. 1).

In the Alpine range the gypsum formations are attributed to the Permo-Triassic and Carnian, while in the Apennines and in Sicily they belong mainly to the Upper Miocene.

The karst topography in the gypsum is marked in the Alpine region by limited areas with many small dolines (average diameter 7 m, depth 1.5 m, maximum frequency 1,100–1,500 to one square kilometre) according to DAINELLI [1908] and MARINELLI [1917].

A typical example is the Moncenisio karst area where, according to the above mentioned authors, the large number and the small dimensions of the dolines are in relationship with the intense and minute fracturing of the gypsum. Karst in the gypsum of the Apennines and Sicily is marked by numerous small blind valleys with swallow holes, generally developed at the contact between gypsum and the underlying clay, and to a lesser degree by large scattered dolines.

At the bottom of these depressions, caves of notable dimensions extend mainly horizontally, often on different levels. Erosional features on the roofs and the walls of the passages are evident, and the caves have often been enlarged through breakdowns. Small pits at the bottom of the dolines are also very frequent. All these features are in great evidence in the gypsum area of the Bologna Apennines [DONINI, 1965] and of central–southern Sicily [MARINELLI, 1917].

We shall now review briefly the gypsum karst areas of Italy.

In the *Western Alps* numerous karst areas with mostly funnel-shaped dolines are to

---

[1] A recent paper by ABRAMI and MASSARI [1968] points out that most of the karstification on the Montello Hill took place during and after the Mindel.

[2] Written by S. Belloni.

be found in the Triassic gypsum and cellular dolomite between the Stura di Demonte and the Val Maira, in the Monginevro–Ambin area, at the Moncenisio Pass, in the upper Aosta Valley and in the upper Toce Valley [CAPELLO, 1955].

In the *Central Alps* there are karst phenomena in gypsum in the Bergamo pre-Alps at Dossena and Oltre il Colle, near the Spluga Pass and at Esine in the Camonica Valley.

In the *Eastern Alps* surface karst forms are frequent in the gypsum outcrops lying between the Fella and Piave rivers, particularly near Treppo Carnico, Mauria Pass, Lorenzago, Valle Sella, Pieve di Cadore, Comelico, and San Pellegrino Pass [MARINELLI, 1917].

In the *Apennines* karst features are mainly to be found along the northeastern slopes and most of all in the Bologna area, in the Val d'Era in Tuscany, and in the Adriatic area of the Marches. Numerous large depressions with subterranean drainage, with an average diameter of 500–1,000 m and a depth of 90–100 m, are found on the Ronzano, Croara and Monte Donato hills in the Bologna neighbourhood (Fig. 7). In the same region are the best known caves in gypsum, such as for example the Spipola-Acquafredda Caves, extending 5,670 m horizontally, 80 m deep, and the Farneto Cave, 870 m long and 44 m deep [BADINI, 1967].

In *Sicily*, where the Miocene gypsum outcrops in many places, karst phenomena are localized in the Agrigento and Calatafimi areas. In the first area the gypsum outcrops widely and constitutes heights rising above the surrounding region. The gypsum lies on clayey–marly formations and the greater part of the karst depressions are developed at the contact of the gypsum with these underlying formations. Near Sant'Angelo Muxaro and Santa Elisabetta dolines are rare, while small blind valleys, terminating in a cave which drains the surface water, are more frequent. The lengthwise profile of these small blind valleys is asymmetric, steep on the gypsum side, gently sloping on the clay side.

Together with the features of the preceding region, there are in the Calatafimi area numerous real dolines completely carved in the gypsum. Near Santa Ninfa about twenty dolines of elongated asymmetric shape are found, with major axis of 300 m and depth of 30 m, developed on a slightly sloping surface. Saucer- and funnel-shaped dolines are rarer, often aligned or grouped, and they almost always have an eccentric swallow hole [MARINELLI, 1917].

HYDROGEOLOGIC CONDITIONS[1]

As we have seen, karst in Italy presents both surface and subterranean features which are extremely variable in character and intensity from one region to another. This obviously is also reflected in the hydrogeologic conditions which cannot easily be synthesized in a few pages. The scarcity of adequate studies, limited generally to

---

[1] Written by B. Martinis.

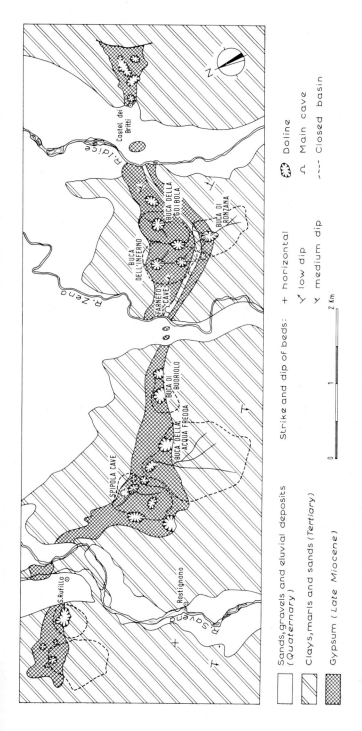

Fig.7. Geologic-morphologic sketch map of the gypsum area near Bologna. (Based on MARINELLI, 1917; SERVICIO GEOLOGICO D'ITALIA, 1963: F° 87 *Bologna*; and BADINI, 1967.)

restricted areas, makes a description of homogeneous characters even more difficult.

For practical reasons and from a hydrogeologic point of view, we can distinguish the karst areas with concentrated subterranean drainage from those with diffuse drainage. In the latter we can further distinguish areas with suspended water from others where the surface waters infiltrate uniformly underground, feeding the groundwater. As stated above we are not able to deal with the subterranean hydrology of all the Italian karst regions, and therefore we review briefly the most typical and best-studied examples, bearing in mind the distinctions made above.

The *concentrated subterranean drainage* as already mentioned, is particularly well developed in the Karst of Trieste, where this phenomenon is chiefly displayed by the famous river Timavo, which begins its subterranean course in the caves of San Canziano (Skocyan) and flows underground throughout the entire region until it rises near San Giovanni di Duino, less than 1 km from the Adriatic Sea, where the Lower Timavo River originates. The imposing resurgences of this river were already described in Roman times, when a famous pagan temple stood near the river bank. At that time the Timavo flowed into the Lacus Timavi, the present Lisert swamp, now almost completely reclaimed. The exploration of the subterranean course began with the first descent into the caves of San Canziano by J.C. Rosenmüller and A. Tillius in 1805; systematic investigations were undertaken by G. Svetina in 1839. Recently, the most modern methods of investigation have been used, in order to identify the underground course of the river, such as geophysical surveys [MORELLI, 1954; MOSETTI, 1954] and successful experiments with the injection of radioactive tracers into the water [MOSETTI et al., 1963].

From the caves of San Canziano, where the Upper Timavo River engulfs, the water flows with steep falls towards the west through a series of imposing caverns for about 1,730 m and disappears from view. The river is reached only in two places by caves, which open onto the surface just above its underground course: the Serpenti abyss, near Divaccia and the Trebiciano Cave, about 5 km to the northeast of Trieste. In this latter cave, at a depth of 329 m, the Timavo is visible for about 200 m, showing a width of 15 m and a depth ranging from 5 to 11 m, at low water. After the Trebiciano Cave and for about 26 km in a straight line, that is to say until the resurgence of the river, its course cannot be checked (Fig. 8).

The numerous caves which open onto the plateau of the Karst of Trieste do not reach the river except for two small natural wells which are, however, near to the resurgences. These rise at the foot of the karst on a 2-km front where eighteen springs can be seen, apart from the small and periodical ones. The three principal springs form the branches or mouths of the Lower Timavo which flow after 300 m into one single course, about 550 long and 42 m wide.

The subterranean course of the river near the resurgences has been studied by MOSETTI [1954] by means of a geoelectric survey, which revealed a greater number of branches than are present on the surface (Fig. 9).

The total discharge of the resurgences varies from 1,000,000 to 13,000,000 m^3/

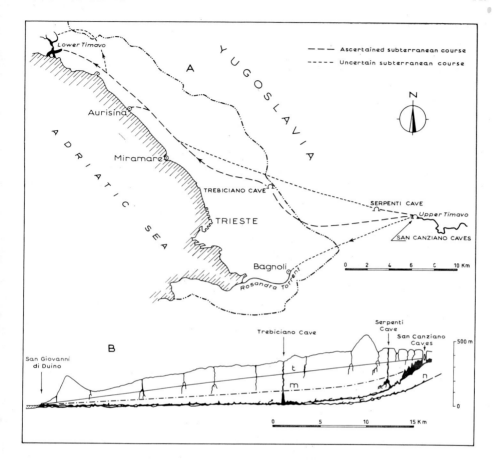

Fig.8. The subterranean course of the river Timavo. A. Plan view. B. Profile: t=theoretical hydrostatic level during exceptionally high water; m=maximum hydrostatic level; n=minimum hydrostatic level. (Based on D'AMBROSI, 1960, 1962; and MOSETTI, 1965.)

24 h, while in the San Canziano caves, where the river begins its subterranean course, flows varying from 11,000 to 8,500,000/24 h have been measured [BOEGAN, 1938]. Along the subterranean course the river increases greatly owing to the water falling into the subsurface catchment area, the surface of which has been estimated at about 500 km². According to D'AMBROSI [1960], the total water falling annually on the entire catchment area of the Timavo equals the quantity rising from the springs of San Giovanni di Duino.

A supposition has been put forward that part of the water of the Isonzo and Vipacco rivers, which flow respectively to the west and the north of the karst plateau, also joins the underground course of the Timavo. However, most authors believe that this contribution is very small.

On the other hand it has been ascertained that the river Vipacco contributes substantially to the lakes present on the Monfalcone Karst. In fact it has been shown

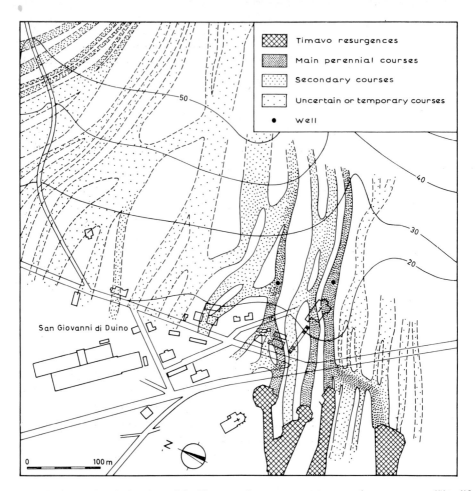

Fig.9. Geoelectrical survey of the Timavo subterranean course near the resurgences. (Simplified after MOSETTI, 1954.)

that the Doberdò, Pietrarossa and Sablici lakes are fed by the water of the Vipacco after a long subterranean route, and that they have no communication with the nearby Timavo rises.

Examples of concentrated subterranean drainage are also noted elsewhere in Italy, but none reaches the importance of the Timavo. Often this concentrated drainage represents a secondary feature to a wider diffuse underground drainage.

It is enough to mention some cases in the Alpine range which lead us to believe that subterranean streams of a certain importance exist, such as the springs of the river Livenza in Friuli, which rise at the foot of the limestone massif of Monte Cavallo, and also the Fiumelatte in Lombardy. This latter, after a subsurface course which has been explored for about 250 m, comes to light from Triassic limestone on the western slope of the Grigne Group at 100 m above the Como Lake. The Fiumelatte flows for only 200 m on the surface, and the subterranean course and the catchment area are uncertain. In the Lombardian pre-Alps there are other examples of subterranean concentrated drainage, such as those at Buco dell'Orso, where the waters are captured, and at Buco della Volpe, where little is known about the catchment area. Further west, in the province of Varese, we can find interesting examples on Monte Campo dei Fiori, a Mesozoic limestone–dolomite massif where the impervious basement is represented by a Triassic clayey-arenaceous formation. The Remeron caves open here, with two communicating lakes of which one is probably fed by a subterranean water course; further, it would appear that there are communications between the cave and lake Varese. The river also flows partly underground through the Cunardo caves, with a fall of 150 m. Finally, the subterranean stream of the Bossea Cave in Piedmont is known; this forms some small lakes and has a discharge ranging from 45 to 300 l/sec.

Subterranean concentrated drainage patterns, inadequately studied, are known in the limestone of the Apennine Mountains, where a clear surface drainage is often lacking.

In the Apuanian Alps, for example, in the Val Secca, the water of the homonymous torrent is engulfed and, after a subterranean course, re-emerges to form the "Pollaccia" Rise, near Isola Santa.

In the Matese area (southern Apennines) similar phenomena are even more imposing: the Sava and Lete rivers, tributaries of the Volturno, have in fact a partly subterranean course. The river Bussento, also in the southern Apennines, disappears into a swallow hole near Salerno and returns to light after a subterranean course of about 5 km [NANGERONI, 1957a]. Further, in the Alburni Mountains, there is a stream which flows through the Pertosa Cave; in the Salerno district there is another stream flowing through the Castelcivita Cave and forming numerous small subterranean lakes.

Finally, it is worth mentioning that in many Apennine "piani", temporary lakes are formed during heavy rains, such as those in the "Quarti" area near Sulmona, or the lake of Piano del Dragone in Irpinia. In these lakes there is generally only one active swallow hole which serves as a drain for the water and not as a rise. According to SEGRE [1948], this is due to the fact that the greater part of the "piani" is rather high on the water table. The engulfment of the water through the swallow hole points to the presence, at least for a certain way, of subterranean concentrated drainage, the character of which, however, is unknown.

The *subterranean diffuse drainage* is developed more or less intensively in the subsurface of all the karst areas. Detailed research in this regard, however, is rare and limited to a few restricted areas. This type of drainage may cause several aquifers or only one sole aquifer which contains the ground water, according to the lithological and structural characteristics and the heights of the karst-bearing limestone mountains. Along the whole pre-Alpine belt this phenomenon is developed in varying degrees. This is particularly evident where a surface drainage pattern is lacking, as is the case in many of the karst areas already mentioned, and which we need not repeat.

Elsewhere, in the pre-Alps, we can cite the Lessini area where the infiltration of the water takes place mainly through the joints and bedding planes, and the diffuse drainage arrives at the Jurassic formations which represent the impervious basement.

Even in Piedmont, in areas where karst phenomena are rare, there is a subterranean flow originating in the outcrops of gypsum, cellular dolomite and calcareous schists. The springs are relatively small, however; the most important geohydrologic phenomena are in the Val del Gesso.

In the peninsular portion of Italy, as in the pre-Alps, this phenomenon displays varying intensity and it is not possible, therefore, to generalize. In the Apennines of Umbria and Marches, cases of diffuse underground drainage can be found in the Sibillini Mountains, chiefly in the "piano" of Castelluccio, where water is drained through numerous dolines and flows towards the Norcia Basin, at the eastern border of the Sibillini. There are many springs in this basin, some of them intermittent.

According to recent investigations in Abruzzi, described by MANFREDINI [1964], the base-level of the underground drainage in the Abruzzi–Campania facies is placed in the Upper Triassic–Lower Liassic poorly pervious sediments. The overlying deposits, made up of over 2,000 m of Mesozoic and Tertiary limestones, contain an active water flow feeding very many springs, for example the Popoli springs with a discharge of about 7,000 l/sec. In the Umbrian facies on the other hand, there is a notable reduction in the thickness of the fissured limestone formations, represented above all by Lower Liassic limestones, feeding the springs of the Aterno Valley and the surroundings of L'Aquila.

Faults and folds often interrupt the geohydrologic continuity of the pervious formations, causing numerous small secondary basins at different heights which give rise to numerous small springs.

Particular attention should be paid to the groundwater flow in Apulia, which is influenced by the sea water as a result of particular geographical position and altitude of the limestone masses. This is an extremely interesting phenomenon which has been the object of particular study during the last few years owing to its importance in water supply.

As already noted, this region is prevalenty made up of Mesozoic limestones, dolomitic limestones and, in a lesser degree, of dolomites, all with an irregular network of intercommunicating fractures and interstices which allow water to flow freely and regularly downwards, forming a wide and thick groundwater reservoir. Also Tertiary and Quaternary sediments, chiefly composed of calcarenites with primary porosity and permeability, easily drain the surface water.

Fractures are often enlarged by karst which, however, as far as we know, does not cause a deep concentrated drainage as opposed to a diffuse flow. Only locally are elements to be found which point out the first type of underground drainage; for example, some littoral and submarine springs caused by groundwater flow under pressure, perhaps through actual deep passages.

In the greater part of the region and particularly in the Salento Peninsula, sea water penetrates deeply into the limestone rocks, aided by the intense coastal karst, so as to produce a flow of salt water; thus the waters of the Adriatic meet the waters of the Ionian Sea [COTECCHIA, 1956].

In addition to the groundwater, small perched aquifers can also be found locally, lithologic conditions permitting. For example, suspended waters are present in the Gargano and also in the Salento where they feed the springs of the Alimini Lakes, near the Adriatic coast to the north of Otranto.

The groundwater emerges locally along the northern side of the Gargano and it feeds both the Lesina and Varano lakes and the springs present in this area; elsewhere it produces submarine springs, particularly along the eastern coast of the headland [COTECCHIA and MAGRI, 1966]. Submarine springs are also to be found more to the south, in Apulia, as far as the coast of Salento. On the whole the discharge of these springs represents a small part of the annual rainfall on the region, and even taking into account the high evapotranspiration, a conspicuous quantity of the water infiltrated into the limestone contributes to the underground water storage.

As a surface drainage pattern is lacking in this region, especially in the Salento area, the recovery of underground water is particularly important, which means the identification of the piezometric surface of the groundwater and the depth of the interface between fresh and salt water. Through numerous wells drilled in the region for this purpose, it has been found that the piezometric surface is near the sea level at the coast and rises slowly inland with a piezometric gradient of $0.3–0.6^0/_{00}$ [COTECCHIA, 1956].

Owing to different densities fresh water floats on salt water, and the interface occurs at a depth according to the Ghyben and Herzberg theory. Research has shown that the contact between fresh and salt water is not abrupt but occurs through a dispersion zone of brackish water, probably caused by fluctuations of the lower level of the fresh water due to the strong seasonal rainfall variations. The interface between fresh and salt water on the continent can be determined bearing in mind the piezometric gradient under static equilibrium conditions. According to COTECCHIA [1955, 1959], the interface is to be found at a depth below sea level, equivalent to about 1/60 of the distance from the coast, which means that this surface is lowered by about 15 m for every kilometre distance from the coast (Fig. 10).

The results obtained in the study of the groundwater flow in the Apulia limestones are very important as they can be applied to many other Italian or foreign karst areas near the coast.

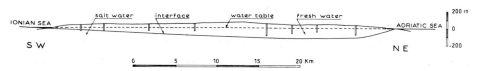

Fig.10. The interface between fresh and salt water in Apulia (Salento Peninsula). (Simplified after COTECCHIA, 1955.)

PRACTICAL PROBLEMS AND THEIR APPROACH[1]

There are various practical aspects connected with karst which have been given attention in Italy, from the use made of karst basins for hydro-electric purposes to the capture of subterranean water courses for water supply, and from the reclaiming by drainage of the major close karst basins to the use of commercial caves for tourist purposes. Further, but less important, can be added the exploitation of fertilizers and ornamental stones from caves and the utilization of caverns for the cultivation of mushrooms, for medical-therapeutic purposes and for scientific research [DELL'OCA, 1962].

The most notable examples of use for *hydroelectric purposes* are the karst lake of Matese with a power of 18,200 kW, and the lake of Canterno with 1,100 kW. Other examples of similar usage are the caves El Fontanon and Böcc della Tuera, in the province of Como which supply a water discharge of 200–400 l/sec to the hydroelectric station of Rescia, with 500 kW of power, the Pertosa Cave near Salerno with a water discharge of 250 l/sec, the Oliero Caves near Vicenza and the cave of Lete in Campania.

Among the best known cases of usage for water supply are the Timavo resurgences captured for the Trieste aqueduct, the perennial subterranean stream of the Buco dell'Orso for the Torriggia (Como) aqueduct, the Buco della Volpe for the Cernobbio (Como) aqueduct, the Fiumelatte for the Fiumelatte and Varenna (Lecco) aqueducts, the San Giovanni Cave for a part of the Cagliari aqueduct, the cave of San Giovanni di Su Anzu, near Dorgali, for the Orosei aqueduct.

Among the examples of *reclaiming by drainage* of karst basins are Pian del Lago near Siena, Piani di Colfiorito in Umbria and the lake of Canterno near Fiuggi.

Among the principal *commercial caves* for tourist purposes (Fig. 2) in Italy are the Toirano Cave near Savona, the Grotta Gigante near Trieste, the caves of Pertosa, Castelcivita and Smeraldo (equipped with a lift) near Salerno, the Grotta Azzurra of Capri, the Castellana caves near Bari, the Su Marmori, Bue Marino and Nettuno caves in Sardinia. The most important of these are the Castellana caves, which extend for about 2 km and are equipped with lifts for the 60 m descent down the entrance abyss. Each lift carries 700 persons per hour and the caves are visited by over 100,000 tourists each year. Near the entrance is a modern speleologic muscum.

With regard to the *exploitation of fertilizers and ornamental stones* from caves, guano is extracted from the Zinzulusa Cave in Apulia (now worked out), from the Pertosa Cave near Salerno (about 13,000 kg), from the Diavolo Cave near Sassari, the Buco del Frate near Brescia and the Spipola Cave near Bologna. Phosphates have been exploited from the San Bernardino Maggiore Cave near Vicenza, calcareous sinter from the Nettuno and Fossa di Pranu Pirastru caves in Sardinia, small blocks of calcite from the Su Marmuri Cave in Sardinia, talc from the Su Gorgovone Cave in Sardinia, and crystalline limestone from the Antro del Corchia in Tuscany.

Among the other uses of caves there can be cited the cultivation of mushrooms,

---

[1] Written by S. Belloni.

the ageing of wines, etc. in the Covoli di Costovizza on the Berici Hills, in the Quinzano Cave near Verona and in the Opreno Cave in Lombardy; the medical utilization of the waters of the Acquasanta Cave near Ascoli Piceno, the Giusti Cave in Tuscany and the Monte Cronio Cave in Sicily, for the treatment of skin diseases, rheumatism, etc.

Finally, an example of scientific usage of a cave is the Grotta Gigante near Trieste, where instruments are installed to measure the earth tides.

REFERENCES

ABRAMI, G. and MASSARI, F., 1968. La morfologia carsica nel colle del Montello. *Riv. Geograf. Ital.*, 75:1–45.
ALMAGIÀ, R., 1959. *L'Italia*. Unione Tipografica Editrice Torinese, Torino, 1324 pp.
ANELLI, F., 1959. *Castellana*. Officine Grafiche De Robertis, Putignano, 157 pp.
BADINI, G., 1967. Le Grotte Bolognesi. Meroni, Albese, 143 pp.
BADINI, G. and GECCHELE, G., 1965. Le più profonde voragini d'Italia. *Mem. Rass. Speleol. Ital.*, 7(2):183–195.
BALDACCI, O., 1957. Per una sistematica ricognizione speleologica nel Gargano. *Boll. Soc. Geograf. Ital.*, *VIII*, 10(9/10):431–456.
BERTARELLI, L. V. and BOEGAN, E., 1926. *Duemila Grotte: Quarant'Anni di Esplorazioni nella Venezia Giulia*. Touring Club Italiano, Milano, 494 pp.
BISSANTI, A., 1966. La dolina Pozzatina nel Gargano. *Riv. Geograf. Ital.*, 73(3):312–321.
BOEGAN, E., 1938. Il Timavo: studio sull'idrografia carsica subaerea e sotterranea. *Mem. Ist. Ital. Speleol., Ser. Geol. Geofis.*, 2:1–251.
CAMPOBASSO, V. and OLIVIERI, C., 1967. Osservatione Preliminari sulla Stratigrafia e sulla Tettonica delle Murge fra Castellana Grotte (Bari) e Ceglie Messapico (Brindisi). Ist. Geol. Paleontol., Univ. Bari, Bari, 16 pp.
CAPELLO, C. F., 1938. Introduzione allo studio sistematico delle plaghe carsiche del Piemonte. *Boll. Soc. Geograf. Ital.*, *VII*, 3(4):298–312.
CAPELLO, C. F., 1950. Il fenomeno carsico in Piemonte: le zone marginali al rilievo alpino. *Consiglio Nazl. Ric., Centro Studi Geograf. Fis., Ric. Morfol. Idrograf. Carsica*, 3:1–90.
CAPELLO, C. F., 1952. Il fenomeno carsico in Piemonte: le Alpi Liguri. *Consiglio Nazl. Ric., Centro Studi Geograf. Fis., Ric. Morfol. Idrograf. Carsica*, 4:1–114.
CAPELLO, C. F., 1955. Il fenomeno carsico in Piemonte: le zone interne del sistema alpino. *Consiglio Nazl. Ric., Centro Studi Geograf. Fis., Ric. Morfol. Idrograf. Carsica*, 6:1–140.
CAPELLO, C. F., NANGERONI, G., PASA, A., LIPPI BONCAMBI, C., ANTONELLI, C. and MALESANI, E., 1954. Les phenomènes karstiques et l'hydrologie souteraine dans certaines regions de l'Italie. *Assoc. Intern. Hydrol.*, 37(2):408–437.
CAPPA, G., 1962. Note di speleologia nel promontorio del Gargano. *Rass. Speleol. Ital.*, 14(1):7–18
CASTALDI, F., 1961. Differenziazione e datazione del fenomeno carsico nella penisola sorrentina. *Mem. Rass. Speleol. Ital.*, 5(2):137–151.
CASTIGLIONI, G. B., 1960. *Atlante Internazionale dei Fenomeni Carsici*, 2. Bosco del Cansiglio (Prealpi Venete), 4 pp.
CASTIGLIONI, G. B., 1964. Forme del carsismo superficiale sull'altopia no del Cansiglio. *Atti Ist. Veneto Sci., Lettere Arti, Classe Sci. Mat. Nat.*, 122:327–344.
COLAMONICO, C., 1951. Alcune caratteristiche del carsismo pugliese. *Rend. Accad. Sci. Fis. Mat. Soc. Nazl. Sci., Napoli*, 18:264–267.
COPPADORO, A., 1903. Contributo allo studio dei fenomeni carsici dell'Altipiano del Cansiglio. *In Alto*, 14(2):19–23.
COTECCHIA, V., 1955. Influenza dell'acqua marina sulle falde acquifere in zone costiere, con particolare riferimento alle ricerche d'acqua sotterranea in Puglia. *Geotecnica*, 2(3):105–128.
COTECCHIA, V., 1956. Gli aspetti idrogeologici del Tavoliere delle Puglie. *L'Acqua*, 34(11/12):168–180.

COTECCHIA, V., 1959. Sulle caratteristiche delle sorgenti e sulle modalità di rinvenimento delle falde profonde nella Penisola Salentina in rapporto alla struttura dei calcari cretacici della regione. *Ann. Fac. Ing. Univ. Bari*, 2:227–245.

COTECCHIA, V., 1963. Geohydrological aspects of the Cretaceous limestone aquifer in Apulia, and their bearing on the practical avoidance of sea water contamination in extraction from wells and springs. *Quaderni Geofis. Appl.*, 24:1–42.

COTECCHIA, V. and MAGRI, G., 1966. Idrogeologia del Gargano. *Geol. Appl. Idrogeol.*, 1:1–86.

CRESCENTI, U. and VIGHI, L., 1964. Caratteristiche, genesi e stratigrafia dei depositi bauxitici cretacici del Gargano e delle Murge; cenni sulle argille con pisoliti bauxitiche del Salento (Puglie). *Boll. Soc. Geol. Ital.*, 83(1):285–338.

DAINELLI, G., 1908. Cavità di erosione nei gessi del Moncenisio. *Mondo Sotterraneo*, 3(3/4):55–68.

DAINELLI, G., 1940. *Atlante Fisico Economico d'Italia*. Touring Club Italiano, Milano, 17 pp., 82 geographical maps.

DAL PIAZ, G., 1907. Le Alpi Feltrine. *Mem., Reale Ist. Veneto Sci., Lettere Arti*, 27(9):1–176.

D'AMBROSI, C., 1956. Paleoidrografia miocenica in Istria e sua successiva trasformazione in rapporto con lo sviluppo del carsismo. *Atti Congr. Nazl. Speleol.*, 6, Trieste—Le Grotte d'Italia, III, 1:144–173.

D'AMBROSI, C., 1960. Sul problema dell'alimentazione idrica delle fonti del Timavo presso Trieste. *Tec. Ital.*, 25(8):5–23.

D'AMBROSI, C., 1961. Sull'origine delle doline carsiche nel quadro genetico del carsismo in generale. *Boll. Soc. Adriat. Sci., Trieste*, 51:205–231.

D'AMBROSI, C., 1962. Le acque del carso ed il problema del rifornimento idrico della città di Trieste e della sua zona industriale. *Tec. Ital.*, 27(1)3–14.

D'AMBROSI, C., 1966. Considerazioni sull'origine e sul periodo di svolgimento del ciclo carsico in atto nella Venezia Giulia con particolare riguardo all'Istria e al Carso di Trieste. *Atti Mem. Comm. Grotte "E. Boegan"*, 5:29–47.

DA SCHIO, A., TREVISIOL, G. and PERIN, G., 1947. *Scienza e Poesia sui Berici*. Tipografia Commerciale Editrice, Vicenza, 305 pp.

DE GASPERI, G. B., 1916. Materiali per lo studio dei fenomeni carsici,2. Grotte e voragini del Friuli, *Mem. Riv. Geograf. Ital.*, 30:1–219.

DE GASPERI, G. B. and FERUGLIO, G., 1910. L'altipiano del Cansiglio: contributo allo studio dei suoi fenomeni carsici. *Mondo Sotterraneo*, 6(3/4):41–59.

DELL'OCA, S., 1962. Note di speleologia economica: utilizzazioni delle grotte. *Rass. Speleol. Ital.*, 14(1):76–109.

DEMANGEOT, J., 1965. *Géomorphologie des Abruzzes Adriatiques*. Centre Natl. Rech. Sci., Paris, 403 pp.

DE MARCHI, L., 1911. *Idrografia Carsica dei Sette Comuni*. Uffic. Idrograf. Magistrato Acque, Venezia, 47 pp.

DONINI, L., 1965. Brevi note sulle grotte dei gessi bolognesi. *Natura Montagna*, II, 5(4):151–167.

FABIANI, R., 1902a. Le grotte dei Colli Berici nel Vicentino. *Antologia Veneta*, 3(2):1–14.

FABIANI, R., 1902b. Fenomeni carsici dei Colli Berici. *Antologia Veneta*, 3(5):1–11.

FERUGLIO, E., 1913. Fenomeni carsici nella Valle dell'Alberone. *Mondo Sotterraneo*, 9(2):37–42.

FERUGLIO, E., 1914. Fenomeni carsici dell'Altipiano di Monte Prat. *Mondo Sotterraneo*, 9(4/5): 90–93.

FERUGLIO, E., 1923. L'altipiano carsico del Ciaorlécc nel Friuli. *Mondo Sotterraneo*, 18–19:1–89.

FERUGLIO, E., 1954. La regione carsica di Villanova in Friuli. *Consiglio Nazl. Ric., Centro Studi Alpini*, 54:1–68.

FORTI, F., 1968. La geomorfologia nei dintorni di Slivia (Carso triestino) in rapporto alla litologia ed alla tettonica. *Atti Mem. Comm. Grotte "E. Boegan"*, 7:23–61.

FORTI, F. and TOMMASINI, T., 1967. Una sezione geologica del Carso triestino. *Atti Mem. Comm. Grotte "E. Boegan"*, 6:43–139.

FUREDDU, A. and MAXIA, C., 1964. *Grotte della Sardegna: Guida al Mondo Carsico dell'Isola*. F.lli Fossataro, Cagliari, 310 pp.

LAENG, G., 1923. Le cavità naturali del bresciano. *Vie Italia*, 29(8):868–874.

LAZZARI, A., 1958. La grotta Zinzulusa presso Castro, provincia di Lecce. *Ann. Ist. Super. Sci. Lettere "S. Chiara"*, Napoli, 8:237–295.

LEHMANN, H., 1959. Studien über Poljen in den venezianischen Voralpen und in Hochapennin. *Erdkunde*, 13:258–289.

LIGASACCHI, A. and RONDINA, G., 1955. Il fenomeno carsico nel territorio varesino (Prealpi Lombarde). *Consiglio Nazl. Ric., Centro Studi Geograf. Fis., Ric. Morfol. Idrograf. Carsica*, 7: 1–119.

LIPPI BONCAMBI, C., 1948. I Monti Sibillini. *Consiglio Nazl. Ric., Centro Studi Geograf. Fis., Ric. Morfol. Idrograf. Carsica*, 1:1–79.

LORENZI, A., 1903. Fenomeni analoghi a quelli carsici nei conglomerati messiniani di Ragogna e Susans nel Friuli. *In Alto*, 14(1):13–14.

MANFREDINI, M., 1964. Schema idrogeologico dell'Abruzzo. *Sorgenti Ital.*, 9:26–41.

MARCACCINI, P., 1961. I fenomeni carsici in Toscana. *Riv. Geograf. Ital.*, 68(3):221–243.

MARCACCINI, P., 1962. I fenomeni carsici del Gargano nelle recenti tavolette dell'Istituto Geografico Militare. *Riv. Geograf. Ital.*, 69(2):186–193.

MARCACCINI, P., 1964. Fenomeni carsici di superficie nelle Alpi Apuane. *Riv. Geograf. Ital.*, 71(1): 35–54.

MARINELLI, O., 1897. *Fenomeni Carsici, Grotte e Sorgenti nei Dintorni di Tarcento in Friuli.* Doretti, Udine, 71 pp.

MARINELLI, O., 1917. Materiali per lo studio dei fenomeni carsici,3. Fenomeni carsici nelle regioni gessose d'Italia. *Mem. Riv. Geograf. Ital.*,34:263–416.

MARTINIS, B., 1955. Osservazioni sull'anticlinale pontica del Montello e rilievo geologico dei Colli di Conegliano (Treviso). *Mem. Ist. Geol. Mineral. Univ. Padova*, 18:1–16.

MARTINIS, B., 1962a Ricerche geologiche e paleontologiche sulla regione compresa tra il T. Iudrio ed il F. Timavo (Friuli Orientale). *Mem. Riv. Ital. Paleontol. Stratigraf.*, 8:1–245.

MARTINIS, B., 1962b. Lineamenti strutturali della parte meridionale della penisola salentina. *Geol. Romana*, 1:11–23.

MARTINIS, B., 1962c. Sulla tettonica delle Murge nord-occidentali. *Accad. Nazl. Lincei, Rend. Classe Sci. Fis. Mat. Nat., VIII*, 31(5):299–305.

MARTINIS, B., 1965. Osservazioni sulla tettonica del Gargano orientale. *Boll. Serv. Geol. Italia*, 85: 45–93.

MARUSSI, A., 1941a. Il Paleotimavo e l'antica idrografia subaerea del Carso triestino. *Boll. Soc. Adriat. Sci. Nat. Trieste*, 38:5–27.

MARUSSI, A., 1941b. Ipotesi sullo sviluppo del carsismo (osservazioni sul Carso triestino e sull'Istria). *Giorn. Geol., II*, 15:1–12.

MAUCCI, W., 1961a. Evoluzione geomorfologica del Carso triestino successiva all'emersione definitiva. *Boll. Soc. Adriat. Sci. Trieste*, 51:165–188.

MAUCCI, W., 1961b. La speleogenesi nel Carso triestino. *Boll. Soc. Adriat. Sci. Trieste*, 51:233–254.

MORELLI, C., 1954. Rilievo gravimetrico alle foci del Timavo. *Tec. Ital.*, 9(2):1–5.

MOSETTI, F., 1954. Rilievo geoelettrico del delta sotterraneo del Timavo. *Tec. Ital.*, 9(2):6–10.

MOSETTI, F., 1965. Nuova interpretazione di un esperimento di marcatura radioattiva del Timavo. *Boll. Geofis. Teor. Appl.*, 7(27):218–243.

MOSETTI, F., ERIKSSON, E., BIDOVEC, F., HODOSCEK K. and OSTANEK, L., 1963. Un nuovo contributo alla conoscenza dell'idrologia sotterranea del Timavo. *Tec. Ital.*, 28(4):157–171.

NANGERONI, G., 1957a. Il carsismo, le grotte. In: *L'Italia Fisica*. Touring Club Italiano, Milano, pp. 284–303.

NANGERONI, G., 1957b. Il carsismo e l'idrologia carsica in Italia. *Atti Congr. Geograf. Ital.*, 17, Bari, 2:83–111.

NANGERONI, G., 1960. "Campi", "piani", "polja" in Italia in una recente pubblicazione. *Riv. Geograf. Ital.*, 67(3):306–313.

ORTOLANI, M. and MORETTI, A., 1950. Il Gran Sasso d'Italia (versante meridionale). *Consiglio Nazl. Ric., Centro Studi Geograf. Fis., Ric. Morfol. Idrograf. Carsica*, 2:1–119.

PASA, A., 1954. Carsismo e idrografia carsica nel Gruppo del Monte Baldo e nei Lessini veronesi. *Consiglio Nazl. Ric., Centro Studi Geograf. Fis., Ric. Morfol. Idrograf. Carsica*, 5:1–150.

PASINI, G., RIBALDONE, G. and DI MAIO, M., 1965. Spedizione 1963 alla "Spluga della Preta". *Mem. Rass. Speleol. Ital.*, 7(2):39–65.

PRACCHI, R., 1939. Geomorfologia dell'Alta Brianza. *Univ. Cattolica S. Cuore, Saggi Ric., X*, 1: 183–241.

PRACCHI, R., 1943. Contributo alla conoscenza del fenomeno carsico in Lombardia. *Univ. Cattolica S. Cuore, Saggi Ric.*, X, 3: 1–105.

ROVERETO, G., 1923. *Geologia Morfologica*. Hoepli, Milano, 1187 pp.

SAIBENE, C., 1951. I fenomeni carsici nel circo di Moncòdeno (Grigna settentrionale). *Riv. Geograf. Ital.*, 58(2):65–78.

SAIBENE, C., 1955. Il Gruppo delle Grigne (note di geomorfologia). *Atti Soc. Ital. Sci. Nat. Museo Civico Storia Nat. Milano*, 94(3/4):255–328.

SAIBENE, C., 1957. Note sul carsismo in Sicilia. *Atti Congr. Geograf. Ital.*, *17*, Bari, 2:137–145.

SEGRE, A.G., 1947. Aspetti antropici del fenomeno carsico nell'Italia peninsulare. *Mem. Geograf. Antrop.*, 1:185–217.

SEGRE, A.G., 1948. I fenomeni carsici e la speleologia del Lazio. *Pubbl. Ist. Geograf. Univ. Roma, Ser. A.*, 7:1–248.

SERVIZIO GEOLOGICO D'ITALIA, 1963–1970. *Carta Geologica d'Italia, 1:100,000:F° 87 Bologna (1963); F° 155 S. Severo (1969); F° 156 S. Marco in Lamis (1970; F° 157 Monte S. Angelo (1965); F° 163 Lucera (1963); F° 164 Foggia (1969); F° 165 Trinitapoli (1968); F° 174 Ariano Irpino (1963); F° 175 Cerignola (1963); F° 176 Barletta (1970); F° 177 Bari (1966); F° 178 Mola di Bari (1963); F° 188 Gravina di Puglia (1966); F° 189 Altamura (1966); F° 190 Monopoli (1969); F° 191 Ostuni (1968); F° 201 Matera (1969); F° 202 Taranto (1969); F° 203 Brindisi (1970); F° 204 Lecce (1968); F° 213 Maruggio (1968); F° 214 Gallipoli (1968); F° 215 Otranto (1968); F° 223 Capo S. Maria di Leuca (1968)*.

SESTINI, A., 1963. *Il Paesaggio*. Touring Club Italiano, Milano, 232 pp.

SOMMARUGA, C., 1952. Problemi scientifici e turistici delle grotte di Capo Caccia (Alghero). *Rass. Speleol. Ital.*, 4(1):7–18.

TONIOLO, A.R., 1907. Materiali per lo studio dei fenomeni carsici, 1. Il Colle del Montello. *Mem. Riv. Geograf. Ital.*, 3:257–393.

TRENER, G. B., 1933. La distribuzione geologica e geografica dei fenomeni carsici nella Venezia Tridentina. *Atti Congr. Nazl. Speleol.*, *1*, Trieste, pp. 90–97.

UFFICIO IDROGRAFICO MAGISTRATO ACQUE VENEZIA, 1963. *Carta Geologica delle Tre Venezie 1:100,000: F° 38 Conegliano (1963); F° 40A Gorizia (1951); F° 53A Trieste (1953)*.

VARDABASSO, S., 1955. Sardegna speleologica. *Rass. Speleol. Ital.*, 7(3):119–134.

VARDABASSO, S., 1957. Il carsismo nella fascia costiera della Sardegna. *Atti Congr. Geograf. Ital.*, 17, Bari, 2:124–136.

VENZO, G.A. and FUGANTI, A., 1965. Analisi strutturale delle deformazioni tettoniche del Carso goriziano (Gorizia). *Studi Trentini Sci. Nat.*, 42(2):335–366.

Chapter 5

Karst of France

J. AVIAS

C. E. R. H., Geological Institute, University of Montpellier, Montpellier (France)

INTRODUCTION

From a morphologic and karst-hydrologic point of view, France is exceptional in the variety it has to offer. Calcareous formations have exceptionally varied lithologies including extremely pure lithographic limestones from the Upper Jurassic, dolomitic limestones, marly limestones and chalks. Structures range from the calcareous tabular plateaus in the Paris Basin to faulted, folded or metamorphosed calcareous formations in the Jura, Languedoc, Pyrenees and Alps regions. The climatic conditions in the French karst areas range from wet continental or wet maritime to Mediterranean or semi-arid climates in southeastern Europe. The palaeoclimatic environment of the karst includes glacial and periglacial palaeoclimates as well as hot or arid palaeoclimates.

For this reason it is easily understood why, after Yugoslavia, France is the leading country where karst observations have been made and theories formulated.

The most important karst area in Europe (Vercors) is located in France. France has three of the ten longest caverns in the world—Reseau de la Dent de Grolles (Grande-Chartreuse), 27,715 m; Reseau de Courry—La Cocalière (Gard, Ardèche), 23,000 m; Goule de Foussoubie (Vagnas, Ardèche), 22,000 m. Moreover, the three deepest caverns—Gouffre de la Pierre Saint-Martin (Pyrenees), 1,152 m (Fig. 9); Gouffre Berger (Vercors), 1,135 m; Reseau Trombe (Pyrenees), 911 m. Furthermore, four of the deepest vertical caverns—Gouffre de la Pierre St-Martin, 333 m; Gouffre S. Bouchet (Basses-Pyrenees), 200 m; Gouffre Jean Nouveau (Vaucluse), 166 m; Aven du Mont Marcou (Hérault), 162 m; and two of the largest underground rooms known in the world, including the second most important—Gouffre de la Pierre Saint-Martin, 200 m long and 200 m high; Aven d'Orgnac, 200 m long and 70 m high.

Studies of French karst are primarily of two different types which are becoming more and more interrelated.

(*a*) Descriptive studies directed towards research, description, inventory and mapping of karst forms, accessible cavities or traceable circulations in limestone massifs. These studies are made chiefly by the following: (*1*) regional speleologic groups and clubs belonging either to the *Fédération Française de Spéléologie* (130, Rue de Saint-Maur, Paris 11°) or to the *Commission de Spéléologie du Club-Alpin Français* (7, Rue de la Boétie, Paris 8°); and (*2*) various university researchers whose activities are coordinated by the *Commission des Phénomènes Karstiques du Comité National de Géographie* (191, Rue Saint-Jacques, Paris 5°).

(*b*) Studies directed more towards the geological, chemical, biological and hydrological aspects of karst phenomena and their application (water resources). These studies are made chiefly by the bodies: the *Centre National de la Recherche Scientifique (Commission de Spéléologie)*—13, Quai Anatole France, Paris 7°—which operates an underground laboratory (biospeleology) in Moulis (Ariège) and publishes the *Annales de Spéléologie;* the B.R.G.M. (*Bureau de Recherches Géologiques et Minières, Service d'Hydrogéologie*, Orléans-la Source, France) which is officially responsible for the speleological inventory of France, and also engages in basic research and applied studies of karsts; and university laboratories which are more or less specialized such as the C.E.R.H. (*Centre d'Etudes et de Recherches Hydrogéologiques* of the University of Montpellier, Faculté des Sciences, Place E. Bataillon, Montpellier 34, France) which studies problems of the evalution and exploitation of karst water reserves, and which established the first French representative and experimental watershed in karstified ground (Saugras Basin). This group also includes the laboratories of the universities of Bordeaux, Besançon, Dijon, Lyon, Lille and Marseille.

Among early French authors famous for their work on karstology are E. A. Martel whose two books *(La France Ignorée)*, published in 1928 and 1930, totalling 596 pages and several hundred figures and photographs, still constitute the basic inventory of karst phenomena in France, and E. Fournier who published a series of volumes on the Jura region entitled *Explorations Souterraines et Recherches Hydrologiques en Franche Comté*.

Among the present French authors who have worked or are still working on karst are: R. Abrard, M. Aigrot, M. Albinet, J. Avias, L. Balsan, P. Barrere, P. Birot, A. Blondeau, G. Boillot, A. Bonnet, A. Bonte, A. Bourgin, R. Caro, N. Casteret, V. Caumartin, A. Cavaille, P. Chevallier, J. Choppy, R. Ciry, J. Corbel, C. Drogue, M. Dreyfuss, P. Dubois, M. Dubois, H. Elhai, P. Fenelon, B. Gèze, J.-P. Henry, R. De Joly, M. Laures, G. De Lavaur, G. Magniez, L. Martin, C. Megnien, G. Mennessier, C. Mugnier, R. Muxart, J. Nicod, H. Paloc, T. Pobeguin, C. Pomerol, C. Pommier, P. Renault, H. Roques, J. Rouire, C. Rousset, H. Salvayre, H. Schoeller, M.-R. Seronie-Vivien, A. Siffre, M. Siffre, T. Stchouzkoy, N. Theobald, H. Tintant, J. Tricart, F. Trombe, A. Vandel, P. Verdeil, G. Waterlot, P. Weydert.

There are also regional speleological researchers and groups whose studies have been published, mainly in the *Spelunca* and *Spelaion Carso* reviews, in the *Annales de Spéléologie*, and in various local bulletins.

Lithostratigraphy

The main calcareous or dolomitic outcrops in France may be summed up as follows:
(*1*) Palaeozoic: nearly always folded, eroded and more or less exposed formations: (*a*) Cambrian limestones (specially limestones with Georgian *Archaeocyathus*);

(b) black limestones of Upper Silurian age; (c) Devonian limestones and dolomites; (d) Carboniferous limestones and dolomites.

(2) Mesozoic: (a) Triassic dolomites; (b) Rhaetian limestones; (c) Sinemurian/Hettangian marly limestones (north) and dolomites (south); (d) Upper Pliensbachian sandy limestones; (e) Aalenian ferruginous limestones; (f) Bajocian oolithic reef limestones (north) and dolomites (south); (g) Bathonian marly limestones; (h) more or less lithographic and more or less in reef-facies, in their upper part sometimes dolomitic (Portlandian), Upper Jurassic limestones; (i) Upper Cretaceous chalky or calcareous reef formations (Urgonian).

(3) Tertiary and Quaternary: marine or fresh-water calcareous beds or lenses from Palaeogene (example: Lutetian) or Neogene formations, in transgression basins of the Hercynian peneplain and Quaternary travertines and calcareous tuffs.

Structure

The types of structures are as follows : (1) tabular formations that are horizontal or affected by slight dips (a few degrees), or large basins (e.g., the Paris Basin); (2) folded formations of the Jurassic type; (3) folded faulted and overthrust formations of a Pyrenean and Alpine style.

Also epeirogenic movements (on the Languedocian coast for example) together with climatic changes caused variations in the relative basal marine or oceanic level. That level after all is the base level of the drainage phenomena controlling karst processes.

Climate

Calcareous formations outcropping at present or joining outcrops, are affected by different climates such as the continental climate in Lorraine, the maritime climates of the Atlantic type, mountainous climates (glacial, periglacial, mountain) and warm climates including the Mediterranean with concentrated precipitations and long dry and hot seasons, sometimes bringing with them semi-arid conditions or the Aquitainian climate which is more humid.

However, as we have stated, one must also bear in mind that hot or cold, glacial or other palaeoclimates may have affected some calcareous formations presently covered (or not) with marine and continental formations. For example, there are the karst in Carboniferous limestones covered with Cretaceous formations in the north of France and the Oligocene karst in Jurassic limestones fixed by transgressive Miocene or Pliocene formations or outcropping again, but with remnants of ancient karst phenomena, e.g., Upper Cretaceous (Isthme Durancien in the Languedoc and the Provence), Eocene, Oligocene, Miocene, Pliocene, and the Glacial Stage (during glaciations or during interglacials).

Enumeration of the preceding factors alone shows how complex the history of some karst phenomena may be.

Karst phenomena in the large natural French units (Fig. 1), their morphologic as well as hydrologic aspects, and their importance notably in the area of water resources, are considered successively in the following section.

Fig.1. Map of the great natural units and general distribution of calcareous areas in France.

KARST PHENOMENA IN PALAEOZOIC, MAINLY HERCYNIAN MASSIFS (ARDENNES, VOSGES, ARMORICAN MASSIF, MASSIF CENTRAL, MONTAGNE NOIRE, PALAEOZOIC MASSIFS IN THE ALPS AND THE PYRENEES)

In these massifs, karst phenomena are relatively slight because of the great extension of schist, sandstone, and graywacke facies (Cambrian, Silurian, Carboniferous flysch) connected with the geosynclinal character of the primitive sedimentation (being opposed to the epicontinental sedimentation limestones rich in most of the Mesozoic or Tertiary deposits), and because of the frequent invasion by metamorphic and crystalline rocks.

From a lithostratigraphic point of view calcareous formations are very much limited (thin beds or lenses), and there has often been more or less pronounced dolomitization of carbonate rocks, and sometimes even marmorization. Examples are the Cambrian (Upper Georgian and Acadian) limestones with *Archaeocyathus* and calcschists well developed in the Massif Central (Cevennes), Armorican Massif, Montagne Noire, Pyrenees, the thickness of which seldom exceeds 200 or 300 m, and Devonian limestones and dolomites (mainly Frasnian and Famennian) and Carboniferous (mainly Dinantian and Visean reefs).

From the tectonic point of view, these limestones are always extremely folded and have always undergone an important fracturation, often more or less obstructed by recrystallization of calcite.

With regard to the phases of karstification, they may be multiple and agree with different phases during which these carbonate rocks were emerging (mainly Permo-Triassic, Upper Cretaceous, Tertiary, Quaternary). But very few traces of ancient phases remain because of the erosion following the uplifts in these regions and because of the intense peneplainization to which they were subjected.

It follows from these facts that in most places carbonate rocks appear as lenses, strings of lenses, or more or less as uninterrupted and very much exposed beds, often projecting into the landscape. In them are generally the only well-marked Pliocene or Quaternary phases of karstification. The network prior to the Quaternary is more often no longer living in its major part (upper part) because of the recent lowering of the base level by erosion (e.g., perched karst in Cambrian limestones in the region of Cantignergues in the Montagne Noire).

Cambrian, Devonian or Carboniferous calcareous dolomitic beds usually give rise only to small hydrologic systems, with karst springs, generally only of local importance. However, the circulation is often very active in these more or less impermeable formations.

In this regard we should cite the exposed and steeply dipping Cambrian marbly limestone flag with a monoclinal structure, on the flanks of Canigou in the Pyrenees. In this formation an important circulation exists notably where the granite is in contact with the calcareous rocks. Corresponding springs give birth to the river Canterrane (Fig. 2A). Another example is the synclinal folded Devonian limestone

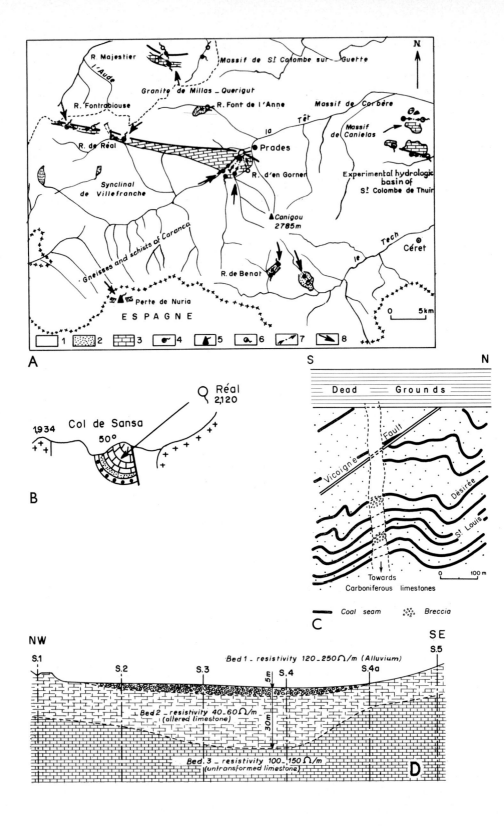

beds in Conflent (from Merins to Villefranche), 35 km long, and in the Pyrenees where the Têt River and its tributaries partly disappear underneath the surface supplying water to resurgent springs (Fig. 2A, B). A portion of the Têt River waters has been captured by the Aude River system. We should also mention the small Devonian limestone remnants in the Pyrenees, some of which at high altitude are fed by melting snow fields or glaciers, for example those giving rise to the Trou du Toro system which is connected with the spring of the Garonne River. We can cite Georgian limestones in Montagne Noire and Cevennes (e.g., Le Vigan). The small outcropping area of the remnants does not allow the development of notable surface karst phenomena, and dolines and lapies are rare.

However, one does find numerous disappearances of water and resurgences there (e.g., the Source du Jaur feeding Saint-Pons in the Montagne Noire). Moreover, the disappearance of brooks into these calcareous formations may, after a more or less long underground flow, a reheating and a mineralization of waters, yield thermal mineral water resurgences, for example the Source d'Avène emerging out of a Devonian limestone outcrop in the Vallée de l'Orb, Montagne Noire.

In the north and northeast of France is an unusual example of pre-Cretaceous karst in Carboniferous limestone covered by deposits of the Upper Cretaceous and Tertiary transgression on the post-Hercynian peneplain (Fig. 2C). Karst in Carboniferous limestones overlain by coal-bearing formations is particularly important and explains the formation of "natural shafts" in coal mines in the north of France (in Belgium as well), such as the one described by Pruvost [1909] in the mine of Vicoigne (Fig. 2C) and those described by Dehee [1926]. These natural shafts, 20–150 m in diameter, cutting through Carboniferous formations, are full of rough breccias from not only Carboniferous rocks but also some of the overlying strata.

KARST PHENOMENA IN MESOZOIC AND TERTIARY FORMATIONS

Introduction

If they occur in Pyreneo-Alpine geosynclines, Mesozoic and to some extent Tertiary rocks are flysch or marly limestones which are not favourable to karstification. However, the sedimentary cover of the eroded Hercynian peneplain deposited, during the Mesozoic and the Tertiary, an epicontinental type of sedimentation, relatively

---

Fig.2.A. Map of the Conflent and adjacent areas. *1* = Metamorphic or primary non-calcareous rocks; *2* = Cambrian limestone; *3* = Devonian limestone; *4* = loss of water; *5* = aven; *6* = resurgence; *7* = underground circulation proved by dye tests; *8* = flow direction. (According to Salvayre, 1960, 1968.)

B. Cross-section of the Conflent synclinal near Real. (According to Salvayre, 1960, 1968.)

C. The natural well (puits naturel) discovered in the coal-mine of Vicoigne, linked to the karstification of Carboniferous limestones underlying the coal strata. (After Dehee, 1926.)

D. Dry valley and subjacent alteration of Upper Jurassic limestones brought to light by electric borings (valley of Sacy, in the southeastern part of the Paris Basin). (After Megnien, 1964.)

poor in detrital sediments and relatively rich in carbonate formations particularly suitable for karstification.

From a lithostratigraphic point of view, the main karst levels are in Middle Jurassic (Bajocian, Bathonian), in Upper Jurassic (Argovian, Rauracian, Sequanian, Kimmeridgian, Portlandian), in Upper Cretaceous (Turonian, Senonian) and in Palaeogene calcareous deposits (Lutetian).

From a tectonic point of view, these formations form two large sets. The first consists of transgressive basins in the Hercynian peneplain (principally the Paris Basin and its annexes, the Aquitanian Basin and Causses) where limestone beds are horizontal or only slightly folded and fractured. The other type found in the Pyreneo-Alpine chains is more or less intensely tectonized with limestone beds often folded and exposed with many large fractures (Jura, Corbières, Languedoc, Provence, sub-Alpine chains, and Pyrenean and Alpine chains.

The phases of karstification naturally correspond to the periods of emergence that followed the end of the Jurassic and the end of the Cretaceous and to the definitive emergence of many regions since the Neogene. The more characteristic and best preserved phases of karstification correspond to the last (Pliocene and Quaternary) cycles. The hot Tertiary palaeoclimates and the alternately cold and temperate Quaternary palaeoclimates played a prime role in karst phenomena. The present climatic differences between the various French regions play an important part in the types of karstification found today in the calcareous outcrops in France. In the northern central and eastern regions most if not all covered karsts morphologies occur. Typical forms of bare karst (karsts nus) occur chiefly in the Mediterranean and Atlantic regions.

From a hydrogeologic point of view, hydrology of more or less karstified calcareous formations is a very important factor in the north (chalk in the Paris Basin, as well as in the south (Tithonian and Urgonian limestones in Provence, Languedoc, and pre-Alps). As discussed in a later section, water in the chalk is often overexploited but in the southern Mediterranean karst the exploitation of water resources generally has not yet begun.

Karst phenomena in Mesozoic and Tertiary formations in the Paris Basin

Mesozoic formations

Lithostratigraphy. The Paris Basin consists of sediments, overlying the Hercynian peneplain, ranging in age from Permian and Triassic (not represented everywhere) to Palaeogene.

From a hydrologic point of view, permeable strata (sands, sandstones, fissured or karstified limestones) and impermeable strata (marls and clays) often alternate. Reef and coralline sediments are very numerous in the Middle Jurassic (Bajocian) and Upper Jurassic (Argovian to Portlandian). In the Upper Cretaceous these special carbonate sediments are chiefly chalk in the outcrops. Due to erosion all these strata outcrop as concentric cuestas with hard beds, limestones for example, rising over talus or gentle slopes of soft beds, such as clays, marls and sands. Table I shows the succession of these strata from Lias to Cretaceous as established by MEGNIEN [1964] in the southeastern part of the Paris Basin between Morvan and Brie.

Structure. From a tectonic point of view, this zone is relatively stable. The strata dip gently, generally 2 or 3 degrees, towards the centre of the basin, affected only by long radial curved warpings, generally not sufficient to form definite anticlinal or synclinal folds, and by faults, usually with small throw.

Warpings and faults are mainly of Oligocene age or younger and follow two principal directions, one of an Armorican type: west-northwest–east-southeast (cf. direction of the anticline of Bray); the other of a Variscan type, east-northeast–west-southwest (e.g., anticlines and synclines in Lorraine).

During Miocene time, the sea left the major part of the Paris Basin and the outcropping formations were exposed to erosion (in fact since the Stampian). This led to the development of a peneplain. During Pliocene and Quaternary the deformation of this "Mio-Pliocene surface" and the variations of the basal level caused its partial clearing as well as the superimposition of some rivers (for example in Lorraine and Barrois). During the Quaternary the Paris Basin was affected by glaciations, as evidenced by various periglacial phenomena and, as for rivers, by the formation of three terraces testifying to the action of at least three periods of erosion followed by a deposition in river valleys.

Phases of karstification. It is possible to distinguish the following phases of karstification:

(1) The Lower Cretaceous surface erosion cut calcareous formations which are distinctly karstified in some places. However, generally the phenomenon cannot be observed because it is covered by sands, clays and more or less continental Wealdian deposits.

(2) During the Mio-Pliocene [TRICART, 1949] hot and wet climates allowed the formation of very highly evolved karst, contemporaneously with the erosion of valleys. This karst was "fossilized" during the Pliocene (a dryer period).

(3) During the Quaternary glaciation, only the zone situated over the "pergelisol" could be karstified (during the summer), forming what CIRY [1959] called *réseaux cutanés* (cutaneous networks).

(4) During Interglacial periods, karstification was again active in the subsurface but was tied to base levels caused by streams and their surface and underground circulation.

These facts, together with alternating impermeable and permeable calcareous strata, and the special characteristics of some carbonate formations (especially chalk) explain all the karst forms that may be observed in the Paris Basin. These forms range

TABLE I

HYDROGEOLOGIC RECAPITULATIVE TABLE, SOUTHEAST OF THE PARIS BASIN (after MEGNIEN, 1964)

Stages	Lithologic levels	Average thickness (m)	Morphology
Upper Cretaceous			
Senonian	compact chalk with silex	220	plateau
Turonian	bedded compact chalk with silex	100	cuesta and plateau
Infra-Turonian	grey marly chalk	60	gentle slope
Cenomanian	bedded hard chalk, with gaize at the bottom	45	cuesta and platform
Lower Cretaceous			
Upper Albian	plastic clays and marls	30	gentle slope
Middle Albian	ferruginous sands and sandy clays	50	small hillocks
Lower Albian	green sands and clays	30	gentle slope
Aptian	black clays	15	gentle slope
Upper Barremian	fine sands and mixed clays	15	small hillocks
Lower Barremian	coquinoid limestones and marls alternating	30	gentle slope
Hauterivian	calcareous sandstone	5–15	platform
Malm			
Portlandian	very fine crystalline limestone	50–80	cuesta and platform
Kimmeridgian	compact limestones and marls alternating	50–100	irregular steep slope
Upper Sequanian	chalky oolitic and very fine crystalline limestones	70	irregular slope
Lower Sequanian	bedded very fine crystalline limestone	35	small cuesta
Marly Rauracian	very fine crystalline limestone with foliated joints and marly levels	150	gentle slope
Reefal Rauracian	recrystallized compact limestone with polyps	150–200	massif cuesta
Argovian	bedded cherty limestone	30	gentle slope
Dogger			
Upper Bathonian	marly and coquinoid limestone	15	gentle slope
Middle Bathonian	bedded oolitic limestone and limestone with debris	100	great cuesta
Vesulian	marly plaquette-like limestone	60–80	gentle slope
Bajocian	compact encrinitic limestone	15	great cuesta
Lias			
Toarcian–Aalenian	hardboard schists and black marls	65	steep slope
Upper Domerian	encrinitic-calcareous sandstone	3–5	small cuesta
Lower Domerian	micaceous foliated marls	60	gentle slope
Pliensbachian	marly limestones and marls	10	gentle slope
Sinemurian and Lotharingian	compact limestones with a kidney texture	5–8	platform
Hettangian	coquinoid limestone and limestones with marly joints	8–10	small cuesta

[1] $+++$ = very permeable; $++$ = permeable; $+$ = slightly permeable; $I-$ = nearly impermeable; I = impermeable.

KARST OF FRANCE 139

relative permeability[1]	Permeability type	Pluviometry (mm/year)	Aquifer levels relative importance	main springs at low water-level total yield ($10^3 m^3/day$)	number
+++	fissuration + karst	650–900	very important	172	9
++	fissuration + karst	650–900	very important		
I	–	650–900	–		
+++	fissuration	650–900	important	1.7	12
I	–	600–700	–		
++	porous interstices	600–700	medium	1.2	8
++	porous interstices	600–700	medium		
I	–	600–700	–		
+	porous interstices	600–700	slight		
I–	–	600–700	–		
+++	fissuration	600–700	slight		
+++	fissuration + karst	650–750	very important	9.7	13
+	–	650–750	slight		
++	fissuration + karst	650–750	little important	5.7	4
+++	fissuration	650–750	very important		
+++	fissuration	650–750	very important	43.5	14
+++	fissuration + karst	650–750	very important		
+++	fissuration + karst	650–750	very important		
+	–	700–800	–		
+++	fissuration + karst	700–800	important	(2.7)	1
+	–	700–800	–		
+++	fissuration + karst	700–800	important	6	38
I	–	750–850	–		
+++	fissuration	750–850	slight	0.2	10
I	–	750–850	–		
+	–	750–850	–		
+++	fissuration	750–850	medium	0.7	17
I–	–	750–850	–		

from exhumed fossil karst to more recent karst, including embryonic forms and "prekarst".

Hydrogeology. The hydrogeologic conditions are as follows:

(*1*) Middle Jurassic karst (Bajocian and Bathonian) lies on Toarcian impermeable marls, and the calcareous plateaus with dolines are more or less alined with fracture networks and the cliffs of the cuesta, the edge of which rises over Liassic layers with many springs.

(*2*) Upper Jurassic karst (lying on Upper Oxfordian) covered by arid and wooded plateaus contrasts with rich regions irrigated by the springs from this plateau.

(*3*) Special karst phenomena consist of the enormous water reservoir constituted by Senonian and Turonian chalk, which exhibits more perfectly than either of the first two areas described, both a true aquifer with a porosity of interstices or fissures (Fig. 2D, 3A) (mainly located in the altered part) and a circulation network with karst channels.

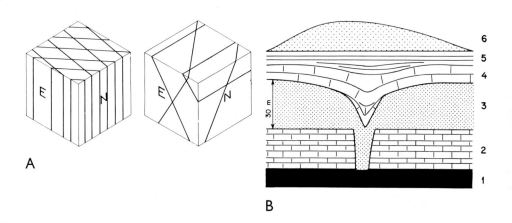

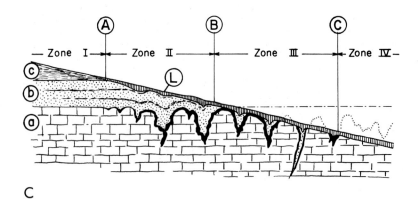

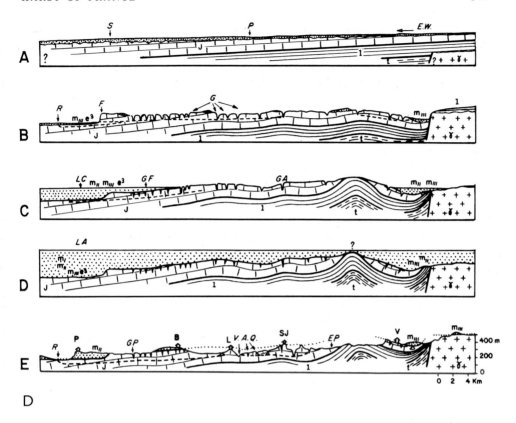

Fig. 3.A. Representation of the two principal sets of joint systems in chalk in the region of Sens, based on statistical measurements. (After MEGNIEN, 1964.)

B. Schematic section of a puisard in a palaeokarst hollowed in Lutetian limestones (2) in Yvillers (Oise). The space draught provoked, during the process of sedimentation, the recession of Bartonian sands (3), limestones (4) and marls (5) 1 = Cuisian sands; 6 = Stampian sands (After POMEROL, 1967.)

C. Formation of dissolution caves (poches de dissolution) under a permeable cover. a = Calcareous; b = permeable layer; c = impermeable layer; L = slimes; zone I = no formation of dissolution caves; zone II = formation of dissolution caves (in black: residual decalcification clays); zone III = destruction of dissolution caves; zone IV = no more dissolution caves. (After BONTE, 1955.)

D. Scheme of the speleologic evolution of the Causses in southern Quercy. A = Eocene (pre-Lutetian) plain covered with decalcification clays and alluvium in a fair way to lateritization solution of the residual phosphate; B = Eocene and Oligocene (Bartonian–Sannoisian) warped and elevated peneplain a prey to karst erosion; C = Oligocene (Upper Stampian) transgression of the Cieurac Lake, consecutive to the folding of Quercy: beginning of the fossilization of sinkholes with phosphorites; D = Miocene (pre-Pontian) transgression of the Agenais Lake entirely fossilizing southern Quercy; E = Pliocene–Quaternary: clearing and superficial erosion (Pliocene), and deepening and karst erosion (Quaternary); reappearing and partial remaking of sinkholes with phosphorite. m' = Aquitanian; m'' = Stampian; m''' = Sannoisian; e^3 = siderolithic; J = oolitic; γ = granite; l = Lias; t = Triassic; P = Puy-la-Roque; B = Bach; L = Limogne; StJ = St.-Jean de Laurs; V = Villeneuve d'Aveyron; $G.P.$ = Eocene sinkholes with phosphorites; $E.P.$ = Pliocene erosion surface; S = siderolithic clays; P = Eocene peneplain; $E.W.$ = great east-west currents; R = resurgence; G = Eocene sinkholes; $L.C.$ = Cieurac Lake; F = edging cliff; $G.F.$ = fossil sinkholes; $L.A.$ = Agenais Lake; $V.A.Q.$ = Quaternary valley and sinkhole. (After GÈZE, 1938.)

In the chalk of Picardy and Artois, the dolines, the *bétoires* and the *creuses* (small, dry, dead-end ravines beginning abruptly, 5–20 m wide, 10 m deep and sometimes more than 2 km long) cut to wider valleys, and are limited downstream by the opposite slope. They may be explained [cf. GOSSELET, 1906] by the infiltration of water along jointed or weak zones of the underlying chalk.

The results of this infiltration is the frequent disappearance of surface flow and the formation of "dry valleys". As seen in the Paris Basin and in Burgundy, these valleys appear to correspond to fossil valleys that may have formed when the periglacial climate, giving rise to a pergelisol, suppressed the general permeability of the subsurface [TRICART, 1949; CIRY, 1959; A. Cailleux, personal communication] or if permeable formations up to the end of the Tertiary were covered with impermeable sediments [PINCHEMEL, 1954].

In the Paris Basin the extent of present karst phenomena is controlled by the possible speed of water circulation across limestones or chalks. When this possibility exists, one may observe quite spectacular phenomena such as the doline field on the plateau of Beaumetz les Aires in Picardy with hundreds of dolines 3–150 m in diameter and 1–8 m deep. This plateau indeed rises 180 m. It is surrounded by very deep valleys and chalky formations overlying impermeable marly formations.

In all these cases, one observes the important part played by these dry valleys or present-day perennial valleys that determine (with their underground flows) the drainage of karstified formations and hence the direction of flow which is seldom parallel to the slight dip of strata. Large fractures may intercept the water, often creating lines of springs.

Dye tests (mainly with fluorescence) have demonstrated subterranean water movement over distances of several kilometres, independent of the topography (except for some small streams following the alignment of streamless valleys) and of the dip of strata. The water generally reaches the permanent flow in the deepest part of the valley nearest to the emission point.

In the chalk BONTE [1955], in his study of "dissolution caves" generally filled with more recent sediments (Landenian sands and grits), showed that the limestone dissolution seems unlikely in the open air under the present climatic conditions, because infiltration is too rapid. On the contrary, the presence of a permeable cover appears to be necessary together with the CO_2 present in soils to allow intense chalk dissolution (cf. Fig. 3C) occurring in these *poches de dissolution* (dissolution caves). Both these surface dissolution phenomena, and chemical and mechanical erosion phenomena in the underground networks, explain the presence of gulfs and "natural wells" which may or may not be filled with subsequent sediments, circulation networks, etc., which have been observed in chalky countries. The part played by the altered surface chalky formations as productive aquifers has also been elucidated (Fig. 2D). Reports of GOSSELET [1906], MARTEL [1921], TRICART [1949], PINCHEMEL [1954], LUTAUD [1959], RICOUR [1959], MEGNIEN [1964] and others contain more details.

Among the famous karst phenomena in the Paris Basin are the following:

(a) Some *muches* (natural underground networks) in Picardy (those of Naours, a little north of Amiens, for example).

(b) Disappearance and reemergence of the Avre River in the Eure chalk (Normandy) which has been collected for the water supply of Paris.

(c) *Vallées absorbantes* (absorbent valleys) and *bétoires* in the Pays d'Othe.

(d) Natural caves, galleries and swallow holes in the region of Andelys (Cretaceous of the Seine).

(e) The Caumont caves in the chalk of Rouen, the ramifications of which are more than 2 km long; underground streams in the Pays de Caux and resurgences in coastal chalky cliffs. Some of these caves were used as underground hiding places as early as Roman times and the Middle Ages. However, most of them are small compared with those known in karstified limestones in the Causses region or the south Mediterranean.

(f) *Poches de dissolution* (dissolution caves) filled with green Albian and Middle Bathonian sands in the region of Aisne and Ardennes.

(g) Pliocene karst fossilized by argillaceous and ferruginous deposits in calcareous Bajocian plateaus in Barrois and Lorraine; these are well-known because formerly they were mined for iron ore.

(h) Fossil karst with Portlandian "strong iron" (fer fort) beds in Haute Marne.

(i) Bajocian karst in the Briey Basin (Lorraine) that have been carefully studied because of the problems they cause in the iron mines of this region.

Tertiary formations

Some karst phenomena are also found in calcareous or gypsiferous Tertiary formations in the central part of the Paris Basin.

The Tertiary calcareous (marine or lacustrine and lagoonal) formations in the Paris Basin are found in Thanetian (Paleocene: limestones of Rilly), in Lutetian (Eocene: *calcaire grossier*—rough limestone—and calcareous with the miliolitic Royal bed, limestones of Provins), in Bartonian (Eocene: limestones of Saint-Ouen, Champigny, etc.), in Stampian (Oligocene: limestones of Etampes and Brie) and in Aquitanian (limestones of Beauce and Orleanais).

These formations present *poches de dissolution* (dissolution caves) here called *puisards* (in Lutetian limestone of Soissonnais and Valois for instance, where they are covered with red clay and filled with more or less ferruginous Bartonian sands—see Fig. 3B; also in Stampian limestone of Champigny and along the eastern cuesta of Ile de France). These dissolution caves are connected with galleries, sometimes giving rise to springs, and true caves may also be located there (the Champlieu Cave, for example). POMEROL [1967] showed that some of these caves were developed before the setting of *cailloutis* (drift-stones) and superficial slimes (end of Pliocene or Quaternary) at Neogene time, and that in the clay that covers them, the major part of the argillaceous, ferruginous and manganese material is not just a dissolution residue, but has been brought "per descensum" and chemical precipitation.

Therefore it seems that the most important karst networks affecting Tertiary limestones in the centre of the Paris Basin began to form from the emergence (after the Aquitanian) of the central part of the Paris Basin and were developed gradually with the lowering of the basal level following the epeirogenic upheaval of the Paris Basin.

In addition let us point out that: (*1*) spectacular dissolution phenomena as well as the disappearance of water (e.g., the Fontis of Montigny, the Trou du Tonnerre in gypsum in the Montmorency forest) occur in gypsiferous Bartonian strata: and (*2*) some sandstone formations (e.g., the sandstone of Fontainebleau) contain caves or galleries which mimic karst forms, but these are nearly always confined to phenomena of mechanical clearing of poorly consolidated or sandy parts by seepage water.

We should also mention that BOILLOT [1964] showed that numerous depressions in the bottom of the English Channel (adjacent to the northern part of the Paris Basin) correspond to karst dissolution basins formed before the submergence by marine water.

Karst phenomena in Mesozoic and Tertiary formations in the Aquitanian Basin

In contrast to the Paris Basin, the Aquitanian Basin presents a marked dissymmetry because its southern edge is the Pyrenean mountain chain which has incorporated in its folds or covered with the debris of erosion the Mesozoic formations that were present there. As a result, towards the south there are no aureoles of calcareous plateaus and hills. The latter are found only in the north, as a belt 300 km long from Quercy to Charentes crossing Périgord, and no more than 60 km wide. These is a lithologic series similar to that of the Paris Basin ranging in age from Triassic to Cretaceous, and showing elevated plateaus cut by rather deep valleys, with rocky-ledged crowned slopes, the altitude of which decreases towards the east.

Quercy

Karstification is developed mainly in Jurassic limestone plateaus in Quercy. Numerous areas have been correctly termed "Causses Mineurs" because of their resemblance to the "Grands Causses" or "Causses Majeurs" in Rouergue, Lozère and Hérault. Here again streams are rare, and arid plateaus are riddled with *cloups* (circular depressions without gulfs at the bottom) and with *igues* (avens) feeding networks of cavities widening into caves and underground streams opening into valleys. Some examples are the Padirac Gulf with its famous underground stream, accessible for tourists, the Saint Sol Lacave Cave, the David Cave with prehistoric paintings, the abyss and cave of the Igue de Bar and those of Marcillac and Combettes.

The part played by faults in the alignment of the site of disappearance and circulation of water is particularly clear here. A fault is responsible for the circulations in Padirac and for the collapse of Bathonian limestone that causes the famous gulf (31.5 m wide, 75 m deep) giving access to an underground stream nearly 3 km long and flowing across several rooms, the highest of which (Grand Dome) is 91 m high.

Among other caverns are the Pech-Merle-Marcenac caves where the first prehistoric stone engravings and paintings of bisons, Equidae, Cervidae and mammoths were discovered in the Lot caves in 1920.

CAVAILLE (1964) showed how underground streams in these "causses" can cut intercalated marly beds and descend from one limestone bed to an underlying one, giving rise to an *étagement* (staging) of the underground circulation.

In the southern part of the Causse du Quercy, mainly between the valleys of the Lot and Aveyron rivers in the so-called Causse de Limogne and Causse de Caylus zones, are many karstified fossil cavities filled with "decalcified clays" rich in iron phosphate and manganese (of the siderolithic type). The filling process probably began during Cretaceous or Eocene time, and ended during Miocene time when southern Quercy was entirely covered by the Aquitanian Lac de l'Agenais transgression. Pliocene and Quaternary erosion exposed these fillings, which were intensively mined between 1870 and 1920 for the phosphates they contained. The phosphates were exported under the name *phosphates de Bordeaux*. Fig. 3D, after GÈZE [1938, 1949], shows how this region evolved from Eocene time to the present. From a palaeontologic point of view, the presence of the Bartonian and Stampian fauna in these *phosphorites de Quercy* is particularly important to the palaeobiological knowledge of the Eocene and Oligocene deposits in these regions.

Périgord and Pays des Charentes

Périgord, drained by the Vezere and Isle rivers shows an attenuation of the karst facies because of the lowering of the limestone erosion surface partly covered by siderolithic deposits and the thick sand beds of Périgord, while a much wider river drainage network was not deeply entrenched. However, the wooded calcareous plateaus are furrowed with a network of dry valleys.

In the Pays des Charentes, this attenuation of the karst facies increases for the same reasons and because the Cretaceous and even Jurassic facies become more and more marly in this region.

Nevertheless there are many caves and places for shelter under rocks, the importance of which in prehistoric archaeology is notable (caves of Cro Magnon, les Eyzies, la Quina, Laugerie Basse, etc.). The bones *(Homo neanderthalensis, Homo sapiens fossilis)*, Palaeolithic artifacts and engravings and cave paintings found there are universally known.

Losses of water are named *soucis* and resurgences *doux, gours, puits, bouillidoux*, etc. Underground networks may be large, more than 4 km, and several gulfs are well-known there, such as that of Proumeyssac.

In the remainder of the Aquitanian Basin, Jurassic or Cretaceous calcareous formations are more or less deeply covered under Palaeogene and Neogene transgressive formations. The karstification that may have affected them is almost unknown, and the only evidence of some aquifers is by borings that show their existence.

The calcareous Tertiary formation in the Aquitanian Basin includes the *calcaires à astéries* (of Stampian age) in Entre Deux Mers and the *calcaires de l'Agenais* (of Aquitanian age) the outcrops of which present some karst phenomena. Moreover the formation contains two developed aquifers (SERONIE-VIVIEN, 1953; LAURA, 1964].

These limestones, and also the Cretaceous formations, extend towards the south under Tertiary overlaps of the Aquitanian Basin and contain relatively uninvestigated aquifers which can be reached only by borings.

Karst phenomena of the Grands Causses

Among the most famous karstified formations in the world are the Grands Causses or Causses Majeurs. Being secondary formations, these still tabular coverings of the southeast Hercynian base of the Massif Central were uplifted to a medium height (800 m) while the massif was rising. Today they tower above Languedoc from which they are separated by the Cevennes fault system which lowers the substratum by successive tiers towards the Oligocene subsidence of the Languedoc and of the Golfe du Lion.

Due to the action of an intense regressive erosion they are subdivided today by three deep and impressive gorges (400–500 m deep) at the bottom of which are more or less perennial rivers, the Tarn, Jonte and Dourbie rivers, with the corresponding plateaus being, from north to south (cf. Fig. 4A,C), the Causse de Sauveterre and Causse de Severac, the Causse Méjean, the Causse Noir, the Causse du Larzac. To the southeast the valley of the Vis River (tributary of the Hérault River) defines another plateau, the Causse de Blandas. On the eastern, northern and western edges they are bordered by the primary and crystalline formations of the Massif Central which tower above the flexure of steep cliffs. The Lot River to the north, the Tarn

---

Fig.4.A. Perspective view of the Grands Causses and their surroundings. Crystalline massifs: *1* = Mont Lozère; *2* = Aigoual; *3* = Aubrac; *4* = Segala. Causses: *5* = Sauveterre causse; *6* = Mejean causse; *7* = Causse Noir; *8* = Larzac causse; *9* = Blandas causse. (After CORBEL, 1954.)
  B. Lithostratigraphic section of the Grands Causses.
  C. Schematic north–south section of the Grands Causses. (After CORBEL, 1954.)
  D. Formation of valleys and terraces in the Grands Causses. (After CORBEL, 1954.)
  E. Formation of the border of the Grands Causses. I = At Pontian times, the crystalline massifs that have just been elevated, are vigorously attacked by regressive erosion. From the calcareous platforms erosion notches the crystalline rocks. Under this climate the crystalline rocks are as friable as limestone. The sheet-like erosion can then develop these surfaces that cut the two rocks indifferently. II = The clearing process of the peripheric depression in contact with limestones. In (*1*) one can see the river flowing from the impermeable (hatched) zone losing itself at the contact with the limestone. The superficial erosion still works at *A*, but is suppressed in *B*. In the foreground (*2*), one can see the result of this difference: the impermeable layer is lowered, regardless of its relative hardness. If limestone is then particularly soluble, for example under a very wet climate, the exposed contact with the two rocks will rapidly retreat. III = Formation of the peripheric depression. In the background a badly-marked depression (as is the case in the eastern part of the Mejean causse); in the foreground a peripheric depression well cleared up by normal erosion. One can recognise the platform on the crystalline rocks that testify to the ancient climatic phase. (After CORBEL, 1954.)

KARST OF FRANCE 147

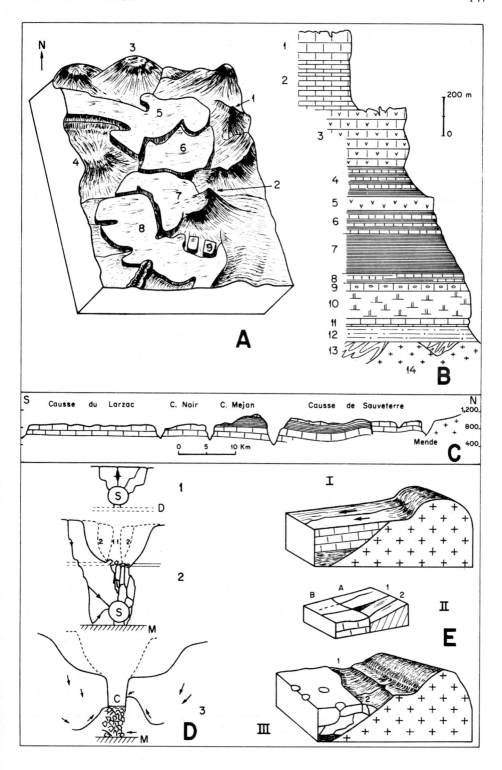

River and its tributaries to the east and to the west more or less closely parallel these cliffs. The process of erosion has isolated several small outlying causses (e.g., Causse de Rodez, Causse de Marvejols, Causse de Mende).

Lithostratigraphy

The complete section (Fig. 4B) is as follows:

(1) At the bottom, Triassic and Rhaetian sandstones and marls. This is overlain by Hettangian dolomites, Sinemurian cherty limestones and Lower Charmouthian marly limestones (the dolomites compose a small autonomous karst system).

(2) Upper Charmouthian and Toarcian marls which with their average thickness of 200 m compose the major impermeable basal level; in a few places there still exists a morphologically continuous passage between the calcareous surface and the crystalline rocks these limestones overlie (e.g., Causse Noir).

(3) The mass of the Dogger (Aalenian and Lower Bajocian marly limestones, Upper Bajocian dolomite, Lower Bathonian marly limestones and Upper Bathonian ruin-shaped dolomite) with a maximum thickness greater than 500 m. This is the most karstifiable zone.

(4) The thin-bedded limestones of Callovian–Sequanian age (300 m).

(5) Finally covering the entire section are grey Kimmeridgian dolomites (150 m), eroded into the shape of ruins.

In more than half of the Causses, the upper formation disappeared leaving the Bajocian–Bathonian strata bare.

Structure

The tabular structure is broken only by some large faults, which for the most part are continuations of fractures in the Hercynian substratum, their major direction being southwest–northeast.

Climate

Situated between the Atlantic and the Mediterranean the Grands Causses are presently influenced by very diversified climates with high rainfall to the east (over 200 cm), moderate or slight rainfall in the centre region, and less than 70 cm to the southwest.

From the palaeoclimatic point of view, according to CORBEL [1954]—relying upon the vegetal palaeoecology—and according to DEMANGEON [1967]—relying upon the mineralogical study of the karst filling—the Grands Causses have successively been under the influence of: (1) a wet tropical climate during the Upper Cretaceous; (2) a semi-arid tropical climate during the Eocene; (3) a plainly arid tropical climate during Oligocene and Middle and Lower Miocene time; (4) a warm and dry continental climate in the summer and a rather cold climate in the winter during the Upper Miocene; (5) a still warm climate, gradually becoming colder, during the Pliocene; and (6) cold climates during the Quaternay Glacial Stages interrupted by Interglacial Stages with a milder climate.

Karstology and hydrogeology

The karstification influenced by the preceding factors is characterized as follows:

(1) An important proportion of dolomites, reducing the normal karstification in numerous places.

(2) The elevation of the basal level (impermeable Liasssic marls) causes a peculiarly efficient drainage of the "perched" karst by the streams running at the bottom of gorges. The nearness of the latter quickens the circulation and causes an intensification of karst phenomena in their neighbourhood (e.g., near the Tarn River the sinkholes and gulfs multiply). The small Causse de Blandas is thus the richest in caves and underground rivers which also abound in the plateau of Guillaumard (the edge of Larzac which parallels the valley of the Vis River).

(3) The existence and direction of fracture networks give an orientation to the major underground circulations causing the formation of backing-spaces *(reculées)* and finally the formation of canyons and valleys by subsidence of the vaults of underground systems and removal (often incomplete) of the subsided materials by water (for example: the gorges du Tarn at the Etroits where the valley is 1,200 m wide at the top, but only 30 m at the bottom). Thus it seems that in numerous instances according to CORBEL [1954] an underground flow was present before the surface flow (cf. Fig. 4D).

(4) Intensity of the rainfall. The parts receiving the most rainfall are the most karstified. Those parts receiving less (east of the Causse Méjean and Causse de Sauveterre, speleologists "purgatory") have few karst cavities or possess only fossil cavities.

(5) Palaeomorphologic and palaeoclimatic periods, which have followed one another since the emergence of the Causses and more particularly since the Upper Cretaceous, have given for the peripheral drainage a prominent part to the variable differential erosion of the limestones and crystalline rocks which surrounded them (Fig. 4E). The granites underwent a more important alteration because of the dry and warm climate. Probably at the end of the Pliocene, the limestone was below the level of these granites. In these conditions, the streams from the crystalline regions had to clear an underground way for themselves in the calcareous masses following the previously fractured zones.

(6) On the whole, the karst phenomena of the Causses Majeurs differ very much from one part to another. On an average, the karst is much less developed than the sub-Alpine karst (as discussed later). Nevertheless we find here:

(a) Plateaus with a sub-structural surface of a wild magnificence, with ruin-shaped dolomites or rubble of plaquette-like limestone residual hillocks (hums). On these plateaus, in various locations there remain either relics of warm climates (coverings of reddish alluvium from the Cretaceous, Eocene, etc.) or relics of cold Quaternary climates (periglacial phenomena: polygonal soils, soils of frost action, etc.). Elsewhere, (especially the southern part of the Grands Causses) are found all of the classical shapes of karst morphology: lapies, dolines of all shapes [MARRES, 1935] showing two different generations, the small ones often being found at the bottom

Fig.5. Temporary overflowing lake of Migayrou aven, November 1, 1963. (Photo H. Salvayre.)

of the largest; dry valleys with old siliceous alluvium; plains looking like more or less complete poljes (e.g., the plain of the Carnac, the plain of St. Maurice de Navacelle) and occasionally becoming the sites of temporary lakes (Fig. 5) because of the relatively impermeable bottoms (due to the decalcified clays) or because of flooding by extravasating water [*eaux d'extravasement*, SALVAYRE, 1964] at the time of heavy rains and engorgement of the underground networks (e.g., Lac des Rives).

Lavognes, small temporary ponds which are artificially made, usually by shepherds, to collect surface run-off must be distinguished from the naturally occurring ones.

(*b*) Moderate underground circulation in the Causse de Sauveterre; slight circulation in the Causse Méjean, developed circulation in the Causse du Larzac, enlarged circulation in the Causse de Blandas or the small plateau near Bramabiau. The circulation connects the particularly numerous streams disappearing and re-emerging in the Gorges du Tarn (over 40) and in the valley of the Dourbie River.

(*c*) Very spectacular *reculées* (backing spaces): backing space of Gourgas (Bout du Monde); backing space of the Sorgue River bound to a Triassic anticlinal axis.

(*d*) Particularly spectacular gorges (e.g., gorges of the Tarn, Jonte, Dourbie and Vis rivers).

(*e*) These gorges seem for the most part to be relatively recent (Quaternary), which explains the frequent disruption between the surface and underground discharge networks and which seems to be corroborated by their history (comparison between the contents of the caves at the bottom of the gorges and archaeologic remains found on the plateaus).

Among the famous tourist sites are the Tarn gorges, the Navacelles cirque, the Dargilan Grotto (above the Jonte River), those of Maleval, the Bramabiau Grotto in the Hettangian with its river, the drainage system of which is nearly 10 km long. The aven Armand which was discovered by L. Armand and explored by E.A. Martel is particularly rich in speleothems with more than 400 stalagmites, some of them of a gigantic size, more than 30 m high (Fig. 6,7).

Fig.6. Map and sections of the Armand aven (Méjean causse). (After MARTEL, 1930.)

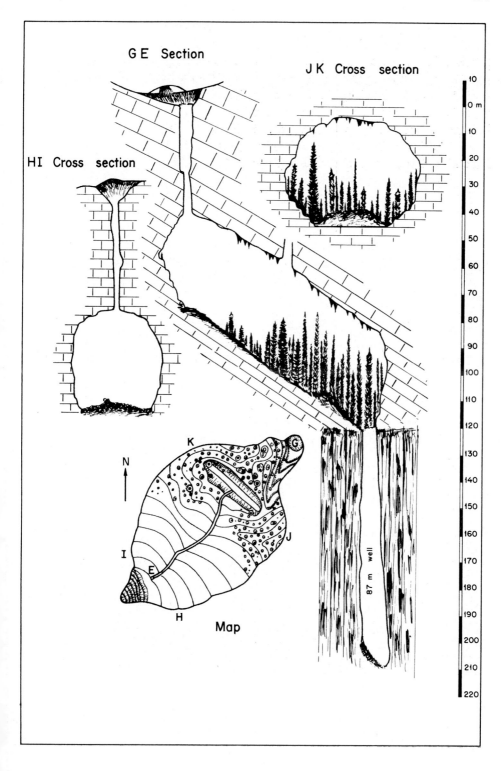

Fig.7. Armand aven: gigantic stalagmite. The scale is given by the standing man.

The Languedocian karst

Setting

Following the Grands Causses towards the east, and composing a more or less continuous belt between the Hérault River and Valence, they border the Cevennes to the west and extend to the Rhône River to the east. The para-Mediterranean or Mediterranean climate affects the Tertiary or Cretaceous and Jurassic carbonate formations which constitute the main part of the Languedocian *garrigues*. They are drained by a succession of small rivers: the Hérault, the Lez and the Vidourle and by three tributaries of the Rhône River: the Gard, the Cèze and the Ardèche. Some of their outlets are at the bottom of ponds or even at sea.

Lithostratigraphy

Four units stand out in the regional Mesozoic and Tertiary:

(*1*) The first dolomitic unit (Liassic—dolomitic Hettangian—and Middle Jurassic —Bajocian, mainly dolomitic Bathonian) with the intermediary marly Toarcian screen is not important, except at the edge of the Cevennes because outcrops are found in only a few places.

(*2*) On the other hand, the second unit composed of the thick Upper Jurassic to Lower Berriasian calcareous formation, which is nearly 800 m thick, is important for it is composed of particularly thick and resistant pure limestones.

(*3*) This formation is followd by a third unit: the Neocomian unit; the Lower Valanginian marls (200 m thick) form an impermeable bed tightly separating the Upper Valanginian limestones (60–100 m thick) from the previous limestones and playing a prominent hydrogeologic role.

Toward the east, along the Rhône, appears the reef calcareous Urgonian facies extending south of Donzère into the marls of the Vocontian Depression. Nowhere can be seen the Upper Cretaceous chalky facies of the Paris Basin.

(*4*) Finally, a fourth weakly karstified unit of negligible importance is composed of lacustrine Lutetian limestones lying above impermeable Lower Eocene marls.

Structure

The region has been intensely fractured, on one hand, by great faults parallel to the edge of the Cevennes (*sous-cévenoles* faults) defining successive tiers, horst zones and graben zones causing contact between fissured and karstified limestones and impermeable calcareous-marly layers.

On the other hand the region has undergone pressures either due to gravity or to the uplift of crystalline masses, or to the orogenic Pyreneo-Alpine movements. Those thrusts, essentially Eocene–Oligocene, caused the very extensive folding of some blocks, in some places overlapping as much as several kilometres to the northeast. The tectonic structure in Hérault and Gard becomes less complex in the department of Ardèche where there are faulted monoclinal belts dipping towards the Rhône Valley graben.

Finally at the even edge of the Cevennes, fragments of Jurassic limestones remain at the primary or crystalline surface or have been kept there by subsidence between faults.

The main structural units, essentially calcareous and almost all limited by faults, are the following: the Montagne de la Sérane immediately below the Causse du Larzac; the Monts de St-Guilhem, the Causse de la Selle and the Montagne de la Sellette which is crossed by the gorge of the Hérault River; the Massif du Thaurac and the Bois de Monié; the Massif de la Fage, the Causse de Pompignan; the Dôme du Coutach (Plate I,A,B,D,E,F); the Causse de l'Hortus, the Causse de Viols-le-Fort and the Massif du Pic St-Loup; the Bois de Paris; the Massif de Logrian; the Causse d'Aumelas and the Pli de Montpellier; the Montagne de Sète and the Montagne de la Gardiole. The anticline de la Vaunage and the garrigue de Nîmes; the Plateau d'Uzès and the Plateau de Lussan, the Montagne de Berg and the Plateau des Gras.

The rivers are in canyons or gorges in the calcareous formations. Examples are the gorges of the Hérault, Vis, Cèze, Gardons and the Ardèche rivers.

Climate

The present climate is essentially Mediterranean, that is to say the rains, though relatively heavy (over 80 cm), are infrequent and concentrated so that the hot summer is often almost rainless. This accounts for the scarcity of the xerophytic vegetation in the garrigue and the frequency of bare calcareous rock. However, the rainfall increases nearer the Cevennes where it can reach more than 2 m and from which the high basins of the streams and rivers mentioned are abundantly fed.

From a palaeoclimatic point of view and considering the phases of karstification, the region of the Languedoc has undergone an almost uninterrupted long continental period from Middle Cretaceous time to the present. During this period there were: (*1*) a succession of varied warm climates (Cretaceous, Eocene, Oligocene, Miocene, Pliocene) or cold climates (Quaternary); (*2*) important variations of the sea level; and (*3*) orogenic and epeirogenic events with major consequences such as the formation of the Golfe du Lion by subsidence. Some subsidence to the south of the fault of Nîmes shows throws at more than 3 km.

If according to these elements we outline the regional karst evolution, we can show, with DUBOIS [1964a], five important karstification periods:

(*1*) While Middle Cretaceous emergence took place, a karstification occurred, the networks of which must have been fossilized by filling in of bauxite. These are mined today on the old area of the Durancian Isthmus. They are covered with Upper Cretaceous continental deposits either more or less completely destroyed by erosion, or subsided and more or less preserved.

(*2*) A Lower Tertiary (Eocene–Oligocene) and Lower Miocene karstification which must have been very important, the corresponding morphologic surface being one of the clearest palaeosurfaces in the Languedoc, and with numerous present-day surfaces of Causses dating from the same period. This surface, partly dislocated at the

PLATE I

A,B,E. Lapies corrosion features in Upper Jurassic limestone (Rieumassel Basin, Gard).
C. Cross-section of the karst formation showed in a trench of the Ceyrac dam foundations. Note: (*1*) the surface regolithic soil; (*2*) the faults and joints; (*3*) the corrosion features; and (*4*) the clay filling.
D. Dissolution meandering channel on the upper surface of a limestone layer, found in the Ceyrac dam foundations (width 10 cm).
F. Dam construction in the Jura karst area (Vouglans Dam).

time of Oligocene subsidences, has been transgressed by the seas of Middle Miocene which sealed numerous karsts which are still covered or exhumed. During the Upper Miocene, the very important Pontian regression, perhaps 200 m below the present sea level, must have caused a particularly active karstification, while the gorges of present rivers were hollowing by superimposition.

(*3*) During Pliocene and Early Quaternary, the Miocene surface was more or less fossilized by the deposits following the rise of the sea level to about 100 m above the present level. It must have been the period during which the high-level grottoes and the hanging dales of the Vis, Vidourle, and Hérault valleys were formed.

(*4*) During the Middle and Late Quaternary the sea level was lowered in a fluctuating manner, from $+100$ m to the present level. The old or contemporary networks were deepened or cleared of deposits. Down-cutting of valleys and underground networks took place, especially during glacial regressions. At the close of the Last Glacial (Würm Glacial), the sea level rose again (Flandrian transgression) from -80 m to the present level. The lower parts or even all parts of some karst networks were drowned, while the lower parts of the valleys were filled.

During the Quaternary cold phases the pergelisol were favourable for the formation of networks, of surface grottoes and of "gelivation" and "solifluxion" phenomena, etc.

Main characteristics

(*1*) It follows from the previous factors that the karstification of the Languedocian zone is particularly complex. Most of the karst was polygenetic and the variations of the base levels were linked to the variations of the level of the Mediterranean Sea and to the tectonic deformations so that the following can be observed: rejuvenated or active, dead, drowned or buried, perched fossil karst, and present-day karst; springs at all levels including the coast where they emerge under the sea or discharge into the middle of ponds, for example the resurgence at -32 m in the abyss in Etang de Thau, and the source of Inversac at the edge of the Gardiole which works alternatively as an emissive or an absorbing gulf.

(*2*) A typical characteristic of this whole region is the very frequent situation of faults joining karstified calcareous masses and impermeable formations. Under these conditions faults in addition to their role in drainage, play a major part as a dam or barrier thus creating *karsts barrés* [DUBOIS, 1961] and causing an orientation of the underground flows with the appearance of springs on the surface. Examples are the south fault of La Serane: Foux de la Buège, Foux de Brissac; fault of Corconne: fontaine de Sauve, the spring of the Lirou River; the succession of faults between Les Matelles and Castries: the spring of the Lez River, the spring of the Fleurette, Gour Noir, etc; faults of La Gardiole: Roubine, Cauvy, Issanka and probably the abyss.

(*3*) Various studies, principally made by local speleological clubs, have shown the existence of numerous underground streams, more than ten of these being 5–20 km long, with the velocity fluctuating between 500 m/h and 14 m/h.

The course of these underground streams is either along faults (e.g., the fault of Corconne or along a synclinal axis as the drainage of the Hortus syncline by the underground river Lamalou) or along the surface course of some streams such as the Vis, Virenque and Vidourle rivers (Fig. 8).

The underground circulation generally runs through well diversified channels, but the existence of actual karst aquifers has been proved either in the dolomitic parts, the cavities of many of these having been filled up with a slightly permeable dolomitic sand [PALOC, 1961], or in fissure networks (mainly located around valleys and gorges in the case of tabular causses, e.g., along the Vis and Hérault valleys) in highly dis-

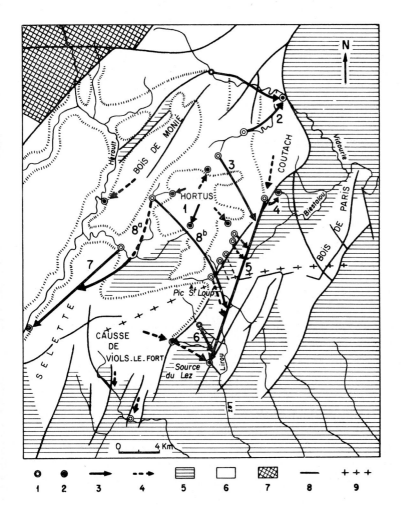

Fig.8. Underground circulation in the north Montpellieran karst. 1 = Loss; 2 = resurgence; 3 = underground circulation proved by dye tests; 4 = probable underground circulation; 5 = post-Valanginian strata: no or very few limestones; 6 = Jurassic–Valanginian limestone zone; 7 = Palaeozoic; 8 = fault; 9 = anticlinal axis of Saint-Loup (Bois de Paris). (After DUBOIS, 1965.)

turbed limestones (e.g., calcareous zones north and west of Montpellier—Dôme de Coutach, Causse de Viols-le-Fort, Causse d'Aumelas, Gardiole, Pli de Montpellier, etc.).

The study of recession curves of the springs flowing from this karst enables us to distinguish the discharge of submerged channel networks from that of fissure aquifers. The comparison between flood hydrographs at the end of a dry period and at the end of a wet period enabled DROGUE [1963a] to calculate the reserves of some of these fissure aquifers.

(4) There are two types of karst springs:

(a) Karstsprings, the impermeable base level of which is above the drainage level (present valleys), e.g., springs running out of the Upper Valanginian limestones of the Causse de l'Hortus.

(b) Karstsprings where the impermeable base level lies under the present drainage level.

These springs are known as "vauclusian fountains" and are particularly important in the north Montpellier karst region where about fifteen of the perennial fountains yield several hundred l/sec to several m³/sec. An example is the spring of the Lez River, which supplies the town of Montpellier. During flood time more than a hundred springs have yield comparable to the fifteen perennial springs. The spring at the head of these fountains is always several tens of metres deep, and some of them rise from depths below sea level (e.g., the spring of the abyss in the Etang de Thau which rises from a depth of 32 m below the surface of the pool). All data seem to show that they correspond to old channel networks of karst circulation, originating when the sea level was much lower than at present, either because of the lowering of the sea level (for instance at the time of the pre-Flandrian regression and of its later uplifting) or because of the general sinking linked to the subsidence of the Golfe du Lion which is still occurring. More likely it is due to a combination of these two phenomena.

(5) Among the exceptional development of fossil caves with no circulation, several have stalactitic and stalagmitic calcite crystallizations. Among the most splendid are the Demoiselles Grotto, the Clamouse Grotto (Plate II A, B), the Aven d'Orgnac, discovered by R. de Joly, the Aven de Marzal, and the network of Courry la Cocalière (23 km long) at the boundary of the departments of Gard and Ardèche.

(6) Among the gorges and canyons are those of the Vis River with the famous meanderings of the Navacelles cirque and those of the Ardèche River, dug in the Urgonian, through which this river runs in canyons 31 km long and which, according to E. A. Martel, is "one among European wonders". This includes the famous Pont d'Arc (a natural archway due to the intersection of a meander) besides spectacular defiles and numerous grottoes.

In conclusion, we must emphasize the enormous complexity of the subject which at the same time constitutes the interest of the Languedocian karst. From now on, different investigators and the *Centre d'Etudes et de Recherches Hydrogéologiques*

A. Aragonite crystal flowers (Clamouse Cave, Hérault).
B. A view of "Le couloir blanc" in the Clamouse Cave, with *gours* and calcite stalactites bearing white aragonite crystals, and magnesian (moon-milk, huntite) on stalagmites.

PLATE II

(C.E.R.H.) of the University of Montpellier will carefully study and advance our knowledge of this karst region and the various karst phenomena.

Karst in Minervois, Corbières and the Pyrenees

The region bordered by the valley of the Hérault River and the Mediterranean Sea to the east, the Montagne Noire to the north, the Palaeozoic and crystalline axial zone of the Pyrenees to the south and the Aquitanian Basin to the west is characterized by the important development of its transgressive Tertiary and Quaternary deposits.

There are, however, some splendid karst phenomena in the folded, uplifted, sometimes overthrust Mesozoic formations (Jurassic and Cretaceous) in the Massif de la Clape, in the Corbières and in the masses of the north Pyrenean zone.

There are also, in Minervois, some outstanding karstifications in the Lutetian flag dominating "in cuesta" the Palaeozoic shaly sandstone formations in the Montagne Noire upon which it transgresses.

Lithostratigraphy

The carbonate formations of this region are the following:
(*1*) Middle Triassic limestones and dolomites; (*2*) Hettangian limestones in "plaquettes" and bedded dolomites; (*3*) Middle Jurassic limestones; (*4*) Upper Jurassic limestones; (*5*) Upper and Middle Cretaceous reef or lacustrine marls facies; and (*6*) Palaeogene limestones, first marine (rich in Foraminifera, especially Lutetian limestones with *Alveolina and Nummulites*), then lacustrine. The whole alternates with detrital deposits (sands, conglomerates, molasse), clayey–marly deposits or, during the Oligocene, with gypsiferous strata.

Structure

The intense palaeotectonics (in particular linked to the different phases of the Pyrenean orogeny) and the changing palaeogeography of this region have caused complex distributions of the different facies and numerous local lacunae. Mesozoic formations are more often highly uplifted or intensely folded and have frequently been overthrust towards the north upon a saliferous Triassic marly sole.

Among other important events were: (*1*) the Aquitanian collapse of the Pyreneo-Provençal range and the transgression of the Miocene sea at the same time; (*2*) the Pontian regression; and (*3*) the opening of the Straits of Gibraltar during the Lower Pliocene and the subsidence which has continued to the present time with the deposition of a thick Plio-Quaternary series in Roussillon.

Karstification phases

As a result of this complex evolution, numerous phases of karstification may have taken place and then disappeared because of aerial erosion and marine transgressions. It also follows that numerous karst networks have disappeared today beneath the present surface of the sea giving rise to numerous submarine springs, or have been covered by Plio-Quaternary formations which can receive water by upward movement from deep karst.

Main natural units

Karst in the Corbières (Cretaceous limestones). The surface of the Corbières shows abysses locally called *barrencs* such as the barrenc d'Opoul where investigators have tried in vain to reach the underground stream which reappears at the submarine coastal springs of Fontdame and of the Rigole in the pond of Salces, springs which at flood time may yield several m^3/sec of slightly brackish water.

Rivers make impressive gorges where they cut through limestones, such as the *clues* of the Agly River in Galamus and that of the Aude River in Pierre de Lys.

Karst in the region of Sault. At the boundary of the Departments of Aude and Ariège should be mentioned the region of Sault with its Lower Cretaceous calcareous plateaus often offering numerous lapies along with losses and closed basins. It is also towards the western boundary of this region that the famous Fontestorbes intermittent spring can be found, the working of which has been partly elucidated by VERDEIL [1958].

Karst in the north Pyrenean ranges. On the one hand, karst affects all the folded or uplifted calcareous ridges running along the northern part of the Palaeozoic and crystalline inner zone of the Pyrenean range; on the other hand karst affects the high calcareous summits of the Pyrenees. The calcareous north Pyrenean ridges (mainly Eocene and Cretaceous limestones) offer not only splendid karst phenomena on the surface, but also numerous gulfs, underground circulations and resurgences. Among these are the Gouffre des Corbeaux, the Caoügno de los Gouffres, the Herm Grotto near Foix, the Lombrive and Sabrat caverns near Ussat-les-Bains.

Information on the circulation of the water comes from many complete borings *(percées)*, for example those in the Lesser Pyrenees and in the ranges of Plantaurel between the Ariège and Garonne rivers. Moreover, many of these caverns have been inhabited by prehistoric men. Among these is the Niaux Cavern near Foix famous for its rock paintings of bisons, wild boars, stags and ibex and its human footprints preserved in clay. There is also the Portel Grotto and its "little red horse". The boring and the cave of the Volp 8 km from St-Girons, where the first clayey statues (bisons) were discovered by Comte Begouen and his sons, should be noted. Not far from there in the Trois Frères Cave is a painting of a "prehistoric sorcerer". The *percée* (boring) of the Arize River at the Mas d'Azil, 410 m long, is famous for its traces of prehistoric man and for H. Filhol's and Abbé Breuil's excavations.

In Haute Garonne, not far from the springs of this stream is the Reseau Trombe (Coume Ouarnede-Arbas; 911 m deep). In the commune of Arbas, the gulf of the

Henne morte should also be mentioned. It opens at an altitude of 1,300 m in the Urgonian limestones fed by dolines almost always full of snow.

In Haute Garonne is the grotto of Marsoulas where the first prehistoric paintings were discovered, the Aurignac and Lespugne grottos famous for their industries and art and the famous statuette of a woman (Venus of Lespugne) which was found there, and also the famous *oubliettes de Gargas* where skeletons of bears, hyenas, wolves etc. were discovered in a heap. Everywhere in Haute Garonne, grottoes, avens, abysses and blowholes are known, as well as numerous circulations such as the underground Labouiche River near Foix. The circulation of water through passages, some almost vertical, frequently causes catastrophic floods.

In the Urgo-Aptian limestones is the Gulf of Poudak (high valley of the Arize River) which shows a remarkable intermittence phenomenon (rhythmic decrease and increase of flow) after rain.

In Lourdes the grotto of Betharram, a famous Roman Catholic pilgrimage site has an underground river 2 km long with five higher levels of cave passages.

High Pyrenean karst. Excluded is the karst that affects pre-Mesozoic dolomites or limestones.

North of Lourdes, in the valley of the Pau River, calcareous masses over 2,000 m above sea level (Mont Perdu, Plateau du Marbore in Spain, Millaris Basin, etc.) show highly developed lapies with crevasses, snow-holes and grottoes more or less obstructed by neves.

In these masses, seepages account for springs (resurgences) on their periphery (e.g., the springs of the Pau Cave) as well as some waterfalls, such as the famous one of the Gavarnie cirque. West of high Pyrenean karst is the karst region of the high valleys of Ossau and Aspe with more downstream resurgences such as the famous resurgence of the Grotte des Eaux Chaudes. Karstified surfaces are also present around the Pic d'Annie (2,504 m).

Karst of the Basque Country. The Basque Country begins at the northwest side of the Pic d'Annie and extends a hundred kilometres as far as the Atlantic coast. The altitude decreases progressively from the high mountain lapies (Pic d'Annie) to the low hills which edge Labourd.

The Mesozoic limestones (principally Cretaceous) outcrop as far as a short distance from the Nive River valley not far from St-Jean Pied-de-Port). Beyond there are few carbonate rocks except in the Palaeozoic formations.

These limestones are cut by canyons (such as the canyons of the Soule region and those of Cacouette, Holcarté and Olhadibie) composing true *clues* which are as much as 300 m deep, with widths of only 3 m in some places. On their surface, lapies can be found (such as that of Braca) as well as variously shaped cavities most of which are blocked by stones or snow and depths called *lecias* frequently more than 100 m

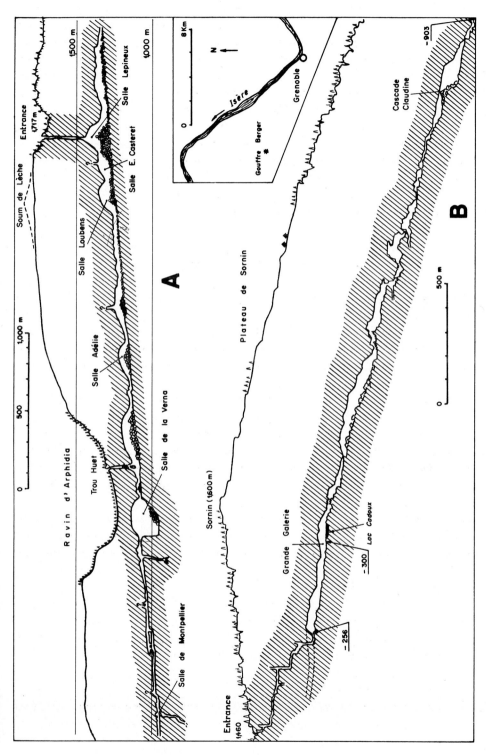

Fig.9. The two deepest gulfs in the world. A. Réseau de la Pierre Saint-Martin (Pyrenees, 1,152 m deep). B. Réseau du gouffre Berger—pro parte—(Vercors, 1,135 m deep).

deep. One of them, the Gulf of Pierre St-Martin near Anette (Basses-Pyrenees), is linked with a network which presently is the deepest system known in the world (gradient: 1,152 m) and which overlaps the French–Spanish border for a distance of 12 km (Fig. 9).

Gulfs include those in: (*1*) the Orion Forest, west of Iraty and the Gulf of the Lecia-Handia in Spain which feed, at least partly, the springs of the Nive River of Beherobie; and (*2*) the Arbailles Forest southwest of Mauleon; there is a Cretaceous limestone mass with dry gullies and numerous gulfs which feeds the resurgences forming the Bidouze River and the source of the Arrangorena River (public baths of Garaïby).

Karst in Burgundy and the Jura Mountains

In the Jura Mountains, the Mesozoic series which are often separated from the substratum at the Triassic level were fractured, then folded into "single covering folds" (anticlines and synclines) or exposed masses that remained tabular (Jura plateaus), divided by folded and uplifted bends which were often fractured and injected with red Triassic marls.

A relatively unimportant erosion affected the folds, leaving their general structure fairly recognizable, but it removed the soft outcrops to the *combes* (dales) and caused the uncovering of hard strata (limestones) which projected as *crêts* (crests) or cliffs.

Lithostratigraphy

Here again are found the same calcareous units as in the Paris Basin, the Jurassic units being the fundamental ones from karst viewpoint.

As far as karst is concerned, the two most important levels are the Middle Bathonian—often with many dolines—and the mass of the Upper Jurassic (Rauracian-Kimmeridgian and Portlandian), which account for the greatest part of the crests and calcareous cliffs overlooking the landscape.

The Cretaceous deposits, on the other hand, appear to be unimportant owing to their often marly or sandstone facies and because erosion has left only limited outcrops. However, the massive oolitic limestones of the Valanginian (Lower Cretaceous) as well as the chalky limestones of the Urgonian, separated by Hauterivian marly limestones and marls, are favourable for karst.

The Tertiary deposits are mainly composed of molasse which is totally unsuited for the development of karst phenomena. To the south, glacial deposits (morainic deposits) covered the Jurassic and Cretaceous formations, and in some places fossilized the preglacial karst phenomena.

Structure

The relative structural simplicity which appears in the morphology must not mask the true complexity of the tectonic history of the Jura, the vertical dislocation of the

Hercynian substratum and of its Mesozoic cover (phases of distension), and different phases of folding (phases of compression) separated by phases of erosion.

The paroxysmal folding phase is the Pontian, which caused recumbent folds and even overlaps in some places. These tectonic facts, together with the marly screens of the stratigraphic series, caused the division of the Jura karst into sections which are more closely related to one another than is apparent at a first glance.

Phases of karstification

The oldest, but insignificant phase, must have corresponded to the emergence at the end of the Upper Jurassic.

The second phase originated at the beginning of the Tertiary (Eocene) and was almost uninterrupted ever since, continuing simultaneously with the deformation and erosion processes. The peri-Alpine Miocene sea was the basal level for a long period, and the Mio-Pliocene surface accounts for the major features of the present morphology together with the consequences of the Pontian Alpine thrust.

This thrust caused the sinking of the hydrographic network and a change in the direction of drainage. Nevertheless intense karstification marked the Pliocene, the climate of which must have been subtropical. This karstification indeed was particularly important in the uplifted parts (anticlinal or tabular blocks). It may partly account for some inversions in relief, for it favoured the erosion of valleys along the long axis of anticlines. There are numerous more or less large karst features from this period, such as dolines, small poljes and so forth. Most of these are filled with terra rossa and covered by vegetation or fossilized by overlying morainic glacial deposits.

This leads us to the Quaternary phases of karstification which were mainly influenced by the Jura glaciers, the action of which was either positive (with abundant water coming from the melting ice, partly "agressive" water) or negative owing to the covering of numerous surfaces by the ice (with mechanical erosion connected with the movement of this ice) or to the Postglacial morainic deposits [AUBERT, 1965].

During Recent time karst processes started again and these continue up to the present, favoured by the prevailing wet climate. The clearing of the morainic deposits by percolating water uncovers old karst surfaces where the superposition of the present karst erosion, still only slightly developed, can be observed rather well on the old karst erosion which is very well developed in many places.

Main characteristics

Finally, the Jura karst shows the following characteristics:

(*1*) More or less polygenetic lapies which developed mainly in the southern part of the Jura (these are particularly abundant on Middle Bathonian and Kimmeridgian limestones).

(*2*) Pliocene, Recent or rejuvenated dolines are especially numerous in dry valleys near passes (e.g., the Col de Richemont) and horizontal surfaces dominating *cluses* (cross valleys) such as those near Nantua. Some also occur on high plateaus.

CHABOT [1927] described in detail the Jura dolines, some of which have become *emposieux* or true uvalas. Some of the uvalas became enlarged and formed small poljes or closed basins.

(*3*) Avens and gulfs *(lanches, cabornes)* are rare owing to the relatively restricted thickness of the limestones; some *bourbouillous* with many fissures situated in the hollows occasionally acting as absorbing wells and sometimes throwing back water that bubbles out (from which comes the name *bourbouillous*).

(*4*) The disappearance of rivers with intercommunications of networks (Doubs, Loue, Andeux, Lizon), drying up the downstream part of a valley (Ravin d'Athose, near Ornans), is a phenomenon which leads to the genesis of dry valleys. Many of these reach the high-yield rivers through dry "waterfalls" or even hang over the main valley (e.g., tributaries of the Loue River, and particularly the dry valley of the Creuse River which is followed by highway N 7).

(*5*) Complex underground circulations are present with cavities abandoned by water after tectonic deformations [FOURNIER, 1923–1928]. These even deformations can explain the creation of new issues, e.g., the Saône Marsh [DREYFUSS, 1957], or as this author [DREYFUSS, 1967] described it: "The chronological succession of networks is often evident in the fact that the most recent ones, composed of underground rivers either impenetrable or hard to explore, are generally linked to old wider cavities, sometimes dry or more or less filled up, or which still have temporary or permanent streams, e.g., Cavottes Grotto, near Montrond, Doubs".

The complexity of the underground circulations is also seen in the multiplicity of resurgences coming from the same basin (e.g., the Saône Marsh) or from interconnections between adjoining basins. Quantitatively variable according to the season, these relations are difficult to pinpoint using dye and can be revealed better by correlation of yields [VANCON, 1965]. When passages of one network lie at different altitudes, the pressure reached in the uppermost parts is often enough to stop the flow in the lower galleries, sometimes causing the submersion of the lower zones (e.g., the Saône Marsh, the Orgelet Basin) or even inversions of the direction of circulation (as in the well of La Brême, near Ornans and Doubs), which is sometimes absorbing and sometimes emissive.

Some borings (e.g., at Novillars, Chalèze and Metabief) have obviously revealed deeply submerged artesian networks which give very different yields according to the local fissures. The synclinal axis often forms the preferential drains. Among the famous resurgences in the Jura is the resurgence of the Loue River, the origin of which (losses of the Drugeon and Doubs rivers) was demonstrated on August 11, 1901, when the Pernod factory in Pontarlier burned and a large quantity of absinthe was discharged into the Doubs River, traces of which were found the following day in the Loue River.

In conclusion, the origin of the karst phenomena in the Jura is linked to the complexity of its tectonic features and to long continental evolution.

Towards the northeast, following the Jura Mountains, Burgundy belongs to the

Paris Basin. However, its western edge is on the Massif Central (Morvan), and the structure was affected (tectonic blocks) by the deformation of the Hercynian substratum lying under Mesozoic and Tertiary deposits (Plaine de la Saône). Surface karst phenomena are almost absent there because of the very important soil cover and vegetation. On the other hand, there are dry valleys, losses and resurgences (generally contact springs, *de débordement*) at the point of contact between limestones and clayey marls (Liassic or Callovian–Oxfordian); the spring of the Laigne is an example. The pergelisol played an important part there during the Quaternary Glacial Stages by decreasing the permeability of the limestones and by developing "cutaneous karst" *("karsts cutanés")*—see CIRY [1959, 1963].

Alpine karst and Provence karst

We shall group together in this chapter the so-called Alpine karst, the pre-Alpine masses (the most important), and the Provence masses which extend towards the southwest and south.

Bordering them on the west, the inner crystalline massifs of the Alps (massifs of Belledonne, Pelvoux, Mercantour) enclose the "high Alpine" karst (Haute-Savoie, Devoluy), the pre-Alpine karst of the mass of the Grande-Chartreuse, the karst of the Vercors mass, the Diois, Ventoux, Vaucluse and Luberon karst, the karst of the zone of Barres and of the pre-Alps of Nice. These latter join the Provence karst of the zône des Plans and of the properly called Provence masses (mass of the Sainte Victoire, Sainte Beaume, Beausset, etc.). They are bordered on the south by the ancient Maures and Esterel masses and by the sea.

Lithostratigraphy

Transgressing upon the granites, gneiss and micaschists, rhyolites and ancient masses are:

(1) The Triassic (mainly Middle and Upper Triassic) composed of grey limestones, dolomites, *cargneules*, gypsum, salt, marls and clay which prevail in the upper part. As a whole the formation appears as an impermeable unit for the overlying formations, but it contains some small karst or pseudo-karst systems and rather frequent springs at the bottom of the limestone, dolomitic, or gypsiferous beds.

(2) The essentially calcareous and dolomitic Jurassic in the south and east (Provence facies) but changing into marly limestones and black marls for all levels except the upper one, the Tithonian, towards the northwest (Dauphinois facies). In the Provence facies the Lower Jurassic (Liassic) is reduced almost everywhere to Rhaetian and Hettangian dolomites and is not separated from Middle and Upper Jurassic by a marly Liassic as in most other regions of France. The Jurassic on the whole may therefore be entirely karstified in all the zones in which limestone and dolomitic facies prevail.

(3) Cretaceous rocks are principally of marly and marly limestone facies, some-

times even of pure limestone, occurring as small reefs both in the lowermost Cretaceous (Neocomian) and in the Upper Cretaceous (Turonian). However, one must distinguish: (*a*) the Provence zone situated southwest of the Donzère–Castellane line in which the marly Barremian limestones change into coralline Urgonian limestones, the white cliffs of which give the Provence masses their characteristic appearance; and (*b*) the sub-Alpine zone situated in the north of the Vocontian region where the Urgonian facies, the cliffs (sometimes more than 300 m high) of which crown the "sub-Alpine border" (Fig. 9A), are still better developed.

(*3*) Tertiary rocks include sandstones or marly limestones followed by marls, schisto-sandstones and sandstone flysch (Annot Sandstone) from Eocene and Oligocene times. The Miocene rocks include the molasse facies. The Pliocene marly or sandy facies is overlain by the alluvial and conglomerate Quaternary facies.

These Cretaceous, Tertiary and Quaternary formations, except the Urgonian limestones, are non-karstified on the whole. However, there are some small caves and some little underground networks in calcareous horizons (Neocomian, Turonian, Lutetian) and some pseudo-karst forms in conglomerates.

In conclusion only the Urgonian facies of the Jurassic and Cretaceous formations in the Alps and Provence are highly karstified.

Structure

The calcareous strata in the Alps and Provence may be either tabular and more or less horizontal (zône des Plans), or somewhat tilted or folded, or inserted into complex imbricated structures or overthrusts.

Where the strata are steeply tilted, the outcrops are reduced to relatively narrow rocky ridges between impermeable soils, and the development of underground networks is limited (e.g., the left bank of the Var River).

On the other hand, where the majority of strata are moderately tilted or tabular karst phenomena may be well-developed (e.g., nappe de Marguareis, zône des Plans) because of the relatively high altitude gradient of the strata from the base levels.

Karstification phases

There are two zones of karst: (*a*) the low altitude zone of the Bas Dauphiné and especially the Provence, where traces of old karstification can still be seen; and (*b*) the high altitude zone where the Late Quaternary (up to the present day) erosion is almost entirely responsible for all the phenomena that can be observed today, or has more or less completely altered the previous phases of karstification.

In the first of these zones appear most of the Provence karst forms. Thus in the zône des Plans, arid plateaus (average altitude 1,100 m) follow one another in waves, overthrusting to the south and joined by lines of cliffs facing south, the dips of which are generally to the north. There are swallow holes or dolines filled with Miocene deposits in situ, demonstrating the presence of pre-Miocene karst. MENNESSIER [1961] for instance, has shown that in the zone of Callian–Montauroux (Var) a karstification had developed at the same time that a pre-Vindobonian surface was forming subsequent to Provençal tectonic deformations (Lutetian–Bartonian). During the Mio-Pliocene this surface underwent the second tectonic Alpine phase when it was dumped, folded and transgressed, while the drainage was reversed and a new karstification (Pliocene) developed. This karstification, which continues to the present, formed dolines, small poljes, closed hollows, avens more or less lined up on joints, faults or borders of impervious beds.

In the second of these zones, at high altitudes (1,500–2,500 m), the Quaternary and present climatic and topographic conditions are important. The scarcity of vegetation and the corrosive power of water resulting from the melting of snow and glaciers caused the formation of spectacular lapies with many dolines, snow-wells and gulfs leading to networks the vertical and longitudinal development of which may be considerable.

Main characteristics

High Alpine karst. With the karst of the Grande Chartreuse Massif the high Alpine karst constitutes what E. A. Martel called the "country of lapiaz". It developed where limestones outcrop either in structural surfaces of panels or folds, or in crests (either in frontal thrust outliers or tectonic windows).

From north to south are:

(1) The Platé Desert (average altitude 2,500 m) with Eocene limestones eroded by the dissolution and which dominates from 2,000 m the Sallanches Basin northwest of the Mont Blanc.

(2) The Parmelan Plateau (average altitude 1,700 m), east of Annecy, famous for its lapiaz with its "snow gulfs" and the resurgences at the bottom of the peripheral walls which begin to yield abundant water when the snow melts.

(3) The karst in Devoluy, the synclinal structural mass of which (average altitude 1,600 m) is characterized by Upper Cretaceous (Senonian) and Neocomian fissured limestones, which are highly inclined towards the north and lie with discordance upon the Upper Jurassic and which swallow rains and melted snows in abysses and crevasses [MARTEL, 1933].

One may see dry valleys and gulfs, which are called *chouruns* in this region, doubtless because of the Saracen incursions there (chourun is the Arab word for gulf). There are many chouruns on the plateau of Aurouze, in the Lure Mountain, on the eastern side of Costebelle (2,380 m) and on the eastern side of the Grand Ferrand (2,761 m) where there are sometimes small underground "ice caves" and "glaciers". The underground water emerges as resurgences and in springs all around the mass.

A whole series of small masses are present with calcareous outcrops, deeply karstified. The most spectacular is the Oucane de Chabrières (altitude 2,200–2,400 m) in the district of Gap which stretches in Jurassic limestones over some ten hectares and exhibits one of the most developed lapies in the world (at regular fissured network with fissures as much as 12 m wide and more than 25 m deep measured to the snow that obstructs them). In many absorption hollows can be seen the waters which reemerge more than 300 m below by means of numerous resurgences.

Lastly, in the Alps above the region of Nice is the Marguareis at an altitude of more than 2,000 m with innumerable lapies, dolines, and snow-holes.

There are many dissolution hollows in the Briançonnais Gypsum.

Pre-Alpine karst in the Grande Chartreuse. Between Chambéry and Grenoble the pre-Alpine mass of the Grande Chartreuse is separated from the high Alpine masses by the Graisivaudan Depression. At a lower altitude (1,500–2,000 m) with a wetter climate, it has a common structure (Fig. 9A). Mainly Urgonian and Upper Jurassic calcareous outcrops form imposing cliffs and determine karstified surfaces which are particularly developed in the region of Les Echelles in the high valley of the Guiers River with the Trou du Glaz network and its subsidiaries (the cavern of the dead Guiers and resurgence of the live Guiers in the Sainte-Même cirque). This network seems to be fed only by condensation water during the dry season [CHEVALIER, 1946].

There are traces of karstification and of ancient surfaces present prior to the deepening of the Graisivaudan Valley.

Pre-Alpine karst in the Vercors. Continuing towards the southwest on the Grande Chartreuse Massif beyond the cluse of the Isère River, one reaches the Vercors, the crests of which are more than 2,300 m and which has an average altitude of about 1,600 m constituted of short-radius curved folds, especially undulating. The main part of its surface is composed of structural outcrops of Urgonian limestone more than 400 m thick. Measuring about 1,000 km^2, it is the largest block of karstified limestone in France and probably in Europe (Fig. 9B). Here CORBEL [1956] distinguished the eastern and western karst.

(*A*) Eastern karst. The precipitation is very great (over 1,500 mm) in this region, and the snow lingers for a long time, principally in the northern part. The surface of limestones is more or less timbered but has no streams and is covered with lapies and dolines interrupted by networks of small dry valleys which always lead to large gulfs that the peasants here call *scialets*. One of the famous scialets is the Combe d'Enfer where the valley through an imposing porch penetrates the limestones as a tunnel. The infiltration water which filters into the tunnel runs down and concentrates in the troughs of synclines. The widened fissures of lapies are generally about 10 m deep and 2 or 3 m wide (Fig. 10A). According to CORBEL [1956], who made a study of this region in 1956, when going down into those fissures, one can almost always perceive at the foot of the wall at the bottom the beginning of a gallery by clearing a joint of stratification. The dolines here are often huge. Some have a diameter of more than 100 m and exhibit gaping gulfs (scialets) over 100 m deep at their bottom. When the snow melts, stupendous torrents of CO_2-rich and particularly "agressive" water fall into those gulfs, causing intense mechanical erosion in addition to the dissolution. It is in the

Fig.10.A. Sornin lapies (Vercors). B. Vaucluse fountain (Sorgue Spring, Vaucluse).

heart of the Vercors that J. Berger discovered in 1953 at an altitude of 1,460 m the opening of the gulf which bears his name and which happens to be the second deepest in the world (Fig. 9B, 11B). Crossing first the Urgonian, more than 250 m thick, this gulf indeed leads to an underground river which meanders along several kilometres in a huge gallery hollowed out in Hauterivian marls and which issues at the resurgence of the Cuves de Sassenage (altitude 290 m) only 5 km from Grenoble. The evolution of the whole of the eastern karst of the Vercors appears to have occurred very recently and seems to date only from the Quaternary. During glaciation the *tjäle* (permafrost) reduced the permeability of the calcareous surface which may have caused the formation of small valleys which nowadays are dry. The lower networks to which the gulfs lead have not had time to develop sufficiently to be large enough to be explored by man, and the underground reservoirs have a comparatively small capacity. Nevertheless the flow of resurgences which emerge at the bottom of the periphery of calcareous blocks is large, and the carbonate content shows [CORBEL, 1956] that the average erosion of the region must correspond to the removal of 0.10 mm/year of calcareous stratum.

(*B*) Western karst. Formed upon three large parallel synclinal depressions interrupted by the Gorge de la Bourne, this is the region of important underground rivers and of closed depressions.

The most western synclinal depression, that of the Basin of La Lyonne, shows beautiful gulfs (such as that of Ambel) and important underground rivers. The valleys [CORBEL, 1956] there offer an alternation of narrow gorges (calcareous strata) and wide valleys (non-calcareous strata). Thus we have a continuation of perched basins joined by very steep canyons. Those depressions, partly covered with Miocene alluvium, give the impression of a topography of poljes recently joined together by the narrow gorges.

The second synclinal depression, that of the Brudoux (in the region of the Forêt de Lente), also shows numerous dolines and snow-holes feeding the underground Brudoux River which drains the synclinal axis and concentrates the waters of *pots* (crater-form dolines) of the forests covering the anticlinal zones. The underground Brudoux River emerges along 3 km in a closed depression, disappears again, and resurges at the bottom of the impressive backing-space of the Cholet by a famous waterfall.

Downstream the karst dissolution was not sufficient to form passages large enough to be explored.

Fig.11.A. Subalpine ledge. 1 = Marls; 2 = muddy limestones; 3 = zoogenic limestones; 4 = sandstones; M = Maastrichtian; G = Albian; U = Urgonian; H_2 = Upper Hauterivian $H1g$ = Lower glauconitic Hauterivian; V_3-V_1 = Upper–Lower Valanginian; P = Portlandian K = Kimmeridgian; S = Sequanian; R = Rauracian. (After GIGNOUX, 1960.)

B. Schematic east–west section of the Vercors. (After DE MARTONNE, 1942.)

C. Vertical section of the Vaucluse fountain. (After MARTEL, 1933.)

D. Vertical section of the Lez source (Hérault)—vauclusian type.

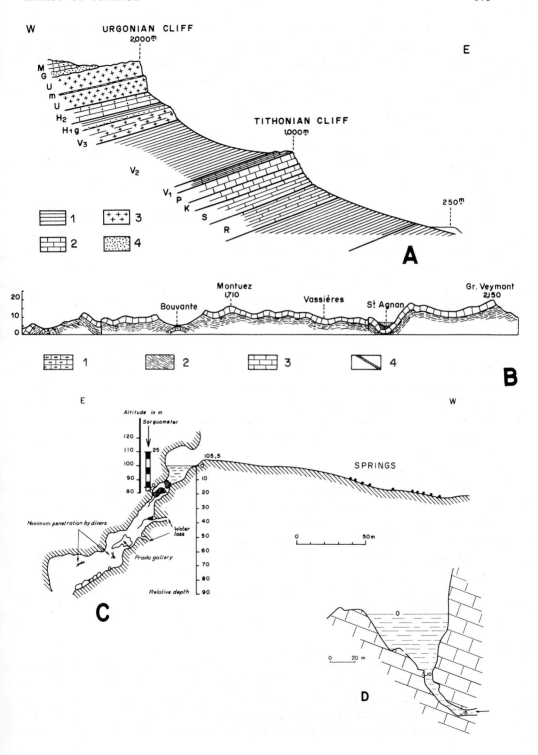

The third synclinal depression is drained by the Vernaison River. Here the rains are less heavy and the climate drier. The dolines are numerous, but gulfs are less important. The surface stream of the Vernaison, flowing on an impermeable covering of Gault, Miocene and morainic deposits, loses a part of its water which form the underground stream of the Bournillon while the other part flows through the gorges of the Vernaison. In the crevées de la Luire there is a temporary rise (at flood time) of the waters of the underground Vernaison, through an open fault drawing away spherical boulders that it moves all around the resurgence. This rise of more than 300 m shows the extent of the loads that can affect the karst channels in these regions.

The underground Vernaison opens in the double resurgence of the Arbois-Bournillon. The roof of the entrance of the Bournillon (Fig. 12), certainly one of the highest in the world, is over 100 m high. The flow of water from here can increase from zero to over 40 m^3/sec within a few hours. The gorges of the Bourne River, which crosses the three preceding units, have evidently led to the discovery of important underground resurgences such as those of Arbois-Bournillon, mentioned above, and the Goule Noire, the volume of which can exceed that of the Bourne.

There are many others such as the Goule Blanche, the Goule Verte and the Grand Réseau des Merveilleuses de Valchevrières with its gulfs and fossil galleries with many speleothems. An extraordinary karst area of the Vercors, the Presles Plateau, on the right shore of the Bourne River, although not exceeding 50 km^2, contains an exceptional group of gulfs and underground rivers (the Chevaline, Goufin, Gournier and Jallifier rivers, etc.). The waters of the latter which are cold and aggressive at their source, deposit calcareous tufa where they issue.

The region of Diois is included with those south of Les Barronies and Mont Ventoux.

Karst in the regions of Diois, Les Barronies, Mont Ventoux, Vaucluse Plateau and Mont Luberon. Continuing south, the pre-Alps of the Vercors, and the regions of Diois, Les Baronnies and Mont Ventoux show the increasing influence of the Pyrenean field of tectonic forces which combines here with the Alpine field of tectonic forces. Therefore in this area are found anticlines crushed in an east–west direction between troughs of synclines projected by inversion of relief, with ellipse-shaped crests having their major axes in the same direction. The north–south Alpine direction can be seen especially in the networks of faults which interrupt the structural units. We shall note, for example, the meridian network of faults which cuts the slope of the mass of the Mont Ventoux (culminating around 2,000 m) into east–west crests.

Moreover in this zone, when coming from the north, we enter the area of the fosse vocontienne (Vocontian Trough) and its marly facies which cause a considerable reduction or even the complete disappearance of the Urgonian facies. The latter is well developed only from the south edge of this trough, i.e., south of the Ventoux in the Vaucluse Plateau, in Mont Luberon, on the southern border of the mountain of Lure and in the Provence masses themselves farther south. Therefore the karst is

Fig.12. Great arch (100 m high) and outlet lake of the Bournillon subterranean river (Vercors). (After MARTEL, 1933.)

very different from the Alpine karst we have just described.

Only slightly developed in the Diois, it is well developed only south of the north side of the Mont Ventoux, particularly in the famous Vaucluse Plateau. At the foot of this plateau is the famous fountain of the same name which is the even type of an ascending spring, having a variable but important flow called "vauclusian", and situated at the foot of a cliff more than 200 m high (Fig. 10B, 11C) 22 km east of Avignon. It issues in a rocky cirque setting, bounded on all sides up-stream, at the head of the Sorgue River valley which is formed by the spring. That is the origin of its name (vallis–clausa). This perennial spring has an average flow close to 30 m³/sec, fluctuating from 4 m³ to over 150 m³/sec. The maximum volume is probably the largest of any limestone spring in the world.

The problem of the drainage basin of this spring and of its hydrogeology has long formed the subject of numerous speleologic and hydrogeologic studies. The first board of studies was appointed in 1873. Presently a concerted action bearing on the period 1966–1967 has been undertaken under the auspices of the D.G.R.S.T. (Délégation Générale à la Recherche Scientifique et Technique) and the B.R.G.M. (Bureau de Recherches Géologiques et Minières) in conjunction with the Universities of Marseille and Lyon. As far as we know now, it seems that the spring originates from the infiltration of water from the outcrops of Urgonian limestones of the southwest through brachy-syncline, the north side of which consists of the south flanks of the Ventoux and Lure Mountain and of the plateau of Vaucluse; the south side is formed by Mont Luberon.

In the axial part of the syncline, Urgonian limestones are underlain by the marly impermeable Neocomian formation and are overlain by impermeable marly Aptian and Albian formations which are covered by Tertiary deposits.

The axis of the brachy-synclinal structure slopes to the west and causes the flow to move in that direction, away from the areas of infiltration. The lateral flow is diverted by relatively impermeable Tertiary (Miocene molasse), faulted down against the limestone, and comes out at a practically unique emergent point, the fountain of Vaucluse. The emission eroding the downstream lip of the exit basin would have caused its lowering to the present position. The probable drainage basin is about 2,000 km². It contains numerous avens, gulfs, gorges and places of water loss. Among them are the avens of Caldaire ($H = 885$ m) and of Jean Nouveau ($H = 831$ m), which are nearly 500 m deep and the aven of Castor. Notable water losses are those at Calavon, Coulon and Nesque. Dye tests made in 1963 by P. Renault, H. Paloc and the Speleologic Federation of Vaucluse in the Nesque loss were the first to show that this stream emerged at the fountain of Vaucluse after a distance of 22 km with an average velocity of 26.20 m/h. Several of the avens of the Vaucluse Plateau become obstructed, then suddenly clear. This occurs for example, in the region of Sault following heavy rain or snowfall.

Provence karst. The sedimentary region of the Provence includes two distinct areas from a geologic point of view: western Provence where the folds are Pyrenean in direction and age; and the Provence pre-Alps to the west, where the folds are of Alpine direction and Miocene age.

(*A*) *Western Provence.* In the western part a succession of small east–west masses dominates the landscape, such as the small range of Alpilles, of Sainte-Victoire, la Sainte-Beaume, of the Nerthe and the Etoile. In those complex tectonic structural massifs, the deeply enclosed strata of calcareous dolomitic Jurassic and calcareous Cretaceous (Urgonian) are raised or overturned and outcrop extensively at altitudes ranging from 100–1,100 m, giving rise to karst phenomena which are only slightly developed compared with those in the calcareous pre-Alps.

Here we find lapies, closed basins, gulfs and resurgences. The greatest among the closed basins is that of Cuges (south of the Sainte-Beaume Massif) which has absorbent fissures and gulfs in its bottom (e.g., the gulf of La Roque). Two smaller basins, the Saint Cassien which probably feeds the resurgence of the Belle Foux de Nans and the Plan d'Aups, are submerged from November to March. In these basins the absorbent gulf of the Tourne (altitude 665 m, depth 25 m) must, according to MARTEL [1921], correspond to the resurgent spring of the Hubeaune and to its two overflow caves, while in the west of the same basins other *embuts* seem to be joined to multiple water exits of the Saint-Pons park (east of Géménos). Among the other closed basins is the Agnis in the Sainte-Beaume Massif.

At the top (994 m) of the Saint-Pilon summit in the Sainte-Beaume there is a *garagai* (gulf in the Provençal language), the garagai of Gaspard de Besse. Near the top of the mountain of Sainte-Victoire there is also another garagai with a double opening and a natural archway.

The dominant position of these massifs is favourable for the development of lapies in ruiniform landscapes (e.g., the aiguilles de Valbelles, 20 km southwest of Brignoles), of absorbent gulfs and of underground streams leading to peripheral resurgences or forming various rivers in gorges (e.g., the vauclusian emergence of La Grand Foux, the waters of which come from the Saint-Cassien lapiaz with dolines in the eastern part of the Sainte-Baume Massif). The Alpilles and their calcareous heights also show a whole succession of emergences, locally called *laurons*, on their circumference.

The coastal massifs and their Urgonian cliffs also exhibit different karst phenomena, for example, north of Toulon, the *rascles* ("lapiaz" in the Provençal language) and the *ragas* ("abyss" in the Provençal language) of Urgonian outcrops above the Dardenne cirque.

In this cirque is the resurgence of the Foux de Dardenne (altitude 102 m) and the famous ragas of Dardenne which opens in the cliff dominating this *foux*, 47 m above and 600 m to the southwest. This ragas corresponds to an almost vertical chimney, which during high waters acts as an emissive gulf.

The gorge-like depressions in these coastal massifs abound near the sea and form the famous *calanques* of the Marseille coast, for example the calanque of Port Miou and its famous submarine resurgence. In the neighbourhood of Marseille, caves and caverns are known such as the Beaume-Loubière, the Monard Grotto and the gulf of Carpiane etc. In the Provence important underground streams and garagai have frequently been encountered, for example in the lignite mines in Fuveau Gardanne and in bauxite mines in the region of Brignoles.

(*B*) Provence pre-Alps. The Provence pre-Alps are essentially composed of an arc of small ranges and of overthrust blocks more or less separated from their substratum at the Triassic level. In this arc, the enormous mass of calcareous Lias disappears progressively from north to south and the Mesozoic Series decreases in thickness from more than 4,000 m to less than 500 m. Their boundaries to the west are the western Provence and the pre-Alpine folds of Baronnies and of Ventoux, to the east the Mercantour and the Hautes Alpes, and to the south the Provence, the Estérel Massif and the sea. To the north they give way to the Devoluy.

Erosion has exposed the hard masses and especially the Jurassic limestone formations which, in crests *(barres)*, in anticlinal arches or in plateaus outcrop with more or less erected edges *(plans)*.

The folds, the barres and the plans, more or less intensively karstified, are interrupted by canyons or deep gorges (called *clues* where they are particularly narrow) by the valleys beginning in the Alps, such as the Verdon (tributary of the Durance), the Artuby (tributary of the Verdon), the Loup and the Var.

The canyon of the Verdon, which is about 50 km long and deeper than 900 m between steep or overhanging cliffs, is justifiably famous (Etroits, Gorge Beaumes-Fères des Passe de l'Imbut and its rocky grating). In 1905, E. A. Martel and his crew first descended into it where the erosion appears to be particularly rapid.

The Upper Jurassic barres and plateaus (between 700 m and 1,100 m altitude) above the floor of the canyons have well-developed karst phenomena favoured by the drainage of the canyons themselves.

The plans include the Plan de Bauduen-Majastre, the great Plan de Canjuers, the small Plan de Canjuers, and the Plan de Caussols. On all these plans are lapies (rascles), dry valleys, and hundreds of gulfs (e.g., gros aven in the great Plan de Canjuers, over 90 m deep; the aven of the Nouguière; the aven of the clos del Fayoum, over 150 m deep). NICOD [1965] made detailed studies and land surveys of karst phenomena in this area (Fig. 13).

On the periphery of these massifs and plans, resurgences can be observed such as the famous l'Evêque Fountain (or Font de Sorps) flowing from Jurassic limestones at an altitude of 410.30 m at the northern foot of the Plan de Bauduen, at the downstream outlet of the depression of Les Salles, just before the beginning of the downstream gorges of the Verdon, 600 m from the left bank of this river. Its flow varies from 2.5 m^3/sec to over 15 m^3/sec.

Dye tests have shown that this system is connected with the losses of the Artuby, a tributary of the Verdon running 30 km from the fountain, with some losses of the Verdon itself and with some avens of the plans de Majastre and Canjuers. The relatively low temperature (12.8°C) confirms tests showing that the water enters the limestone at high altitudes.

After the Verdon Valley project is completed the resurgence of l'Evêque Fountain will be covered by the water contained by the dam of Les Salles (about 80 m high). Some important observations will certainly be made on the effects of this increase of "outlet pressure" on the underground circulation in the area.

To the west of the plans zone is the zone of the barres and the pre-Alps of Nice where one finds less developed (owing to the reduction of calcareous outcrops)

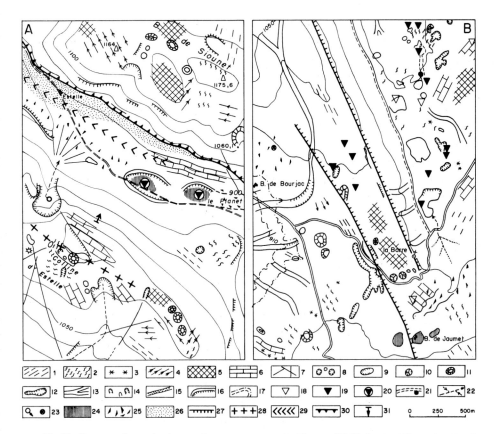

Fig.13. Extracts from the karst phenomena maps in NICOD (1965) in the Plans du Verdon. A. Region of the Bois de Siounet and the colline d'Estelle. B. Region of the small Plan de Canjuers and its border towards Bourjac. *1* = Normal lapies; *2* = small lapies; *3* = ruiniform lapies; *4* = sharp-crested lapies; *5* = crossed each other lapies; *6* = table of lapies; *7* = fissures and joints; *8* = dolines; *9* = dissymetric doline; *10* = circ-like doline; *11* = trough-like doline; *12* = stretched doline; *13* = dry valley; *14* = open dolines; *15* = canyon; *16* = polje (morphologic contour); *17* = polje (closed morphologic curves); *18* = obstructed aven; *19* = aven; *20* = ponors; *21* = underground network and inner wells; *22* = temporary stream; *23* = spring, well; *24* = terra rossa; *25* = cone with angular material; *26* = non-karstifiable covering; *27* = known throw fault; *28* = anticlinal axis; *29* = synclinal axis; *30* = overthrust; *31* = dip. (After NICOD, 1967.)

karst phenomena but nevertheless gorges and impressive clues. Examples include the clue of Trigance, the clue of Daluis (on the Var), the clue of Riolans only a few metres broad and 200–400 m high, the clues or gorges du Loup, and the clue of Sisteron on the Durance.

From hydrogeologic and karst points of view, the Jurassic plays the essential part in all of the karst of this region. The series is characterized as follows by CREAC'H [1967] in the speleologic inventory of the Maritime Alps:

"From top to bottom, Portlandian (J^{8-6} of geologic maps in 1 : 80,000) generally forming the surface of the high plateaus and of the Cheiron, with masses, lapies but few gulfs. In some places in a chalk zone there are meanders. The explorer in his downward progress is almost always stopped by highly dolomitic Kimmeridgian, which produces usually impenetrable splits or narrows. The Sequanian–Rauracian forms some lapies and is a zone of holes. More marly calcareous Argovian–Oxfordian and Callovian are pierced by small clayey gulfs or caverns and with irregular sections; few large galleries such as the one at Revest can be seen. Bathonian composes the cliffs bordering the plateaus. There are numerous gulfs which end near Bajocian, generating some horizontal cavities. Finally it is almost always in Lias, at the base of Hettangian or at the base of Rhetian above Triassic marls that underground waters reappear, often as important rivers but almost always showing wetting vaults *(vôutes mouillantes)* near their outlet.

Apart from Jurassic karst there are small cavities in Neocomian, Turonian, Lutetian and calcareous horizons and pseudokarst forms (broad fissures, rocky shelters) in the Oligocene sandstones, in the calcareous molasses (for example in Tourette sur Loup) or in the Miocene to Quaternary conglomerates".

We have seen above in the karst zone of Callian–Montauroux (Var) how intricate the evolution of all these karst systems was all during the Tertiary and Quaternary, with the special feature of a particularly rapid and important tectonic evolution of this region.

CONCLUSIONS: SCIENTIFIC AND PRACTICAL INTEREST, RESOURCES AND EXPLOITATION OF FRENCH KARST

Scientific interest

From the scientific point of view in France, as in other karst countries, karst affords an opportunity for interesting scientific studies in the fields of physics (water circulation, cavitation problems, etc.). Chemistry (dissolution and deposition problems, etc.), and biology (nature of underground flora and fauna, evolution and adaptation of fossil species, physiological problems, etc.). In the same way, karst studies are important from a geological point of view (erosion problems, the relation of karst phenomena to lithology, tectonics, historical geology, etc.), especially in regard to hydrogeologic problems (problems of water resources, of the structure and operation of karstic aquifers, etc.).

We shall note here only the main fields in which French karstologists have been able to obtain original results in recent years.

Geologic and hydrogeologic problems

From a geologic and hydrogeologic point of view preceding sections show that France contains an unusually broad variety of lithological, structural, climatic, palaeoclimatic and karstification phases conditions. The main known karstifications were during Carboniferous, Cretaceous, Eocene, Oligocene, Miocene, Pliocene and

Quaternary times in addition to the karstification continuing to the present.

Corresponding climatic episodes interfered with erosion and changes of base levels and drainages linked to orogenic and epeirogenic movements. The result is that often, notably in areas with a complex tectonic evolution, there exist *polygenic karst*, the history of which is often unsuspectedly complicated. However, the reconstruction of the chain of events is of great importance for the general geologic history of France, especially since the Cretaceous.

Systematic studies in the field of karst phenomena are carried on by "COMITÉ NATIONAL FRANÇAIS DE GÉOGRAPHIE", by the C.E.R.H. and by the B.R.G.M. These led to the publication of the first karstologic map in regular sheet (*"Hydrogeologic Map of the Languedocian Karst"* by PALOC, 1964; and *"Karst Phenomena Maps in Provence"* by NICOD, 1965]. GEZE [1958, 1965a] gave excellent discussions about water penetration and circulation problems in karst systems.

Erosion and concretion mechanisms

Among specific studies of karst in France, we should cite those covering erosion and deposition processes in karst.

As for carbonate dissolution problems, we should mention: CORBEL's [1954, 1956, 1957, 1965] studies (action of cold water, melting snow water and ice); AUBERT's [1960, 1965, 1967] studies about quantitative aspects of karstic erosion phenomena; AVIAS' [1956, 1966] studies about polarized dissolution and concretion phenomena in calcareous formations with a discontinuous structure and in limestone pebble alluvial deposits; the studies by POBEGUIN [1957], SCHOELLER [1965a], CARO [1965], ROQUE [1965] and others of the chemical processes of limestone dissolution by carbonic acid rich water.

As for the deposition phenomena, studies by GÈZE and POBEGUIN [1962], among others, on the mineralogy and genesis of carbonate deposits in caverns include calcite formations of stalactites, stalagmites, curtains, floating calcite, calcarenitic sands, calcitic bubbles, sidewalks, *gours*, pearls of caverns (muriform, cubic, disc-like, top-like, etc.), budding forms (*choux fleurs* or cauliflowers), *plumets* (plumes), excentric concretions, aragonitic *houpettes* (powder puffs), *sapins d'argile* (clayey firs), *fleurs de gypse* (gypsum flowers) (Plate II).

The minerals are essentially calcite, but may also be aragonite, dolomite, huntite, hydromagnesite, nesquehonite, giobertite, tricalcic phosphates or alumina phosphates (minervite), gypsum or epsomite.

Biologic and prehistoric questions

In France special attention is also given to the biological study of the underground environment (microflora, microfauna, fauna) (A. Vandel's laboratory in Toulouse, the Underground Laboratory in Moulis, Pyrenees, part of C.N.R.S., and so on).

We must mention, among others, the exceptional interest of caves in continental palaeobiology. Caverns constitute shelters, traps, fossilization places for living (or dead) beings, including man, who penetrate them. Deposits and remains preserved in caverns are most valuable in reconstructing the early history of mankind.

France is particularly rich in famous prehistoric underground sites, some of which gave their names to stages of prehistoric man's evolution (the caves of Moustiers, Aurignac, La Madeleine, Solutré, etc.).

Practical interest

Tourism

From an economic and practical point of view, karst formations are generally poorly suited for agricultural purposes, because they are often covered with forests in the north of France. They are usually covered with only slight vegetation, or are bare in high mountains and in the southern zone of the country.

As tourist attractions, however, they are regions of exceptional interest. In France, there are at present only 26 caverns open to the public; among them are: the Aven Armand (Aveyron), Betharram Grotto (Hautes-Pyrénées), Clamouse Grotto, Demoiselles Grotto (Hérault), Grand Roc Grotto and the prehistoric site of Laugerie Basse (Les Eyzies, Dordogne), Lascaux Grotto (Dordogne), Aven d'Orgnac (Ardèche), Gouffre de Padirac (Lot). Some karst sites, such as the gorges du Verdon, de l'Ardèche, du Tarn, the ruinform chaos in Montpellier-le-Vieux, and the Bois de Paiolives, are more and more frequently visited.

Water supply

However, the main interest of these formations is that they contain the most important under-developed freshwater resources in France. The water supply for the industrial zone in the northern part of France comes almost entirely from Carboniferous limestones and chalk.

In the southern part of France relatively bare karsts constitute particularly efficient water traps and reservoirs under Mediterranean climatic and vegetal conditions (rare but intense rainfall and slight ground covering). They contain the most important water resources of the French Mediterranean.

Most of the time only overflow is used neglecting the water stored underground. Among the studies made for more detailed survey and development of the stored water are, in Languedoc, the works of the C. E. R. H. (J. Avias and C. Drogue), and in Provence, the works of the B.R.G.M. In the Languedoc, pumping from the vauclusian spring of the Lez River, begun in 1967, will furnish the city of Montpellier with its entire water needs, without using polluted, treated Rhône River water. Water obtained from underground rivers in Larzac will allow some abandoned regions to flourish again. In the Reg d'Argent in the Massif de la Clape (Aude), the first French attempt to store water underground using an underground dam has been made. Several avens have also been used (e.g., Aven de la Soeur near Sauve, Hérault) for catching water to fill the drinking water supply of nearby communities.

Several dams in karst fields have been built or are being built in France to effect either storage or *écrêtement de crue* (flood control). These include the famous Génissiat dam on the Rhône River, a short distance from branch to Lac Léman, completed in 1948, the Charmine-Moux and Vouglans dams in the Jura (Plate IF) and the Ceyrac, La Rouvière and Conqueyrac dams in Languedoc.

Studies have also been undertaken concerning the problem of collecting the freshwater emerging as submarine springs (e.g., the work of the C.E.R.H. in Fontdame, of the B.R.G.M. in the spring of Les Abysses, Languedoc, and in the Port Mion resurgence, Provence).

Further aspects

We should also mention the present prospect of some karst systems isolated by impermeable beds in the Languedoc and the Provence able to store petroleum products. From the military point of view, the advent of the atomic age gave a new interest to the underground cavities. These have always been used by men as shelters against bad weather and adversity, but their interest has never been as great as it could become in the case of an atomic conflict. The underground world could then offer almost the only escape from destruction and atomic contamination. As GÈZE [1965b] recently wrote: "Espérons tout de même que les cavernes qui ont été à l'origine de l'apparition de l' *Homo sapiens* sur terre, ne seront pas un jour les derniers lieux qui assisteront à la disparition de l'espèce": Let us hope, anyway, that the caverns which were present at the appearance of *Homo sapiens* on earth will not one day be witnesses of the extinction of the human species.

REFERENCES

AIGROT, M., 1967. L'écoulement de la fontaine de Vaucluse (étude hydrogéologique). *Trav. Lab. Hydrogéol. Géochim., Fac. Sci., Univ. Bordeaux*, 15:97 pp.
ANONYMOUS, 1964. Liste des clubs (de spéléologie) membres de la Fédération Française de Spéléologie. *Spelunca, Bull.*, 4(1):42–43.
ANONYMOUS, 1967. *Le Laboratoire Souterrain de Moulis (Ariège)*. Centre National de la Recherche Scientifique, Paris, 54 pp.
ARCHAMBAULT, G., 1967. Les dépressions fermées au nord-est d'Orléans. *Bull. Inform. Géol. Bassin Paris*, 2:273.
AUBERT, D., 1960. Structure, activité et évolution d'une doline. *Bull. Soc. Neuchatel. Sci. Nat.*, 89: 113–120.
AUBERT, D., 1965. Calotte glaciaire et morphologie jurassiennes. *Eclogae Géol. Helv.*, 58(1):555–758.
AUBERT, D., 1967. Estimation de la dissolution superficielle dans le Jura. *Bull. Soc. Vaudoise Sci. Nat.*, 69:366–374.
AVIAS, J., 1956. Sur des phénomènes de corrosion et de concrétionnement polarisés de galets des formations alluviales et de calcaires en plaquettes de l'est de la France. *Bull. Soc. Géol. France*, 6:275–288.
AVIAS, J., 1963. *Méthodes d'études et Mise en Valeur des Ressources en Eaux Karstiques du Midi Méditerranéen Français*. Food and Agriculture Organisation (United Nations), Athens, 29 pp.
AVIAS, J., 1965. Sur la recherche des ressources en eaux karstiques sous recouvrement mio-plioquaternaire dans le bassin méditerranéen. *Bull. Assoc. Intern. Hydrol. Sci.*, 1965:410–419.

AVIAS, J., 1966. Sur l'utilisation des phénomènes de dissolution et de concrétionnement polarisés des formations calcaires en paléohydrogéologie. *Mém. Assoc. Intern. Hydrol. Sci.*, 6:79–80.
AVIAS, J., 1968. Sur la méthodologie en hydrogéologie karstique. (d'après l'exemple du Languedoc méditerranéen). *Mém. Centre Etudes Rech. Hydrogéol.*, 1:5–20.
AVIAS, J. and DUBOIS, P., 1963. Sur la nappe aquifère des karsts barrés par failles du Bas-Languedoc. *Spelunca, Mém.*,3:68–72.
BALSAN, L., 1950. *Grottes et Abimes des Grands Causses.* Fédération Française de Spéléologie, Millau, 150 pp.
BARON, G., CAILLERE, S., LAGRANGE, R. and POBEGUIN, T., 1959. Étude du Mondmilch de la grotte de la Clamouse et de quelques carbonates et hydrocarbonates alcalino-terreux. *Bull. Soc. Franç. Mineral. Crist.*, 82:150–158.
BARRERE, P., 1964. Le relief karstique dans l'ouest des Pyrénées centrales. *Rev. Belge Géol.*, 88:9–62.
BIROT, P., 1966. *Le Relief Calcaire.* Centre de Documentation Universitaire, Paris, 236 pp.
BOILLOT, G., 1963. *Géologie de la Manche Occidentale: Fonds Rocheux, Dépôts Quaternaires, Sédiments Actuels.* Thesis, University of Paris, Paris, 220 pp.
BONNET, A., DU CAILAR, J., COUDER, J. and DUBOIS, P., 1962. Le massif calcaire du Mont Perdu. *Congr. Intern. Spéléol., 2me, Bari, 1958, Actes,* 1:135–142.
BONTE, A., 1954. Poche de dissolution, argiles de décalcification (karst dans la craie). *Ann. Soc. Géol. Nord,* 74:67–94.
BONTE, A., 1955. Sur quelques aspects de la dissolution des calcaires. *Comt. Rend., Congr. Soc. Savantes, 80me, Lille, 1955,* pp.109–116.
BOURGIN, A., 1942. *Le Dauphiné Souterrain.* Arthaud, Grenoble, 82 pp.
BOURGIN, A., 1945. Hydrographie karstique. La question du niveau de base. *Rev. Géograph. Alpine,* 33:99–107.
BOURGIN, A., 1946. La Luire et la Vernaison souterraine. *Ann. Spéléol.*, 1:31–42.
BRAQUE, R., 1966. Les mardelles du Nivernais. *Bull. Assoc. Franç. Etudes Quaternaires,* 3(8):167–179.
CAILLETEAU, P., CARTIER, G. and DREYFUSS, M., 1965. Étude géologique et hydrogéologique du bassin fermé d'Orgelet (Jura). *Terres Eaux,* 46:31–33.
CARO, R., 1965. La chimie du gaz carbonique et des carbonates, et les phénomènes hydrogéologiques karstiques. *Chron. Hydrogéol.*, 7: 51–77.
CASTERET, N., 1931. Le problème du Trou du Toro. Détermination des sources du Rio Esera et de la Garonne occidentale. *Bull. Soc. Hist. Nat. Toulouse,* 61:89–131.
CAVAILLE, A., 1953. Le karst des gorges de l'Aveyron. *Congr. Intern. Spéléol., 1er, Paris, 1953,* 2:91–103.
CAVAILLE, A., 1961. Morphologie et karst des Causses du Quercy. *Ann. Spéléol.*, 16:113–131.
CAVAILLE, A., 1964. Observation sur l'évolution des grottes. *Intern. J. Speleol.*, 1:71–100.
CHABOT, G., 1913. Le Revermont, étude sur une région karstique du Jura méridional. *Ann. Géograph.,* 22;224:399–416.
CHABOT, G., 1927. *Les Plateaux du Jura Central.* Thesis, University of Paris, Paris, 351 pp.
CHEVALLIER, P., 1946. Le réseau souterrain de la Dent de Crolles (Isère). *Ann. Spéléol.*, 1:15–29.
CHEVET, B., 1964. Le plateau karstique de Jaut (Basses-Pyrénées). *Soc. Prehist. Bordeaux,* 12/13: 85–90.
CHOLLEY, A., 1923. Évolution du relief karstique du Parmelan. *Compt. Rend.*, 177:276–278.
CHOPPY, J., 1965. Les types de cavités du Vercors. *Spelunca,* 4:50–56.
CIRY, R., 1959. Une catégorie spéciale de cavités souterraines: les grottes cutanées. *Ann. Spéléol.*, 14:23–30.
CIRY, R., 1963. Aperçu des principaux types de cavités souterraines de la Bourgogne, Côte d'Orienne. *Spelunca, Mém.*, 3:73–80.
COLIN, J., 1966. *Inventaire Spéléologique de la France, 1. Le Jura.* Fédération Française de Spéléologie, Paris, 276 pp.
COMBES, J.-P., 1965. Dissolution karstique sous une couche bauxitique. Remarques sur l'origine des gisements en poches. *Compt. Rend. Soc. Géol. France,* 4:123–124
CORBEL, J., 1954. Les phénomènes karstiques dans les Grands Causses. *Rev. Géograph. Lyon,* 29(4): 287–315.
CORBEL, J., 1956. Le karst du Vercors. *Rev. Géograph. Lyon,* 31(3):223–241.
CORBEL, J., 1957. Karsts Haut-Alpins. *Rev. Géograph. Lyon,* 32(2):135–158.

Corbel, J., 1965. Introduction au karst des Pyrénées françaises. *Spelaion Carso*, 2:34–54.
Corroy, G., Gouvernet, C., Chouteau, J., Sivirine, A., Gilet, R. and Picard, J., 1958. Les résurgences sous-marines de la région de Cassis. La Fontaine de Vaucluse. *Bull. Inst. Océanog. Monaco*, 1131:35 pp.
Creac'h, Y., 1967. *Inventaire Spéléologique de la France, 2. Alpes Maritimes.* Fédération Française de Spéléologie, Paris, 187 pp.
Dalmasso, E., 1964. Quelques observations sur les phénomènes karstiques dans le massif du Marguareis. *Méditerranée*, 5(3):257–264.
Daveau, S., 1965. Vallées sèches des plateaux du Jura. *Rev. Géograph. Est*, 5(4):46–47.
Dehee, R., 1926. Note sur le puits naturel dans le terrain houiller de Carvier, résultant d'un karst dans le calcaire carbonifère. *Ann. Soc. Géol. Nord*, 51: 32.
De Lavaur, G., 1953. Hydrologie souterraine du Causse de Cramat (Lot). *Congr. Intern. Spéléol., 1er, Paris, 1953*, 2:87–90.
Demangeon, P., 1967. Premiers résultats d'une étude minéralogique des remplissages karstiques des Causses au sud du Tarn. *Compt. Rend.*, 264:2721–2724.
De Martonne, E., 1933. L'hydrographie du karst. *Ann. Géograph.*, 237:1–580.
De Martonne, E., 1942. *La France Physique.* In: *Géographie Universelle de Vidal de la Blache et L. Gallois.* Collin, Paris, 6:451 pp.
Dreyfuss, M., 1950. Cavités anciennes et cavités récentes dans la chaîne du Jura. *Bull. Assoc. Spéléol. Est*, 3:45–51.
Dreyfuss, M., 1957. Contribution à l'étude hydrogéologique du Marais de Saône. *Ann. Sci. Univ. Besançon, Sér. Géol.*, 2(7):3–20.
Dreyfuss, M., 1967. *Le Karst Jurassien.* 2 pp. (unpublished).
Dreyfuss, M. and Damidot, L., 1957. Observations hydrologiques et morphologiques aux environs de Gennes (Doubs). *Ann. Sci. Univ. Besançon, Sér. Géol.*, 2(7):49–54.
Drogue, C., 1963a. Méthode de détermination de la capacité de rétention par fissures et microporosité des massifs karstiques à partir des variations saisonnières des donnes hydrométriques. *Compt. Rend.*, 256:5377–5379.
Drogue, C., 1963b. Sur l'utilisation des ions naturels comme traceurs dans les aquifères calcaires. *Ann. Spéléol.*, 18:405–408.
Drogue, C., 1964. Contribution à l'étude géologique et hydrométrique des principales résurgences de la région nord-Montpelliéraine. *Mém. Centre Etudes Rech. Hydrogéol.*, 1:62–121.
Drogue, C., 1965. Un bassin témoin en terrains calcaires: le bassin de Saugras (Languedoc). *Bull. Assoc. Intern. Hydrol. Sci.*, 1:383–386.
Dubois, M., 1959. *Le Jura Méridional: Etude Morphologique.* Thesis, University of Paris, Paris, 633 pp.
Dubois, P., 1961. Les circulations souterraines dans les calcaires de la région de Montpellier. *Bull. Bur. Rech. Géol. Minières*, 2:31 pp.
Dubois, P., 1962. Étude des réseaux souterraines des rivières Buèges et Virenque. *Congr. Intern. Spéléol., 2me, Bari, 1958, Actes*, 1:167–175.
Dubois, P., 1964a. Les circulations souterraines dans les karsts barrés du Bas-Languedoc. *Congr. Intern. Spéléol., 3me, Vienne, 1963*, 2: 167–174.
Dubois, P., 1964b. Esquisse de l'hydrogéologie du Massif de la Gardiole (Bas-Languedoc). *Spelunca, Mém.*, 4:57–67.
Dubois, P., 1965. Sur la répartition des eaux souterraines karstiques entre trois cours d'eau du Bas-Languedoc: le Vidourle, l'Hérault et le Lez. *Ann. Spéléol.*, 20(1):1–14.
Dubois, P. and Griosel Y., 1963. La source sous-marine de l'Abysse (Hérault). *Spelunca, Mém.*, 3:87–91.
Durozoy, G. and Paloc, H., 1966. *Projet de Captage Experimental de la Résurgence Sous-Marine de Port-Mion.* Bureau de Recherches Géologiques et Minières, Paris, 11 pp.
Durozoy, G. and Paloc, H., 1967. Recherches sur les ressources en eaux karstiques du littoral méditerranéen français. *Publ. Assoc. Intern. Hydrol. Sci.*, 72:457–465.
Fenelon, P., 1965a. Cartographie des phénomènes karstiques quaternaires et tertiaires. *Bull. Assoc. Franç. Etudes Quaternaires*, 2:5–8.
Fenelon, P., 1965b. Rapport quadriennal (1960–1964) de la commission des phénomènes karstiques. *Norois*, 12(45):63–75.

FORKASIEWICZ, J. and PALOC, H., 1965. Le régime de tarissement de la Foux de la Vis (étude préliminaire). *Bull.Assoc. Intern. Hydrol. Sci.*, 1:213–226.

FOURNIER, E., 1923–1928. *Explorations Souterraines et Recherches Hydrologiques. 1. Grottes et Rivières Souterraines; 2. Gouffres; 3. Les Eaux Souterraines; 4. Phénomènes d'Erosion et de Corrosion; Applications Scientifiques et Pratiques.* Imprimerie de l'Est, Besançon, 159, 207, 222, 350 pp.

GALLE-CAVALLON, H. and PALOC, H., 1967. *Étude de la Fontaine de Vaucluse.* Bureau de Recherches Géologiques et Minière, Paris, 11 pp.

GALLOCHER, P., 1952. Introduction à l'étude hydrospéléologique du Massif de la Ste Beaume. *Ann. Spéléol.*, 12:115–141.

GÈZE, B., 1938. Influence de la tectonique sur la localisation des sources vauclusiennes. *Spelunca*, 9:10 pp.

GÈZE, B., 1949. Les gouffres à phosphate du Quercy. *Ann. Spéléol.*, 4(2):89–107.

GÈZE, B., 1953. Le genèse des gouffres. *Congr. Intern. Spéléol., 1er, Paris, 1953*, 2:11–23.

GÈZE, B., 1957. *Les Cristallisations Excentriques de la Grotte de Moulis.* Centre National de Recherches Scientifique, Paris, 16 pp.

GÈZE, B., 1958. Sur quelques caractères fondamentaux des circulations karstiques. *Congr. Intern. Spéléol., 2me, Bari, 1958, Actes*, 1:3–22.

GÈZE, B., 1965a. Les conditions hydrogéologiques des roches calcaires. *Chron. Hydrogéol.*, 7:9–39.

GÈZE, B., 1965b. *La Spéléologie Scientifique.* In: *Collection Microcosme.* Du Seuil, Paris, 22:188 pp.

GÈZE, B. and POBEGUIN, T., 1962. Contribution à l'étude des concrétions carbonatées. *Congr. Intern. Spéléol., 2me, Bari, 1958, Actes*, 1:396–414.

GIGNOUX, M., 1960. *Géologie Stratigraphique.* Masson, Paris, 735 pp.

GIGOU, R. and MONNIN, J., 1966. Inventaire spéléologique du sud-est du Département du Doubs. *Ann. Spéléol.*, 21:269–355.

GLANGEAUD, L., PÉZARD, R., FRANÇOIS, S., PERRENOUD, M.-J. and TOITOT, M., 1956. Les nappes phréatiques et artésiennes du Jura septentrional (Département du Doubs); leurs relations avec les réseaux karstiques. *Bull. Soc. Géol. France*, 6(6):531–546.

GOSSELET, J., 1906. Observations sur les creuses de l'Artois et de la Picardie et réflexions sur l'importance de la dissolution souterraine de la craie. *Ann. Soc. Géol. Nord*, 35:237–243.

JEANNEL, R., 1926. *Faune Cavernicole de la France.* Lechevalier, Paris, 334 pp.

JEANNEL, R., 1943. *Les Fossiles Vivants des Cavernes.* Gallimard, Paris, 323 pp.

JEZEQUEL, F. and CUNHAC, R., 1963. Étude hydrogéologique du Bassin de Lantouy (Lot-Aveyron). *Spelunca, Mém.*, 3:114–126.

JULIAN, M., 1965. Le modelé quaternaire des versants de la bordure des Grands Causses (région de Millau). *Méditerranée*, 3:161–168.

LAURA, R., 1964. Hydrogéologie et géochimie du calcaire à Asteries (Stampien) de l'entre deux mers dans le Département de la Gironde. *Trav. Lab. Hydrogéol. Géochim., Fac. Sci., Univ. Bordeaux*, 13:77 pp.

LAURES, M. and PALOC, H., 1962. Note sur les concrétions de la Clamouse. *Congr. Intern. Spéléol., 2me, Bari, 1958, Actes*, 1:423–427.

LERICHE, M., 1928. Sur un curieux effet produit dans les terrains tertiaires de la Picardie par la dissolution de la craie sousjacente. "Les Demoiselles de Lihus (Oise)". *Compt. Rend., Congr. Soc.-Savantes, 61er, Lille, 1928*, pp. 170–173.

LEROUX, E., RICOUR, J. and WATERLOT, G., 1958. Variation du niveau des nappes aquifères du nord de la France. *Bull. Soc. Géol. France*, 8:191–206.

LETOURNEUR, J., 1964. Sur les entonnoirs d'ablation: les folletières du Brionnais (Saône et Loire). *Bull. Serv. Carte Géol. France*, 60:545–550.

LUTAUD, L., 1959. L'hydrologie superficielle des plateaux crayeux du nord-ouest de la France. *Mém. Assoc. Intern. Hydrol., Sci.*, 2:147–149.

MAHEU, J., 1906. *Flore Souterraine de la France.* Masson, Paris, 644 pp.

MARGAT, J., MOLINARD, L., PALOC, H. and BONNET, A., 1967. Instructions pour l'inventaire des cavités naturelles de la France. Mode d'emploi du dossier de cavité naturelle. *Rappt. Bur. Rech. Géol. Minières*, D.S. 67–A–11:34 pp.

MARRES, P., 1935. *Les Grands Causses: Étude de Géographie Physique.* Arrault, Tours, 213 pp.

MARTEL, E. A., 1894. *Les Abîmes.* Delagrave, Paris, 275 pp.

MARTEL, E. A., 1921. *Nouveau Traité des Eaux Souterraines*. Gaston Doin, Paris, 824 pp.
MARTEL, E.A., 1930–1933. *La France Ignorée, 1. Des Ardennes aux Pyrénées; 2. Sud-est de la France.* Delagrave, Paris, 302,294 pp.
MARTEL, E.A., 1936. *Les Causses Majeurs*. Fédération Française de Spéléologie, Millau, 520 pp.
MEGNIEN, C., 1964. Observations hydrogéologiques dans le sud-est du Bassin de Paris. *Mém. Bur. Rech. Géol. Minières*, 25: 288 pp.
MENGAUD, L., 1926. Sur quelques sondages profonds dans le Bassin d'Aquitaine. *Bull. Soc. Hist. Nat. Toulouse*, 54: 122–142.
MENNESSIER, G., 1961. La zône karstique de Callian Mautauroux (Var). *Ann. Spéléol.*, 16:267–276.
MENNESSIER, G., 1968. *Introduction Morphologique et Géologique. Inventaire Spéléologique de la France: Var.* Bureau de Recherches Géologiques et Minières, Orléans. (In press.)
MUGNIER, C., 1965. Les karstifications antépliocènes et plioquaternaires dans les Bouges, la Chartreuse septentrionale, et les chaînons jurassiens voisins (Savoie, Haute-Savoie, Isère). *Ann. Spéléol.*, 20:15–46; 167–208.
MUXART, R. and STCHOUZKOY, T., 1967. Contribution à l'étude de la dissolution des calcaires dans les eaux naturelles. *Ann. Spéléol.*, 24(4):639–651.
NAPIAS, M., 1963. Étude hydrogéologique du Massif des Arbailles, Basses-Pyrénées. *Trav. Lab. Hydrogéol. Géochim., Fac. Sci., Univ. Bordeaux*, 12:220 pp.
NICOD, J., 1965. Karst du gypse dans les Alpes et en Provence. *Actes Congr. Natl., Soc. Savantes, Nice, 1965, Sect. Géograph.*, pp.87–104.
NICOD, J., 1967. Recherches morphologiques en Basse-Provence calcaire. In: *Revue Géographique des Pays Méditerranéen*. Ophrys, Gap, 5:573 pp.
PALOC, H., 1961. *Hydrogéologie de la Région Viganaise*. Thesis, Centre d'Études et de Recherches Hydrogéologiques, Montpellier, 96 pp.
PALOC, H., 1962. Contribution à la connaissance des circulations karstiques dans une région type du midi-méditerranéen. Observations sur le comportement aquifère des dolomies. *Mém. Assoc. Intern. Hydrol. Sci.*, 5:243–248.
PALOC, H., 1964. Normes de représentation adoptées par le Bureau de Recherches Géologiques et Minières pour la réalisation à titre expérimental d'une carte hydrogéologique en zône karstique. *Spelunca*, 4:41–43.
PALOC, H., 1964–1968. *Carte Hydrogéologique de la Région Karstique Nord-Montpelliéraine*. Bureau de Recherches Géologiques et Minières, Paris, 227 pp.
PALOC, H., 1965. Les recherches hydrogéologiques et les captages et aménagements hydrauliques en milieu calcaire: enseignements tirés de quelques exemples récents. *Chron. Hydrogéol.*, 7:87–109.
PARDE, M., 1966. Les circulations souterraines dans la région d'Orléans et quelques sources existant dans le monde. *Norois*, 13(51):344–354.
PETREQUIN, P., 1964. Les grandes cavités du Jura français (Département du Doubs). *Spelunca, Bull.*, 4:25–37.
PINCHEMEL, P., 1954. *Les Plaines de Craie du Nord-ouest du Bassin de Paris*. Colin, Paris, 502 pp.
POBEGUIN, T., 1957. Reproduction expérimentale de concrétions de carbonate de calcium. *Ann. Spéléol.*, 12:5–12.
POMEROL, C., 1967. Genèse, datation, remplissage de cavités karstiques dans le Tertiaire du Bassin de Paris. *Actes, Coll. Néogène Nordique, Paris, 1965*, 7 pp. (In press.)
PRUVOST, P., 1909. Sur un puits naturel dans le terrain houiller. *Ann. Soc. Géol. Nord*, 37:437.
RENAULT, P., 1963a. Granulométrie des dépôts de caverne. *Spelunca, Mém.*, 3:55–67.
RENAULT, P., 1963b. Quelques réalisations de spéléologie expérimentale: vermiculations argileuses, corrosion sous remplissage. *Spelunca, Mém.*, 3:48–54.
RENAULT, P., 1967. Contributions à l'étude des actions mécaniques et sédimentologiques dans la spéléogenèse. *Ann. Spéléol.*, 22(1):1–16.
RENAULT, P. and SAUMADE, P., 1963. Étude de la radioactivité des sédiments souterrains de la grotte laboratoire de Moulis (Ariège). *Spelunca, Mém.*, 3:127–129.
RIBIS, R., 1965. *Contribution à l'Étude Géologique du Crétacé Supérieur de la Haute Chaîne, dans la Région de la Pierre St-Martin (Basses-Pyrénées)*. Thesis, University of Paris, 210 pp.
RICOUR, J., 1959. Inventaire des ressources hydrauliques du nord et du Pas de Calais. *Mém. Assoc. Intern. Hydrol. Sci.*, 2:45–48.

Roques, H., 1965. Contribution à l'étude statique et cinétique des systèmes gaz carbonique–eau–carbonates. *Ann. Spéléol.*, 19:255–484.

Rouire, J., 1963. Quelques idées générales sur le géospéologie des Grands Causses. *Spelunca, Mém.*, 3:130–143.

Rouire, J., Caubel, A. and Fabry, J., 1966. Recherches spéléologiques dans les Grands Causses. *Spelunca, Bull.*, 6:162–170.

Rousset, C., 1965. Tectonique et géomorphologie du plateau de Caussol (Alpes Maritimes). *Actes Congr. Natl. Soc. Savantes, 90me, Nice, 1965, Sect. Géograph.*, 13 pp.

Salvayre, H., 1960. Hydrogéologie du Conflent (Pyrénées Orientales). *Ann. Spéléol.*, 15:3.

Salvayre, H., 1963. Carte des écoulements souterrains et des faciès du Larzac et de son avant-causse. *Spelunca, Mém.*, 3:144–150.

Salvayre, H., 1964. Les lacs temporaires du Larzac et de son avant-causse. *Spelunca*, 4:79–103.

Salvayre, H., 1968. Étude hydrogéologique du synclinal de Merens à Villefranche. *Ann. Spéléol.* (In press.)

Schlumberger, R., 1965. Observation sur la morphologie des falaises sénoniennes du secteur d'Etretat-Yport (Seine Maritime). Rôle des karsts. *Compt. Rend. Soc. Géol. France*, 3:73–74.

Schoeller, H., 1965a. Sur la limitation de la circulation en profondeur dans le karst. *Chron. Hydrogéol.*, 7:43–49.

Schoeller, H., 1965b. Hydrodynamique dans le karst (écoulement et emmagasinement). *Bull. Assoc. Intern. Hydrol. Sci.*, 1:3–20.

Séronie-Vivien, M.R., 1953. Guide de l'excursion dans l'entre deux mers. *Congr. Natl. Spéléol., 7me, Bordeaux, 1966*, 5:160–163.

Siffre, M., 1962. Morphologie souterraine et hydrogéologique des formations plio-quaternaires de la région de Nice. *Congr. Intern. Spéléol., 2me, Bari, 1958, Actes*, 1:176–186.

Theobald, N., 1968. *Aperçu sur la Géologie et la Géographie Physique du Département de Haute Saône.* Bureau de Recherches Géologiques et Minières, Orléans. (In preparation.)

Tricart, J., 1949. *La Partie Orientale du Bassin de Paris. Étude Morphologique* (2 volumes). Thesis, University of Paris, 774 pp.

Trombe, F., 1952. *Traité de Spéléologie.* Payot, Paris, 370 pp.

Vancon, J.-P., 1965. Étude quantitative des pertes du Doubs et du Danube. *Bull. Bur. Rech. Géol. Minières*, 4: 151–167.

Vancon, J.-P., 1966. Méthode d'étude des débits par correlation. *Bull. Assoc. Intern. Hydrol. Sci.*, 3:17–25.

Vandel, A., 1964. *Biospéléologie.* Gauthier et Villars, Paris, 621 pp.

Vandenberghe, A., 1960. Obstacles à la circulation des eaux dans les terrains calcaires. Notion de limite d'un réseau karstique—application aux calcaires carbonifères du nord de la France. *Ann. Soc. Géol. Nord*, 80:147–155.

Verdeil, P., 1958. Les phénomènes d'intermittence dans les réseaux karstiques. *Congr. Intern. Spéléol., 2me, Bari, 1958, Actes*, 1:62–78.

Verdeil, P., 1968. Sur la rivière intermittente de Fontestorbes. *Mém. Centre Études Rech. Hydrogéol.*, 1:29–35.

Vila, G., 1963. La grotte de Clamouse (Hérault). *Spelunca, Mém.*, 3:151–162.

Wahl, J.-B., 1967. Le gouffre-grotte du Leubot à Gousans (Doubs). *Spelunca*, 7(2): 88–96.

Weydert, P., 1965. Les relations entre les failles et les dolines sur le plateau de St. Christol (Vaucluse). *Bull. Musée Anthropol. Préhist. Monaco*, 12:5–16.

Chapter 6

Karst of Germany

D. PFEIFFER AND J. HAHN

Niedersächsisches Landesamt für Bodenforschung, Hannover (Germany)

INTRODUCTION

A description of the karst and its phenomena cannot be restricted to purely morphological problems but must also consider geology, sedimentology, hydrology, tectonics, soil science and engineering geology. It seems appropriate within the scope of this work to describe those kinds, forms and mechanisms of leaching which are especially important and characteristic in Germany, and to touch also upon the forms and manner of dissolution of evaporates (chlorides and sulphates). The phenomena and problems of the limestone karst in humid and semi-humid climatic zones are, however, treated more briefly because they are similar in most central European countries, and are described repeatedly in the present work. Data on karst conditions in the German part of the Alps will be found in the section "*Karst in Austria*".

The eventful stratigraphic and tectonic phases in the geologic past are reflected also in the present landscape of Germany. From a morphological point of view, there are clearly three zones. There is in the first place a lowland zone, the "north German Plain" (Norddeutsche Tiefebene) which is covered by Quaternary deposits. Its width is up to 300 km, and its altitude is mostly less than 50 m above sea level. Adjoining it in the south is an upland zone (altitude generally 200–600 m above sea level) which was subject to strong tectonic influences and which has a complex geologic composition. It consists on the one hand of remnants of mountains folded during the Caledonian and Variscan, and on the other hand of younger, mostly Mesozoic sediments which, due to the consolidated underground, reacted to prevailing tectonic stress with fracturing (germanotype). The upland zone borders on the Bavarian molasse basin. Filled with Tertiary and Quaternary deposits, this basin extends in front of the young folded mountains of the Alps which in Germany attain altitudes of up to 3,000 m.

Germany is situated in the zone of transition from maritime to continental climates. This zone is characterized by the lack of extremes. The monthly mean temperatures vary from 16°C (summer months) to 4°C (winter months). The precipitation amounts, on an average, to 780 mm/year.

Karst regions in Germany occupy only a relatively small area. According to the geologic and stratigraphic facts, these regions are either isolated occurrences of massive limestone or calcareous and sulphate and chloride salinar deposits, usually not very thick and which are karstified in the zones of their outcrops but also locally at greater depths.

In the study of the karst areas, the north German Plain may be disregarded. The generally loose Quaternary rock deposits are penetrated only at a few closely confined places by Mesozoic or Palaeozoic sediments subject to karstification; these are predominantly diapirs (300 such diapirs have been located below the earth's surface) or blocks dragged upwards by salt diapirs (e.g., Bad Segeberg, Lüneburg and Rüdersdorf near Berlin; see Fig.1). In the upland zone the situation is different. It is in this heterogenously composed area—the combination of the individual regions is only justified by the general mean altitude—that German's most important and most extensive karst regions are located (Fig.2). The oldest karstified rocks of a greater extent are Devonian limestones. These occur in locally considerable thicknesses, especially in the Rhenish Schiefergebirge, and, on a smaller scale, in the Harz Mountains, Thuringia and Saxony.

The Lower Carboniferous in Germany has a largely clastic facies (Culm); in the Aachen district, however, this formation still occurs in the northwest European "carbonate limestone facies", and, with a thickness of up to 300 m, represents an important karstification horizon in this area.

Strong leaching phenomena are observed in the areas of the Upper Permian (Zechstein) salinar sulfate and chloride deposits outcropping in a small strip on the southern and western margins of the Harz Mountains, in the fringe of the Thuringian Basin and on the eastern margin of the Rhenish Schiefergebirge as well as in parts of the hilly country of southern Lower Saxony (Südniedersächsisches Bergland).

Major karst phenomena are also shown by the deposits of the Muschelkalk (Triassic), which are present today, in larger coherent planes at the surface in the Thuringian Basin, the northern foreland of the Harz Mountains, the Hessian Depression, Franconia, Baden-Württemberg, Palatinate, Saarland, and occasionally in the Eifel.

The largest continuous karst regions are the outcrop areas of the Upper Jurassic in the Swabian and Franconian Alps. Cretaceous sediments are almost exclusively found in the northern part of the upland zone (Münsterland Basin, the mountainous country of southern Lower Saxony, the foreland of the Harz Mountains). Karst phenomena are generally confined to the Cenomanian and Turonian as only these formations have a calcareous facies; on the island of Rügen, however, the entire Upper Cretaceous is calcareous, and actually developed as chalk. The Tertiary in Germany seldom has a calcareous facies in larger areas. Only in the marginal regions of the Upper Rhine graben and on the fringe of the Mainz Basin are some limestone horizons of small extent and very little karstification found at the surface.

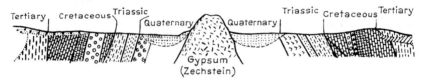

Fig.1. Geological cross-section through a salt-gypsum diapir in the north German Plain.

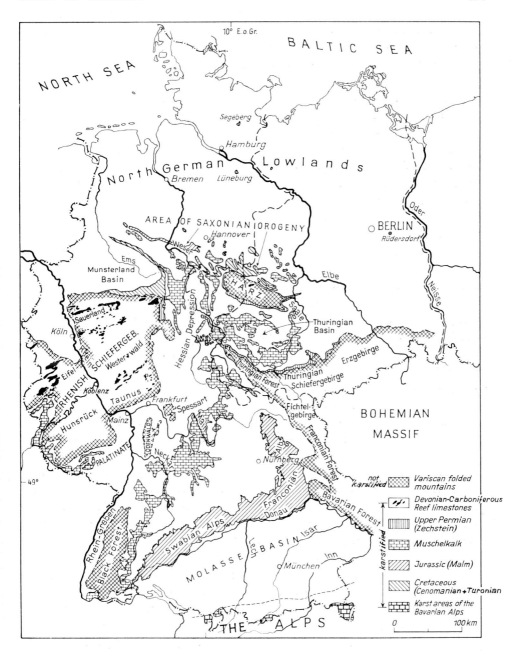

Fig.2. Distribution of major karst rocks in Germany.

STRATIGRAPHIC AND SEDIMENTOLOGIC ASPECTS

Geologic development of Germany

The geologic development of Germany is many-sided and full of changes. Therefore, only the essential trends will be dealt with here.

Precambrian

The occurrence of rocks of this age is mainly restricted to the central cores of younger mountain massifs, such as the Fichtelgebirge, Thuringian Forest and Erzgebirge. The Assyntian orogenesis which begins at the end of the Algonkian is the oldest orogenesis proved so far in German territory.

Palaeozoic

During the Cambrian, Ordovician and Gotlandian, the formation of a depression zone took place which strikes in an east–west direction and is a part of the north European Caledonian geosyncline. This zone can be established in Germany by the presence of deposits in the Rhenish Schiefergebirge, Fichtelgebirge, Thuringian Forest and Lausitz. During this time, Germany was largely transgressed by a shallow sea; limestones are not common and occur only locally. Thus the Lower Cambrian of the Lausitz has an *Archaeocyathus* Limestone facies, but in the layers of the Ordovician and Gotlandian intercalations of thin limestones are also encountered locally.

The sediments of the Caledonian trough were essentially subjected to the young Caledonian folding which began at the end of the Gotlandian; resulting in a varying degree of folding.

The Devonian is characterized by the subsidence of the Variscan geosyncline slowly proceeding from the west to the east. In Germany this geosyncline is bordered in the north by the Old Red continent (line Ardennes–Harz Mountains), and in the south by the Alemannic–Bohemian island (south Germany). At least in the Lower Devonian, a separation into a sandy–clayey (Rhenish) and a more calcareous (Hercynian) facies area is still possible. However, a division of the basin, which began in the Middle Devonian, and which was due to synsedimentary movements and later to foldings, resulted in a strong differentiation of the facies. In ridge zones or in shallow-water zones, limestone reefs were formed, or limestones of locally considerable thickness were deposited. Basin regions mostly show slate facies marginal changing into the reef limestone facies of the ridge regions. The Devonian limestones are strongly karstified in the various parts of the Rhenish Schiefergebirge, in the Harz Mountains, and, to a smaller degree, in the Thuringian Forest and Saxony.

Of special importance with regard to the geologic structure and consolidation of central Europe, is the Variscan orogeny. This begins at the boundary Devonian/Carboniferous with the Bretonic phase, continues in the Carboniferous with the remarkable Sudetic, Erzgebirgic and Asturian phases, and ends in the Permian with the Saalic and Palatinate phases.

In the Lower Carboniferous, in the German part of the Variscan geosyncline, two facies regions may be distinguished: (1) the west European, marine "carbonate limestone facies" which in Germany extends from Aachen to Düsseldorf; and (2) the Culm facies. The thickness of the "carbonate limestone facies" is 300 m near Aachen, decreases toward the east, and finally wedges out. The Culm sediments are erosion products of the Bretonic uplift, and consist of a thick alternation of greywackes, conglomerates, clay slates and "Kieselschiefer". The Upper Carboniferous sediments of the fore-through and of the inner depressions of the Variscan mountains (where considerable amounts of coal are developed) are subjected to the Asturian folding. After this folding, the Variscan mountains were largely levelled as early as the Lower Permian (Rotliegendes). The disintegration of the mountain massifs was accompanied by the formation of depressions and graben zones, in which the terrestrial waste products were deposited under arid climatic conditions.

The Upper Permian (Zechstein) began with a transgression, which came from the north and advanced several times into Germany to form fjords. This shallow-water encroachment covered the north and centre of Germany up to the edge of the Variscan mountains. As only a relatively small quantity of fresh water was supplied through narrow connections with the ocean, and due to the arid climatic conditions, evaporations repeatedly occurred, in the course of which limestones, anhydrite and

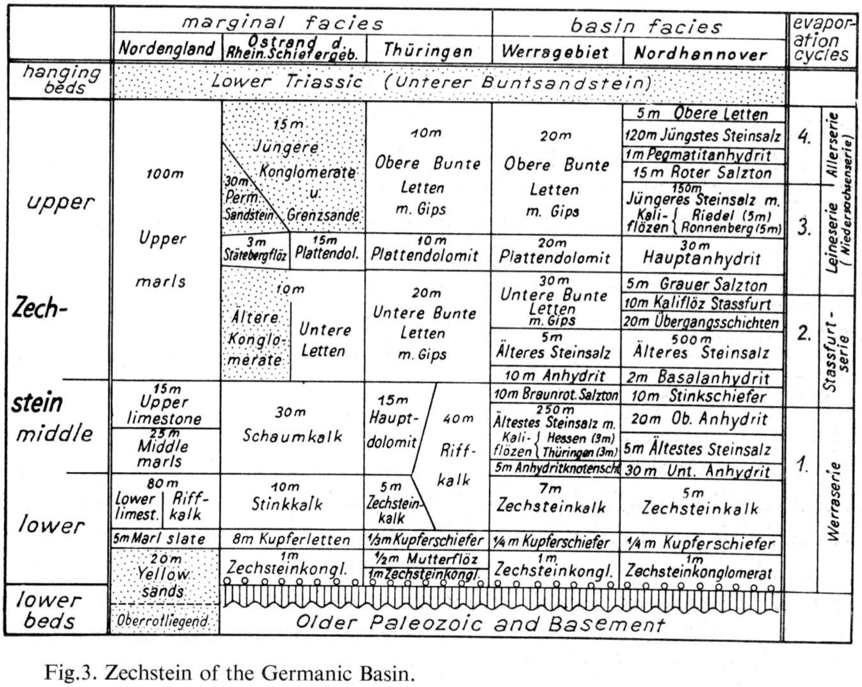

Fig.3. Zechstein of the Germanic Basin.

Na- and K-chlorides were precipitated in chronological order. Four cycles of this kind may be distinguished, viz., the Werra, Strassfurt, Leine and Aller Series, in which saline sediments of a maximum of more than 700 m thickness were deposited (Fig.3.). In the outcrop zones of these series, manifold karst phenomena are encountered.

The salt masses which had become mobile during the Saxonian movements, rose up owing to their low specific weight and concentrated in the so-called salt domes or diapirs which, on their rise, dragged upwards rock blocks of younger formations. More than 300 salt domes have been located in the north German Plain. Many of these reach the zone of mobile groundwater and are marked by suberosion phenomena.

Mesozoic

During the Mesozoic, there was a time of relative tectonical quietness; the sediments lie flat to undulating and extend over wide areas. The sedimentation, compared with the other parts of Europe, is marked by a special facies ("Germanic facies").

The Germanic Interior Basin which appears in outlines as early as in the Zechstein, widens in the Triassic. The sedimentation area of the Buntsandstein (Lower Triassic) is bounded on the west by the Gallic uplift, and on the south by the Vindelician Ridge and the Bohemian Massif, both of which separated the Germanic Basin from the Alpine Tethys area. The mostly fluviatile-terrestrial fillings in the sedimentation basin came from the Bohemian Massif and the Gallic uplift. Calcareous sediments have been deposited only locally.

In the Upper Buntsandstein marine influences are observed; thus, in the uppermost Buntsandstein (Röt) dolomites, gypsums and clays were deposited. A second shallow-water encroachment from the Tethys via Silesia, which comprised the entire Triassic sedimentation area in the Germanic Basin, led to the deposition of Muschelkalk (Middle Triassic) with a thickness of up to 250 m. During the Middle Muschelkalk, an interruption of the connection to the Tethys resulted in the formation of salts, which was ended, however, by a further marine transgression from the Tethys. The carbonate sediments of the Muschelkalk are important karstification horizons. In the uppermost formation of the Triassic, the Keuper, rather limnic-terrestrial features are observed, owing to the interruption of the connection of the Tethys. These are marked by large sandstone fillings, and by deposits of clays, marls and dolomites; in the Middle Keuper, however, separations of gypsums also took place.

In the Jurassic there were new transgressions, advancing from the north and southwest into the Germanic Basin which, owing to the young Cimmerian movements, was then divided into a northern and a southern part initially connected by the Hessian Strait. While in the Liassic (Lower Jurassic), dark clays, marls, bituminous shales, and, to a small degree, also limestones and sandstones were deposited, brown sandstones, clays, and secondarily limestones were chiefly deposited in the Dogger.

At the beginning of the Malm (Upper Jurassic), the division of the sedimentation area into a northern and a southern basin was completed. The deposits of these basins increasingly show differences in the facies, while in the southern basin (Swabia and Franconia) chiefly limestone was deposited. Very much clayey–marly material, in addition to limestones, was deposited in the north German Basin; in the Portland there even occurs a saline series in the Münder Marls.

During the Cretaceous, the sedimentation was rather changeable, owing to the Saxonian movements. South Germany was largely a mainland, while in north Germany an increased subsidence and hence the deposition of thick sediments took place. In south Germany—with the exception of the Alps and the western margin of the Bohemian Massif—Cretaceous deposits are found only locally. In north Germany, in the Lower Cretaceous, chiefly clays and sands were deposited. In the Upper Cretaceous, limestones and limestone marls from the Cenomanian transgression, with thicknesses of about 800 m predominate, and they attain 2,500 m in marginal depressions. In addition, the chalky facies are widespread in northwest Germany. The areas of karstification are chiefly to be found in the Münsterland Basin (see p. 198), in the hilly country of southern Lower Saxony and in the foreland of the Harz Mountains (see p. 199).

Tertiary

The boundary between the Tertiary and the Cretaceous is largely characterized by a stratigraphic break and a transgression. In the course of the Tertiary the sea gradually retreated into its present boundaries; the present morphology of Germany was also essentially developed during this time.

In central and northern Germany, the germanotype orogenesis continued, while in western and southwestern Germany grabens were formed (e.g., Upper Rhine Graben).

The Tertiary sediments are mostly loose and clastic. In the valley of the upper Rhine, coquinas as well as gypsum and saline deposits are found.

The karstification of the limestones in the Swabian–Franconian Alps, as well as in the region of the Variscan mountains, had its maximum in the humid climate of the Tertiary. In the cavities and sinks of karstified limestones terra rossa accumulated which may be associated with iron and manganese ore deposits.

Layers subject to karstification are found, to a small extent, in the Mainz Basin and in the valley of the upper Rhine (Upper Oligocene/Lower Miocene). The *Hydrobia* Beds, *Corbicula* Beds, *Cyrena* Marls (Mainz Basin) and freshwater limestones (valley of the upper Rhine) are made up of marls which have a variable content of limestone and a total thickness of about 150 m.

Rocks subject to karstification

A considerable number of rocks is subject to karstification. A common factor to these rocks or to the individual rock components is their water-solubility. These rocks include: (*1*) sedimentary rocks: limestone (also "Quellkalk", "Kalktuff", "Kalksinter", travertine), dolomite, anhydrite, gypsum and some rock salt, as well as calcareous rocks such as those of the marl series, calcareous sandstone, limestone conglomerates, limestone breccias and dolomitic greywacke (Zellendolomit); (*2*) metamorphic rocks: marble and calcareous mica schists; and (*3*) volcanic rocks: calcareous tuffs.

Important rock types showing karst phenomena are the limestones and dolomites, as well as anhydrite, gypsum and rock salt; the other rocks mentioned above are only of local importance. The post-Variscan formations of extra-Alpine Germany contain, for instance, 100,000 km^3 of salts, an amount corresponding to that of the limestone deposited in this area.

In general it can be said that deposits subject to karstification are widespread in Germany, since the deposition of calcareous or saline sediments occurred repeatedly in the various formations. The degree of karstification, however, is quite different and depends on various factors. The chief factors are the kind of the rock, its chemical composition (purity), the pore space, the jointing, and in particular the climate; furthermore a part is played by the stratification of the rocks, the tectonic stress, the altitude of the rocks above sea level, the hydrological conditions, the vegetation and the covering soils. For the intensity of karst phenomena, the time of the beginning of karstification is not the least decisive factor. The distribution areas of calcareous and saline deposits respectively, are shown in Fig. 2. These deposits, however, are not always so strongly karstified that, on the basis of their morphological features, they might be regarded as typical karst areas.

Carbonate sediments

Limestone and dolomite. The shape of the calcareous deposits is manifold, and the various deposits are subject to karstification in a different degree. Karstified rocks are, above all, the pure (up to 98% $CaCO_3$) to dolomitic massive reef limestones of the Devonian (max. 1,000 m massive limestone—Massenkalk, Stringocephalenkalk—, max. 300 m Dorper Limestone, max. 200 m Iberg Limestone, and others), of the Zechstein (max. 60 m Bryozoa reefs and Stromaria reefs) and of the Upper Jurassic (max. 150 m Korallenoolith, max. 200 m Massenkalk and Frankendolomit, respectively). These rocks mostly reacted to tectonic stress with fracturing and jointing, whereby karstification was promoted. In addition, this process is supported by the lack of detritus of the reef limestones, and by the lack of clayey inpermeable horizons in these largely unstratified deposits. In the case of impure limestones, for instance, Devonian limestone "Knotenschiefer" (i.e., highly clayey sediments, where the $CaCO_3$ content is concentrated in nodules), and "Pläner" (limestones of the Upper Cretaceous which are rich in clay and sand) or generally the members of the marl series or calcareous sandstone series, the karstification is essentially dependent on the petrographic properties of the sediments, so that often only small karst forms are developed.

Saline sediments (sulphate and chloride)
The formation of saline deposits (sulphate and chloride) which is due to the precipitation of salts dissolved in the sea water, took place under arid climatic conditions, in an ocean basin largely cut off from the open sea. In the ideal case, the precipitation of the sea salts proceeded according to their solubility, i.e., in the succession $CaCO_3$–$MgCO_3$–$CaSO_4$–$NaCl$–K-Mg-chlorides. In Germany, two saline types may be distinguished:

(1) "Red formations" of a continental-terrestrial origin. These include the salines of Upper Rotliegendes, Zechstein 4, Röt and Middle Keuper. The salts of this type are largely mixed with clastic sediments (mostly pelites); anhydrite is present in small thicknesses, whereas carbonates are lacking (the "red formations" were formed under arid climatic conditions in the centre of continental depressions).

(2) Marine saline series. These include the salines of the Zechstein 1–3, Middle Muschelkalk, Portland and Oligocene. The salts of this type are accompanied by carbonate rocks (limestone and dolomite) and mostly by thick anhydrite masses. The precipitation took place in genuine sea basins which, however, were distant from oceans; drifting of detritus did not occur.

Of these formations only the marine deposits are essentially of importance for karstification and leaching phenomena of a large areal extension.

Anhydrite and gypsum. In Germany anhydrite ($CaSO_4$) and gypsum occur essentially in Permian, Triassic and Jurassic deposits. These include the Zechstein 1 (max. 200 m anhydrite), Zechstein 2 (max. 25 m), Zechstein 3 (max. 20 m), Upper Buntsandstein (max. 50 m), Middle Muschelkalk (max. 80 m.), Middle Keuper (max. 10 m) and Portland (max. 15 m). Anhydrite, owing to its low solution rate, is subject to karstification in a very small degree; only its hydration to gypsum ($CaSO_4.2H_2O$) which begins at the contact with water makes the solution and karstification possible. The volume increases by 62 % as a result of the hydration. Although the solubility of anhydrite in water is higher than that of gypsum (at $18°$:2.6 g gypsum/l H_2O), its solution rate is 3–4 times less. The solubility of gypsum, however, is especially high on contact with waters containing chloride; a 3.5 % NaCl solution, for instance, dissolves 6.4 g gypsum/l at $21°C$. At a concentration of 17.5 % NaCl and at $17°C$ as much as 9.3 g gypsum/l of brine are dissolved. As the deposits of sulphates are often overlain or partly underlain by Na-K-Mg chlorides, leaching phenomena of both rocks often occur simultaneously; here the leaching phenomena in the gypsum can be promoted by suberosion water coming from the rock salt.

Rock salt (halite). In Germany the precipitation of rock salt[1] took place essentially during the Zechstein. Rock salt was also precipitated in a smaller degree and less widespread, in the Röt (max. 200 m), Middle Muschelkalk (Ladinian, max. 40 m)

---
[1] When we speak about "rock salt" here, we always mean a naturally occurring association of different salt minerals where, however, NaCl largely prevails.

and Münder Marl of the Upper Portland of northwest Germany (max. 200 m), as well as in the Oligocene of the graben of the upper Rhine Valley. The deposition of thicker rock salt layers took place chiefly in the lower parts of the evaporation basins. In the marginal regions of these basins, the rock salt deposits are substituted by limestone, dolomite and anhydrite, and, near regions of denudation, by terrestrial sediments such as conglomerates, sandstones or clays. Rock salt is highly soluble; a concentrated NaCl solution contains 356 g NaCl/l H_2O at $10°C$.

GEOLOGIC AND TECTONIC SETTING OF THE KARST AREAS

Variscan basement

The geologic unit "Rhenish Schiefergebirge" is geographically subdivided into the uplands of Eifel, Hunsrück, Taunus, Westerwald, Siegerland, Bergisches Land and Sauerland as well as into a few smaller regions. It is bordered in the north by the lower Rhine embayment and the Münsterland Basin, in the east by the Hessian Depression, and in the south by the Mainz Basin and the southern margin of the Hunsrück (Fig. 4).

The largely Devonian sediments of the Rhenish Schiefergebirge are subject, as to their facies, to great local variations. They are composed of slates, greywackes, quartzites and magmatic rocks, with intercalations of a large number of larger and smaller isolated limestone deposits of a locally considerable thickness (up to 1,000 m). The sediments of the Rhenish Schiefergebirge were folded by the Variscan orogeny into wide-spanned, mostly asymmetrical anticlines and synclines which are in their turn dissected by faults.

The limestones are mostly massive reef limestones (Massenkalk) formed in ridge zones. They are represented, in the direction of the interior of the basin, by bedded, platy limestones passing into phacoidal ("Schlieren"-like) clayey limestones which in their turn are interspersed with clayey deep-water sediments.

Apart from these Devonian limestones, at least on the left side of the Rhine (e.g., near Stolberg, see Fig. 4), there exist Lower Carboniferous carbonate limestones of a rather great thickness. All limestone ranges are karstified and distinguished by dolines, sinks and caves. They are important ground-water aquifers, as they are transversed by numerous cavities and joints. In the karst caverns of the Tertiary, iron–manganese ores were formed (Hunsrück, Lindener Mark near Giessen). In a Pleistocene cave of the Dorper Massenkalk, in the Neandertal near Düsseldorf, the remains of a pre-human being, the *Homo neanderthalensis*, were found. Larger karst regions are encountered in the Eifelian limestone troughs, in the area of the Lahn River as well as in the Sauerland (anticlines of Remscheid, Warstein, Brilon, and Attendorn).

The individual karst occurrences cannot be described here because of their large number; the most important ones are shown in Fig. 4.

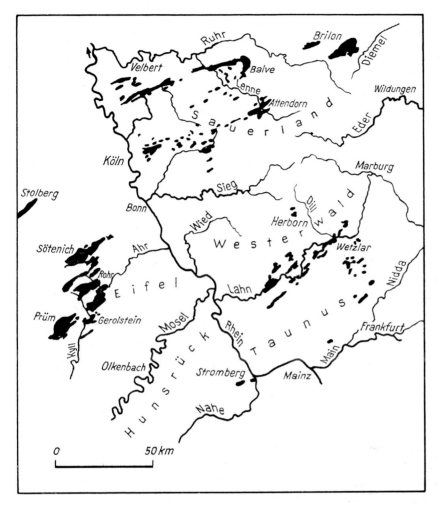

Fig.4. Areas of distribution of the Devonian reef and massive limestones (black) in the Rhenish Schiefergebirge. (After JUX, 1960.)

The Harz Mountains form another Variscan block which, similarly to the Rhenish Schiefergebirge, is largely built up of Devonian and Carboniferous sediments, as well as by magmatic rocks. Karst phenomena in the Harz Mountains are in the main encountered only in two regions: on the Iberg near Bad Grund and in the district of Elbingerode and Rübeland. The limestones of these regions correspond, as to their genesis and structure, to those of the Rhenish Schiefergebirge; they are massive reef limestones which are intensely karstified, a usual feature with tectonically highly stressed massive limestones. In the small complex of the Iberg (1.8 km²) alone, more than 20 caves are known. In the karst region of Elbingerode, which has an extension of about 15 km², Iberg Limestone and Massenkalk outcrop at the earth's surface. The Baumannshöhle and the Hermannshöhle, which are situated in the Iberg Limestone (lowest Upper Devonian) at Rübeland, are well known. The surface of the limestone is largely covered by residual loam; Tertiary deposits are found in sinks.

The Thuringian Forest, the Thuringian/Saxonian Schiefergebirge and the Erzgebirge are made up of Palaeozoic and Precambrian rocks. Karst phenomena are restricted to the calcareous–clayey *Tentaculites* Limestones (Lower Devonian) and to the Upper Devonian Knollenkalk (limestone Knotenschiefer) which essentially occur in the Schiefergebirge. It is true that in the Knollenkalk (limestone intercalations in a clayey matrix, up to about 100 m thick) caves, dolines and ponors exist locally; this horizon, however, has a very changeable facies over short distances (clayey, calcareous or volcanic).

The remaining parts of Germany which were affected by the Variscan folding (Black Forest, Odenwald and Spessart as well as the regions situated on the margin of the Bohemian Massif: Fichtelgebirge, Fränkisches Mittelgebirge, Oberpfälzer and Bayerischer Wald) are largely built up of rocks which are not subject to karstification. This also applies to smaller basement areas. Karst phenomena are restricted to several marble ranges having only a local importance.

Thuringian Basin

The Thuringian Basin is an important karst region. In this depression, which has a Hercynian longitudinal axis (i.e., directed west-northwest) and which is surrounded by the Harz Mountains, the Thuringian Forest, the Thuringian Schiefergebirge and partly by the Erzgebirge, there is a continuous sequence of deposits from the Rotliegendes (Lower Permian) up to the Keuper (Upper Triassic) as well as locally from the Lower Jurassic (Fig. 5,6). During the Zechstein, a cyclic large-scale separation of saline series took place in the Thuringian Basin, which shows, for instance, on the southern edge of the Harz Mountains a thickness of 350 m, and in the Werra area bordering on these mountains in the west a thickness of 400 m; these series are composed of limestones, dolomite, anhydrite and gypsum, respectively, as well as of Na-Mg chlorides.

The overlying formations show karstified rocks in the Upper Buntsandstein (Röt gypsum), Middle Triassic (Muschelkalk) and Upper Triassic (Keuper gypsum). The egg stone-oolitic limestone in the Lower Buntsandstein is also karstified to a smaller degree. The Thuringian Basin generally shows a curving strike of the layers. The margins of the basin are marked by the outcrops of the Zechstein layers in which karstification phenomena occur to a considerable extent. Of special importance are the underground leaching phenomena in the Zechstein salts which characterize the surface forms in large parts of the basin. The frequent occurrence of brine springs is documented by a large number of names of places and waters, and gives evidence of a still efficient leaching of salts in the underground.

On the basin's southern margin (Thuringian Forest, Thuringian Schiefergebirge),

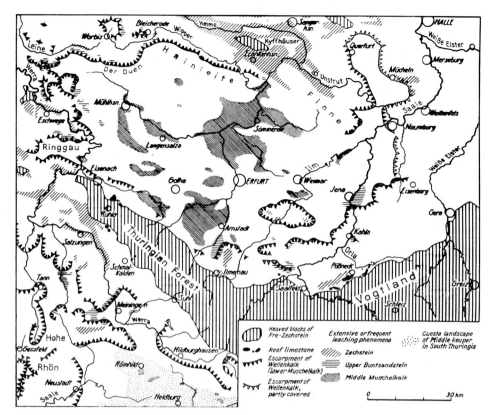

Fig.5. Outline of the main forms of the Thuringian Zechstein and Triassic landscape. (After WEBER, 1955.)

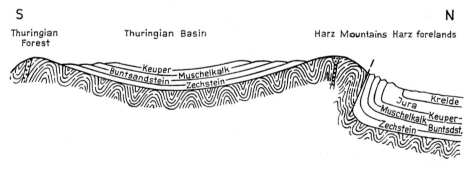

Fig.6. Schematical cross-section through the Thuringian Basin and the northern Harz foreland. (After HAASE, 1951.)

which is largely identical with the southern shore zone of the Germanic Basin, the outcrop and karstification zones of the Zechstein deposits are only small and irregular. The former nearness of land may be recognized by a changing facies of the strata, for instance, by terrestrial sediments or by shallow-water formations, such as algal and

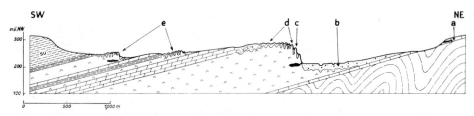

Fig.7. Schematical cross-section through the southern foreland of the Harz Mountains. a = Caves in limestones and dolomites of an isolated Zechstein block lying above the rocks folded by the Variscan orogeny (Harz Mountains); b = Quaternary valley filling within the range of a suberosion depression; c = escarpment of the Werra Gypsum; d = karst phenomena in the top region of the Werra Gypsum (sinks, dolines); e = sinks and dolines in "younger gypsums" which have been largely dissolved (Basalanhydrit and Hauptanhydrit); su = Lower Buntsandstein with wide-area salt leaching depressions (striking rim syncline).

bryozoan reefs, as well as by a primarily smaller thickness of the strata which was caused by fluctuations of the ocean level. Thus in the area between Gera and Saalfeld, there are today highly karstified reef limestones up to 60 m thick in the former shallow-water zone.

Similar limestone reefs are found in many places in the former coastal region at the northern and southern margins of the Thuringian Forest (Bad Thal, Bad Liebenstein, Asbach–Altenstein); they dominate the landscape with their bizarre karst forms. The margin of the Thuringian Basin bordering on the Harz Mountains is characterized by a widely outcropping zone of saline deposits (Fig. 7). Thus along the southern margin of the Harz Mountains, there extend the white rock walls of the Zechstein gypsum. The high solubility of the gypsum in water caused the intense karstification of this region. More than 80 caves are found here, such as the Himmelreichhöhle (dimensions 170 × 80 × 15 m) and the Heimkehle with a length of more than 4 km. Furthermore, karst springs with large discharges (Rhumequelle, 4 m^3/sec), ponors, bulge caves ("Quellungshöhlen"), small poljes (Nixsee Basin) and large dolines areas (Fig. 8–11) which have an almost exclusively Zechstein gypsum facies are found. In addition, 18 gypsum caves ("Mansfelder Schlotten") are known to exist in the Mansfeld Trough, a Zechstein depression separated from the Thuringian Basin by a ridge zone.

The salt solution in the depths had important consequences. The fertile plains of the Goldene and Diamantene Aue, situated north and south of the basement uplift of the Kyffhäuser, are the result of large-area subsidences which were due to the leaching of deep-seated salt deposits. Gypsums showing karstification phenomena are also exposed in larger areas of the Kyffhäuser block itself (Barbarossa Cave).

The Röt, the topmost formation of the Buntsandstein, though having a largely clayey facies, may contain saline intercalations subject to karstifications. Thus karstification effects are observed in some areas east and south of Jena, west of Rudolstadt as well as in the western part of the basin between the Werra and Leine rivers. The Röt clays generally form the underlying bottom layer for the waters of the superposed, intensely karstified Muschelkalk horizons.

Fig.8. Natural slope of gypsum at the Trogstein (southern Harz Mountains).

Fig.9. Bulge cave (Quellungshöhle), Zwergenschmiede near Walkenried.

Fig.10. The Werra Gypsum at the Röseberg near Walkenried which shows big sinks and which is covered with detritus of dolomite.

Fig.11. View of the Beierstein south of Osterode (Harz Mountains) toward the step of Lower Buntsandstein. In the foreground are dolines.

The Muschelkalk, with its escarpment of hard limestones (Wellenkalk), striking around the basin, is a dominating morphological element in the Thuringian Basin which is largely filled with soft sediments. The dissolution of the gypsum in the underlying Röt clays which is caused by the fissure and crevice water of the Muschelkalk leads to considerable falls at the escarpment of the Muschelkalk. These phenomena are especially intense in the Plauenscher Grund, and on the northern margin of the Ringgau. The flat dip of the layers towards the basin's interior results in the occurrence (in the marginal zones of the basin) of a belt of karstified plateau surfaces, such as the Ihn–Saale Platform, the Gossel Plateau southwest of Arnstadt, the Ringgau, the area between the Hainich, Dün, Hainleite etc. The vegetation of these areas which are often lacking in water and have mostly shallow soils, has a steppe character ("Steppenheide").

The interior of the Thuringian Basin is not uniformly covered with Keuper deposits; owing to Saxonian movements, it is divided by faults and undulations of a Hercynian direction into wide Keuper synclines and Muschelkalk anticlines, so that again and again long, karstified Muschelkalk ranges also occur in the interior of the basin.

The cuesta landscape of Swabia and Franconia

In the zone of the south German block ("Süddeutsche Grossscholle"), in the Mesozoic Era, a wide-area deposition of Triassic and Jurassic sediments took place. These deposits later subjected only to a minor tectonic stress. This complex is partly horizontal and partly slightly dipping southeastward. It consists of alternating sandstone, clay, marl and limestone horizons of differing resistivity, and it thus represents a typical example of a cuesta type of topography (Fig. 12).

This area is bordered in the west by the Black Forest, the graben of the upper Rhine Valley, and the Odenwald, in the north by the Hessian Depression and the Thuringian Forest, and in the east by the crystalline rocks of the Fichtelgebirge, Franconian Forest and Bavarian Forest. In the south the beds of the south German block plunge beneath the Tertiary and Quaternary deposits of the molasse basin.

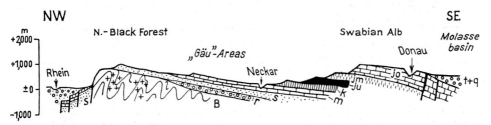

Fig.12. Schematical cross-section through the south German cuesta landscape. B = Basement; r = Rotliegendes; s = Buntsandstein; m = Muschelkalk; k = Keuper; ju = Liassic; jm = Dogger; jo = Malm; $t+q$ = Tertiary and Quaternary. (After GELLERT, 1958.)

The south German cuesta landscape, exclusive of the Buntsandstein, consists of two or three great steps and two plateaus; in addition, there is a number of smaller steps and terraces. Seen from the north to the south, a Muschelkalk escarpment is followed by the Muschelkalk–Lower Keuper plateau (Gäu areas) which ends at the step of the Keuper sandstone. The adjacent Keuper plateau (Keuper land) includes the Lias, above which the Dogger and Malm rise in the form of steps (Alb). The following horizons are subject to karstification: Muschelkalk (in Swabia about 200 m, in Franconia 110 m), Keuper (80 m of clay with gypsum intercalations, only in Swabia) as well as Malm (about 400 m).

The zone of the Gäu areas corresponds to the outcrop region of the Muschelkalk and the Lower Keuper, and extends, in a changing width, from the southeastern Black Forest to the north; it bends off from the Neckar river, towards the east, and then strikes northeast in the direction of the Thuringian Forest (see Fig. 2). The most important karst horizon is the hard, thick-bedded Upper Muschelkalk (Hauptmuschelkalk) with a thickness of up to 90 m. It is particularly susceptible to karstification, as the Middle Muschelkalk shows intercalations of 10–40 m of rock salt and of about 40 m of gypsum, the leaching of which results in a strong jointing and disruption of the overlying Upper Muschelkalk plate. Large parts of this outcrop area are lacking in water, they have little fertility and are, therefore, thinly populated. Karst springs, ponors, sinks, dolines, caves and dry stream beds are widespread. The relation of valley density to river density which indicates the displacement of the run-off into the underground, amounts to 10/1–100/1. While in the Black Forest (valley of the Murg River) there are 5 km of waterway to 1 km², the amount in the eastward adjacent Muschelkalk area is 0.11 km, in the Keuper area, however, it is 1.84 km (Fig. 13).

The ground-waters of the Muschelkalk area are often very hard, and in the case of active gypsum solution in the underground, are not suited for drinking purposes. The Keuper deposits superimposed on the Muschelkalk, are largely clayey–marly, and, especially in the east of the area, show thicker sandstone horizons. Despite of their relatively small thickness of 255 m (Swabia) to 350 m (Franconia), these deposits cover the major part of the south German cuesta landscape, and outcrop in a strip attaining a width of 100 km (mostly about 50 km). This so-called Keuper land has a relatively high river density. Karst phenomena first occur in a broad boundary zone of Muschelkalk–Keuper. The karst phenomena of the Middle and the Upper Muschelkalk are traced here as far as into the overlying Keuper layers. Secondly they occur in the western part of the area (for instance, northwest of Stuttgart) in the Keuper deposits themselves, where a clay–gypsum series as much as 80 m thick is intercalated here in the Middle Keuper. The overlying Jurassic layers, owing to their slight south, southeast and east dip, respectively, also outcrop widely, forming several cuestas, the most important of which are the Swabian–Franconian Alps (Alb) built up by the Malm limestones.

The largely clayey–marly Liassic deposits which are 80 m thick (35 m in Franconia), constitute the wide foreplan of the Alb which is utilized for agriculture. The

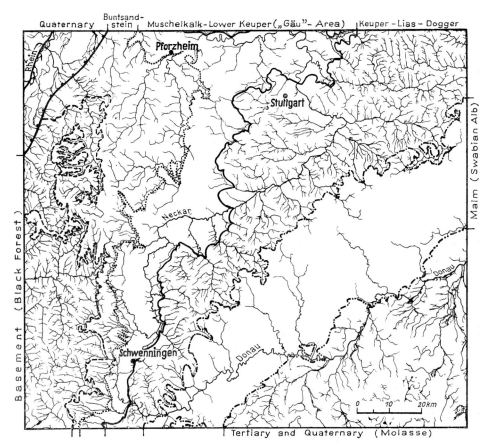

Fig.13. The differing drainage density in the areas with crystalline basement, Triassic and Jurassic layers. (Section from the hydrographical map of southwest Germany.)

Dogger (thickness 240 m and 150 m, respectively) forming the slope of the Alb is covered by meadows and woods, while the sediments of the Lower Malm form the escarpment ("Albtrauf") towards the Alb Plateau (maximum altitude 1,015 m).

The mountain range of the Swabian–Franconian Alps (width about 40 km, length more than 450 km, altitude 600–900 m above sea level) is composed of an alternating series of limestones and dolomites (secondarily also of marls) of the Malm (White Jurassic), which in the Swabian part attain a thickness of more than 400 m, and in the Franconian part about 200 m. This area is the largest coherent karst region of Germany (in the Franconian Alps alone 6,400 km^2 are karstified, 1,242 caves of all kinds have been discovered in this area).

The marls are practically impermeable; their outcrop is sometimes accompanied by springs. The limestones and dolomites, as a rule, are permeable owing to their jointed nature and their karstification. The degree of karstification is different. While the regionally occurring dense, bedded limestones and rock limestones are not very susceptible to karstification, the porous, jointed limestones and dolomites of the many former sponge reefs (Stotzen) are subject to karstification to a higher degree, and are also partly morphologically accentuated. The limestones of the Upper Malm especially have a largely reef facies (Massenkalk, Frankendolomit), and are very much subject to karstification.

In large parts of the Alb Plateau, underground drainage prevails which takes place in fissures, joints or channels. In many areas water can only be obtained from the alluvium of the valleys or from cisterns. There are no springs on the plateau. The drilling of wells is risky, furthermore since there is danger that the groundwater recovered may be bacteriologically objectionable.

For instance, within the plane-table map Münsingen, scale 1:25,000, there is not a single perennial water course. In the region of Neresheim, in an area of 233 km^2, only 6 km of river course are found. Dry stream beds and ponors or river sinks are widespread. Dolines sometimes of a larger extent (in the form of karst depressions), are also found, such as the undrained area (7.7 km^2) of Battenau east of Geislingen. Intermittent springs and temporary lakes (for instance, the Eichener See near Dinkelberg) are evidence of the fluctuating karst water table. If the latter is cut by the earth's surface, which happens in many river valleys, strong karst springs often appear (for instance, Brenztopf, averaging 1.3 m^3/sec and Blautopf averaging 2 m^3/sec). The yield of the karst springs generally fluctuates very much; a ratio of minimum to maximum yield of 1:50 is no rarity. A typical example is the Blautopf, where a minimum of 0.35 m^3/sec and a maximum of 26 m^3/sec have been measured. Exceptional are those springs which are fed by a karst water reservoir, such as the Egau Spring with a minimum yield of 0.6 m^3/sec and a maximum of 1.2 m^3/sec.

An important karst phenomenon is the Aach Spring north of Singen. Large amounts of water from the Danube which soak in below Immendingen (654 m above sea level), after an air-line wandering of 12 km, issue here in the giant Aach Spring at an altitude of 475 m above sea level. The Aach Spring, the greatest karst spring of Germany, which has an average yield of 9 m^3/sec (maximum 22 m^3/sec), drains into the Rhine River.

The sediments of the Muschelkalk, which are so important for the constitution of the Swabian–Franconian cuesta landscape, continue also on the left side of the Rhine. They are superposed as residues on the Buntsandstein of the Palatinate and the Saarland, and they subsequently encroach northward into the area of the Eifel (Trier Basin). The profile of the Muschelkalk deposits is incomplete in these areas, it has a reduced thickness and shows gypsum intercalations. Essentially small karst forms are developed of which only a small sinter cave near Niedaltdorf/Saar is remarkable.

Westfalian Cretaceous embayment (Münsterland Basin)

This basin-like structure which is bordered in the north and northeast by the Teutoburg Forest, in the east by the Eggegebirge and in the south by the Haarstrang

Fig.14. Schematical cross-section through the highland of Paderborn.

(Rhenish Schiefergebirge), owes its formation to Upper Cimmerian movements (boundary Jurassic/Cretaceous) which caused a subsidence in this region.

While the deposits of the Lower Cretaceous (sandstones and marls) very distinctly bear the stamp of the near continent and do not show any sediments subject to karstification, limestones ("Pläner") were deposited after the transgression of the Cenomanian and Turonian. These limestones at present outcrop on the margin of the basin and are overlain in the centre by younger sediments.

In the north, on the margin of the Teutoburg Forest, the "Pläner" dip relatively steeply and accordingly outcrop in a relatively small zone. On the other hand, in the south and southeast there is a broad outcrop of the "Pläner"-layers, especially on the Paderborn Plateau, owing to the flat dip of 3°–5° in the direction of the basin's interior (Fig. 14). The "Pläner" beds are often faulted and jointed, and the run-off of the karst water is mostly bound to fault or joint zones; velocities of ground-water movement of 135–280 m/h have been measured here.

The plateau presents here the aspects of a covered karst region; dry stream beds, intermittent river courses, intermittent springs, ponors and a large number of sinks are to be observed. Caves are seldom found, but dissolution of limestone in the underground is considerable. About 26,000 m^3 of limestone per annum are dissolved by the karst waters flowing to the Pader Spring alone.

At the foot of the plateau, at the contact of the limestones with the younger superposed marls, a number of strong springs emerge, such as the Pader Spring at Paderborn with an average yield of 4 m^3/sec, the Heder Spring at Upsprunge with 2 m^3/sec, and the Lippe Spring at Lippspringe with 1 m^3/sec.

The area of the Saxonian orogeny

This area comprises the uplands of northwest Germany which are situated between the Münsterland Basin and the Harz Mountains, and is built up by Mesozoic rocks. The northern border is formed by the Wiehengebirge, Wesergebirge, Bückeberg, Deister as well as by an imaginary line running via Hannover, Lehrte, Peine, Braunschweig to Wolfsburg. There is no natural southern border, as the tectonic movements slowly decline here. The approximate boundary towards the south extends from the southern end of the Eggegebirge via Holzminden and Northeim to the Harz Mountains. Common to all these regions is the fact that they belong to a zone in which the post-Variscan movements caused block-faulting tectonics ("Bruchschollentektonik"), which in its turn involved the upward movements of the Zechstein salts which were highly plastic under pressure, a process which additionally complicated the tectonic conditions. Consequently the tectonic structures and the distribution of the individual Mesozoic sedimentary series at the earth's surface are manifold and varied.

A detailed description of the many small limestone regions of this zone, the karstification of which is mostly limited, is not possible within the scope of these pages. It is generally a green karst, and agriculture and forestry are widespread. Rocks subject to karstification are the following: Zechstein gypsums (eastern margin of the Rhenish Schiefergebirge, Stadtoldendorf, western margin of the Harz Mountains), clays of the Upper Buntsandstein (Röt) which are rich in gypsum, Muschelkalk (Lower and Upper Muschelkalk: calcareous–marly, Middle Muschelkalk: marly with gypsum intercalations) and Middle Keuper (gypsum). In the Upper Jurassic there are two karst horizons with several locally limited karst-water layers: a lower predominantly marly–calcareous series (Heersum layers to platy limestone of Eimbeckhausen), and an upper calcareous–marly series with saline intercalations: the Münder Marls (lower serpulite). Of the Cretaceous deposits, practically speaking, only the "Pläner" of the Cenomanian and Turonian are subject to karstification.

Though karst phenomena frequently occur—for instance, in the area discussed about 6,500 dolines were counted—this area cannot be regarded as a pronounced karst region. As a result of the geologic and tectonic diversity of the area, the factors favouring karst formation may operate in a different degree.

Karst regions or leaching regions, respectively, are encountered, in a more or less pronounced form, in the following zones: Teutoburg Forest and Eggegebirge (Muschelkalk and Upper Cretaceous "Pläner"), on the eastern margin of the Rhenish Schiefergebirge (limestones and dolomites of the Zechstein), on the northern slope of the Wiehengebirge and Wesergebirge as well as in the Deister and Süntel (Malm limestones, Münder Marls), in the Lower Hessian–Lippische upland between the Eggegebirge and the Weser River as well as near Bad Pyrmont (Muschelkalk). Furthermore karst phenomena are found among others in the Hils syncline (Malm limestones and "Pläner"), in the Göttinger Forest, on the Dransfeld Plateau (Muschelkalk), on the southwestern and northern margins of the Harz Mountains (Zechstein, Muschelkalk, Jurassic and Cretaceous), in the Rhüden anticline (Muschelkalk), on the western margin of the Ringelheim syncline (Cretaceous limestones) as well as in the Elm, Fallstein, Huy and Hackel (Muschelkalk).

PHASES OF KARSTIFICATION

Karst types

Karst is the sum of all those processes and phenomena which come into existence on and in certain rocks owing to the capacity of these rocks for being dissolved by water. Accordingly, a precondition for the development of these phenomena is sufficient precipitation and easily soluble rocks. The extent of karstification is influenced by a great variety of factors. In addition to the solubility and composition of the rocks, a part is played by the jointing, the layering of the rocks, the morphology of the area, the position of the hydrostatic base level, soils, vegetation, climate (especially temperature and precipitations as well as their seasonal distribution) and duration of karstification.

From the point of view of solution phenomena, a distinction of the different karst types is not justified in itself. The nature and distinctness of karst phenomena are a function of the intensity and duration of karstification, as well as of the kind of the factors influencing solution. Thus, under arid climatic conditions, rock-salt deposits outcropping at the earth's surface can show the same or at least similar solution phenomena as limestone or gypsum rocks under humid conditions.

In the present report a distinction is made, according to the parent rock, between a carbonate karst (limestone, dolomite) and a saline karst (sulphates and chlorides); the limestone karstification is based on a chemical process—mostly according to the formula $CaCO_3 + H_2O + CO_2 = Ca(HCO_3)_2$—and the karstification of sulphate and chloride rocks on physical dissolution.

Limestone karst

The regions of limestone karst (as well as those of gypsum karst) are marked by characteristic solution phenomena of the rock surface and the underground. Certain differences between both rock types are caused by their different solubility and resistivity. A peculiarity of a karst landscape is the increased displacement of the surface drainage into the solution cavities of the underground (see Fig. 13 "water net"); this process is due to the partial dissolution of the underground. As regards the forms of solution, it must be stated that there is probably no karst phenomenon which is not encountered in the German karst regions in a more or less wide distribution; even tropical karst forms are no exception, such as the "Kegelkarst", fossil residues (originating in the Tertiary) of which have been preserved in the Swabian Alps.

As the individual forms of limestone karst are treated in the other chapters of the present work, their description is not necessary here.

The karst regions of Germany, according to the altitudes and to the climate of central Europe, are generally to be regarded as covered karst and green karst, respectively. They show a more or less continuous cover of vegetation which comprises a part of the karst phenomena, especially the small forms. The original characteristic type of vegetation, the "Steppenheide", has largely disappeared owing to anthropogenetic influences (cultivation, exploitation of forests, etc.).

Bare karst is essentially developed only in the Alps (for instance, Karrenfelds of the Gottesacker Plateau). In these regions erosion predominates, and the residues of weathering and soils have been washed off.

Gypsum karst

The karstification of gypsum is the result of a low but rapid water-solubility of gypsum. The karst phenomena are identical with those occurring in limestone karst regions (dolines, ponors, karst springs, sinks, caves, karren etc.).

The karstification of gypsum is based upon the conversion of the parent rock, anhydrite, into gypsum. The depth of conversion varies from a few centimetres to 30–40 m; it depends on the degree of jointing of the anhydrite deposits and their inclination towards the earth's surface. The hydration of the $CaSO_4$ is generally effected from the earth's surface. The steeper the dip and the stronger the jointing, the faster, more complete and more thorough-going is the conversion of the anhydrite into gypsum. The hydration is made difficult by the fact that the volume increases during the conversion, as a result of which small water-bearing cracks and fissures in the anhydrite become closed again. Solution phenomena are widespread on the surface of gypsum rocks, they are lacking only on steep slopes, where erosion predominates. The surface of the uncovered gypsum rocks is mostly pitted by small sinks, the depth of which amounts to no more than a few decimetres. Similar features are shown by gypsum occurrences underlying permeable layers (gravel, sand). Under uniformly permeable layers (limestone, sandstone) the punctiform to linear action of the surface water leads to intensified solution processes in the gypsum rock, as a result of which sinks (up to 25 m deep) tapering off downward are formed. Under more or less impermeable (clayey) layers, the solution action is the smallest.

If the covering layers sag into the solution cavities, dolines are formed. The following types of dolines may be distinguished:

(1) Steep-walled dolines which have been formed by the breakdown of underground cavities.

(2) Pointed funnel-shaped dolines as a superficial form of solution.

(3) Funnel-shaped dolines as a normal form of the covered karst.

(4) Bowl-shaped dolines as a special form of the covered karst; *(3)* and *(4)* have been formed by superficial solution as well as by the breakdown of underground cavities.

(5) "Fossil" dolines with a more or less even bottom; the latter often contains datable Quaternary deposits.

Here and there fault systems and groundwater passages may be recognized by a serial arrangement of dolines.

As to the caves composed of gypsum, two types are distinguished which, however, can merge into each other:

(a) Joint caves. These have their origin in larger joints, where the formation of the gypsum and the dissolution of the rocks by percolating waters, respectively, begins. Channels and flowing water, operating as a forming (eroding) factor are infrequent.

(b) Leaching caves. Characteristic form elements of this cave type are leaching roofs (even horizontal cave roofs and roof parts, comparable to the "sky" in the sink works of Alpine salt mines) as well as facets (oblique, lateral walls starting under the

Fig.15. Scheme of the leaching cave formation. (After BIESE, 1931.)

leaching roofs, see Fig. 15). The facets are only formed in more or less undrained caves filled with water, where a differentiation of the waters according to the degree of saturation, i.e., according to the specific weight, takes place (Laugenschichtung). Near the bottom, where the saturated solutions have concentrated no further dissolution of the gypsum occurs; in the direction of the roof of the cave the degree of saturation of the brine continually diminishes which, in the progress of the dissolving process, leads to an increased bevelling of the cave's walls.

The facets and the roof always form acute angles; in the case of many caves several facets can be identified—example: Himmelreichhöhle near Walkenried at the southern margin of the Harz Mountains (maximum length 170 m, width up to 80 m, height up to 15 m).

A third cave type is the so-called bulge cave ("Quellungshöhle"), its formation being based upon the increase in volume of hydrated anhydrite. In the case of bedded anhydrite lying more or less parallel to the earth's surface, the increase in volume during the hydration may lead to a swelling-like lifting of the top layers converted into gypsum. These caves are small ("dwarf holes", see Fig. 9) and bound to the earth's surface—example: Waldschmiede at Blumenberg on the southern margin of the Harz Mountains (diameter 7,5 m, height 2 m).

Saline karst

Although the morphological phenomena due to rock salt leaching are not related by some authors to the karst phenomena, they will be mentioned here because of their importance and extent in Germany.

The saline karst areas of Germany, unlike those in arid climatic zones, are not areas with outcropping salt deposits, where the phenomena of karst morphology might be studied. These areas are rather characterized by underground leaching processes; the loss of substance in the underground caused by the dissolution of the salt leads locally to the formation of wide-spanned depressions, and to some extent of sinks. With regard to the stratification, two types of rock salts may be distinguished in Germany:

(*1*) Deposits in which the rock salt, having originally been deposited in layers, has accumulated to form salt diapirs, and has penetrated younger, overlying layers (see Fig. 1). The occurrence of salt domes is essentially restricted to the area of the Saxonian orogeny, and to the north German Plain.

(*2*) In the case of these deposits, the rock salt is intercalated, in the form of seams, in other sediments (Fig. 16). In deposits of this kind lie mainly rock salts of the Röt, Muschelkalk, Keuper and Jurassic, as well as the Zechstein beds of different regions, for instance, of the Thuringian Basin (see p. 200) and of the Werra Basin.

The salt diapirs generally lie at depths at which no groundwater circulates, and accordingly no or only a little leaching takes place. Only in the case of diapirs which have gone up into the ground-water area do the solution phenomena make themselves also felt morphologically. Generally trough-like depressions are formed which in the case of solution phenomena in the residual gypsum (gypsum cap, caprock) may also be accompanied by sinks. As the salt domes mostly occupy only a small area, the effects are inconsiderable (however, an exception is made, for instance, by the city of Lüneburg).

Stronger effects are shown by the leaching of rock salt deposits having the form of layers, as major areas are concerned.

According to the kind of ground-water attack, the following leaching types may be distinguished (see Fig. 16):

(a) "Regular salt leaching". This is a salt leaching which attacks more or less regularly from the outcrop in the direction of the layer's dip, and which is adjusted to the salt level which depends on the ground-water conditions. The limit of leaching is accompanied by a striking "salt slope". Subsidences caused by loss of substance in the underground come to pass slowly (maximum: some millimetres per annum) and with a varying intensity. The leaching depressions in most cases take the form of basins or dolines; they can, however, change and enlarge their shape in the progress of leaching. Sinks, thus, do not occur in the case of regular salt leaching.

(b) "Irregular salt leaching". By this is meant the leaching of a more or less intact deposit lying at a greater depth; here the solution of the salts is effected by waters circulating in zones of deformation and disruption. These local regions of solution, which may often be situated a great distance from the outcrop of the salt layers, are leaching basins sinking more or less continuously. According to the resistivity of the rocks, larger sinks and marginal tensional gashes also occur.

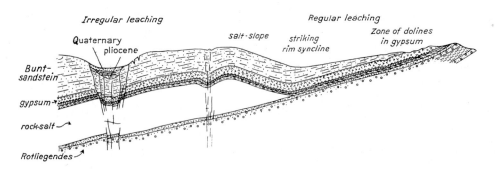

Fig.16. Schematical pattern of regular and irregular leaching. (After WEBER, 1952.)

The age of karstification

Both denudation and karstification begin as soon as rocks are exposed to the effects of the atmosphere. In the major part of the German karst regions this happens at a relatively early date.

Limestone karst

The denudation of the Variscan mountains began in the Carboniferous. The major part of the central and south German upland has been dry since Jurassic times. Phenomena of these early karstification periods naturally no longer exist. The oldest karst phenomena are observed in the Franconian Alps, where karst sinks with Cenomanian deposits have been discovered. There is an abundance of karstification phenomena from the Tertiary, during which both an intensive chemical decomposition and a karstification of the limestones took place under moist and warm climatic conditions. In the Swabian Alps, as well as in the Rhenish Schiefergebirge, there is a great number of sinks and caves in which remnants of mammals from all stages of the Tertiary have been found. In many regions of the Alb relict soils from the Tertiary are encountered. Karstification of the limestones continued during the Pleistocene and Holocene, as may be inferred from the many findings of Glacial and Recent faunas in the caves.

Gypsum karst

The karst phenomena in gypsum regions seem to be essentially of Holocene age. Probably similar forms were produced in the Pleistocene as well, which later, however, fell in.

With regard to the salt karst formation, the first solution phenomena probably had occurred by the beginning of the Mesozoic, as may be inferred from primary variations of the thickness of the Mesozoic sediments. The climax of salt solution was in the Tertiary. In the regions of solution, the gradual subsidence led to the formation of important lignite deposits (Geisel Valley near Halle/Saale, Frankenhausen/Thuringia, amongst others). From the fact that brine springs still outcrop today at many localities at the surface, it may be concluded that salt solution continues. Several ore occurrences of an essentially Tertiary age are confined to fossil karst areas of Germany. Thus, in karst hollows of the Rhenish Schiefergebirge, concretions of hydroxyl apatite are found; phosphoric acid was formed on the decomposition of Devonian eruptives containing apatite, and it was precipitated at its contact with limestones. Residual ore deposits containing concretions of iron and manganese ore are found in troughs, pockets, funnels and cave galleries in Devonian, Zechstein, Jurassic and Cretaceous limestones of many parts of Germany. The most important German manganese deposit, the Lindener Mark near Giessen, is also related to this type. In the Swabian and Franconian Alps, the formerly economically important "bean ores" (Bohnerze), brown haematite concretions in the residual clays occurring in funnels and sinks, are found. In addition ochreous earths (a filling of sinks) occur which are still exploited for use as a basic substance for colours.

Soils in karst areas

The characteristic soil type in the German karst regions is the rendzina (with its subtypes), shallow A–C-soils rich in $CaCO_3$ and $CaSO_4$; these soils are formed on carbonate and gypsum rocks respectively, by solution weathering. The soil, which is heavy owing to its clay content (solution residue), has a good humus form (much limestone humate); humus is present mostly in the form of mull. Mould formation prevails in the case of dryness and an especially shallow soil as well as in the case of the rendzinas lying on sulphate rocks. The humus content seldom amounts to less than 5%. The following subtypes are found: syrozem-rendzina, mull-like rendzina, brownish rendzina, Braunerde-rendzina, terra-fusca-rendzina. The brown subtypes prevail, corresponding to the general mean altitude and to the existing climatic conditions. In addition, some other soil types exist in the German karst regions; thus, in areas of periglacial loess drifts, loess soils (Parabraunerde, chernozem) are developed, and pseudogleys with impeded internal drainage are often developed on the clay horizons which are frequently intercalated in limestones. Soil formation on talus and in the valley flats is influenced by many factors, and accordingly the soil types formed are manifold.

In old karst regions (Swabian and Franconian Alps, Devonian and Carboniferous limestones of the Rhenish Schiefergebirge) pure rendzinas are relatively uncommon and brown carbonate soils and terra fusca are predominantly developed.

In addition, the residues of dislocated terra-rossa or terra-fusca soils are found at localities protected from erosion (often dolines or fissures).

HYDROLOGIC CONSEQUENCES

Though the bulk of the ground-water catchment in Germany originates from the alluvial sands and gravels of the river valleys, from the Buntsandstein and from the Pleistocene deposits of the north German Plain, the amount of waters caught from karstified horizons by borings or the tapping of springs is not unimportant. In the karst regions at least these waters share considerably in the total catchment.

The waters which are found in the karst regions are generally not very favourable, firstly because their often high degree of hardness is detrimental and secondly, because the very rapid infiltration and the high velocity of flow of the karst waters in joints and solution channels, allow only an insufficient biological auto-purification since the time underground necessary for the decomposition of organic matter is mostly too short. The lack of water in many karst regions, however, necessitates the use of these water reserves to ever increasing extent, even though the water hardness is 25–30° of German hardness (1° of German hardness $= 2.5$ mg $CaO/l\ H_2O$)

and precautionary measures have to be taken against eventual or existing bacteriological contamination.

We shall mention here as examples the hydrological conditions in the areas of the Eifelian limestone troughs (Devonian massive limestone), of the southern Harz Mountains (Zechstein) and of the Swabian and Franconian Alps (Jurassic).

The Devonian reef limestones of the Rhenish Schiefergebirge are important aquifers. Intercalated in more or less impermeable rocks such as argillaceous schists, greywackes and quartzites, they represent the only natural water storage in these regions, apart from Quaternary loose sediments.

The areas of karstification in the Eifel region are essentially confined to the "Eifelian north–south-zone" (width: 30–40 km), in which there are several Middle–Upper Devonian limestone troughs striking in a northeast–southwest direction. Among the sediments in these troughs there is a tectonically much disturbed, dolomitized coral limestone having a thickness up to 500 m and which forms an important groundwater horizon owing to its strong karstification. Furthermore karstified Upper Devonian and Lower Carboniferous limestones on the northern edge of the Eifel (anticline of Aachen) are good aquifers, and form the basis for the drinking water supply in the industrial area of Aachen.

The dolomites of the Eifelian limestone troughs are very pure, hard rocks, almost without bedding and with only local intercalations of thin clays. The troughs are surrounded by more or less impermeable Lower Devonian schists, greywackes and quartzites which constitute the aquiclude of the groundwater present in the limestones.

The massive limestones have no hydrologically effective pore space, i.e., the rocks are primarily dense and consequently impermeable. Thus, water can only flow in these rocks if the systems have secondary openings. Such hollows may be joints, fissures or solution-formed karst cavities starting at these joints.

The groundwater in the Eifelian limestone troughs generally has a free water table depending on the local hydrostatic base level. The direction of ground-water flow varies locally, depending on the morphology, altitude, stratification of the rock layers and on the direction of drainage.

Springs rise at those places, where the ground-water table is cut by the earth's surface, or where relatively impervious beds of Lower Devonian outcrop and are overlain by limestone beds (contact springs), which is chiefly the case at the margins of the Eifelian limestone troughs. The water supply of many villages in the Eifel is based upon such springs. Strong springs are found, for example, at: Birgel (Kreis Daun) with 0.1 m³/sec, Müllenborn (Kreis Daun) with 0.125 m³/sec, and Dreimühlenquelle near Weyer with 0.07–0.09 m³/sec.

Water supply by means of borings (water wells) seems possible at those places, where the Eifelian limestones outcrop in a great thickness below the hydrostatic base level, where these limestones show a large catchment area, and where a strong jointing and karstification exists. At such places up to 50 m³/h per well can be obtained.

In the Eifelian limestones, as a residual phenomenon of Tertiary and Quaternary volcanism, there are local occurrences of carbonic acid; the existing ground and spring waters locally form acidulous springs. The solution of the carbonate rocks is considerably increased by the water charged with carbonic acid. The result thereof is an intensified karstification which has locally attained so high a degree that the limestones are present in a slacklike form.

In an effort to mobilize all useful ground-water storages for water supply one proceeds sometimes, despite certain risks, to catch water also from Zechstein horizons. This is done chiefly at the southern margin of the Harz Mountains (see pp. 199–202). The main water storage is the jointed Plattendolomit, of 20 m thickness, which is intercalated in saline sediments. The chief difficulty of water exploration results from this fact: the sealing of the wells to prevent the invasion of waters of a high sulphate or chloride content. The clay layers underlying and overlying the Plattendolomit are also not always impermeable, so that useless water can flow to the store itself.

When drilling a well, poor water in minerals is often at first encountered in the dolomites. After a certain time of delivery, clayey deposits present in fissures and cavities of the underground are washed out, and thus connections are established with other karst channels, in which waters containing gypsum and salt often circulate. The water delivered thereby obtains a higher content of dissolved matter. Waters with a total hardness of more than 40°, at a sulphate content of 700 mg/l, are generally not suited for drinking purposes (1° of German hardness corresponds to 2.5 mg $CaO/l\ H_2O$).

The possibilities of exploiting waters which are usable from a chemical point of view, are often subject to rapid local changes. Thus the town of Bad Lauterberg did not succeed in obtaining good groundwater in a sufficient quantity till a number of borings had been made, all of which fall within a relatively small area.

The towns of Seesen, Pöhlde, Bad Sachsa and others are supplied from wells in Zechstein beds.

The Jurassic of the Swabian and Franconian Alps does not show any continuous ground-water reservoir, rather there are several karst-water reserves, with their own karst circulation. In some regions the karst-water flows in larger joints and other secondary openings; in other regions, however, water circulates only in the fine fissures. The latter regions show a low "water conductivity", and act as a wall between the catchment areas of larger karst springs, preventing a compensation between the different reserves so that the individual karst-water bodies have their own drainage systems.

It appears that in the Alb there are areas of a differing karst development; definite karst-water systems and spring types are associated with these areas. These conditions, in a modified way, are also valid in the other karst regions.

Shallow karst

Here, the impermeable layer of the ground-water storage lies above the hydrostatic base level. The waters soaking into the underground flow quickly and practically unfiltered through the joints, fissures and caves of the strongly karstified rocks, and they issue from contact springs in those areas where the aquiclude is cut by the earth's surface (mostly in valleys). The water storage capacity of these areas is small, and the discharge of these karst springs is subject to great fluctuations.

Deep karst

In this type of karst region the impermeable layer below the karst water is below the level of the valleys. Different zones may be separated (Fig. 17). In the "outer zone" the storage reservoir is relatively small, as the bottom of the aquifer lies only a few metres below the hydrostatic base level. The springs of this zone have a greater discharge than those of the shallow karst, but they, too, show greater fluctuations of the discharge; the water is often objectionable as regards bacteriology.

In the "inner zone" of the deep karst, however, karstified limestones occur far below the hydrostatic base level. Thus the karst-water reservoir is large and often has a considerable extent. The fluctuations of the karst-water table and of the discharges of the springs are only small. Decomposition of organic matter is possibly due to the long time during which the waters are underground; hence the waters of this zone are mostly not objectionable as regards hygiene.

The Jurassic layers of the Swabian Alps and to some extent those of the Franconian Alps, plunge in the south below the impermeable Tertiary and Quaternary deposits of the molasse basin; they are karstified to a depth of 2,000 m, and borings have revealed that they contain fresh water at this depth. This zone is designated as a

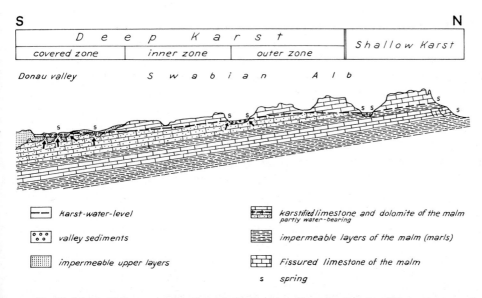

Fig.17. Schematical cross-section through the Swabian Alps, showing the different karst-water zones. (After WEIDENBACH, 1954.)

"covered zone". The recharge is from the area of the "inner zone" of the deep karst. The groundwater is confined; its movement is extremely slow and probably amounts, at most, to a few metres per day.

Large amounts of hygienically satisfactory groundwater may be obtained from this zone.

Because of the general lack of water on the Swabian Alps, already 100 years ago, people began to supply the localities from central water systems. According to their geographical situation, hydrological points of view, and the planned capacity of the pumping station, villages and farms are connected to a "water supply group" and supplied from central water sources through pipe lines. The waterworks are mostly in valleys, and there they take the groundwater from Quaternary gravels and sands or from limestone beds. In these valleys springs are often utilized for the water supply.

Today, in the Swabian Alps, there are 23 groups of waterworks, supplying an area of 5,000 km^2 (140,000 inhabitants). Difficulties are caused, however, by the increasing amount of waste water. The still normal practice of disposing of the waste water in sinks involves local danger to the sanitary quality of the central water supply, due to the quick percolation and the migration in limestone joints.

PROBLEMS OF ENGINEERING GEOLOGY IN KARST AND LEACHING AREAS

The limestone, owing to different factors, may represent a particularly unstable foundation soil. Engineering structures, especially in bridge building, tunneling and dam building as well as blankets for dams require, therefore, extraordinary precautionary measures.

The typical foundation problems of limestone karst regions will not be described in detail here. These problems can better be treated when taking as an example countries with great typical karst regions. Only some aspects of engineering geology to which attention must be paid in Germany, and which concern above all areas of leaching, will be mentioned here. In areas with a conspicuous, flat to undulating stratification of the Mesozoic limestone, many slides occur owing to the frequent intercalation of clay horizons (aquicludes) in the limestone sequence, especially at slopes or at artificial slope cuts; these slides take place on the clay horizons between the karstified limestone layers. Moreover, the great increase in volume during the swelling of the clays leads to a differential settling, which causes danger especially in the case of the structures with a shallow foundation, as well as in the case of railways and streets.

In bituminous limestones and clays, the oxidation of the pyrite they contain gives off sulphate to the groundwater, which from a certain concentration onward, attack concrete.

Isolated dolines do not generally prevent building. However, special foundations based upon special local investigations are indispensable.

Settling occurring below the loading soil is caused by the loss of substance of the deeper underground; this loss is due to the leaching of the soluble salts by groundwater. In the wide areal extent of the regions of Germany with saline (rock salt) deposits engineering structures are required in these areas of potential subsidence.

A criterion for judging the degree of danger and the presumable development of the subsidence is the "youthfulness" of the latter which can be recognized by the surface forms. The amount of settlement in salt areas is generally from fractions of millimetres up to some millimetres per annum. Areas especially endangered can be determined by observation of the surface forms, geologic mapping, tectonic investigations, hydrologic and hydrochemical observations, measurements of the gravity and of subsidence, etc. Building is possible even in areas of active subsidence (if there is no leaching of gypsum with the subsequent formation of dolines), because the subsidence due to the suberosion of salts are continuing processes which lead to the formation of constantly progressive, basin-shaped or striking zones of subsidence of a wide areal extension. Relatively safe building is possible in the centre of the basin unless there is a danger that the dolines are formed due to the karstification of gypsum. Only the margins of the depressions should be excluded from building, as in this tensional area the subsidences occasionally have a jerky character, so that fissures may be formed. Important industrial plants or buildings sensitive to differences of subsidence should not be constructed in depression basins, especially if the leaching in the underground has not been completed.

REFERENCES

BAUER, F., ZÖTL, J. and MAYR, A., 1958. Neue karsthydrographische Forschungen und ihre Bedeutung für Wasserwirtschaft und Quellschutz. *Wasser und Abwasser*, 1958:280–297.
BELSER, E., 1936. *Die Abwasserversorgung in ihrer Geographischen Bedeutung*. Thesis, Becht, Tübingen, 117 pp.
BIESE, W., 1931. Über Höhlenbildung, 1. Entstehung der Gipshöhlen am südlichen Harzrand und am Kyffhäuser. *Abhandl. Preuss. Geol. Landesanstalt, Neue Folge*, 137:71 pp.
BIESE, W., 1933. Über Höhlenbildung, 2. Entstehung von Kalkhöhlen (Rheinland, Harz, Ostalpen, Karst). *Abhandl. Preuss. Geol. Landesanstalt, Neue Folge*, 146:170 pp.
BOLSENKÖTTER, H., 1963. Ein Färbeversuch in der Rohrer Kalkmulde (Devon der Eifel). *Gas-Wasserfach*, 104(40):1156–1158.
BRINKMANN, R., 1959. *Abriss der Geologie, 2. Historische Geologie*. Enke, Stuttgart, 360 pp.
CRAMER, H., 1928. Untersuchung über die morphologische Entwicklung des fränkischen Karstgebirges. *Abhandl. Naturhist. Ges. Nürnberg*, 22(7):244–325.
CRAMER, H., 1934. Einige Zahlen über fränkische Höhlen. *Fränkische Berg- Wintersport Z.*, 2.
DORN, P., 1960. *Geologie von Mitteleuropa*, 2nd edition. Schweizerbart, Stuttgart, 488 pp.
ECKERT, M., 1902. Das Gottesacker Plateau, ein Karrenfeld im Allgäu. Studien zur Lösung des Karrenproblems. *Z. Deut. Österreich. Alpenver., Wiss. Ergänzungshefte*, 1(3):108 pp.
ENDRISS, G., 1936. Die Karsterscheinungen der Schwäbischen Alb. *Z. Erdkunde*, 4(14):628–632.
FRICKE, W., 1965. Zum Karstwasserproblem des norddeutschen Oberjura. *Geol. Jahrb.*, 83:641–666.
FRIESE, H., 1935. Die Karsthohlformen der Schwäbischen Alb. *Bl. Schwäbischen Albver.*, 47:121–127.
GELLERT, J., 1958. *Grundzüge der Physischen Geographie von Deutschland,1. Geologische Struktur und Oberflächengestaltung*. VEB Deutscher Verlag der Wissenschaften, Berlin, 586 pp.

GERECKE, F. and NEUNHÖFER, H., 1964. Beobachtungen natürlicher Bodenerschütterungen im Gebiet des Südharzkalibergbaus und der Mansfelder Mulde. In: *Beiträge zur Regionalen Geologie Thüringens und Angrenzender Gebiete sowie zu Anderen Problemen—Abhandl. Deut. Akad. Wiss. Berlin, Kl. Bergbau, Hüttenwesen, Montangeol.*, 2:399–405.

GEYER, O. F. and GWINNER, M. P., 1962. *Der Schwäbische Jura—Sammlung Geologischer Führer, 40*. Borntraeger, Berlin, 452 pp.

GRÄBE, R., 1964. Die Karsterscheinungen im oberdevonischen Knotenkalk bei Schleiz (Ostthüringisches Schiefergebirge). In: *Beiträge zur Regionalen Geologie Thüringens und Angrenzender Gebiete sowie zu Anderen Problemen—Abhandl. Deut. Akad. Wiss. Berlin, Kl. Bergbau, Hüttenwesen, Montangeol.*, 2:443–452.

GRAHMANN, R., 1958. Die Grundwässer in der Bundesrepublik Deutschland und ihre Nutzung. *Forsch. Deut. Landeskunde*, 105(2):198 pp.

HAASE, H., 1936. *Hydrologische Verhältnisse im Versickerungsgebiet des Südharz-Vorlandes*. Thesis, Univ. of Göttingen, 218 pp.

HAASE, H., 1951. *Das Gesicht der Heimat. Erdgeschichtliche Deutung der Harz-Landschaften*. Krösing, Osterode/Harz, 81 pp.

HAEFKE, F., 1959. *Physische Geographie Deutschlands*. VEB Deutscher Verlag der Wissenschaften, Berlin, 357 pp.

HENKEL, L., 1902. Karsterscheinungen im Thüringer Muschelkalk. *Arch. Landes- Volkskunde Prov. Sachsen*, 12:8–9.

HERRMANN, A., 1964. Gips- und Anhydritvorkommen in Nordwestdeutschland. *Silikat-J.*, 3(6): 442–466.

HERRMANN, A., 1966. Vergipsung und Oberflächenformung im Gipskarst. *Intern. Congr. Speleol., 3rd, Vienna, Sect. C*, pp.99–108.

HUBER, F., 1959. Das "Höhlenkataster Fränkische Alb". Geschichte, Anlage, Bedeutung, Stand. *Geol. Bl. Nordost Bayern*, 9(2):49–84.

HUNDT, R., 1942. Erdfälle als landschaftsgestaltender Faktor in Thüringen. *Thüringer Fähnlein*, 11(1/3):13–14.

JUX, U., 1960. Die devonischen Riffe im Rheinischen Schiefergebirge, 1. *Neues Jahrb. Geol. Paläontol., Abhandl.*, 110:186–258.

KAMMERER, F., 1962. Ingenieurgeologische Methoden in Erdfall- und Senkungsgebieten. *Freiberger Forschungsh., C*, 127:49–109.

KEIL, K., 1954. *Ingenieurgeologie und Geotechnik*, 2nd edition. VEB Knapp-Verlag, Halle/S., 1132 pp.

KEILHACK, K., 1935. *Lehrbuch der Grundwasser- und Quellenkunde*. Borntraeger, Berlin, 575 pp.

KESSLER, H., 1966. Wasserhaushalts- und Wasserwirtschaftsfragen in Karstgebieten. *Intern. Congr. Speleol., 3rd, Vienna*, 5:31–34.

KOEHNE, W., 1913. *Geologische Geschichte der Fränkischen Alb*. Piloty and Loehle, Munich, 42 pp.

KOSACK, H.-P., 1954. Beiträge zur Kenntnis der Karstgebiete in Deutschland. *Ber. Deut. Landeskunde*, 12(2):292–298.

KRANZ, W., 1957. Eine 170 m tiefe Wünschelruten-Fehlbohrung auf Wasser in Bopfingen und weitere Bohrungen in der Umgebung. *Jahresber. Mitt. Oberrhein. Geol. Ver., Neue Folge*, 39:17–23.

LEHMANN, O., 1932. *Die Hydrographie des Karstes—Enzyklopädie der Erdkunde*. Deuticke, Leipzig, 212 pp.

LINDNER, H., 1922. *Kritische Erörterung der Hypothesen über die Karsthydrographie*. Thesis, Technische Hochschule München, Munich, 143 pp.

MÜCKENHAUSEN, E., 1957. *Die Wichtigsten Böden der Bundesrepublik Deutschland*. Kommentator, Frankfurt am Main, 146 pp.

MÜCKENHAUSEN, E., 1962. *Entstehung, Eigenschaften und Systematik der Böden der Bundesrepublik Deutschland*. DLG-Verlag, Frankfurt am Main, 148 pp.

PAL, A. M., 1912. *Karstphänomene auf der Reutlinger und Tübinger Alb*. Laupp, Tübingen, 32 pp.

PENCK, A., 1904. Über das Karstphänomen. *Schr. Ver. Verbreitung Naturwiss. Kenntnisse Wien*, 44:1–38.

PENCK, A., 1924. Das unterirdische Karstphänomen. In: *Recueil de Travaux, offert à M. Jouan Cvijic*. Beograd, pp. 177–197.

PFEIFFER, D., 1961. Zur Definition von Begriffen der Karst Hydrologie. *Z. Deut. Geol. Ges.*, 113:51–60.

PFEIFFER, D. 1963. Die geschichtliche Entwicklung der Anschauungen über das Karstgrundwasser. *Geol. Jahrb. Beih.*, 57:111 pp.

Pietzsch, K., 1956. *Abriss der Geologie von Sachsen*. VEB Deutscher Verlag der Wissenschaften, Berlin, 200 pp.
Reuter, F. and Reuter, R., 1962. Ingenieurgeologische Beurteilung und Klassifikation von Auslaugungserscheinungen (Hinweise für das Bauen in auslaugungsgefährdeten Gebieten). *Freiberger Forschungsh.*, C, 127:5–47.
Richter, M., 1966. *Allgäuer Alpen—Sammlung Geologischer Führer, 45*. Borntraeger, Berlin, 189 pp.
Richter–Bernburg, G., 1955. Über salinare Sedimentation. *Z. Deut. Geol. Ges.*, 105:593–645.
Rutte, E., 1962. Karst- und Überdeckungsbildungen im Gebiet von Kelheim-Donau. *Quartär*, 14:69–80.
Scheffer, F. and Schachtschabel, P., 1966. *Lehrbuch der Bodenkunde*, 6th edition. Enke, Stuttgart, 473 pp.
Scheuermann, R., 1954. Die Solstellen am Kyffhäuser und ihre Pflanzenwelt in Vergangenheit und Gegenwart. *Ber. Naturhist. Ges. Hannover*, 102:39–47.
Spöcker, R.G., 1935. Der Karst des oberen Pegnitzgebietes und die hydrographischen Voraussetzungen für die Wassererschliessung bei Ranna. *Abhandl. Naturhist. Ges. Nürnberg*, 25(1):1–83
Spöcker, R. G., 1950. *Das Obere Pegnitzgebiet. Die Geologischen und Hydrogeologischen Voraussetzungen für eine Wasserschliessung im Fränkischen Karst*. Nürnberg, 140 pp.
Spöcker, R.G., 1952. Zur Landschaftsentwicklung im Karst des mittleren und oberen Pegnitzgebietes. *Forsch. Deut. Landeskunde*, 58: 53 pp.
Stille, H., 1903. Geologisch-hydrologische Verhältnisse im Ursprungsgebiete der Paderquellen zu Paderborn. *Abhandl. Königl.-Preuss. Geol. Landesanstalt Bergakad., Neue Folge*, 38:1–129.
Stolberg, F., 1925. Die Südharzer Zechsteinhöhlen. *Der Harz*, 8:596–599.
Von Bubnoff, S., 1956. *Einführung in die Erdgeschichte*, 3rd edition. Akademie-Verlag, Berlin, 808 pp.
Von Knebel, W., 1906. *Höhlenkunde mit Berücksichtigung der Karstphänomene*. Vieweg, Braunschweig, 222 pp.
Weber, H., 1952. Pliozän und Auslaugung im Gebiet der oberen Werra. *Geologica*, 8.
Weber, H., 1955. *Einführung in die Geologie Thüringens*. VEB Deutscher Verlag der Wissenschaften, Berlin, 243 pp.
Weidenbach, F., 1950. Wasservorräte und Wasserversorgung in Nordwürttemberg. In: *Raumforschung und Raumordnung*. Akad. Raumforschung Landesplanung, Cologne, 4:5 pp.
Weidenbach, F., 1954. Über einige Wasserbohrungen im Jura. *Jahresber. Mitt. Oberrhein. Geol. Ver., Neue Folge*, 36:54–73.
Weidenbach, F., 1956. Über Wasservorräte in Württemberg und Möglichkeiten der Erschliessung. In: *25 Jahre VEDEWA, 1931–1956*.
Weidenbach, F., 1957. Altes und Neues vom Brenztopf. *Jahresber. Mitt. Oberrhein. Geol. Ver., Neue Folge*, 39:25–36.
Weidenbach, F., 1960. Trinkwasserversorgung aus Karstwasser in der östlichen Schwäbischen Alb. *Jahresh. Karst-Höhlenkunde*, 1:169–191.
Wiefel, J., 1966. Auslaugungserscheinungen in der oberen Trias im südöstlichen Teil des Thüringer Beckens. *Ber. Deut. Ges. Geol. Wiss., Reihe A, Geol. Palaeontol.*, 11(3):295–304.

Chapter 7

Karst of Austria

F. BAUER AND J. ZÖTL

Speleological Institute, Vienna (Austria)
Technical University, Graz (Austria)

INTRODUCTION

About one-sixth of the area of the Austrian State is built of karstifiable rocks, nearly all of which are situated in the Alpine region. About 80% of the Austrian karst area is occupied by the Northern Limestone Alps (Nördliche Kalkalpen) alone (see Fig. 1). The marked differentiation of the rock-qualities, the mountain formation, the morphogenesis, and the climatic factors have brought about the formation of different karst types, which cannot be comprehensively described here. In order to give, however, at least a survey of the Austrian karst, its main types only are described here in greater detail by means of some characteristic examples, whose selection has been made with a view to the present state of their investigation.

Austrian karst research was initiated in the Dinaric region, then part of the Austro-Hungarian Monarchy. A member of the Vienna school of geographers

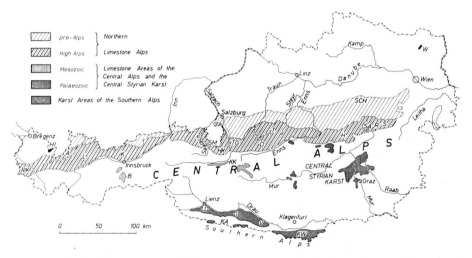

Fig.1. Significant karst areas of Austria. A = Allgäuer Alpen; B = Brenner Mesozoic; D = Dachstein; G = Gailtaler Alpen; H = Hochschwab; HI = Hoher Ifen; HK = Hochkönig; K = Karwendel Mountains; KA = Karnische Alpen; KK = Klammkalk zone; KW = Karawanken; L = Lechtaler Alpen; LD = Lienzer Dolomiten; R = Rax; RK = Raetikon; S = Schneeberg; SA = Schneealpe; SCH = Schlagerboden; SM = Steinernes Meer; T = Totes Gebirge; TG = Tennengebirge; V = Villacher Alpe; W = Waschberg.

headed by Albrecht Penck, GRUND (1903) was the first to treat monographically the hydrography of the karst and to coin the term "karst groundwater". On the other hand, KATZER [1909], the geologist, and BOCK [1913], the speleologist, stressed the importance of independent karst waterways, Bock presuming the existence of large cave-rivers. TERZAGHI [1913] studied the formation of dolines in detail. The economic problems posed by the karst were even then clearly realized and were dealt with by the Imperial-Royal Ministry of Agriculture by means of large-scale afforestation and melioration.

BOCK [1913a,b] was the first to transfer the opinions formed in the classic karst to the Alpine karst areas, which after 1918 became the main field of Austrian karst research. At that time speleology flourished under G. Kyrle. LEHMANN [1927] studied the zonal distribution of lapies and dolines in the high karst areas and, through his work *Die Hydrographie des Karstes* [1932], influenced the general opinion on karst-water movement for decades.

After 1945 karst research received a considerable impetus all over the world. In Austria the Association of Austrian Speleologists (Verband Österreichischer Höhlenforscher) worked towards the systematic registration and study of the Austrian karst caves [SCHAUBERGER and TRIMMEL, 1952]; the surface forms of the Alpine karst were the object of studies by university institutes and the Vienna Speleological Institute. Through the development of new techniques, entirely new ways were opened to karst hydrographic research. New knowledge in this field was gained, above all, through large-scale tracing experiments in coordinated cooperation between the Vienna Speleological Institute and the Graz Association for Hydrogeological Research (Vereinigung für Hydrogeologische Forschungen in Graz).

THE NORTHERN LIMESTONE ALPS

General description

Stratigraphy and lithology

Triassic. The Triassic sequence of the Northern Limestone Alps is developed in several facies, which are connected by manifold transitions [SPENGLER, 1951].

Carbonate rocks are represented in all stages of the north-Alpine Triassic. Thin limestones are occasionally interbedded in the Scythian "Werfener Schichten". The Anisian is represented by mostly dark or black limestones and dolomites (Gutensteiner Kalk und Dolomit). Gray, silex-bearing limestones of the Anisian and Ladinian are called "Reiflinger Kalk". In the Ladinian a sequence of light-coloured, partly bedded limestones (Wettersteinkalk) reach a thickness of up to 1,500 m. In large areas of the Limestone Alps, this limestone is substituted for by white, semi-crystalline dolomites (Wettersteindolomit = Ramsaudolomit). In the western part of the Limestone Alps, the Wettersteinkalk is replaced by the "Arlberg Schichten" (bedded limestones with marl-layers). The upper part of the Karinthian shows carbonate facies (Opponitzer Kalk and cellular dolomites = Rauhwacke). The Norian is developed in the "Hauptdolomit" and "Dachsteinkalk" facies. The greyish, brittle and partly thick-bedded Hauptdolomit is up to 1,000 m thick and covers extensive areas of the Limestone Alps. The Dachsteinkalk, reaching a thickness of up to 1,500 m, is a light to greyish limestone, reaching stratigraphically into the Rhaetian Stage. It is partly well-bedded (Geschichteter Dachsteinkalk), partly homogeneous reef-limestone, cut by numerous vertical joints (Dachstein-Riffkalk). In the Upper Rhaetian, in the western section of the Limestone Alps, the "Kössener Schichten" are overlain by the light, massive to rough-bedded Upper Rhaetian "Oberrhätkalke".

Argillaceous and sandy beds divide the carbonate rocks stratigraphically. The Scythian, representing the base of the Mesozoic limestone sequence, consists of sandstones (Buntsandstein) in the west, and predominantly of micaceous, argillaceous sandstones (Werfener Schichten) in the east. In the Lower Ladinian of the western section there occur dark, thinly bedded marls (Partnach Schichten). The rocks of the Carnian Stage (sandstones, marls of the Lunzer and Raibler Schichten, partly coal-bearing) form an important impermeable layer. In the Rhaetian of the Kössener Schichten, developed only in the west and in the eastern pre-Alps, there are beds of marls and shales.

Locally important deposits of gypsum or anhydrite are found in the Scythian and in the Carnian Stage of the Northern Limestone Alps. Rock salt, tectonically intensely mixed with gypsum, other salt minerals and clay, occurs at the base of the Triassic sequence; this so called "Haselgebirge" dominates in the facies of Hallstatt (developed within the Hallstatt nappes), which cannot be dealt with in this report.

Jurassic. In the Northern Limestone Alps, Jurassic sedimentary rocks are far less thick and less frequent than in the Triassic. The Liassic shows the facies of "Hirlatzkalk" (white to pink crinoidal limestones) and of "Lias-Fleckenmergel" (marls, extensively developed in the western and northern areas of the Limestone Alps). In some areas, the Liassic is represented by multi-coloured cephalopodal limestones (Bunter Cephalopodenkalk). Above the Hirlatzkalk follow the Middle Jurassic (Dogger) "Laubensteinkalk" and "Vilserkalk" (crinoidal and Brachiopoda-bearing limestones), and above the cephalopodal limestones the "Klaus-Schichten" (brown and red limestones). The lower part of the Upper Jurassic (Lower Malm) contains radiolarites. The Upper Malm is developed in the facies of the 'Aptychenkalk" and "Oberalmer Schichten" (platy, marly limestones, often cherty), and also in the facies of "Plassenkalk" (white reef-limestone). The "Sulzfluhkalk" of the Lower East Alpine nappes of the Raetikon Mountains is lithologically similar to the Plassenkalk.

Cretaceous. Marls dominate in the Neocomian of the Limestone Alps. The Cenomanian shows conglomerates, sandstones and marls. The transgressive Gosau Formation (Turonian–Danian) is formed by conglomerates, sandstones and marls, rich in fossils. It contains local coal seams and bauxite deposits.

Lithological factors of morphology. The dominant lithologic elements of the sequence of the Northern Limestone Alps are limestones and dolomites of the Ladinian and Norian stages of the Triassic. Wettersteinkalk and Wetterstein dolomit but also Hauptdolomit and Dachsteinkalk, reach thicknesses of over 1,000 m in the Limestone High Alps. In comparison, other Triassic and Jurassic limestones and dolomites are thinner and less numerous. Wettersteinkalk and Dachsteinkalk are the most important rock formations of the Limestone Alps.

The limestones generally form escarpments, mountains with steep rocky flanks and characteristic plateaus, but they also build the summits of mountain chains. The strong karstification and the predominantly subterranean drainage reduces the erosion and dissection of the limestone massifs. This has often caused the preservation of Upper Tertiary denudational plains.

The dolomites of the Triassic, because of their lithology (often densely fractured and jointed), are easily eroded. Steep rocky walls are rare and never reach the steepness or height of similar landforms in limestone areas. The dolomite areas of the pre-Alps are often morphologically characterized by round-shaped hills and ridges. In the argillaceous and sandy strata which are easily eroded, soft landforms are developed.

Tectonics

The tectonic structure of the Northern Limestone Alps originates from pre-Cenomanian, pre-Gosavian and Early Tertiary orogenetic phases. It is characterized by extensive nappe-type folding in which overthrusts within the Limestone Alps reach distances of over 50 km. The main direction of overthrusts is always northward. The stratigraphic sequences of the various nappes, generally beginning with Scythian strata, show normally different facies. The stratigraphically youngest beds are often eroded.

The nappes of the Voralpen (pre-Alps), forming the northern parts of the Limestone Alps, mostly have a complicated fold and imbrication structure. The nappes of the Hochalpen (High Alps), with their important limestone sequences, form the widespread plateau-mountains, situated in the southern parts of the Limestone Alps. This area is particulary important for its intensive karstification.

The morphologic development of the Limestone Alps

The nappe-type mountain area, folded in the Early Tertiary, has been denuded to a hilly terrain on which hardrock gravels were deposited in Early Miocene times. The gravels probably were derived from the Central Alpine areas. The gravel blanket, originally several hundred metres thick, was eroded again almost completely at the end of the Miocene. Relics of this gravel blanket, named "Augensteine", can be found at present almost everywhere in the Northern Limestone Alps, especially on the plateaus of the eastern Limestone High Alps but also in fissures and caves where they were deposited by water action after having undergone manifold resedimentation. With the uplift of the Alps from Upper Miocene to Pliocene times, widely extended denudational plains were formed, which probably originally covered a high percentage of the Northern Limestone Alps. These denudational plains can be divided into various systems of plains showing a certain degree of interfingering. During the continued uprising of the Alps, these systems of denudational plains were preserved only in areas of extensive limestone sequences (dominantly Wettersteinkalk and Dachsteinkalk) with low dip. In areas where the carbonate rocks show minor thickness or where an intensive intercalation of steep-dipping limestone–dolomite sequences with marls and shales prevails, but also in predominantly dolomitic areas, these old denudational plains were destroyed by erosional processes in connection with the development of a later drainage net. The Upper Miocene–Pliocene denudational plains whose higher elements are generally termed "Raxlandschaft" [Rax-landscape, LICHTENECKER, 1926] form the landscape of the large plateau-mountains of the Northern Limestone Alps.

The levels of the Raxlandschaft were dissected and thrown against each other by younger tectonic block-movements. During the following incision of valleys into the old denudational plain-systems, the general drainage direction, which had originally a northern course from the Central Alps in the south over the Limestone Alps to an erosion base in the north, was interrupted. In the border area between the Central Alps and the Limestone Alps, the important longitudinal valleys of the rivers Inn, Salzach and Enns were formed. This process of valley-development had been completed during the latest Pliocene. At the beginning of the Pleistocene, most of the valleys had been incised to a depth corresponding to that of the present.

Main types of the North Alpine karst

The important differences of facies, thickness of strata and tectonic structure within the Northern Limestone Alps have caused the development of different morphological types. The latter also differ with regard to the intensity and type of karstification and to their altitude (High Alps with summits of around 2,000 m and more and pre-Alps with summits of around 1,500 m and less).

The limestone-plateau type is most expressively well-developed in the regions of the Limestone High Alps east of the Saalach River. Basically, this type is represented by extensive masses of limestones (Triassic Wettersteinkalk or Dachsteinkalk), forming more or less isolated limestone massifs with often impressively steep flanks. The dominant morphological element is the presence of plateau-type denudational plains. The plateau type in its classic form begins in the west in the area of Steinernes Meer and extends over Hochkönig, Tennengebirge, Dachstein, Totes Gebirge, Hochschwab, Schneealpe, Raxalpe and Schneeberg to the Vienna Basin. The development of typical plateaus becomes less distinct toward the north (pre-Alps) and the plateaus decrease in size. At the northern fringe of the Limestone Alps, the plateau type is again represented in the area of Untersberg and Höllengebirge.

The limestone mountain-chain type occurs in its classical form in the limestone ranges of the northern Tyrolean Alps (Wettersteingebirge, Karwendel), with mostly steeply dipping masses of Wettersteinkalk for its main constituent.

In the pre-Alps of lower Austria intensively folded and faulted sequences of limestones, dolomites, argillaceous rocks and sandstones are predominant.

The extreme west of the Northern Limestone Alps is characterized by high-Alpine dolomite ranges (Raetikon, Lechtaler Alpen, Allgäuer Alpen). Because of the predominantly surficial drainage and of the great amount of erosion in this area, no significant karst topography is developed there. In the dolomite mountains of the pre-Alps, which are mostly covered by forest, karst phenomena are likewise rare.

The types of karst as mentioned above are characteristic examples; between them there exist all the transitional stages.

The development of the North Alpine karst until the end of the Glacial Epoch
Indications of karstification in the Northern Limestone Alps have been observed already within the Mesozoic sequence (Triassic–Jurassic border): Liassic Hirlatzkalke are deposited in fissures of the Triassic Dachsteinkalk, in cavities which were probably formed by karstification processes. However, these fossil remnants of karst phenomena have no bearing on the younger karstification.

The principles of Tertiary karstification can best be studied in the plateau-massifs of the Northern Limestone Alps.

During Early Miocene times (Augensteinzeit), no more than a few hums or inselbergs may have risen above a gravel plain and undergone a locally restricted karstification.

The karst cycle responsible for the present landforms began with an uplift of the Northern Limestone Alps during the Middle Miocene which caused a denudation of the Augenstein Beds. The first stage of karstification may well have occurred already underneath the overlying gravel blanket of "Augensteine", at a time when these were elevated above a first local erosional base level.

We cannot exclude the possibility that a (subsurface) karstification, having occurred below the probably permeable gravel blanket, has contributed much to the formation of many karst phenomena (karst depressions and dolines with diameters of several hundred metres), found entirely within the plateaus of the High Alps. This would easily explain the origin of the above karst phenomena for which mainly climatic–morphologic reasons used to be given.

Repeatedly resedimented remains of Augensteine are found today (often in dolines and caves) in areas of Late Tertiary denudational plains. During the continuing uplift of the limestone plateaus, the karstification process became more intensive and caused the transport of Augenstein gravels into fissures and cavities from whence further redeposition into cave systems took place.

At further stages of the uplift, which are reflected in several systems of denudation plains, and during the incision of the first valleys (at a transitional stage between sheet-like denudation and plateau-forming processes and the linear incision and erosion), important subterranean drainage systems were developed. Waters originating in the Central Alps, whose run-off at this stage was not directed into the longitudinal valleys of later days, found their way through the limestone massifs. We have to assume that several valleys and depressions (their bottoms probably as much as 1,000 m above their present levels) in the southern sections of the Limestone Alps functioned as poljes during Pliocene times [ZÖTL, 1963]. Areas of the central Enns Valley (drainage via the limestones of the Gesäuse Mountains, partly through the eastern part of the Dachstein Massif and partly through Dachsteinkalk of the Dachsteinhöhlen) and, at a later stage, also the Aussee Basin (drainage through the limestones of the Saarstein-Zinken area) must be considered as old polje areas.

The concentration of large cave systems in elevations of about 1,500 m is considered to be an indication of a main phase of subterranean flooding of the karst massifs of the High Alps. In sections of the cave systems of the Dachstein, [Petrefakten Cave, Rieseneis Cave, Mammut Cave; BAUER, 1954b], deposits of crystalline sands and scallops point to a local northwestward directed flow of streams originating in the Central Alps. This subterranean flooding represents the latest stage of a hydrological influence of the waters from the Central Alps on the North Alpine karst.

After completion of the valley-forming processes in Late Pliocene times, the Limestone Alps were hydrologically isolated from the Central Alps. For the further development of their subterranean karst systems, only precipitation within these karst areas was available. The old cave-rivers no longer functioned. New, local drainage systems developed, with the newly incised valley bottoms as their base level.

In areas consisting of limestones and dolomites, less extensive or folded and containing intercalated shales or marls, the older denudational plains were largely destroyed by erosion. Regional subterranean drainage from the Central Alps to the north of these areas was not possible because of the interbedding of limestones and impermeable layers. Therefore, phenomena as described for the plateau massifs of the High Alps can hardly be observed here.

During the Pleistocene, the higher elevations of the Limestone Alps were covered by glaciers. On the plateaus extensive glaciers were formed, reaching thicknesses of up to 400 m (e.g., on the Dachstein Massif). Dolines already formed during Late Tertiary times, but also old depressions and tectonic zones of weakness were widened, eroded and incised. Those plateaus that were once covered by Pleistocene glaciers, are characterized by a distinct "Gassenlandschaft": more or less straight, narrow ravines which always occur on fault lines and joints, reach lengths of over 100 m and show an average depth of several metres. Morphologically they represent the network of the main faults and joints. Deposits of Augenstein gravel and terra rossa-like soils which occur quite often on the bottom of these "Karstgassen" indicate their pre-Glacial disposition or origin. Their present forms derive from the selective action of the glaciers.

The "Plattenlandschaft" is also restricted to areas of former glaciation and is found entirely within the limestone areas with distinct bedding and flat dip-angles of the strata. The Plattenlandschaft is a landscape showing ice-polished bedding planes and step-like scarps of the thickly bedded limestones [Schichttreppenkarst; BöGLI, 1964].

Old karst lapies were denuded by glacial erosion on the limestone plateaus and only relics of deeply incised "Kluftkarren" may have been preserved. The karstification may also have been partially effective underneath the glaciers. Frequently jamas and karst shafts occur on the formerly glaciated plateaus; their origin is partly ascribed to the action of glacial waters. Moraine debris has been deposited in depressions of the plateaus. Moraine deposits in caves, for example in the Dachsteinhöhlen, show that direct connections existed between the surficial sinkholes and the cave systems below. Non-glaciated top-ridges within the plateaus show distinctly rounded forms (probably periglacial landforms), in which only the larger and old karst depressions may have been preserved.

In the Interglacial stages a surficial karstification was possible in areas laid open after the regression of the glaciers. The few indications of such phenomena do not permit a stratigraphic determination.

Principles of hydrology

The principal parameters for the development of a subterranean drainage are the thickness, extension and stratification of the bedrock, the morphology of the surface, the amount of water available for penetration into the ground, the orographic position of the base level of drainage and the duration of the karstification. The tendency towards subterranean drainage is largest in areas where the surface of the bedrock is horizontal or only gently inclined, thus especially in limestone plateaus. An important factor is the increase of precipitation with increasing orographic height: in the karst massifs of the High Alps, where the yearly precipitation is up to 2,500 mm, the karstification is strongest. Another factor is the vegetation: in the predominantly wooded pre-Alps (with a smaller annual precipitation), an important amount of water, moreover, is lost by evapotranspiration. On the other hand, in these regions covered by soil and vegetation the infiltration water shows a higher content of CO_2 than the rainwater infiltrating in the zone bare of vegetation, thus having a higher corrosive activity on the rock surface.

The development of subterranean drainage nets in limestones and dolomites normally begins along zones of weakness of the bedrock (joints, fissures, faults, bedding planes, lithologic boundaries). It proceeds more rapidly in the more soluble limestones than in the dolomites. The subterranean drainage directions depend solely on the geologic structure of the karst massif and on the altitude of the local base level. They are, therefore, largely independent of the morphology of the infiltration areas.

Within extensive limestone plateaus, the precipitation penetrates in a sheet-like manner along the manifold fissures and bedding joints, preferably at the junctions of such elements. Using the faults and cracks in the bedrocks, the water gradually develops a widely extended, three-dimensional drainage net which may reach to considerable depths. The drainage takes place in a concentrated form along certain fault directions but mostly in a diffuse way, making use of all available fissures and bedding joints, generally in the direction of the base level. In cases where a massif drains towards several local base levels, the karst-water run-off moves towards them radially from a central culmination of the karst-water table. In cases of a single local base level only, the water flows centripetally toward the former. Where large amounts of surface waters penetrate at certain points only (e.g., in swallow holes of glacial streams) the subterranean run-off may follow a main direction (development of more or less isolated subterranean water channels). The results of tracing experiments in karst regions of the High Alps during the last decade show that the subterranean water channels often may intersect or overlap each other. Therefore within a karst region, the recharge areas of the karst springs may often largely overlap each other; generally they deviate considerably from the orographically defined recharge areas.

In steeply dipping limestone ranges (Kettencharakter) which are often bordered laterally by poorly permeable or impermeable strata, the karst drainage may be limited to the longitudinal axis of the mountain chains.

In dolomitic rocks the development of subterranean drainage is determined by the smaller degree of solubility and the strong cleavage of these rock types. In most cases, a diffuse network of very thin drainage passages and veins is developed and larger cavities are much rarer than in limestones (caves are less frequent in dolomites). In dolomitic rocks, the flow of subterranen water is remarkably slower and the storage capacity is greater.

The orographic position of karst springs does not indicate the depth of the subterranean drainage system, which can also reach below the local base level. The subterranean percolation is limited only by the depth of the karst rocks (base level of karst erosion).

The more developed the subterranean drainage of a karst massif is, the fewer are the important karst springs (with a discharge of up to several m^3/sec) that discharge the larger part of percolating waters of the area. Smaller springs, situated above the larger karst springs mostly have local recharge areas only and frequently may go dry as they are not connected with the main karst-water body. The considerable variations of discharge are characteristic of karst springs. The amount of discharge of perennial karst springs during snow melting in springtime or after heavy rains may reach more than 20 times the minimum amounts.

Relations between the water temperature of springs and the elevation of the recharge area were determined by ZÖTL [1961] and were confirmed by deuterium analyses of the spring waters.

Within a karst massif three depth zones of water flow can be distinguished:

(1) A zone lying above the karst water table into which waters from the surface infiltrate. These waters penetrate solely according to the laws of gravity along open fissures and joints and have vertical drainage directions (vadose zone).

(2) A deep zone of the karst-water body which is limited below by the depth of the karstified rocks and above by the lowest level of the karst water table. Water moves

there under hydrostatic pressure (phreatic zone). In the upper sections of this zone one can expect higher velocities of waterflow than in the deeper.

(3) Between both zones there is an area in which the water movement takes place according to the two principles mentioned above and corresponding to the elevation of the karst-water level as resulting from the variations of seasons and precipitation. This third zone shows the highest percolation rate and, correspondingly, the highest degree of widening of cavities by solution. Because of the increased dissolution of the rocks in this zone, the subterranean cavities are often found to be concentrated at certain levels.

Tracing experiments in the Northern Limestone Alps gave indications of the average velocity of the water flow measured in a straight line between sinkhole and spring, ranging from 10 to 250 m/h. The actual velocities must be many times higher than these average velocities, because the actual subterranean watercourse is much longer than a straight line. Furthermore, hydrologic factors such as dry periods or periods of heavy precipitation will certainly influence the flow velocity. No conclusions are possible as to the actual velocities of subterranean water movement from the amount of water penetrating into swallow holes and from the discharge of the springs.

Caves

The origin of the North Alpine karst caves and of their development to their present form derive from formerly active subterranean drainage systems. In many cases, Late Tertiary, Glacial and post-Glacial stages of development of the caves can be distinguished. The original shape of the cavities was often changed by crumbling, occurring mostly during the Quaternary. The various horizontal or vertical cave sections are mainly confined to fissures and bedding joints (and, especially, to junctions of the above) and to lithologic borders between karst rocks and impermeable strata. The development of the caves differs widely, depending on the specific lithology and the tectonics of the various karst massifs.

Caves consisting chiefly of horizontal passages were derived in most cases from a water flow in the uppermost levels of a karst water body, directed towards the local base level. Caves with a chiefly vertical extension (shaft caves) were formed by waters infiltrating vertically towards the karst-water body. The dimensions of the caves depend on the amount of water flow and the duration of the flow.

In the plateau massifs of the High Alps, large cave systems (niveaugebundene Grosshöhlen) are concentrated at altitudes of 1,300–1,800 m; they are relics of an extensive Late Tertiary drainage system. These caves were mainly formed by allochthonous waters running from the Central Alps towards the northern foreland. The formation of these caves took place under palaeohydrographic conditions (which cannot be reconstructed in detail) during intermittent cessations or retardations of the Late Tertiary uplift of the Alps. The present altitude of the large cave systems of the various karst massifs can only be compared under certain restrictions; different stages of the uplift and local karst-hydrographic conditions must be taken into account. Local differences of permeability within one cave system may cause differences in the altitudes of the tunnels amounting to as much as a hundred metres and more. In many cases one can discern several cave levels within one system, connected by shafts or steep passages. In some sections of a cave system large tunnels, halls and domes with diameters of up to several tens of metres alternate with narrow passages. Of the original cavities corresponding to the stage of Late Tertiary flooding, only exceptional relics have been preserved, and these show characteristic current marks [BAUER, 1954b,1961a]. Deep shafts cutting through horizontal tunnels belong to a younger genetic phase with a dominantly vertical drainage trend (see Fig. 2). The deepest shaft cave of the Northern Limestone Alps is the Gruberhornhöhle (Hoher Göll, Salzburg) with a difference in altitude of 854 m.

Of the cave systems originating in Late Tertiary times, only the explored sections are being investigated scientifically. Notwithstanding this incomplete knowledge. the distinct horizontal trend of the main tunnels of the large cave systems and their dependence on certain levels can be clearly recognized. The storey-like structure of certain systems of large caves indicates their formation by gradual uplifts of the Alps.

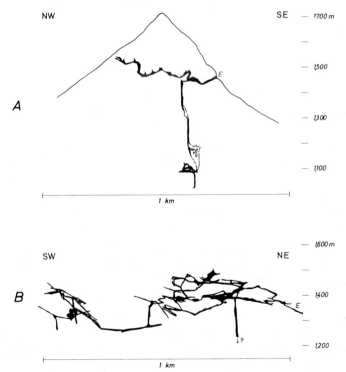

Fig. 2. Schematic sketches of important caves in the Northern Limestone Alps, showing more or less horizontal gallery systems cut by deep shafts (E = entrance). A. The "Geldloch" in the Ötscher (Lower Austrian Limestone Alps). Total length of the galleries is 1,800 m, depth of the shaft is 432 m. B. The gallery system of the Mammut Cave (Dachstein). Total length of the galleries is 16.5 km. (After TRIMMEL, 1966.)

The cave systems of the plateau massifs of the Northern Limestone Alps, being restricted to certain altitudes, must therefore be considered as parts of a larger-scaled, Late Tertiary percolation system, whose altitude was determined by the position of the base level of erosion (see Table I).

TABLE I

THE MOST IMPORTANT CAVES OF THE NORTHERN LIMESTONE ALPS

Cave	Height above sea level (m)	Total length (km)
Eisriesenwelt Cave (Tennengebirge)	1,600–1,800	42
Mammut Cave (Dachstein Mountains)	1,250–1,500	16.5
Tantal Cave (Hagengebirge)	1,300–1,700	16
Frauenmauer–Langstein–Tropfstein Cave (Hochschwab)	1,400–1,600	10.5

A large part of the karst caves of the Northern Alps owe their origin and formation to local speleologic factors. The formation of these caves, taking place at various times, was caused chiefly by the available amount of infiltration water deriving solely from precipitation within the cave area. General coincidence with a certain elevation cannot normally be ascertained.

The tendency for drainage towards the respective local base levels can be recognized in some active drained caves (e.g., Lamprechtsofen near Lofer, and the Koppenbrüller Cave in the Dachstein Massif).

Surface karst phenomena

Sheet corrosion and the formation of lapies within limestone areas are the main elements of the Postglacial karst landscape development. The possibility of the development of karst forms in distinct climatic zones depends on lithology, cleavage and jointing and on the inclination of the surface. The climatic factors determine the altitude of the snow line and the timber line and thus the altitude of specific lapies forms. Changes of climate shift these zones and are responsible for a remodelling of older karst forms or their destruction. The surface-karstification is strongest on little-inclined surfaces of pure, fine-grained limestones and under the influence of a large amount of precipitation, that is, above all, on the large limestone plateaus of the High Alps. The coarser or the more argillaceous the limestones are, the more indistinct becomes the development of definable lapies forms. Fissure and bedding joints are the starting points of intensified, vertically directed solution processes (Kluftkarren). Lapies are rarely formed in dolomites that weather easily into debris.

Lapies. The conditions of the recent formation of lapies can best be studied within the extensive plateaus of the Limestone High Alps at altitudes of 1,600–2,000 m. At these altitudes lapies are dominant karst phenomena, covering wide areas (e.g.,

Karrenfelder of the Dachstein Massif and the Totes Gebirge). There, older lapies had been eroded by the action of Quaternary glaciers. The present lapies were therefore formed—with few exceptions—in Postglacial times.

The following principal lapies forms can be discerned [BAUER, 1958; Fig. 3]: lapies forms which originate on exposed limestone surfaces (such as channeling and "Trittkarren") and lapies forms originating under an overburden of soil (such as Rinnenkarren, up to 1 m deep, with rounded contours).

The various overlapping generations of forms quite often depict the Postglacial climatic oscillations. Broad, well-rounded lapies were partly remodelled by a younger solvent action, creating channeling at the crests and narrow drainage rills at the bottom of the older forms. The rounded lapies originated during the "Postglacial Thermal Maximum" under overlying soils; the younger rills were formed after the erosion of the overburden on the now denuded surface (partly within the last centuries). The rounded lapies, formed in the Dachstein Massif at an altitude of 2,100 m during the Postglacial Thermal Maximum under soil cover, were partly denuded by young and brief advances of glaciers [BAUER, 1961b].

Classification of Postglacial karst forms according to their respective altitudes must therefore take account of the climatic oscillations in Postglacial times. At the time of the Thermal Maximum, the timber line was approximately 300–400 m higher than it is today. The lapies-forming processes which can be studied at present at an altitude of about 1,800 m formerly went on at elevations of about 2,100–2,200 m. The mutual overlapping of different generations of lapies can therefore be identified only within the zone of the fluctuating timber lines.

"Kluftkarren" and "Schichtfugenkarren" are forms determined by lithologic factors. They are probably mainly deepened under soil cover. Old Kluftkarren may, under favourable conditions, have outlasted the Quaternary glaciation. At elevations above 2,200 m the frost weathering becomes more and more important and prevents the development of significant lapies.

In the predominantly wooded pre-Alpine karst areas (and also in the lower parts of the karst massifs of the High Alps) the formation of lapies is determined by the smaller amount of precipitation and the thicker and more argillaceous soil cover. In places where the surface of limestones is exposed by recent soil erosion, irregularly shaped, more or less rounded rills and lapies of different depth are found. Lapies forms, such as channelings and "Trittkarren", are hardly ever developed below 1,600 m.

The amount of chemical erosion of the sparsely vegetated karst areas of the High Alps, caused by the dissolving action of water in Postglacial times, is assumed to be about 15–20 cm of limestone [BAUER, 1964].

Dolines. Large karst depressions and large dolines doubtlessly originated in pre-Glacial times. Only smaller depressions may be of Postglacial origin. Dolines in "classic" pattern, being characteristic for the Dinarides, have a limited distribution

A

B

Fig. 3. Typical lapies forms. A. Channelings originated on exposed limestone surfaces (Dachstein, 1,900 m). B. Rinnenkarren, up to 1 m deep, originated under an overburden of soil (Totes Gebirge, 1,600 m).

in the Northern Alps. Such forms as pre-Glacial remnants reach diameters and depths of several tens of metres, and sometimes of more than 100 m, e.g., at the plateaus of the Hochschwab and in the Sengsengebirge [BAUER, 1952].

In the plateau areas of the High Alps, which have been glaciated during the Pleistocene, pre-Glacial karst depressions and the "Gassenlandschaft" predominate.

Within the calcareous moraine debris of the High Alps there occur funnel-shaped solution sinks of a few metres in diameter. They are commonly caused by solution processes which took place after the Postglacial Thermal Maximum and under a soil cover which was mostly eroded later on [BAUER, 1958b].

"Erdfalls" occurring mostly in the pre-Alps, are mainly caused by a denudation of an impermeable soil cover into karst hollows of the subsurface. They often develop into local swallow holes and small dolines.

Solution depressions on gypsum (gypsum dolines) are confined to the sparse outcrops of gypsum in the Limestone Alps. They often show strong deepening in Recent times [GÖTZINGER, 1957]. Large gypsum dolines (e.g., around Lünersee, Raetikon) are certainly older.

The sparse well-developed dolines of the pre-Alpine areas are mostly of only local importance.

Regional examples of North Alpine karst

In this report only some examples of significant karst areas and karst problems of the Northern Limestone Alps can be mentioned. Because a general description has been given above, here only some local, prevailing hydrological problems are described.

Schneebergalpen

The Schneebergalpen comprise the plateau massifs of Schneeberg (70 km^2, 2,075 m), Raxalpe (65 km^2, 2,007 m) and Schneealpe (110 km^2, 1,903 m) which consist mainly of Wettersteinkalk. During Glacial times the plateaus have only locally been glaciated. Lapies are rare. The plateaus are mostly vegetated (pastures and dwarf pines). Relics of Late Tertiary soils are abundant. The morphological features of the limestone plateaus in this area have been described by LICHTENECKER [1926] as "Raxlandschaft" (Fig. 4). Most of the large springs of this region are used for the water supply of the city of Vienna (Erste Wiener Hochquellwasserleitung).

The first, large-scale experiment in the Alps employing colour tracers was undertaken on the Schneeberg Massif [DOSCH, 1956]. These tests showed a radial drainage pattern of the subterranean waters percolating in the center of the massif, directed towards almost all the springs around the massif.

During 1966–1968, a tunnel 9,7 km long was driven through the Schneealpe to get spring water from the southern part of the Schneealpe into the Erste Wiener Hochquellwasserleitung. The tunnel runs for a good part along the base of the limestones, above the Werfener Schichten, and has a maximal overburden of an average

Fig. 4. The Rax Massif from the west.

of 900 m Wettersteinkalk. This tunnel permits important observations of the hydrologic conditions in the depths of a karst massif. Large quantities of water have been encountered in the tunnel, issuing at numerous joints, faults and in cavities (total discharge in winter time about 400 l/sec). According to the relations between discharge in the tunnel and the precipitation and melting of snow on the plateau, three distinct hydrologic zones have been distinguished, in spite of the short time available for observations.

Beginning with the northern tunnel entrance, there are:

(*1*) The outer zone (0–1,000 m tunnel length): the waters in the tunnel react strongly with the surface precipitation in amount of discharge, temperature and chemical composition. Within this outer zone the karst water table sank gradually to the tunnel level.

(*2*) The transitional zone (about 1,000–3,000 m): discharge is almost constant in temperature and quantity and shows only changes in its chemical compostition.

(*3*) The central zone (from 3,000 m onward): the partly plentiful discharge shows occasionally only slight changes in chemical composition in relation to the precipitation.

The three zones described above clearly indicate a tri-section of the karst-water body of the Schneealpe. Zone (*3*) corresponds to the deepest, poorly flooded core of the karst-water body, lying so deep below the mean karst water table that fluctuations of the latter do not cause detectable changes in discharge. Zone (*1*) characterizes the surface layer of the karst-water body which also feeds the karst springs. Zone (*2*) represents a transitional stage, lying already below the mean water table so that the amount of its discharge is no longer influenced by fluctuations of the water table, although it will be reached by waters from the surface layer within few days.

The largest cavity encountered in the tunnel was in the central part of the Schneealpe. Out of this cave about 1,500 m^3 of sand flowed into the tunnel. The visible part of the cavity had a volume of at least 1,000 m^3. This proves that the deep karstification of Alpine karst areas can reach down to the base of the limestones, even if the flow-speed of waters in this zone is small.

Measurements of water pressure in the southern limb of the tunnel showed that the karst-water table within the massif, in a horizontal distance of 500 m from the point of measurement to the spring, lies up to 30 m higher than the corresponding vauclusian springs (Siebenquellen).

In a debouchure, at 3,040 m from the northern entrance of the tunnel, water with a content of about 0.06 mg uranine/m^3 was found in May, 1968. This tracer comes from experiments conducted in the Schneealpe area in June, 1963—that is five years earlier (F. Dosch, unpublished report)—during which only 33 % of the 15 kg of uranine injected were recovered in the spring water.

About 37,000,000 m^3 of potable water from the karst massifs of the Schneeberg, Raxalpe and Schneealpe areas was brought to Vienna in 1967 by the Erste Wiener Hochquellwasserleitung.

Totes Gebirge

The Totes Gebirge, extending over 590 km^2, is the largest karst massif of the Northern Limestone Alps. Its plateaus (with altitudes of 1,500–2,300 m) consist in the eastern part of flat-lying Dachsteinkalk; in the western part the Dachsteinkalk is overlain by Jurassic limestones. The main massif is separated from the Tauplitz Plateau in the south by an important fault zone. Tracing experiments [MAURIN and ZÖTL, 1964] showed that tracers fed into the sinkhole of the Steyrer See (which is situated on the northern edge of the fault zone) were mainly transported through the complete massif itself over a distance of 30 km to springs at the northern edge of the Totes Gebirge; only small amounts were found in nearby springs in the south. A radial subterranean drainage from a central point of the massif was observed (Fig. 5).

Dachstein

The Dachstein Massif (ca. 400 km^2) is the best known karst massif of the Northern Limestone Alps. The presently glaciated summit area (Hoher Dachstein, 2,996 m) rises above a wide plateau system. The massif has steep rocky flanks of up to 1,500 m in height. The main structural element of the Dachstein is the slightly northward-dipping Dachsteinkalk of up to 1,000 m thickness (Fig. 6).

The hydrology of the Alpine karst was studied here for the first time in numerous tracing experiments [ZÖTL, 1957; BAUER et al., 1958]. The subterranean drainage pattern shows, on the one hand, a somewhat diffuse, radial flow from small sinkholes on the plateau towards a number of springs. On the other hand, water from large glacial sinks flows in an almost linear drainage system towards vauclusian springs, discharging at the foot of the massif (Fig. 7).

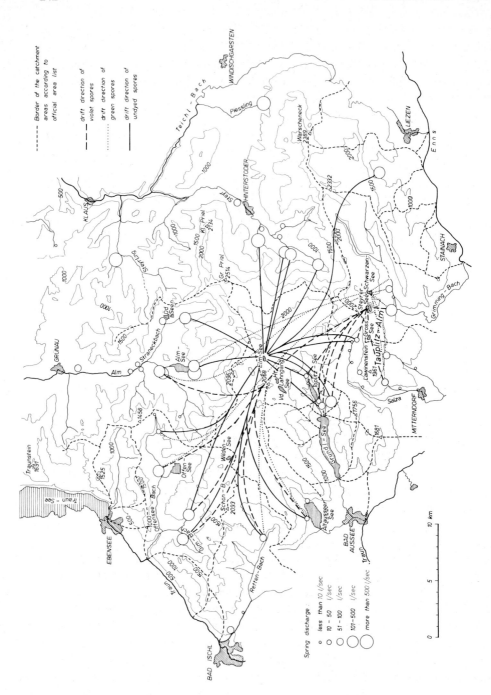

Fig. 5. Results of tracing experiments in the Totes Gebirge. The spores injected into the lake water ponors of the Steyrer See and Schwarzer See crossed the main massif towards the northwest, overlapping the tracers injected into the centre of the massif (Elm

KARST OF AUSTRIA 243

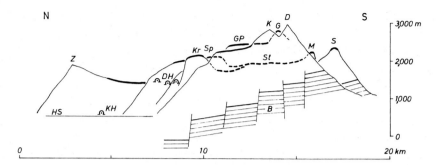

Fig. 6. Schematic geologic section of the Dachstein Massif. The thick lines indicate the old plateau levels. B = Impermeable base of the karstified Dachsteinkalk; HS = level of the Hallstatt Lake; DH = Dachstein caves (Mammut Cave, Rieseneis Cave and Petrefakten Cave as relics of a Late Tertiary subterranean drainage system); KH = Koppenbrüller Cave (as a part of the recent drainage system); D = Hoher Dachstein; G = Hoher Gjaidstein; K = "Hohes Kreuz"; GP = Gjaidstein Plateau; Sp = Speikberg; Kr = Krippenstein; M = Miesberge; S = Sinabel; Z = Zinken; St = plateau "Am Stein".

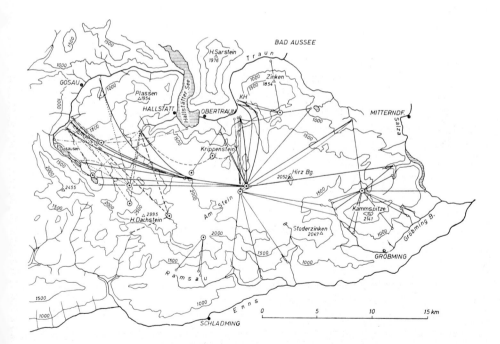

Fig. 7. Results of tracing experiments in the Dachstein area. Large circles with point indicate the injection point of the spores; small circles are springs and sampling points; KH = Koppenbrüller Cave.

The development of the Late Tertiary plateau systems can be recognized quite clearly within the Dachstein Massif. The plateau levels around 1,700–2,000 m, covered mostly by *Pinus mugo* are characteristic of most of the karst plateaus of the Northern Alps (Fig. 8).

The development of karst forms and their relationship to local climatic factors as well as the water balance of the plateau areas are being studied at the observation station Oberfeld of the Speleological Institute at the northern edge of the Dachstein Massif at 1,830 m.

The large cave systems of Rieseneis Cave and Mammut Cave, situated at the northern border of the massif at 1,200–1,500 m and in which strong indications of Late Tertiary flooding have been found, are one of the main fields of work of the "Verband Österreichischer Höhlenforscher" (Fig. 9).

Hochkönig
The Hochkönig Massif (70 km^2, 2,941 m) consists principally of northward-dipping Dachsteinkalk. On the plateau, covered partly by relics of a strongly regressing glacier, different tracers were injected into five sinkholes in August, 1966 (F. Bauer, unpublished report). Three of them, which were fed into small sinks (1–2 l/sec; see Fig. 10,*A,B,C*) were found after 1–3 days in two springs. Two of the tracers, which were fed into big sinks, where large amounts of meltwater (10–30 l/sec; see Fig. 10,*D,E*) disappeared, were found in only one vauclusian spring, after 16–19 days.

The long time interval between injection of the two tracers in the big sinks and their discharge in the vauclusian spring indicates that the tracers, together with large amounts of meltwater, passed into deeper levels of the phreatic zone, because heavy precipitation with snowfall in the plateau area occurred three days after the injection, and the ice-melting and the subterranean drainage from the meltwater sinks to the spring was interrupted. Additionally, the large amounts of rainwater precipitating in the lower, marginal parts of the Hochkönig Massif probably functioned as back-storage for the subterranean waters.

The three tracers from the small sinks must have been fed into the topmost karst-water level which flows with considerable speed to the springs.

Steinernes Meer
The Steinernes Meer is the most extensive karst massif between the Saalach and Salzach rivers (280 km^2). It is formed mainly of Dachsteinkalk; Dachstein dolomite and Jurassic deposits occur in the north and northwest. The extensive karrenfelder of the plateau are surmounted in the south by summits of over 2,600 m; in the north, the Watzmann (2,714 m) is in German territory.

High valleys, traversed by Pleistocene glaciers and descending in mighty steps to the recent valley bottoms, are characteristic of the Steinernes Meer.

The Diessbach Basin (Fig. 11) is a Tertiary valley, which today extends into the transverse valley of the Saalach in the shape of a hanging valley. This Tertiary valley,

Fig. 8. The eastern Dachstein Plateau from the Krippenstein. In the righthand part of the picture is the glaciated so-called "Kargebirge" with the summit of the massif.

Fig. 9. The "Paläotraun", the most important gallery of the Dachstein Mammut Cave. (Photo by K. Mais.)

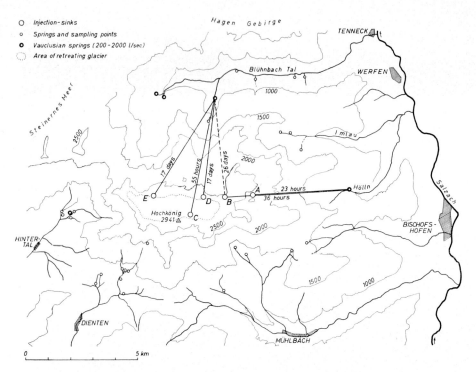

Fig. 10. Results of tracing experiments in the Hochkönig Mountains. A, B, D, E = injections of dyed spores; C = injection of uranine.

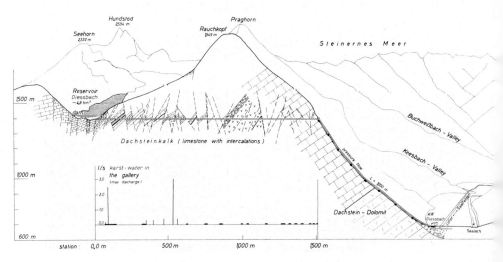

Fig. 11. The Diessbach reservoir and power plant (Steinernes Meer). (After BRANDECKER et al., 1965.)

about 1 km long and up to 200 m wide, is situated at an altitude of about 1,400 m, which is more than 700 m higher than the Saalach. This difference in altitude suggested the following experiment. By means of a 30-m high dam, a reservoir was built in the high-situated Diessbach Basin for 4,800,000 m³ of water, the water being fed into a power plant through a tunnel and down the valley slope, which has a slope ranging from 45° to 60° (!).

Large-scale karst hydrologic studies [BRANDECKER et al., 1965] were necessary with respect to the water-tightness of the reservoir, which is situated exclusively in the Dachsteinkalk. Tracing experiments showed that there exist indeed several karst-waterways in the Dachsteinkalk leading into the deep Saalach Valley but for the main part of the eastern karst-water body it is still the Late Tertiary high valley that acts as the local base level of erosion—a fact which had been suggested by the existing springs and by the Diessbach feeding the reservoir.

By means of injection-boring totalling 7,500 m in length and involving the use of a total of 862.8 tons of concrete, the reservoir could in fact be made completely watertight. The cost of these injections amounted to 3% of the total cost of the project. At a drop of 718 m the present capacity of the plant is 12,000 kW and its annual production of energy 25,800,000 kWh; further extension, which is now in progress, will increase its capacity to 24,000 kW and its annual capacity to approximately 40,000,000 kWh.

Karwendel Gebirge

The Karwendel Gebirge (Karwendel Mountains) is clearly divided into four ranges, the "Vordere Karwendel-Kette", the "Hintere Karwendel-Kette", the "Gleiersch-Kette" and the "Solstein-Kette" (Fig. 12). The numerous peaks rising to above 2,500 m emphasize its character as a range of high mountains. They are mainly very compact mountain ranges, whose slopes are carved with numerous cirques of Pleistocene origin. The longitudinal valleys separating the several ranges of the Karwendel Mountains are situated up to 900 m above the Inn Valley.

Tectonically speaking, the basal series of the Karwendel Mountains are part of the Lechtaldecke (Lechtal nappe), over which the Karwendel nappe has been thrust [AMPFERER and HAMMER, 1898; AMPFERER, 1903, 1928]. Towards the east, large parts of the Karwendel nappe have been denuded and wide areas of the basal series (Lechtal nappe) laid bare (Fig. 13, profile *B*). On the southern slope of the Solstein-Kette there extends, at an altitude of about 1,000–1,100 m, the overthrust plane between the basal Lechtal nappe and the overlying Karwendel nappe along the Inn Valley (Fig. 13, profile *C*). Here again, the difference between the two tectonic units, which principally consist of the same Triassic rocks, is very marked: whereas the strata of the lower unit repeat themselves in sequences of varying thickness, the Karwendel nappe has, on the whole, its strata in the order of their original deposition. Furthermore there are, in the basal unit, Late Triassic strata (e.g., Raibler Schichten, Hauptdolomit), which have already been eroded in the Karwendel nappe.

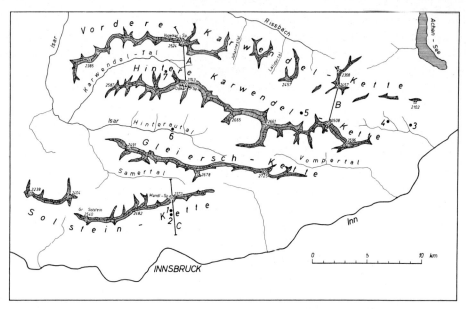

Fig.12. The Karwendel Mountains. 1–7 = Large karst springs; 1,2 = Würm and Klammbach springs; 6 = spring horizon "Bei den Flüssen".

The mountain forms are determined, apart from the overthrust planes and geotumor structure, by numerous longitudinal as well as cross faults, which mainly influenced the formation of valleys.

Of the rocks of Triassic age that exist in either nappe, the Wettersteinkalk, which is here up to 1,500 m thick, is extremely karstified. Also the stratigraphically lower-situated Muschelkalk, whose thickness runs to several hundred metres, and the Partnachkalk are highly liable to karstification. Wettersteindolomit and Hauptdolomit are doubtlessly less karstifiable than the above-mentioned limestones; the detritus often prevents the formation of lapies; dolines also occur less frequently. But karstification is nevertheless going on here, too; subterranean drainage, one of its characteristic accessories, can be observed. The redclay-sandstones of the lowermost Triassic (Buntsandstein) are not liable to karstification; neither is the subsequent cellular dolomite of the "Reichenhaller Schichten", a strongly decomposed limestone–dolomitic rock of usually high clay content. Similar is the reaction of the (later) Raibler Schichten, where they are marls and sandstones. The Liassic limestones and the Aptychenkalk of the Jurassic Period are easily karstifiable, but are comparatively rare, as are the Neocomian strata.

Surface karst mainly expresses itself by the distribution of lapies and dolines.

The distribution of lapies in the Karwendel Mountains was investigated by LECHNER [1948], who found that lapies are dependent on certain land forms in the especially karstifiable rocks (Wettersteinkalk, Muschelkalk) and that they are massed at certain altitudes. Steep slopes with dips up to 80°, and the slopes of glacial *roches*

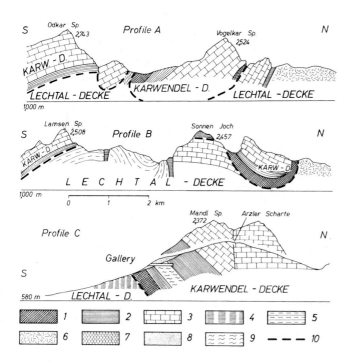

Fig.13. Geologic sections through the Karwendel Mountains. *1* = Buntsandstein; *2* = Muschelkalk; *3* = Wettersteinkalk, Wettersteindolomit; *4* = Raibler Schichten (mainly marls and sandstones); *5* = Hauptdolomit; *6* = mylonite of the Hauptdolomit; *7* = Jurassic and Cretaceous sediments; *8* = Höttinger Breccia; *9* = "Rauhwacke"; *10* = border between the Lechtal Decke (Lechtal nappe) and the Karwendel Decke (Karwendel nappe).

moutonées and of solitary blocks, are the land forms favoured by "Rinnenkarren", whereas "Kluftkarren" are numerous chiefly in the bare Wettersteinkalk and Plattenkalk of the hilltops and passes as well as in the terrain covered with dwarf-pines. The zone of the most active lapies formation in the Karwendel Mountains is between 2,000 and 2,200 m. Solitary lapies occur even at an altitude of 2,600 m. The formation of dolines (and jamas) in the denuded rock of flat tops, crests, hills and peneplains is frequent, chiefly in connection with steep bedding-joints, conjugated joints, and faults. The zone of the most active doline formation in the bare rock, as established by LECHNER [1948], for the total area of the four Karwendel ranges, is the altitude of 1,980–2,260 m. The reason for this frequency lies not in climatic altitudes; this is clear from the fact that the frequency of the dolines in rock at a certain altitude is connected with the concentration of dolines in the area of cirques (cirque threshold and subsequent cirque floor); cirques, however, are bound to a certain zone of altitude.

According to Lechner, dolines are numerous in areas of detritus (Schuttdolinen). Preglacial dolines may have favoured the formation of Glacial cirques [FELS, 1929].

Most of the precipitation and meltwater sinks into numerous sinks in the entire limestone area and comes to the surface in a few large karst springs in the valleys (Fig. 12).

Among the most important karst-water risings are the spring horizons "Bei den Flüssen" in the Hinterau Valley and the Würm and Klammbach springs north of Innsbruck. The spring horizon "Bei den Flüssen" is a karst-water rising at the base level of the bottommost longitudinal valley within the Karwendel ranges. The Würm and Klammbach springs, whose geological situation has been made widely known by the investigations of AMPFERER [1903,1928] and their subsequent opening-up for the water supply of Innsbruck, are depression springs, caused by the northward dip of the unkarstifiable strata of the Lower Triassic of the Inntaldecke (Buntsandstein, cellular dolomite; see Fig. 13, profile C) and also by the water-damming bar formed by the (Interglacial) Höttinger Breccia, which extends in front of the southern slope of the Solsteinkette. In the process of their formation, however, a considerable quantity of water flowed into cavities or below the breccia and thus was lost. Following a suggestion by Ampferer, a system of galleries was driven through the Höttinger Breccia and the cellular dolomite right into the karst-water body in the Muschelkalk (Fig. 13, profile C). The system of galleries consists of three window galleries and a collecting gallery of a total length of 1,663 m. The tunnelling was severely handicapped by irruptions of water on the borderline between the cellular dolomite and the Muschelkalk (up to 1,000 l/sec).

The average yield of 1,500 l/sec in summer and 800 l/sec in winter as well as the altitude of the galleries of 1,140 m (Innsbruck 760 m), suggested the combination of water supply for the town of Innsbruck together with the working of a power plant, whose average annual energy production amounts to 35,000,000 kWh (annual average capacity 4,000 kW). To the drop of 445 m, there corresponds a capacity of 3.58 kW per l/sec and a work value of 1.0 kW/m^3.

This opening-up of a large karst-water body is not accompanied by the usual drawbacks of using karst-water. Its bacteriological quality is impeccable (much less than 1 germ/cm^3!), the total hardness varies in the different galleries from 5 to 9 German degrees, its temperature from 4.1° to 5.6°C. The variation in the rate of the yield is exceptionally low (below 2). These conditions, exceptionally favourable in karst areas, must be ascribed to the absence of any settlement, the preservation of the natural flora up to the vegetation border, the long-lasting snow at the higher altitudes, and also to the quality of the rock in the recharge area. The Wettersteinkalk, which forms the main complex, is here slightly dolomitized, so that in spite of numerous dolines [LECHNER, 1948, p.107] the ways into the interior of the mountain have remained narrow. In the Muschelkalk, however, karstification has created large cavities.

One of the galleries hit on a cave, in which there is a subterranean lake, about 30 m long, 25 m wide, and up to 15 m deep, which bears unimpeachable evidence of active subterranean karstification.

The number of the known cavities is small. The Austrian Catalogue of Caves lists under the heading Karwendl Gebirge only 42 caves, of which only seven have a total length of more than 50 m. Most of them are jamas, in accordance with the structure of the mountains.

Pre-Alps

In the karst areas of the pre-Alps the subterranean drainage systems are determined by the often intensive folding and faulting. The locally developed karst drainage generally has only small dimensions. Caves with tunnels longer than 500 m are rare [FINK, 1967].

An example of a pre-Alpine karst drainage system in lower Austria is the Schlagerboden near Frankenfels, a small polje which is drained by two main sinks. Both ponors are separated by an east–west striking horizon of "Lunzer Schichten", which determines the drainage directions (Fig. 14). Ponor A was investigated on May 14, 1965 with uranine [FINK, 1965], and on June 28, 1965 with uranine and *Lycopodium* spores (F. Bauer, unpublished report). During the first experiment, which was conducted under high-water conditions, the tracer moved from the sink to the Höll spring in 24 h. During the second experiment, conducted at medium (and constantly decreasing) discharge the uranine in the Höll Spring was detected on July 5, seven days after injection. The spores, fed into the ponor during the second experiment about 30 min after the uranine, were found in the Höll Spring for the first time after extreme precipitation on July 18 (20 days after injection!). In the water sample from July 19 (maximum), over 15,000 spores were counted; however, not a single spore was detected in all samples taken before July 18. In contrast to the uranine, the spores must therefore have been transported into a drainage system which ceased to function because of the decreasing water level and which was reactivated only after heavy precipitation on July 18. The large time difference between injection and discharge of the tracers indicates a deep-lying subterranean drainage system, in which different parts are activated under various hydrologic conditions. A few of the spores placed

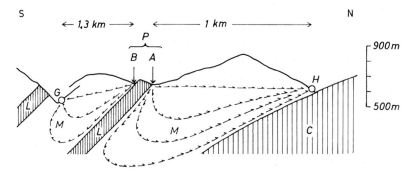

Fig.14. Schematic geologic section through the Schlagerboden area. C = Cretaceous; L = impermeable Lunzer Schichten; M = Muschelkalk (karstified); A,B = ponors; H = Höll Spring; G = Gabauer Spring; P = polje area; the arrows indicate the possible subterranean water movement. Differences in elevation: $A–H$ = 60 m; $B–G$ = 110 m.

in ponor *B* on June 28 were found in the Gabauer Spring already after 2.5 days; here also the main discharge of spores took place on July 18, 1965.

THE CENTRAL STYRIAN KARST

The term "Central Styrian Karst" [coined by BOCK, 1913a,b] is applied to the area of the Palaeozoic carbonate rocks of the Grazer Bergland. The columnar section of the "Grazer Palaeozoicum" comprises mainly Silurian and Devonian slates, sandstones and (partly half-metamorphous) limestones and dolomites, with occasional intercalations of diabases and tuffites [FLÜGEL, 1961]. Of the carbonate rocks, the rather frequent, relatively pure, and well-jointed "Schöckelkalk" (Devonian) is especially liable to karstification.

Variscan as well as Alpine phases have contributed to the Central Styrian Karst's orogeny, which is a complicated nappe tectogenesis. Its morphogenesis, hydrological development, and karstification were influenced by a Tertiary fracturing tectogenesis and by Late Tertiary upheaval phases [WINKLER-HERMADEN, 1957]. Relics of Late Tertiary peneplains are to be found in the whole area. The "Trahüttener Niveau"

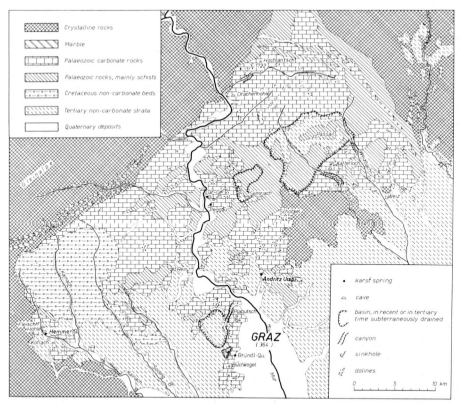

Fig.15. The Central Styrian Karst.

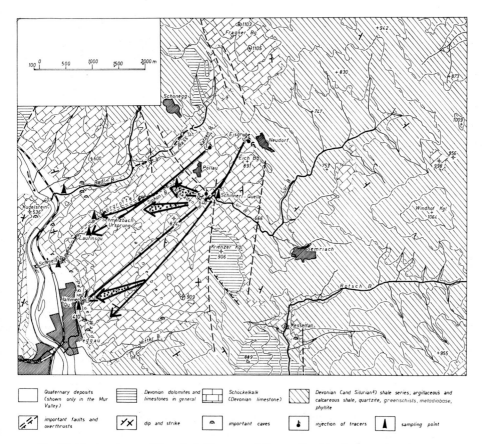

Fig.16. The karst area of Semriach–Peggau and the results of tracing experiments.

(Pannonian) was the first to influence the formation of level-bound karst-water systems. In this connection, mention should be made of the gigantic tunnel of the Dragon's Cave (Drachenhöhle) near Mixnitz (Fig. 15), which is like a number of other caves in the Hochlantsch area, at an altitude level of 950–980 m. Of morphogenetic importance is also the "Hochstradner Niveau" (Dacian), which is at an altitude of 700–750 m, and which has been preserved, apart from other phases, in the region of Tanneben (Fig. 16) and in the Semriach Basin. The Tertiary fracturing tectogenesis, moreover, tectonically outlined the formation of smaller basins within the Grazer Bergland (Semriach and Passail basins; Fig. 15).

Dolines and lapies occur in the Central Styrian Karst up to elevations of more than 1,000 m. More than half of the dolines are, however, bound to levels of 660–780 m, i.e., to that of the Hochstradner Niveau [VORMAIR, 1943]. This level-bound accumulation has probably been caused by the existence in the Hochstradner Niveau of over-developed wide valleys which were covered with permeable Tertiary gravels and thus offered the essential prerequisites for the formation of dolines. This large number of dolines in the Hochstradner Niveau is not limited to the Palaeozoic limestones of the Grazer Bergland, but also characterizes these levels in the Leitha Limestone (Tortonian) of the "Wildoner Buchkogel". The degree to which climatic influences may have contributed to the formation of the dolines has not yet been determined. Lapies rarely occur as morphogenetic factors in this mostly wooded area. Where they have been laid bare by soil erosion, they are of the type of the Kluftkarren, rounded under the cover of the soil. The numerous caves of the Grazer Bergland are, as was established by MAURIN [1952,1956,1961], at the same levels as the several peneplains. Karst systems formed, reaching more or less extensively into the body of the mountain system according to the duration of the (surface-)degrading phase. This process was repeated in the formation of the adjacent lower morphologic levels of the land surface. In the process, the subterranean headward migrating erosion often formed vertical connections with a step-like dip between different cave stages. Thus, the step-like structure of the landscape was to a certain extent reflected in the subterranean drainage of the mountain body. Where tectonic or climatic conditions caused the aggradation of valleys, as was the case in the Upper Pannonian or in the Glacial periods, there followed the inundation and obstruction of cave passages at lower levels. These were, however, partly reopened by later phases of erosion [BOCK, 1913b]. Today inundated cave passages are still to be found in the area of the Late Pleistocene gravel terraces (area of Peggau).

Of the numerous caves (some of which are very interesting from the prehistoric point of view) the following have become well known: the Lur Grotto for its extension (5 km) and its wealth of forms; the Katerloch near Weiz for its wealth of dripstones; the Dragon's Cave near Mixnitz [ABEL and KYRLE, 1931] for innumerable finds of bones of cave-bears.

Today all of the important karst springs discharge at the level of the valley. As the karst-waterways reach down below the Late Pleistocene valley filling (e.g., 40 m near Peggau), some of the springs discharge (with rising branches) at the bottom of the valleys (Hemmer Spring at Köflach, Andritzursprung). Some are connected with karst tubes at higher levels which become active only during high-water periods, when they function as outlets for the main spring. Such high-water outlets are the Fleischhacker Cave (Fig. 15) for the Hemmer Spring, the Frauen Cave for the Andritzursprung, and the Lur Grotto for the Hammerbach Spring. In the Lur Grotto backwater, risings of up to more than 100 m, caused by high water, could be observed before the cave was organized for the admission of the public.

Extensive karst hydrological investigations are being conducted, mainly in the larger area of the Lur Grotto. This cave system is, in its whole extension, situated in the Tannebenstock, which is built up of "Schöckelkalk". This limestone forms a tectonic upfold, which is, considered in detail, wide-folded with faulted slices, but which generally dips more and more steeply towards the west and northwest, whereas its southeastern fringe is horizontal. The Eichberg, adjacent in the northeast to the Tannebenstock, also consists of Schöckelkalk. This limestone area is almost completely surrounded by steeply-dipping structures, whose directions are reflected in the rather regular joint systems of the whole region. This marked jointing of the Schöckelkalk (which is otherwise rather massive and chemically fairly pure), renders the area highly liable to karstification. The prevailing joint directions have also favoured the development of the high and sheer cliffs which characterize the landscape. The basement of the Schöckelkalk is formed by a series of phyllites, green slates, clay slates, and limestone slates. In parts the basis of the Schöckelkalk is to a large extent connected with this basal series of slates by dislodged slices.

A recharge area of 22 km^2 is drained underground through the Lur Grotto system. The waters of the northern Semriach Basin (16 km^2) reach the system through the Lurbach via the Lur Grotto, while the waters from the plateau areas of Pöllau, of the Eichberg, and northwest of Neudorf (2 km^2) disappear in the swallow holes of the Eisgrube and the Katzenbach. The plateau of the Tannebenstock (4 km^2) is also in the recharge area.

Along the last kilometre of its surface course, the Lurbach permanently loses water owing to the perforation of the underlying Schöckelkalk. On this stretch a loss of 28% has been ascertained at low water (66 l/sec). The degree of permeability of the basement varies; at times complete disappearance of the low water has been ascertained in various places. The brook disappears in the Semriach Lur Grotto after about 200 m of subterranean course, into different swallow holes of varying capacity. Larger high waters cannot be absorbed by these swallow holes; they overflow a bar and flow into the otherwise dry cave tract of the Lur Grotto for almost 5 km, which is passable from Semriach to Peggau. The Schmelzbach, a permanent underground stream, flows through the western lower 1,200 m stretch of this cave. In various places in the passable part of the Lur Grotto there appear smaller streams that either disappear quickly (e.g., the Kaskadenbach in the upper part of the cave) or flow into the Schmelzbach (e.g., the Laurin Spring).

The discharge measurements for two years (1966, 1967) at the Lurbach at Semriach and at the Schmelzbach and the Hammerbach at Peggau are summarized in Table II.

TABLE II

THE DISCHARGE OF THE LURBACH, HAMMERBACH AND SCHMELZBACH IN 1966 AND 1967 (IN l/sec)

	1966			1967		
	Average	Highest measurement	Lowest measurement	Average	Highest measurement	Lowest measurement
Lurbach	118	1,115	45	124	4,670	27
Hammerbach	212	637	103	177	754	93
Schmelzbach	80	1,035	43	91	5,390	52

The investigation of the subterranean drainage of the Lur Grotto system is part of the history of karst hydrology. Speleologists of the pioneer generation [BOCK, 1913b] surmised a connection between the Lurbach disappearing east of the Tannebenstock and the Hammerbach surfacing at the foot of the Peggauer Wand in the Mur Valley. Later, in 1920, a dyeing and salting experiment conducted by KYRLE [1928] brought negative results which were used as a fundamental argument in favour of the theory of independent karst tubes, and with which LEHMANN [1932] influenced the prevailing doctrine for decades, indeed up to the present. In 1952 a salting experiment directed by MAURIN [1952] first established the subterranean connection of the Lurbach with the Hammerbach Spring. Tracing experiments conducted in 1959 by MAURIN and ZÖTL [1959] confirmed this result and also showed that the Hammerbach Spring is fed by waters that disappear in the Eisgrube. Light was also thrown on the connection of subterranean waterways [Siphonbach; MAURIN and ZÖTL, 1959]. But the opinion still prevailed that the Schmelzbach and other debouchures rising in the Lur Grotto (Laurins Spring) collected their waters only from the recharge from precipitation at the Tanneben and that therefore the Hammerbach and the Schmelzbach were independent, non-connected karst tubes. Only by means of the large-scale combined tracing experiments of 1966, which included the Lurbach swallow hole as well as the Katzenbach [BATSCHE et al., 1967] was it demonstrated that the Lurbach, Schmelzbach, and Laurins Spring are parts of a large closed karst-water body that generally drains towards the southwest in low-water and middle-water conditions (Fig. 16). With the rising water supply, the drainage shifts to the west on account of the subterranean accumulation until, in periods of high-water, there is a direct overflow from the Lurbach to the Schmelzbach.

The large-scale experiment of 1966 had another important result. Most of the tracers that were injected into the Lurbach swallow hole for comparative purposes reappeared at average rates ranging from 55 to 68%. Comprehensive observation suggested the conclusion that the missing percentage of the tracers (the absorption factor having been taken into account) and therewith that part of the Lurbach water which does not discharge in the Hammerbach, flowed directly into the groundwater body in the loose rock filling of the Mur Valley. This is possible in the Peggau area, as the karstified mountain body extends below the Quaternary gravel filling of the Mur Valley. This, however, furnishes the important proof of the fact that in calculations of the water system of groundwater fields bordering on karstified mountains, not only the volume of the loose deposits but also subterranean feeding must be taken into account.

Attention is finally drawn to the compensating effect of the karst system in the Tannebenstock. Whereas the Lurbach rises quickly during heavy rainfalls, the rise of the Hammerbach can usually be observed only 10–15 h later, which suggests extended narrow-profiled siphons, with reservoirs behind them, for the upper reaches of the subterranean system also.

In the course of the latest extensive investigations (1966), an attempt was made at estimating the water volume between the injection point and the Hammerbach. Calculations based on the results of tracing experiments usually furnish minimum values. According to the gravity centres of the passage curves of the various tracers established in the Hammerbach, the subterranean karst-water volume in the recharge area of the Hammerbach, at a middle yield of 134 l/sec, ranges from 24,600 to 36,000 m^3.

Numerous tracing experiments were also carried out in the karst massif of the Buchkogel, west of Graz. The injection of all kinds of tracers into different swallow holes situated in the subterranean drained basin west of the range (Fig. 15), made possible the testing of those tracers in ideal experimental territory and showed that the subterranean drainage of this area is entirely by the Bründl Spring east of the mountain range. This spring has an average yield of 9.4 l/sec; the water volume between swallow hole and spring was estimated as ranging from 2,130–2,850 m^3. Comparison of the results of tracing experiments in the areas of the Buchkogel and of Peggau shows that the larger water quantities in the karst system of the Lurbach do, in spite of the lesser fictive slope (77% as against 88.7% in the Buchkogel area), cause much higher flow velocities. The flow velocities were, under the given middle water conditions and referring to the maximum of the tracer passage, 62–64 m/h as against 13–18 m/h in the Buchkogel. On the other hand, the only important recharging point of the water from the respective subterranean systems is the Bründl Spring rather than the Hammerbach Spring, as is shown by the generally higher percentage of reappearance of the tracers injected into the former. The high percentage of reappearance and the quick reaction of spring-yield and tracer passage even to light rainfall suggest that in the Buchkogel there exists a simpler subterranean karst system than in the Semriach–Peggau area.

Notwithstanding the disadvantages of the use of karst-water for drinking water supply (varying yield, possibility of pollution), this use is for some parts of the Grazer Bergland the only way of securing larger quantities of water. Thus, apart from smaller places, the towns of Weiz and Köflach depend on the use of karst-water, and the groundwater plant at Graz-Nord uses the discharge of the Andritzursprung for the artificial recharge of the groundwater field.

KARST OF THE SOUTHERN ALPS

In the Karnische Alpen Palaeozoic rocks are abundant. Limestones are found in the Upper Silurian, Devonian and Permian. Triassic rocks (e.g., the Ladinian "Schlerndolomit") are of minor importance. Small limestone massifs, partly with plateaus, are built of Devonian reef limestone (400–1,000 m thick) and of Permian "Trogkofel Kalk" (thickness: 400 m). Karst phenomena have been mentioned by GRESSEL and PICHLER [1966].

The Drauzug (Lienzer Dolomiten, Gailtaler Alpen with Villacher Alpe and Karawanken) consists almost entirely of Triassic rocks, showing facies relationships with the Triassic of the Northern Alps. Limestones and dolomites of the Ladinian Stage (Wettersteinkalk and Wettersteindolomit of thickness up to 1,000 m and more), and dolomites and limestones of Norian age (Hauptdolomit and Dachsteinkalk) predominate. The widely distributed dolomites are easily eroded, and sharply accentuated mountain chains and steep peaks are characteristic of mountains like the Lienzer Dolomiten. Well-developed karst features are restricted almost entirely to limestone areas.

The Villacher Alpe (Dobratsch, 55 km^2, summit 2,166 m) is one of the most important karst massifs of the Southern Alps. It consists of Ladinian Wettersteinkalk and shows a step-like arrangement of plateau levels. On these, dolines of up to 50 m in diameter can be observed. Lapies are found on all plateau levels but are never so well developed as on the large plateaus of the Northern Limestone Alps (e.g., Dachstein, Totes Gebirge). The lower intensity of karstification on the plateau of the Villacher Alpe (comparable with that of Schneebergalpen in the Northern Alps) is caused [LICHTENBERGER, 1954] by the more continental climate (lesser precipitation) and denser vegetation (pastures even on the summit areas).

At the eastern foot of the Villacher Alpe, at Warmbad Villach, important thermal springs (max. 28°C) discharge up to 100 l/sec. The thermal springs are closely connected with adjacent intermittent cold karst springs [STINY, 1937; RÉMY-BERZENCOVITCH, 1963]. In the lead–zinc mine of Bleiberg (situated at the northern edge of the massif) a thermal spring (27°C, 57 atm., 40 l/sec) was discovered in 1951 at a depth of 650 m and 287 m below sea level [JEDLICKA, 1956]. The catchment area of the strong cold springs at Obere Fellach (which are employed for the water supply of the city of Villach) probably also lies in the massif of the Villacher Alpe. However, more detailed investigations are needed to explain the mechanism of this karst-water system, especially the spatial relation between the cold karst waters and the thermal waters.

The intensive subterranean karstification of the Villacher Alpe is illustrated by the fact that in spite of the few investigations carried out so far, already 90 caves are known in this area.

Wide parts of the Karawanken are also strongly karstified (e.g., the Petzen Massif).

KARST OF THE HELVETIAN NAPPE SYSTEM

The Helvetian nappe system in Vorarlberg, exposed as a culmination under the overlying ultra-Helvetian flysch, contains Jurrasic and Cretaceous limestones which are partly karstified [WAGNER, 1950].

The Aptian "Schrattenkalk" (an 80–100 m thick sequence of thick-bedded, pure limestone, rich in organic remains) builds the karst massif of the Hochifen area [SCHMIDT-THOMÉ, 1960, 1961]. This strongly karstified limestone is underlain by Drusberg beds (Barremian shales and marls alternating with limestone bands), and

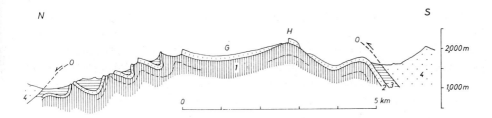

Fig. 17. Geologic section through the Hochifen Massif. *1* = Lower Cretaceous beds; *2* = Schrattenkalk; *3* = Upper Cretaceous and Lower Tertiary beds; *4* = flysch; *O* = overthrust of the flysch nappe over the Helvetides; *H* = Hoher Ifen; *G* = Gottesacker Plateau. (After SCHMIDT-THOMÉ, 1960.)

overlain by Brisi Sandstone (Gault) and Upper Cretaceous "Leistmergel". The Schrattenkalk passes gradually into the limestone bands of the Drusberg beds, and replaces in some areas the facies of Drusberg beds. In any case, the Drusberg beds form the base of the karstified Schrattenkalk (Fig. 17).

In the Hochifen Massif (Hoher Ifen, 2,232 m), the flat-lying Schrattenkalk is exposed over 25 km^2, forming the well-known Karrenfeld (about 10 km^2) of the Gottesacker Plateau [ECKERT, 1902]. At elevations of 1,700–2,200 m the bare surface of the Schrattenkalk is intensely dissected by lapies, which are generally developed along joints and cleavages. The forms of the lapies are the same as on the plateaus of the Northern Limestone Alps.

The present surface of Schrattenkalk mostly represents its stratigraphic top: even in places where the overlying Brisi Sandstone has been eroded, the Schrattenkalk is preserved in its complete thickness because of its karstification. The Drusberg beds form the lower border of the karst-water movements. Where the limestone forms arches, the impermeable Drusberg beds are also upfolded: the culmination of the only 100-m thick limestone here also marks the subterranean water divide, in this case between the Rhine River (North Sea) and the Danube (Black Sea), a very rare case in karst areas.

The best-known caves of the Hochifen Massif are the Schneckenloch near Schönebach (over 1,000-m long channels; TRIMMEL, 1953) and the Hölloch at Riezlern (Kleines Walsertal). The Hölloch can only be reached through a 77-m deep shaft in a sink. Some 300 m of partly water-bearing channels spread out from the bottom of the shaft. The channels of both caves are at the base of the Schrattenkalk, above the impermeable Drusberg beds [SPÖCKER, 1961].

The extremely intensive karstification of the Hochifen Massif is presumably caused by the dense jointing of the pure limestone and the high amount of precipitation in this area (more than 2,500 mm/year).

On the Kanisfluh (2,047 m) and the Mittagsfluh, lapies are well developed in the Quintnerkalk (grey, thick-bedded, fine-grained, crystalline Malm limestones).

KARST OF THE WASCHBERG ZONE

The Waschberg zone ("Äußere Klippenzone") in lower Austria lies in front of the flysch zone north of the Danube and is overthrust over the extra-Alpine molasse zone. Two facies of the Tithonian (the more marly Klentnitz beds and the light-coloured, dense Ernstbrunn Limestone) build the Leiser Berge (492 m, about 9 km^2 of limestone area), and the low but conspicuous hills of Staatz, Falkenstein and Kleinschweinbarth. RIEDL [1957,1958] described numerous small dolines from this area (up to 2 m deep), small caves (maximum length 20 m) and two strong karst springs. The annual precipitation in this shallow green karst terrain is between 600 and 700 m.

KARST OF THE CENTRAL ALPS

The Central Alps consist mostly of granites and metamorphic rocks such as gneisses, mica-schists and phyllites in which karstification is impossible.

Only within the so-called "Schieferhülle" of the Hohe Tauern Mountains are there widely extended calcareous schists (calcareous micaschists, calcareous phyllites) which may show distinct lapies forms [HORNINGER, 1959].

Within siliceous metamorphic rocks south of the Salzach River, steeply dipping laminated, probably Mesozoic limestones (Klammkalk) outcrop. During the excavation of tunnels for a hydroelectric power plant, some tunnels crossing these limestones were heavily flooded. The tunnels opened a communication system of water-filled karst channels with diameters of up to several metres [HORNINGER, 1959].

More or less isolated slightly metamorphosed Mesozoic limestones and dolomites occur in the Brenner area, in the Radstädter Tauern and in the Semmering district. No important karst features have been reported from there nor from the numerous lenses of limestones and dolomites of the "Grauwacken" zone.

ECONOMIC KARST PROBLEMS

Forestry and agriculture

In the high Alpine regions (above 1,500 m), the pasturage of cattle and sheep has been predominant for centuries. In areas of Pleistocene glaciers it is mainly limited to the depressions filled with moraine deposits. In areas without Pleistocene glaciers it is, as a rule, confined to depressions and relatively flat areas with old weathering soils. With careful use, the level of production of these pastures can be preserved at its present state, and may even be increased by appropriate cultivation (manuring etc.).

At altitudes ranging from 1,600 to 2,000 m the dwarf-pine is the prevailing growth. It prefers the rock ridges interspersed between the depressions and it roots mainly in humus soils of up to 1 m thickness that were formed in more favourable conditions during the Postglacial Thermal Maximum. Underneath these soils there are mostly lapies of over 1 m depth, whose crests may perforate the surface of the soil. Disturbances of the vegetation (chiefly the clearance of the dwarf-pines) lead, as a rule, to the erosion of the soil and to the denudation of the lapies. In an area of denuded deep lapies, however, any remarkable formation of new soil is almost impossible: they are to a large extent sterile. In order to preserve the covering of vegetation, any disturbance of it must therefore be avoided in these regions. The utilisation of dwarf-pines in these areas, e.g., as firewood for the Alpine pastures, which was largely the custom in former times, must lead to the permanent unproductivity of these areas as is shown by numerous examples.

Even short-term climatic oscillations such as the period around 1850, may lead to considerable damage to the vegetation in the region of the dwarf-pines and may introduce local erosive processes [BAUER, 1958a].

Erosive processes conditioned by karstification decrease with the decreasing altitude above sea level. Under 1,700–1,500 m forestry prevails. At lower altitudes, thicker weathering soils and more favourable climatic conditions allow, in places, agricultural use of the soil, which may be more or less intensive according to local conditions. The value of produce of these areas may be locally diminished by dolines.

Water supply and hydro-electric power plants

The Alpine karst areas are important water supply sources. Thus karst springs with average discharges of 100–500 l/sec in the Schneebergalpen and the Hochschwab area cover, by means of the first and second Wiener Hochquellwasserleitung (Fig. 18), up to 70% of the requirements of Vienna. The main catchment areas of these springs are at altitudes of more than 1,000 m and are only in use as Alpine pastures or for tourism. Also the cities of Innsbruck and Salzburg are mainly supplied by karst water.

As was shown by tracing experiments with *Lycopodium* spores in Alpine karst regions, the percolating waters are not filtered: spores (diameter 35 μ) drifted through the subterranean channels for distances of up to 30 km. This shows that morbific agents, such as typhoid-fever germs (length 7 μ) can also be transported from the catchment areas into the springs. With the rapidly increasing motor traffic on newly built roads, liquid fuels such as oil and gasoline can also become a serious danger to the purity of springwater.

To preserve the purity of spring water, it is necessary to delineate the borders of the catchment areas and to prevent all contamination within them. In Austria, the "Wasserrechtsgesetz" provides the necessary regulations against water pollution.

For the planning of hydro-electric plants in karst areas, knowledge about the permeability of bedrocks within the proposed reservoir area and of the size of the catchment area are of essential importance. It must be kept in mind that even apparently impermeable bedrock may become permeable under rising pressure by the opening of joints formerly closed by weathering residues. Also, oscillations of the lake

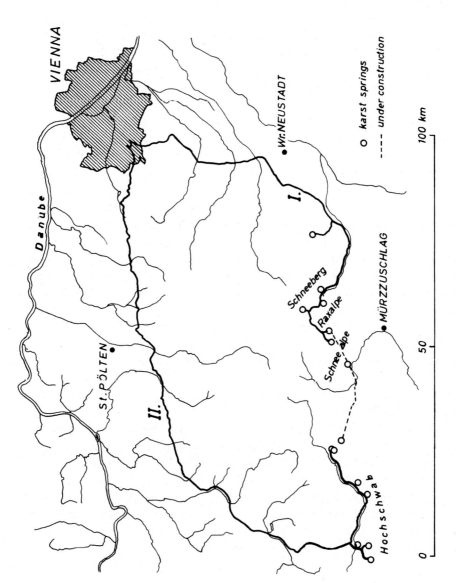

Fig.18. The "Wiener Hochquellwasserleitung". Length of I. Hochquellwasserleitung is 118 km; length of II. Hochquellwasserleitung is 200 km.

level may cause the denudation of more or less thick impermeable layers and the opening of joints in the bedrock along the shore line.

Furthermore the influence of the rising lake level on the local base levels of the surrounding karst region might divert the subterranean watershed towards the reservoir, thus diminishing the inflow.

CONCLUSIONS

The Austrian karst is almost entirely situated in the Alpine region. Therefore the multiplicity and characteristic features of its phenomena are most closely connected with the specific history of the development of the Alps. Relics of older (Tertiary) phases of development are replaced by recent karst phenomena, which prevail in the present-day picture. The mostly high altitudes of the karst plateaus and the climatic factors connected therewith (high precipitation, long-term covering of snow) lead to the origin of specific forms that do not occur in deep-situated non-Alpine karst areas. The karst massifs that were uplifted in the Late Tertiary and are mostly isolated, show subterranean drainage patterns which, as a rule, differ greatly from non-Alpine ones.

The special conditions prevailing in the Alpine karst areas require special methods of investigation. Austrian karst research aims at the development of appropriate methods and the systematic investigation of the karst areas.

REFERENCES

ABEL, O. and KYRLE, G., 1931. *Die Drachenhöhle bei Mixnitz—Speläologische Monographien.* Österreichische Staatsdruckerei, Vienna, 7/9:953 pp.
AMPFERER, O., 1903. Geologische Beschreibung des nördlichen Teiles des Karwendelgebirges. *Jahrb. Geol. Reichsanstalt*, 53:169–252.
AMPFERER, O., 1928. Die Reliefüberschiebung des Karwendelgebirges. *Jahrb. Geol. Bundesanstalt (Austria)*, 78:241–256.
AMPFERER, O. and HAMMER, W., 1898. Geologische Beschreibung des südlichen Teiles des Karwendelgebirges. *Jahrb. Geol. Reichsanstalt*, 48:289–374.
BATSCHE, H., BAUER, F., BEHRENS, H., BUCHTELA, K., HRIBAR, F., KÄSS, W., KNUTSSON, G., MAIRHOFER, J., MAURIN, V., MOSER, H., NEUMAIER, F., OSTANEK, L., RAJNER, V., RAUERT, W., SAGL, H., SCHNITZER, W.A. and ZÖTL, J., 1967. Vergleichende Markierungsversuche im mittelsteirischen Karst, 1966. *Steir. Beitr. Hydrogeol.*, 18/19: 331–403.
BAUER, F., 1952. Zur Verkarstung des Sengsengebirges in Oberösterreich. *Mitt. Höhlenkomm.*, 1952: 7–14.
BAUER, F., 1954a. Aufgaben und Gliederung einer Karstuntersuchung. *Mitt. Höhlenkomm.*, 1954 (1):2–6.
BAUER, F., 1954b. Zur Paläohydrographie des Dachsteinstockes. *Höhle*, 5:46–49.
BAUER, F., 1958a. Vegetationsveränderungen im Dachsteingebiet zwischen 1800 und 1950. *Centralbl. Gesamte Forstwesen*, 75:298–320.
BAUER, F., 1958b. Nacheiszeitliche Karstformen in den österreichischen Kalkalpen. *Congr. Intern. Spéléol., 2me, Bari, 1958, Actes*, 1:299–328.
BAUER, F., 1961a. Sedimentation und Konvakuationserweiterung in aufsteigenden Syphonstrecken. *Rass. Speleol. Ital., Mem.*, 5(2):171–175.

BAUER, F., 1961b. Kalkabsätze unter Kalkalpengletschern und ihre Bedeutung für die Altersbestimmung heute gletscherfrei werdender Karrenformen. *Z. Gletscherk. Glazialgeol.*, 4:215–225.

BAUER, F., 1964. Kalkabtragsmessungen in den österreichischen Kalkhochalpen. *Erdkunde*, 18: 95–102.

BAUER, F., 1970. Aufgaben der karsthydrologischen Forschung in Österreich *Österreichische Wasserwirtschaft*, 22:127–138.

BAUER, F., ZÖTL, J. and MAYR, A., 1958. Neue karsthydrographische Forschungen und ihre Bedeutung für Wasserwirtschaft und Quellschutz. *Wasser Abwasser*, 1958: 280–297.

BOCK, H., 1913a. Der Karst und seine Gewässer. *Mitt. Höhlenk.*, 6(3):1–23.

BOCK, H., 1913b. Charakter des mittelsteirischen Karstes. *Mitt. Höhlenk.*, 6(4):5–19.

BOCK, H., LAHNER, G. and GAUNERSDORFER, G., 1913. *Höhlen im Dachstein*. Verein für Höhlenkunde. Graz, 147 pp.

BÖGLI, A., 1964. Le Schichttreppenkarst. *Rev. Belge Géograph.*, 88:63–82.

BRANDECKER, H., MAURIN, V. and ZÖTL, J., 1965. Hydrogeologische Untersuchungen und baugeologische Erfahrungen beim Bau des Diessbach Speichers (Steinernes Meer). *Steir. Beitr. Hydrogeol.*, 17:67–111.

DOSCH, F., 1956. Färbeversuch Hochschneeberg, 1955. *Gas, Wasser, Wärme*, 10:1–6; 39–45.

ECKERT, M., 1902. Das Gottesackerplateau. *Z. Deut. Österreichischen Alpenver., Wiss. Ergänzungsh.*, 1(3):108 pp.

FELS, E., 1929. Das Problem der Karbildung in den Ostalpen. *Petermanns Mitt., Ergänzungsh.*, 202:1–84.

FINK, M. H., 1965. Der erste erfolgreiche Färbeversuch am Schlagerbodenpolje (Niederösterreich). *Höhle*, 16:67–73.

FINK, M. H., 1967. Tektonik und Höhlenbildung in den niederösterreichischen Kalkalpen. *Höhle, Wiss. Beih.*, 11:128 pp.

FLÜGEL, H., 1960. *Geologische Wanderkarte des Grazer Berglandes (1:100,000)*. Geologische Bundesanstalt, Vienna.

FLÜGEL, H., 1961. Die Geologie des Grazer Berglandes. *Mitt. Museums Bergbau, Geol. Tech. (Landesmuseum Joanneum)*, 23:1–212.

GÖTZINGER, G., 1955. Der voralpine Karst und seine Gesetzmässigkeiten. *Mitt. Geol. Ges. Wien*, 48:33–48.

GÖTZINGER, G., 1957. Beobachtungen im Gipskarst der niederösterreichischen Kalkvoralpen. *Mitt. Höhlenkomm.*, 1955(2):33–37.

GRESSEL, W. and PICHLER, H., 1966. Karst- und Höhlenforschung in Kärnten im Jahre 1965. *Carinthia 2*, 76:158–163.

GRUND, A., 1903. Die Karsthydrographie. *Geograph. Abhandl. (Penck)*, 7(3):103–300.

HORNINGER, G., 1959. Auslaugung an Karbonatgesteinen. *Geol. Bauwesen*, 24: 159–164.

JEDLICKA, F., 1956. Thermalwassereinbruch am zwölften Lauf des Rudolfschachtes in Bleiberg (Österreich). *Z. Erzbergbau Metallhüttenwesen*, 9(5):1–10.

KATZER, F., 1909. Karst- und Karsthydrographie. *Kunde Balkanhalbinsel*, 8·1–88.

KYRLE, G., 1928. *Kombinierte Chlorierung von Höhlengewässern—Speläologische Monographien*. Österreichische Staatsdruckerei, Vienna, 12:94 pp.

LECHNER, A., 1948. *Die Verkarstung des Karwendelgebirges*. Thesis, University of Innsbruck, 185 pp. (Unpublished.)

LEHMANN, O., 1927. Das Totes Gebirge als Hochkarst. *Mitt. Geol. Ges. Wien*, 70:201–242.

LEHMANN, O., 1932. Die Hydrographie des Karstes. In: *Enzyklopädie für Erdkunde*. 212 pp.

LICHTENBERGER, E., 1954. Beobachtungen über Karstformen auf der Villacher Alpe (Kärnten). *Höhle*, 5:63–68.

LICHTENECKER, N., 1926. Die Rax. *Geograph. Jahresber. Österreich*, 13: 150–170.

MÄRZ, CH., 1904. Der Seenkessel der Soiern, ein Karwendelkar. *Wiss. Veröffentl. Ver. Erdkunde (Leipzig)*, 6:213–316.

MAURIN, V., 1952. Ein Beitrag zur Hydrogeologie des Lurhöhlensystems. *Mitt. Naturwiss. Ver. Steiermark*, 81/82:169–180.

MAURIN, V., 1956. Tertiäre, pleistozäne und rezente Verkarstung im Köflacher Becken. *Mitt. Höhlenkomm.*, 1955(2):37–39.

MAURIN, V., 1961. Hydrogeologie und Verkarstung des Grazer Berglandes. *Mitt. Museums Bergbau, Geol., Tech. (Landesmuseum Joanneum)*, 23:173–195.
MAURIN, V. and ZÖTL, J., 1959. Die Untersuchung der Zusammenhänge unterirdischer Wässer mit besonderer Berücksichtigung der Karstverhältnisse. *Steir. Beitr. Hydrogeol.*, 1959:184 pp.
MAURIN, V. and ZÖTL, J., 1964. Karsthydrologische Untersuchungen in Toten Gebirge. *Österreichische Wasserwirtschaft*, 16:112–123.
PENCK, A., 1904. Über das Karstphänomen. *Schr. Ver. Verbreitung Naturwiss. Kenntn. Wien*, 44:1–38.
RÉMY-BERZENCOVICH, E., 1963. Hydrologie des Gebietes der Thermalquellen von Warmbad Villach. *Mitt. Hydrograph. Dienst Österreich*, 35:4–7.
RIEDL, H., 1957. Die Verkarstung der Juraklippen in der niederösterreichischen Waschbergzone. *Höhle*, 8:1–8.
RIEDL, H., 1958. Die Verkarstung des mesozoischen Bereiches der niederösterreichischen Waschbergzone (Leiser Berge). *Höhle*, 9:80–84.
SCHAUBERGER, O. and TRIMMEL, H., 1952. Das österreichische Höhlenverzeichnis. *Höhle*, 3:33–36.
SCHMIDT-THOMÉ, P., 1960. Zur Geologie und Morphologie des Ifenbergstockes (Allgäu). *Erdkunde*, 14:181–185.
SCHMIDT-THOMÉ, P., 1961. Geologie des Hölloches und seiner Umgebung im Bereiche des Hohen Ifen und der Gottesackerwände. In: P. SCHMIDT-THOMÉ (Editor), *Das Hölloch bei Riezlern im Kleinen Walsertal*. Wagner, Innsbruck, pp.13–32.
SPENGLER, E., 1951. Die nördlichen Kalkalpen, die Flyschzone und die helvetische Zone. In: F. X. SCHAFFER (Editor), *Geologie von Österreich*. Deuticke, Vienna, pp.302–413.
SPÖCKER, R.G., 1961. Das Hölloch als geographisches Element. In: P. SCHMIDT-THOMÉ (Editor), *Das Hölloch bei Riezlern im Kleinen Walsertal*. Wagner, Innsbruck, pp. 33–53.
STINY, J., 1937. Zur Geologie der Umgebung von Warmbad Villach. *Jahrb. Geol. Bundesanstalt (Austria)*, 87:57–110.
TERZAGHI, K., 1913. Beitrag zur Hydrographie und Morphologie des kroatischen Karstes. *Mitt. Jahrb. Königl. Ungar. Geol. Reichsanstalt*, 20:225–369.
TRIMMEL, H., 1953. Das Schneckenloch (1,270 m) bei Schönebach (Vorarlberg). *Mitt. Höhlenkomm.*, 1953(2):42–46.
TRIMMEL, H, 1964. Dobratsch-Alpenstrasse und Karst im Gebiete des Dobratsch (Kärnten). *Höhle*, 15:35–39.
TRIMMEL, H., 1966. Österreichs längste und tiefste Höhlen. *Höhle, Wiss. Beih.*, 14:64 pp.
VORMAIR, F., 1943. Die Dolinenwelt des mittelsteirischen Karstes. *Z. Geomorphol.*, 11:123–166.
WAGNER, G., 1950. *Rund um Hochifen und Gottesackergebiet*. Rau, Öhringen, 146 pp.
WINKLER-HERMADEN, A., 1957. *Geologisches Kräftespiel und Landformung*. Springer, Vienna, 822 pp.
WORM, G., 1927. Beiträge zur Geographie und Morphologie der Kare. *Mitt. Ver. Erdkunde (Dresden)*, 1927:49–97.
ZÖTL, J., 1957. Neue Ergebnisse der Karsthydrologie. *Erdkunde*, 11(2):107–117.
ZÖTL, J., 1961. Die Hydrographie des nordostalpinen Karstes. *Steir. Beitr. Hydrogeol.*, 1960/1961: 54–183.
ZÖTL, J., 1963. Zur Morphogenese des Ennstales. *Mitt. Naturwiss. Ver. Steiermark*, 93:155–160.

Chapter 8

Karst of Hungary

F. DARÁNYI[1]
Budapest (Hungary)

INTRODUCTION

In Hungary there are two significant karst regions: one is the Transdanubian Central Mountains (Dunántuli Középhegység) extending northeastward, parallel to the northern shore of the Balaton Lake, up to the Danube River, i.e., to Budapest; the other is the Northeast Range. Rocks susceptible to karsting also occur in both the Mecsek and Villány Mountains west of the southern part of the Hungarian stretch of the Danube, but these areas are of limited significance from the point of view of karst phenomena.

The Transdanubian Central Mountains area can be subdivided into four parts: its western part is the Bakony, joined by the Vértes on the northeast, the northern continuation of which—the Gerecse—extends up to the Danube; the fourth member—the Pilis—connects, like a chord, the arch of the Danube Bend, extending as far as Budapest.

The Northeast Range consists of two members, of which the Bükk with the associated Uppony Block lies entirely within Hungary, whereas the Aggtelek karst to the north-northeast is intersected by the Hungaro-Czechoslovakian boundary (Fig. 1).

Although different in lithofacies and structure, a common characteristic of Hungary's mountains is that they are made up of Mesozoic deposits overlying a Palaeozoic basement and underlying Palaeogene and Neogene strata in the intramontane basins, whose fill grows thicker away from the mountain borders.

LITHOSTRATIGRAPHIC AND SEDIMENTOLOGIC ASPECTS

Transdanubian Central Mountains

Palaeozoic rocks can be traced along the southeast border of the Transdanubian Central Mountains, where they outcrop discontinuously within a narrow belt (Fig. 2). Oldest of these rocks are the phyllites whose lowermost member is believed by ORAVECZ [1964] to be of the Silurian Period. Their whole stratigraphic range, however, is not exposed. For this reason, it can be merely supposed that a part of them may pass

---
[1] Present address: Strohgasse 21/11, Vienna 1030, Austria.

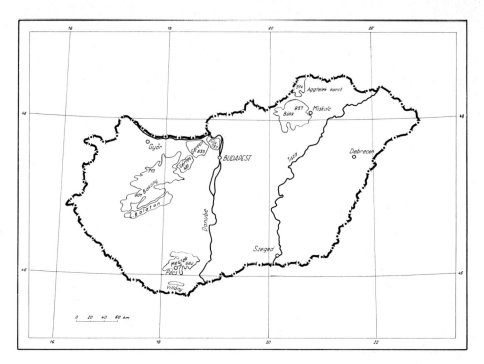

Fig.1. A general map of the Hungarian karst regions.

into the Lower Carboniferous. Sporadic occurrences—geographically confined to the area between lake Balaton and lake Velence on the northeast—of dark grey, sandy limestones belonging, as suggested by KISS [1951] and FÖLDVÁRY [1952], to the Visean Stage of the Lower Carboniferous in a tectonic contact with shaley, crystalline limestones, are also known. On the other hand, as is mentioned by VADÁSZ [1960], an absolute age of 217 m.y $\pm$ 40 m.y. for the granite pluton north of lake Velence, was determined by A. Földváry-Vogl (1959), using the strontium–rubidium method. This would presumably coincide with the Carboniferous Period, an evidence that confirms earlier geological opinions, including the hypothesis which suggested that equivalents of the granite occurred in the Alps.

The Permian red quartzose sandstones, up to 380 m thick and entirely representative of continental deposits, occur in a number of good exposures along the north shore of lake Balaton. The precise age of this sequence, underlain by the phyllites and overlain by Lower Triassic sediments, cannot be determined.

The strata of these Palaeozoic systems, more or less incompletely known, are superimposed the one upon the other, with marked tectonic and stratigraphic unconformities and hiatuses, respectively. They form the basement of the mainly Mesozoic–Tertiary Transdanubian Central Mountains, though their northward extension deeply beneath this mountain range is so far unexplored, and we know of only some traces thereof along one or two large thrust sheets, like those at Litér and

Fig.2. A rough geologic map of the Transdanubian Central Mountains (scale 1:1,250,000). *1* = Early Palaeozoic rocks; *2* = granite; *3* = Permian sandstone; *4* = Lower and Middle Triassic sediments; *5* = Upper Triassic "Hauptdolomit" and "Dachsteinkalk"; *6* = Jurassic sediments; *7* = Cretaceous sediments; *8* = Eocene sediments; *9* = Middle and Upper Tertiary and Quaternary sediments.

Kádárta, where diabasic tuffs associated with phyllites are also exposed.

The complete sequence of the Triassic deposits can be encountered in an easily traceable Alpine development, as was shown by an exhaustive study as early as by LÓCZY et al. [1913]. This sequence is so complete and so undisturbed, compared to the Alps, that every problematic detail of the Alpine Triassic can be convincingly settled here.

The Werfenian Stage of the Lower Triassic can be subdivided into six horizons, three of which are composed of 110 m thick Seisian laminated sandstones, massive dolomites, shaley clays, calcareous dolomites, and marls including the *Pseudomonotis clarai* and *Pseudomonotis aurita* horizons. On the other hand, the Campilian deposits of some 500 m thickness, with three subdivisions, begin with *Pseudomonotis*-bearing sediments resembling the Seisian; they then grade into calcareous marls, characterizable by *Tirolites cassianus* and *Natiria costata*, and finally into laminated and thin-bedded limestones with cellular dolomites. The Werfenian Stage is a gradually transgressive succession of sediments which, however, remains neritic over the whole range of the sequence.

The Anisian Stage of the Middle Triassic is initiated by the up to 200 m thick Megyehegy Dolomite and is terminated by 10–20 m of shell-limestone of South Alpine nature; stratigraphic indices being *Dadocrinus gracilis*, *Rhynchonella decurtata* and *Ceratites trinodosus*. The *Protrachyceras reitzi* horizon of the Ladinian consists of 15 m of marl, with distinct traces of diabasic tuffs along the bedding planes. It is overlain by up to 50 m of red, cherty, massive limestones *(Daonella lommeli)* and then by some 20 m of slightly dolomitic limestone *(Trachyceras aon)*. All these features indicate the existence of an open, though shallow, sea environment in Anisian time.

The Carnian Stage of the Upper Triassic is made up of a very thick sequence (500–600 m), called Upper Marl, with a few interbedded massive limestone and dolomite layers and overtopped by the so-called Sándorhegy Limestone. Its index fossils are *Trachyceras aonoides* and *Tropites subbulatus*. The Norian Stage is entirely filled by the Hauptdolomite, 1,000 m thick or more; the Rhaetian by 400 m of Dachsteinkalk underlain, though not uniformly, by limestones and marly limestones of the Kössen facies.

Accordingly, the Triassic deposits of the Transdanubian Central Mountains, bearing both South- and North Alpine features, have been developed with no break in sedimentation in a total thickness of about 2,900 m, of which more than 2,000 m are accounted for by carbonate rocks—mainly dolomites and subordinate limestones. A complete sequence is known, however, in the Bakony Mountains only, as in the rest of the Transdanubian Central Mountains the pre-Upper Triassic rocks have been buried much deeper and are so far inaccessible to exploration.

The Jurassic deposits have lagged in importance far behind the Triassic, being represented only by sporadic remnants. On the west the Lias developed, with no break in sedimentation from the Triassic, but farther northeast a hiatus appears on the Triassic–Jurassic boundary. An additional characteristic feature of the Jurassic of the Transdanubian Central Mountains is its marked lithologic variety, so that even quite closely spaced bore holes may exhibit somewhat different lithologic logs. It is known best in bore holes, as at Urkut (A. Barabás, personal communication, 1958), where the Lower Liassic, pink and locally cherty, crystalline limestone with joints of crinoid stalks, brachiopods and ammonites, is 30–50 m thick, overlain by 70–100 m of Middle Liassic red, very cherty, nodular, crinoidal limestone with *Pygope aspasia*. The Middle Liassic ends with 2–10 m of massive cherty calcareous marl. The Upper Lias is represented by a 30–45 m thick sequence of alternating radiolarian and Mn-carbonate-bearing clayey marls.

The Lower Dogger consists of a thin, 0.15–0.30 m thick, basal chert layer, overlain by 1–2 m of green-mottled, brown, ammonitic, clayey marl; the Middle Dogger is represented by 90–120 m of ammonite- and Radiolaria-bearing calcareous marl, with interbedded *Posidonomya* Clay; whereas the cherty, siliceous marls of the Upper Dogger, about 70 m thick, grade as suggested by VADÁSZ [1960], into the Malm, although the red, nodular, ammonitic limestones of the Middle Malm, as well as the Tithonian limestones and calcareous marls, are usually absent.

The total thickness of the Jurassic of the Bakony is 300–600 m; it decreases, however, northeastwards, to a thickness of less than 100 m in the Gerecse along the Danube. Typical of the entire Jurassic are: the well-aerated, bathyal marine sediment, the common occurrence of cherts, the red colour of the limestones, the marly and calcareous-marly facies of the Upper Lias and Dogger, and, in general, the North Alpine nature of the sediment.

The Cretaceous System has developed, partly with no break in sedimentation and partly with a hiatus, from the Tithonian; although the Cretaceous deposits, like the Jurassic, have been markedly worn away by erosion, thus being reduced to isolated areas.

As shown by NOSZKY [1934,1943] and TELEGDI-ROTH [1935], the Lower Cretaceous begins with Valanginian–Berriasian laminated, chert-nodular limestones *(Pygope diphyoides)* 60–80 m thick, grading into 60–70 m of Hauterivian thin-bedded, crinoid- and brachiopod-bearing limestone with chert layers. The Barremian is characterized by an emergence, during which 0.1–20 m thick accumulations of bauxite were formed: a number of deposits of varying size are known in some parts of the Transdanubian Central Mountains. In the Gerecse, the bauxite horizon of the Barremian is known, but the more than 150 m thick sandstone sequence of the Hauterivian, containing a marine fauna, passes, as believed by FÜLÖP [1961], into the Barremian which grades upwards into a conglomerate member.

The Aptian Stage of the Middle Cretaceous is represented by a clay-and-marl sequence with calcareous algae, orbitolinae, and oysters in the Bakony Mountains (60–100 m in the northern part and 0–10 m in the south). On the other hand, the Albian Stage is made up of *Requienia*- and *Orbitolina*-bearing limestones which are 30 m thick in the north and 100 m thick in the south. The Cenomanian is represented by 150 m of *Turrilites* Marl in the northern Bakony Mountains and the Vértes. The Turonian witnessed an overall emergence during which, as shown by DARÁNYI [1957], a bauxite horizon was formed and the subsequent submergence, that preceded the Upper Cretaceous transgression, was locally connected with the accumulation of thick gravel beds. The same author also showed that the 60 m thick coal-bearing formation of Ajka belongs to the Senonian, being overlain by a sequence of marls and limestones up to 100 m thick, containing representatives of *Lima, Hippurites,* and *Gryphaea*.

At the end of the Cretaceous and Eocene an overall emergence took place.

What has been preserved of the Cretaceous deposits in Hungary is more complete on the west than it is on the east. In the western part of the Bakony the Cretaceous is terminated by the Senonian, in the eastern part and the Vértes by the Cenomanian, and in the Gerecse by the Aptian. Its total thickness is 500–600 m, one half of which is composed of limestones, the other half by clayey, marly sediments.

Tertiary deposits occur in the intramontane basins and on the mountain borders.

Being usually continental, the Eocene often begins with redeposited bauxites and variegated clays, possibly with accumulations of detrital material at the Cretaceous–

Eocene boundary. The overall submergence began after the Palaeocene. Its freshwater argillaceous sediments attain 70 m in thickness. On the borders of the Central Mountains and within the basins the Lower Eocene is a commercial coal-bearing formation, overlain first by brackish-water sediments and then by strata with Miliolinae and *Alveolina*. During the transgression that advanced northeastwards limestones with *Nummulites* (or their various marly equivalents), representing the Lutetian Stage of the Middle Eocene, were laid down. Their thickness is 100 m, but it exceeds even 200 m in many cases. The Upper Eocene is initiated by a tuff eruption which is followed by the deposition of nummulitic *(Discocyclina)* limestones and clavulinoid-bearing marls, whose uppermost horizon, represented by siliceous and bryozoan calcareous marls, is known in Budapest. The total thickness of the Upper Eocene exceeds 200 m.

Because of the gradual subsidence, the Eocene is characterized by shallow-water sediments of a total thickness of 500–600 m, deposited in a gradually more pelagic environment. The clayey–marly strata are locally associated with huge limestone masses, resulting in the prevalence of carbonate rocks within the Eocene.

The Oligocene manifests itself in the northern foreland of the Transdanubian Central Mountains by a hiatus and several hundred metres of accumulated debris, whereas in the Pilis Mountains it is characterized mainly by coarse-grained sandstones followed by foraminiferal clays and finally by marine sands with *Glycymeris*.

The Miocene occurs, like the Oligocene, on the mountain borders and within the intramontane basins. Sedimentation begins with the Helvetian marine, calcareous sands and sandstones up to 300 m thick with interbedded Leitha Limestones, overlain by a Tortonian coal seam and then by diatomaceous shales with bentonites in a thickness of 60–80 m. The Sarmatian sandy-gravelly clays with boulders (100 m), grading into the Pannonian–Pliocene deposits, are of little importance in the karsted mountain areas of Hungary, but far away from these areas they fill the large basins to a thickness of 2,000–3,000 m. The small basalt sheets occurring on the north shore of lake Balaton also belong to the Upper Pliocene.

Reduced to spots of varying size, the Pleistocene loess, 10–15 m thick, is common in the karstified mountain ranges.

Northeast Range

The oldest formation represented by crystalline limestones, sericite schists, and sandstones (Fig. 3) is placed by BALOGH [1964] in the Devonian, and its upper member in the Carboniferous, being so far undetermined in many cases. The Namurian argillaceous and calcareous shales are conformably overlain by Moscovian shales and sandstones, supporting the conformable Uralian shales, sandstones and bituminous limestones with lenses of *Fusulina* Limestone. The Carboniferous strata grade with no break in sedimentation into the following Permian sequence: 800 m of Lower Permian shale; up to 150 m of Middle Permian terrestrial, red, green, grey sandstone and sandy

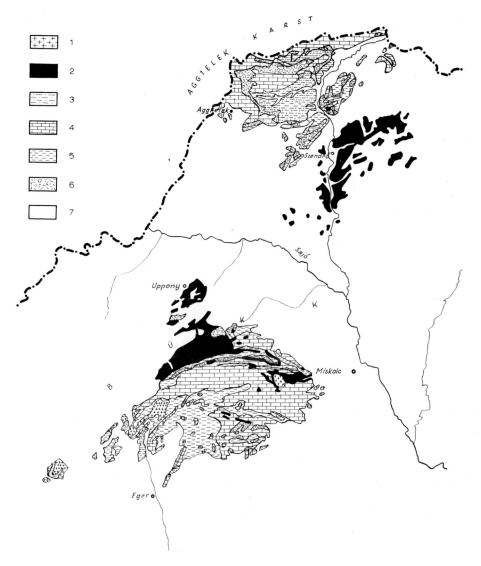

Fig.3.A rough geologic map of the Northeast Range (scale 1:680,000). *1* = Igneous rocks; *2* = Palaeozoic; *3* = Lower Triassic sediments; *4* = Middle Triassic limestone and marl; *5* = Middle Triassic shale, flintshale and sandstone; *6* = Middle Triassic dolomite, and limestone alternating with dolomite; *7* = Tertiary and Quaternary sediments.

shale; and Upper Permian clay shales and bituminous limestones.

Palaeozoic deposits occur mainly in the northwest foreland of the Bükk and in the southeast foreland of the Aggtelek karst. As believed by SCHRÉTER [1959], the Werfenian beds of the Lower Triassic have developed with a 100–150 m thick transition from the Upper Permian.

The Seisian member of the Werfenian Stage is made up of thick-bedded shaly limestones containing interbedded sandstone layers with *Pseudomonotis clarai*. It is overlain by Campilian shales and oolitic limestones with massive dolomites in the upper horizon. Characteristic fossils are *Natiria costata* and *Tirolites cassianus*. On the northeast, along the frontier, the Seisian contains serpentine, anhydrite and gypsum.

The Anisian Stage of the Middle Triassic is represented by stratified dolomites, followed by porphyrites, diabases, diabasic tuffs, and tuffaceous limestones and finally by limestones with chert and dolomite lenses. The Ladinian consists of black shales, radiolarian siliceous shales, limestones with chert, diabases and their tuffs, quartz porphyry and well-stratified limestones. It is possible that a part of this sequence may pass into the Upper Triassic, but this hypothesis has been but partly proven as yet. In the area of the Aggtelek karst, however, the presence of the Upper Triassic can be proved.

The Triassic of the Bükk Mountains bears Southern Alpine features, all but the lower member being pelagic sediments. In the lower 300–400 m of the 1,000 m thick Triassic sequence, the subordinate carbonate rocks are dominated by clays and siliceous shales, volcanic rocks and their tuffs, whereas the upper member shows the predominance of carbonate rocks, particularly so within the Aggtelek karst.

The Jurassic is completely absent in the Northeast Range, while the Senonian of the Upper Cretaceous is represented by isolated occurrences of the Hippurites Conglomerate. The Eocene is also of limited significance, as its transgressive upper stages, composed of sandstones and limestones are, when present, confined to the southern and southeastern borders of the Bükk. Similarly is the case of the Oligocene and Miocene, and, in some place, with the Pannonian sediments which, though widespread in the basins, extend but to the border of the Northeast Range. Loess is common, covering areas of varying size.

Mecsek Mountains

According to WEIN [1952] and SCHMIDT et al. [1962], in the southeastern and eastern forelands of the Mecsek Mountains (Fig. 4), there are outcrops of Palaeozoic shales, phyllites, and granites; in the western part of the mountains these are overlain by Permian deposits, of which the Lower Permian is represented by 500 m of well-stratified, red siltstone and shale with quartz porphyry and the floral remains *Ullmanites* and *Voltzites*; the Middle Permian by 450–800 m of sandstone and conglomerate with subordinate shales; and the Upper Permian by 400–500 m of red sandstone with the floral elements *Walchia* and *Dadoxylon*. Accordingly, the about 1,200–1,700 m thick sequence of the Permian consists almost entirely of arenaceous (sandstone or sand) members.

To the east, the Permian is overlain by the Triassic.

The Seisian member of the Lower Triassic sequence is composed of an alternation, about 100 m thick, of sandstones, gypsum beds, and variegated shales with

Fig.4. A rough geologic map of the Mecsek Mountains (scale 1:800,000). *1* = Early Palaeozoic rocks; *2* = granite; *3* = Permian sandstone; *4* = Lower Triassic sediments; *5* = Middle Triassic sediments; *6* = Upper Triassic sediments; *7* = Lower Jurassic sediments; *8* = Middle Jurassic sediments; *9* = Upper Jurassic sediments; *10* = Cretaceous; *11* = Tertiary and Quaternary.

Pseudomonotis clarai. The Campilian member includes 130 m of marl and 100 m of thin-bedded to laminated black limestone characterized by *Myophoria costata*.

The Anisian member of the Middle Triassic sequence is made up of limestones and dolomites, 400 m thick, with an abundant marine fauna. The Ladinian member consists of clayey marls and thinly laminated marls, and clayey sandstones 180 m thick, with plant fossils.

The Carnian and Norian stages of the Triassic System seem to lack the littoral facies, whereas the transgressive conglomerate sequence of the Rhaetian Stage is 600 m thick (littoral facies).

Of the total 1,500 m thickness of the Triassic deposits, 500–600 m are constituted by carbonate rocks.

Farther east, consistently epicontinental, shallow-water Jurassic sediments can be traced in a total of 3,000 m thickness, of which more than 2,000 m of the Lower Jurassic consists of sandstone, shale, coal-formation, clayey marl, marl, and subordinate limestone.

The Dogger is represented by 250–300 m of marls.

The Malm is made up of limestones and cherty limestones, 200 m thick, indicating a southward increase in sea depth.

The Cretaceous System is represented by volcanics (trachydolerites).

The Miocene is composed of conglomerates, sands, clays, and tuffs occurring on the mountain borders and filling the basins to a thickness of 1,200–1,300 m.

The last members of the stratigraphic succession are the Pannonian sediments and Pleistocene loesses.

The strata of the individual systems and stages have been preserved or worn away depending on local features of geologic history. This is eloquently shown by the geological maps (Fig. 2–4) which indicate the degree of reduction of the particular stages and the repeated action of the concomitant karst processes of recurring phases of emergence and erosion in the geologic past.

TECTONIC SETTING OF THE AREA

Transdanubian Central Mountains

The Transdanubian Central Mountains have a structure of marked complexity, since they have been affected in the various epochs by tectonic agents of different orientation and nature, whose final chronology was established by DARÁNYI [1966b]. The Transdanubian Central Mountains, as well as the Bükk Mountains, emerge along the same southwest–northeast trending major fault which can be traced from the Medvednica and Gorjanci (Žumberak), i.e., from southwest of Zagreb in Yugoslavia, across lake Balaton, to the northeastern Carpathians. SZALAI [1958] describes this fault as a Lower Palaeozoic structure which has repeatedly rejuvenated.

Though of great importance, the tectonic setting of the Palaeozoic cannot be explored because of the limited occurrence of these deposits. In our case this is not even essential, as the Transdanubian Central Mountains are built, for the most part, of Mesozoic and Tertiary formations. These are intensively faulted, with subordinate manifestations of folding. The numerous uplifts and submergences played an important role in the modelling of their surface.

A striking tectonic feature is that the most marked differences can be shown to exist partly between the Jurassic and the Cretaceous, partly between the Middle and Upper Cretaceous and finally between the Tortonian and Sarmatian structures. Accordingly, in terms of H. Stille's nomenclature, the Late Cimmerian, Austro-sub-Hercynian, and Late Styrian movements were the most efficient agents of crustal deformation. These stresses, however, largely differed in orientation, nature, and chronology and they were associated with a number of additional, minor but distinct, tectonic stresses.

After the uninterrupted sedimentation of the Triassic, the sequence of the Jurassic, though continuous in many places, exhibits some local breaks, as is shown by

SZABÓNÉ-DRUBINA [1956] in the manganese ore mine of Eplény, where the Middle Liassic limestone footwall of the Upper Liassic manganese carbonate ore deposits was eroded, whereas at Urkut it has been completely preserved. This uplift of the crust, devoid of any other associated tectonic features such as folds and faults, was connected with the Early Cimmerian orogeny and contemporaneous with the formation of SZALAI's [1951] "Cimmerian ridge", i.e., with the emergence of the south border of the Bakony Mountains.

Erosion and sedimentation, alternating with repeated uplifts and subsidences, make it very difficult to trace the original limits of transgressions and the chronology of the geological events; in other words, to determine the geographic range and duration of transgressions and regressions.

A very strong tectonic feature due to compressive stresses from two different directions, one from the north, the other from the west occurs at the Jurassic–Cretaceous boundary. It is the fold pattern of the Jurassic and older rocks that most clearly shows the tectonic contrast with the Cretaceous and post-Cretaceous deposits in which faults are predominant. Along with the folded forms due to northerly compression, the westerly stresses also produced a number of folds, perpendicular to the former, as well as numerous thrust sheets of eastward trend.

During the Late Jurassic a partial emergence of the Gerecse Mountains took place, but the Valanginian and Hauterivian sedimentation did not completely cover any of the members of the Transdanubian Central Mountains. In Barremian time an emergence of the entire area took place, giving rise to marked erosion, to formation and redeposition of bauxite, to oxidation and partial redeposition of manganese carbonate ore.

The Middle Cretaceous transgression was followed by an emergence in Turonian time when the Austrian orogeny began. For lack of Turonian sediments, this orogenic phase, however, cannot be distinguished from the sub-Hercynian one, so that the two are together referred to as Austro-sub-Hercynian movements. We believe these movements were the most efficient tectonic agents, whose mechanism fits in with SCHMIDT's [1957] theory if generalization is disregarded. According to Schmidt's theory, these tectonic movements ought to be conceived as a southwest–northeast-trending torque which would have given rise to compression along northeast–southwest fracture planes and to tension, with resultant blockfaulting along northwest–southeast lines. Certain phenomena, however, suggest that some thrust sheets and reverse faults may have been produced by stresses that came from the northwest. All in all, it appears that the Austro-sub-Hercynian forces tilted and thrust the last major blocks and also forced the Transdanubian Central Mountains to change their strike from southwest–northeast in the Bakony to south–north in the Vertes-Gerecse.

In Turonian time the entire area was an emergent land which is indicated by evidence of erosion, and by a distinct bauxite horizon. If the Laramian orogeny had affected the complete sedimentary cycle that began with a subsidence and resultant transgression in Senonian time and ended with an emergence in the Late Cretaceous (a hypothesis insisted upon by various authors), so some differences in structure between the Upper Cretaceous and Eocene ought to be present. No difference can be found where the two formations are superimposed (e.g., in the Ajka Coal Basin). This would mean that the Laramian movements were manifested merely by synorogenic uplifts.

After the complete sedimentation cycle of the Eocene a minor attack came from the west. Its manifestations are slightly flexing and some thrust sheets in the Eocene and older deposits. These movements may have taken place in the Oligocene testified by an overthrust of Lower Oligocene over the "Hauptdolomit" north of Banhida in the Gerecse along a north–south axis of several kilometres.

In the following discussion we may disregard the post-Mesozoic emergences which, after the Eocene, gave rise to almost steady shifts of the coast lines and, consequently, to independent, local sedimentation cycles. In fact, any attempt to dwell on their chronology would confuse tectogenetic interpretations.

Much more significant are the very pronounced tectonic movements referable to the Late Styrian phase. Their peculiarity consists in that they have produced—in contrast with the tectonic disturbances hitherto discussed—tension faults acting in two different directions: northeast–southwest and northwest–southeast. The faults vary in size from inconspicuous dislocations to several hundred metres of throw. All in all, they have produced relief differences of some thousand metres in magnitude. The well-known graben–horst-imbrication structure of the Transdanubian Central Mountains was almost completely formed when the earlier compressive stresses were replaced by tensile ones that acted in both the aforementioned main directions. The precise date of these movements is shown by the fact that their manifestations, present throughout the Tortonian deposits, are absent in the Sarmatian.

Younger movements are also known. Their manifestations are tectonic features produced by northerly stresses. These can be readily studied in some Cretaceous, Eocene, and Miocene coal-fields. Here, as in the case of the Oligocene movements, we also have to do with gentle synclines and anticlines in the coal seams and with thrust sheets in the less flexible Mesozoic bedrock; but these latter show here an eastward trend, in contrast with the southward direction observed in connection with the Oligocene movements. Minor thrust sheets and gentle folds can be observed within the Miocene in the south foreland of the Bakony. Manifestations of piling-up and squeezing-out have been recorded in the Sarmatian also. According to J. Kókay (personal communication, 1965), on the north shore of lake Balaton the Lower Triassic was thrust from the north over the Pannonian deposits, so that the movement can be ascribed to the Rhodanian phase.

As a result of all these movements, the Transdanubian Central Mountains have obtained a complex block fault–reverse fault–minor thrust–gently folded pattern with main trends in northeast–southwest and northwest–southeast directions. SCHMIDT [1957] interpreted this structure as a cratosyncline.

Strikingly different in mechanism of movement, the orogenic forces are less distinguishable from one another as far as their final results are concerned. This is quite natural, as in particular aspects, the various stresses may have some common resultant features, even though in their complexity the results differ following the cases.

All the tectonic movements quoted were, as shown above, of different magnitude. They originated from different directions and can be traced from the end of the Triassic to the Quaternary. Of these, the Late Cimmerian stresses proceeded first southeastwards and then—as can be readily demonstrated—eastwards; the Austro-sub-Hercynian orogenic movement was from the southwest; the Oligocene movement from the west; the Late Styrian movements acted in longitudinal and transversal directions; the Rhodanian movement came from a northerly direction. All but the Late Styrian movements were compressive, whereas the latter brought about tension faults in all directions. Large-scale crustal deformation was caused by the main phase of the Late Cimmerian movements, then by the Austro-sub-Hercynian and the Late Styrian phases, the rest of the tectonic phases had a less significant effect.

Northeast Range

As shown by Balogh [1964], the Palaeozoic structure of the Northeast Range developed in Carboniferous time, whereas the tectogenesis of the Mesozoic formations began with the Cretaceous phases of the Alpine movements. Compressed between the rigid Palaeozoic blocks, on the one hand, and the crystalline block, dropped to great depths in the southern foreland of the mountains, on the other, the Northeast Range underwent overall folding, thus being basically different from the predominantly faulted Transdanubian Central Mountains. In the Bükk Mountains the overturned limbs and thrust sheets do not attain the size of the Carpathian nappes (overthrusts), but both normal and overturned folds occur along with plication features, marked strike-slip faults and manifestations of rolling. The products of submarine volcanism have been dynamometamorphically altered and the thick shale sequences have obtained a transversal cleavage. The axes of the folds show a west–east trend and the edge of the folded sequence has been transgressed by Eocene sediments with no manifestation of folding.

Dating of the tectonic movements which moulded the Northeast Range is difficult; the Jurassic and Cretaceous are almost completely absent and the Upper Eocene appears only on the edges, being traceable in a limited number of places; whereas the Oligocene and Neogene, as basin deposits occurring outside the mountain range, can provide but indirect information.

The movement of the Northeast Range was not always uniform, the Aggtelek karst bordering on south Slovakia, as well as the other areas north of the Bükk, sometimes differed in movement the one from the other.

Just before the Senonian, the Bükk was compressed into simple upright folds and its northern part was thrust over the edge of the Uppony Block which, consolidated as it was, acted as a quasi underthrust block.

The Senonian basin which resulted from a collapse of the piled-up sediment masses was filled with erosion waste.

Prior to the Eocene, the upright folds of the Bükk Mountains were overturned and re-imbricated. In this phase the fold-inducing masses interchanged their roles, so that it was the southern massif which acted as an underthrust block, orienting towards itself the motion of the thrust sheets.

In Palaeogene and Neogene times the more or less consolidated Bükk subsided or uplifted in its totality rather than displaying differential movement of its various portions. Early in the Miocene a northwest-trending overlap occurred in the northwest foreland of the Bükk Mountains.

Predominant in the sediments of the northeast and east forelands, the Miocene tension faults are less manifest within the mountains themselves. Thus, in contrast with the Transdanubian Central Mountains, the Bükk Mountains have not been dissected, nor disintegrated.

It is a fact that the Bükk Mountains owe their tectonic pattern to southward movement, but because of contrasts with the northward orientation of the Carpathian nappes (overthrusts), tectonically they prove to have "Dinaric" features, not to speak of sedimentary features.

In the areas lying northeast of the Bükk Mountains, e.g., in the Aggtelek karst, the first imbrication, in Middle Cretaceous time, was oriented southeastwards; the later movements acted northwestwards and brought about slight plication features, thrust sheets, and reverse faults.

Mecsek Mountains

According to VADÁSZ [1960] and WEIN [1952], the tectonic setting of the Mecsek Mountains has been controlled by the situation of the crystalline basement to the north and south. The movements were predominantly directed from the north to the south, but the Villány Block, south of the Mecsek Mountains, underwent northward imbrication. The structural pattern of the Mecsek is of plicate nature, but instead of forward-thrust nappes, it consists of folds and thrust sheets still more or less connected with their original roots. The fault-folded nature of the mountains is conspicuous. The main axes of the Mesozoic folded structures are of west–east orientation, which direction is followed by the later structure elements also.

In the structural units produced by the superposition of deformations of different ages, the individual tectonic phases and features are difficult to discern. The southward dislocation of the Triassic chain, i.e., of the western part of the mountains, may have taken place before the Middle Miocene. A period of heavy orogenesis of the Mecsek Mountains is believed to have been the Austrian phase of the Cretaceous, when the folded chains and their periclinal arrangement, as well as the products of volcanism observable in the southern Mecsek, were brought about.

The detection of certain detail phenomena of orogenic activity is handicapped by the break in sedimentation from the Lower Cretaceous to the Helvetian. All that is known for sure is that some of the movements began very late, a fact warranted by the folding of the Miocene strata, the marked unconformity between the Lower and Upper Pannonian, and the pronounced dip of the Upper Pannonian.

PHASES OF KARSTIFICATION

Transdanubian Central Mountains

It is evident from the preceding chapters that the Transdanubian Central Mountains, though supported by a Palaeozoic basement, are essentially Mesozoic and Tertiary.

Continuous sedimentation in the Triassic Period made impossible the development of contemporaneous karst processes, but later, in post-Triassic times, it was the more than 1,000 m thick Norian "Hauptdolomit" and the Rhaetian "Dachsteinkalk" of the Upper Triassic Series which were most intensively karsted.

In studying the phases of emergence, erosion, and karstification, some difficulties are faced, for we do not know whether the lack of a stratum at a given locality is the result of erosion or is due to non-submergence of the area under consideration. Thus it is difficult to draw the geographic and time limits of one-time transgressions.

According to SZALAI [1951], the earliest Mesozoic emergence was connected with the Early Cimmmerian orogeny, a fact proved by the local uncompleteness of the Middle Liassic sediments between the Lower and Upper Lias. According to DARÁNYI [1966b], there is a distinct manifestation of karstification in the Lower Cretaceous of the Ajka Coal Basin, where numerous dolines are known, overlain by Middle Cretaceous Apto-Albian sediments deposited in the transgressive sea that covered the area after the Barremian continental period. Highly karstified Lower Liassic limestones are known at Urkut, where redeposited manganese ore, overlain by Lower Eocene sediments, occurs in a 30–40 m deep vertical-walled sink-hole. The karstification can be dated, here too, from the Cretaceous, although it cannot be determined to what extent the Barremian, Turonian, and Danian emergences were involved in the process.

During the entire sedimentation cycle of the Eocene the karst processes were inactive, though no information is available on what may have happened beyond the coast lines of the Eocene sea, whose original distribution is unknown due to large-scale erosion. Later on, until the end of the Pliocene, many transgressions and regressions alternated with each other. In addition, steady shifting of the coast lines gave rise to a number of minor sedimentation cycles, during which karstification outside the sedimentation basin was always possible.

Northeast Range

The phases of karstification of the Northeast Range are more uniform than in the Transdanubian Central Mountains, as in the hanging wall of the Triassic deposits, the Jurassic and almost all the Cretaceous—of which merely the Senonian is represented by a limited occurrence—are absent. The older members of the Eocene are not known either, and even the younger ones extend only to the borders of the Northeast Range. The same holds true of the rest of the Tertiary and the loess blanket of the Pleistocene —which has been markedly eroded in the meantime—has only slightly changed this continuous break in sedimentation. Consequently, in the Northeast Range, karstification was at that time an uninterrupted process and has remained so up to the present time.

Mecsek Mountains

When proceeding from the west to the east in the Mecsek Mountains, we encounter younger and younger deposits. On the west, Permian deposits associated with Triassic sediments are predominant. Toward the centre of the Mecsek both Triassic and Jurassic deposits can be encountered, whereas on the east predominantly Jurassic sediments are exposed, overlain by scattered, isolated occurrences of Cretaceous sediments and volcanics. After the deposition of these rocks the area emerged and remained emergent until a transgression came in Helvetian time. Accordingly, the process of karstification lasted for long geologic epochs, being associated with erosion, and this continental regime could be scarcely changed either by the appearance of Miocene and Pliocene deposits, or by the deposition of loess in Pleistocene time.

MORPHOLOGIC FEATURES

The Transdanubian Central Mountains, the Northeast Range and the Mecsek are mountains of middle elevation. The Bakony Mountain is 713 m, the Bükk 957 m, and the Mecsek 683 m high. All are emergent peneplains of a hilly surface, with a steep edge at least on one side and with locally deeper-incised valleys and basin depressions, which are usually of tectonic origin. They are in motion even at present. As shown by BENDEFY [1964], the central part of the Bakony rises 10–13 mm each 10 years, in contrast with the rate of subsidence of 16 mm/10 years on the southeast and 6 mm/10 years on the northwest. In the Northeast Range more marked hypsometric changes—a rise of 13–32 mm/10 years— can be observed, whereas in the Mecsek these figures are –3.8 to +9.6 on the south and –13.0 to +1.9, 6.6 on the north flank. In the Villány Mountains, south of the Mecsek, the rate of rise varies between 0.75 and 6.4 mm/10 years.

Karst surfaces in Hungary occupy an area of some 3,000 km². The contributions of the various epochs are, however, very different, being dependent on stratigraphic position, thickness, tectonic setting, and karstifying agents. Therefore, first of all, the karst rocks most characteristic of the area under consideration have to be studied.

In the case of the Bakony Mountains, the thickness of more than 1,000 m of the Norian "Hauptdolomit" and the Rhaetian "Dachsteinkalk" of the Upper Triassic overshadows the remainder of the karstified rock masses.

In the Northeast Range the several hundred metres thick Anisian and Ladinian limestone sequences constitute the peculiar karst topography.

On the present-day surface of the western half of the Mecsek, the 400 m thick Anisian sequence has been karstified, whereas in the eastern half only the 200 m thick Malm limestones show karst features.

Of course, some questions may also arise in connection with karstification. Dolomite in the original sense is not affected by karstic attack in the same way as are limestones. No caverns like those known in limestones are formed. According to DARÁNYI [1966a], the dissolution of $CaCO_3$ bordering the rhombohedral crystals of dolomite may cause pulverization of this rock and result in a marked increase of its permeability. In addition, the calcite fillings deposited in the meantime are also dissolved, thus contributing to the development of water ducts. The water-storing capacity of dolomites is increased by its low resistance to tectonic stresses and by its consequent liability to tetra- or hexahedral fracturing over large areas. Dolomite is heavily attacked by thermal waters, so it was that while drilling for thermal water at the eastern edge of Budapest (E.R. Schmidt, personal communication, 1965), it was observed that as the drill reached the Hauptdolomit, it sank for 300 m easily, finding no resistance!

The limestones are all subject to karstification, but they are not karstified at the same rate. We do not know, however, why limestones under apparently equal conditions are more karstified at one place than they are at another. Moreover, to make it clear, we should say that at present we have no standard unit of measurement for expressing the degree of karstification of a rock.

Another crucial and still unsettled problem is the determination of the void volume of dolomite and limestone. The idea of laboratory experiments has to be automatically rejected, as a sample can by no means embody the characteristics of cubic kilometres of rock. No doubt, void volume shows large scattering, as evidenced by literature giving values varying between 0.2 and 16%. The fluctuations may, of course, be even greater if a small, very intensively karstified area is compared to deeper horizons, lying beyond the reach of the karstifying agents.

The depth range of karstification itself cannot be precisely determined. In the Transdanubian Central Mountains it usually does not exceed 100 m, in the Northeast Range, however, several deeper seated manifestations of karstification are known.

The degree of karstification itself seems to be illustrated to some extent by the number of caves and their total length. According to BERTALAN [1962], the caves known in the Transdanubian Central Mountains number 450, their total length being 20,508 m. Most of these caves, 107 with a total length of 16,220 m, are concentrated on the south flank of the mountains, in the vicinity of Budapest. This can be ascribed to two factors: on the one hand, to intensive thermal water activity, and on the other hand, to the greater variety of limestones dominated by the well-karstified Upper Eocene limestone strata. Least important of all are the caves in the Vértes Mountains, 10 in number with a total length of 77 m, which is due to the predominance of dolomite in their lithologic composition.

The caves of the Northern Range number 314, with a total length of 40,215 m, of which the Aggtelek karst comprises 89 caves with a total length of 31,727 m. Here the Aniso-Ladinian limestones are karstified to such an extent that one of the world's largest and most beautiful caves has developed within the karst. Its total length, inclusive of the Czechoslovak sector, exceeds 40,000 m.

In the Mecsek 69 caves, concentrated predominantly in the western half of the mountains, are known with a total length of 2,390 m.

The caves listed above do not include caverns which are inaccessible below the hydrostatic level of the karst waters. Cave systems like these are known northwest of Budapest, on the south bank of the Danube, and in the Dorog Coal Basin, where there are huge, contiguous karst cave systems comparable in size to the Aggtelek karst, but inaccessible. They are known in general only insofar as can be determined from hydrologic and other data. This same phenomenon substantially reduced in magnitude is also manifested in other parts of the Transdanubian Central Mountains such as the Dudar Coal Basin within the Bakony.

HYDROLOGIC CONDITIONS

Every karst region is characterized by its peculiar hydrogeology which is a function of lithology, degree of karstification, and their locations. The karst nature of an area is the most readily defined by its hydrogeology.

In a study of the karst springs of Hungary, KESSLER [1959] surveyed data on the reliability of the springs, their qualitative relationships, the percentage ratio of infiltration, the rates of infiltration, and underground run-off. He came to the conclusion that fluctuations in yield, temperature, and ionic concentration are intimately linked with one another, being, however, a function of the lithology of rocks penetrated by the channels supplying the springs. In the case of narrow channels, e.g., in dolomites, the results obtained for yield, temperature, and ionic concentration alike are more reliable, i.e., less variable, than in limestones with wide channels, particularly so if these latter outcrop high above the base level. It can usually be demonstrated that a decrease of the Ca/Mg ratio is characteristic of springs of steady yield and temperature with a high ionic concentration.

A close relationship can be found between total hardness and specific resistivity. The specific resistivity of a water with a hardness of 5 German degrees is 6,600 Ohm/cm, that of one with 10 German degrees being 3,000–3,400, and that of one with 15 degrees about 2,300.

In a karst region the determination of the percentage ratio of infiltration is very important. As noted by KESSLER [1959], there is not even an approximately linear relationship between the annual amount of precipitation and the yields of springs, but the amount of precipitation of the first four months of the year is representative for then the vegetation is dormant before large-scale consumption of water by plants begins and before the evapotranspiration is felt. Naturally, the total annual amount of precipitation cannot be neglected either. The percentage ratio of the representative amount of precipitation to the annual total is the representative percentage precipitation:

$$\frac{\text{precipitation of 1st to 4th month}}{\text{precipitation of 1st to 12th month}} \times 100$$

Essential for the rates of infiltration is the amount of precipitation late in the year, when both evaporation and evapotranspiration are markedly reduced. As shown by empirical data, the end of the year precipitation does not substantially influence the spring yields of the same year, but is primarily consumed by filling, through capillary action, the hair-joints and other narrow crack systems dried during the summer. This process does, however, make it possible for almost all of the precipitation early in the next year to enter the supply channels of the springs. Thus the representative percentage of precipitation has to be adjusted by a correction for the precipitation of the last part of the preceding year. This is done in such a way that we examine the difference in percentage between the amount of precipitation of the last four months of the preceding year, on the one hand, and the long-term average of the same four months on the other. As shown by empirical data, the coefficients of correction pertaining to these correction percentages are those given in Table I.

The mean monthly infiltration varies according to the above conditions; as shown by a preliminary survey, the long-term national average has its lowest value, 12.8, in October and attains its highest value in March when 113.0% infiltrates as a

TABLE I

RELATIONSHIP BETWEEN CORRECTION PERCENTAGES OF PRECIPITATION AND COEFFICIENTS OF CORRECTION (AFTER KESSLER, 1959)

Correction percentage of precipitation	Coefficients of correction
0– 5	0
6–15	1
16–25	2
26–35	3
36–45	4
46–55	5
56–60	7
61–65	10
66–70	13
> 70	15

result of thawing of the snow. In one of the Bükk caves the rate of infiltration measured in March at a single place increased by 525 cm^3/h as a result of 31 mm of precipitation whereas at the same place in August 33 mm of precipitation caused an increase of only 83 cm^3/h.

According to KESSLER [1959], in swallet-free areas of the Aggtelek karst, after 85 mm of artificial rainfall, the velocity of infiltration in 80 m of Middle Triassic limestone was found to vary between 6.7 and 9.1 m/h, while near Budapest, in 25–40 m of Upper Eocene limestone, 100 mm of artificial rainfall could cause velocities of 2.4–4.7 m/h.

Because of its varied constituents the horizontal flow is such a complex phenomenon, that DARÁNYI [1966a] believes that its details can hardly be assessed. The laws governing the behaviour of watercourses in caves are similar to those of the surface streams, while a flow occurring below the hydrostatic level is very intricate, being dependent on continuous changes of the complex channel systems, their size, roughness, interconnections, the angles formed with hydrorelief (the most favourable direction of flow) etc. Since they undergo continuous changes in both horizontal and vertical senses, all these factors result in a labyrinth of varying conditions difficult to comprehend.

Accordingly, in spite of the great number of observations and measurements, many questions remain unanswered. The amount of precipitation falling on the exposed karst surfaces and non-karstified overlying formations, or depression-causing overburden is well-known, but available information on the precise horizontal and vertical distribution of these formations and their water-absorbing capacity in different places is—apart from some local information—far from being adequate to permit us to give a clear-cut answer to these questions.

Essentially the same holds true for the run-off and evaporation. The amount of water running off over a large area cannot be measured, nor can the loss due to evaporation be determined, as some of the run-off may be swallowed by rock cracks and voids while travelling down its path.

In some places water entering the rocks may increase the discharge of springs, whilst the evaporation may be initially higher, if the rock surface is heated, but it will then decrease as the surface cools; so, for the same reason, evaporation shows a seasonal variation as a function of temperature and air moisture; but water may display steady evaporation when exposed to the atmosphere in ponds, pools and swamps around the mountains. Additional difficulties are due to the fact that we cannot separate waters of different origin in the swamps and pools, i.e., the contributions of surface run-off, swallowed or spring water, and direct precipitation on the one hand, and the periodic discharge and the variable surface area of the swamps and pools on the other. In both run-off and percolation we have to take into account the degree of fracturing of the rock, the angle and length of the slope, the amount of precipitation per unit time, and the processes taking place between snowfall and thawing, such as evaporation, snowdrift and rate of thawing.

In the majority of cases the void volume of dolomites and limestones is not known either. At present, there is no method which would permit the determination of their void volume over a large area. Information on the shape and the range of the cone of depression resulting from withdrawal of water, is not satisfactory for this purpose, because in a karstified rock body the cone of depression is irregular, its configurations being controlled by the fracture systems as well as by the karst caverns and channels. Its range of fluctuations observable along fracture planes may be many times that observed in the direction perpendicular to it, and in fact its shape resembles an irregular star rather than a funnel. The area of a stabilized cone of depression in an artesian aquifer is defined by the amount of recharge from precipitation—the only form or recharge occurring in Hungary. The stabilized cone extends over an area where the amount of infiltration or recharge equals the amount of water withdrawn.

Difficulties are also faced with the amount of water flowing below the hydrostatic level, since our knowledge about the factors of flow themselves is more or less incomplete in dependence on location and time of measurement. The problem will be even more complicated if the sizes of the water-bearing cavern network are examined, for in the karstified upper levels these seem to be a multiple of the decreasing figures of the lower levels. The ratio of flow, i.e., the water yield which can be obtained at a given place, is defined by water pressure and the water-conducting capacity of all the channels and caverns connecting that place with the drained area. It follows from the above, however, that all this can only be calculated by using a method of approximations since the yield may change by leaps and bounds within a short distance, in dependence on the water-bearing channel tapped.

All the variability mentioned concerns both limestone and dolomite. Many authors suggest the void volume of dolomite to be larger than that of limestone, although there is no sound evidence that would warrant this suggestion. It can not be doubted, however, that the distribution of water in dolomite is more even than it is in a karstified limestone. Therefore, its water discharge is more balanced also.

The void volume of a rock, i.e., the volume of its water-bearing fracture system, may vary greatly along faults. In the case of a large fault, whose vertical displacement exceeds the thickness of the karstified zone, all or part of the water-bearing beds may be faulted to a position opposite to the relatively impervious beds. This would practically mean that such vertical displacement along major faults produces a marked damming effect on the circulation of the water, i.e., the water yield to be expected will be lower than calculated on the basis of hydrostatic pressure. However, in the case of step-faults, with sizes of individual steps smaller than the downward decrease in void volume, the damming effect, if any, will be very much reduced.

Since some factors of the karst water regime such as precipitation, run-off, infiltration, void, lithoclase system, depression area, flow conditions, fault-induced disturbances, etc. are known only approximately and since they are always inferred from other approximate data, the calculations always involve two or more unknowns. In the case of karstified rocks, any conclusion on hydrology can be theoretically approached, as a rule, with less certainty than in unconsolidated sediments. Practical experience is always very important for karst hydrology, as the individual constituents of a theory deduced from karst water phenomena may be arbitrarily changed according to the given circumstances.

Transdanubian Central Mountains

The area of the Transdanubian Central Mountains is nearly 5,000 km^2, of which some 2,400 km^2 are exposures of Mesozoic, for the most part Upper Triassic sediments. Uniformity of the so-called main karst aquifer of the mountains is assured by a more than 1,000 m thick sequence of the Upper Triassic Hauptdolomit and Dachsteinkalk, characterized by good water-bearing properties. The mean annual precipitation is estimated at some 650 mm.

The higher karst aquifer is represented by limestones referred to the Albian Stage of the Middle Cretaceous. In the northern part of the Bakony Mountains, the 100 m thick Aptian clay sequence lying between the main karst and the higher-sited Cretaceous limestone aquifer ensures the complete isolation of the two, so that both of them are characterized by independent hydrogeologic regimes. In the southern Bakony the 60 m thick, Upper Cretaceous coal-bearing marl-and-clay sequence overlying the main karst provides a hydraulic system for both the main karst aquifer and the Eocene water-bearing limestones.

Both the Middle Cretaceous and Eocene aquifers differ basically from the hydrologic regime of the main karst. The difference consists mainly in the fact that they do not form a continuous hydrologic system, also in that, because of their higher elevation and more reduced thickness, they are more or less—in some places even completely—worn away because they are separated by faults with displacements exceeding the thickness of the aquifers.

In addition to stratigraphic control, the hydrogeology of the Transdanubian Central Mountains is defined by the smaller ratio of the exposed portion of the very thick water-bearing main karst sequence compared to the subsurface portion accounting for 1/2–3/4 of the total, so that the karst is bordered on the south and north by impervious or relatively impervious sediments (e.g., lowermost Triassic or Palaeozoic on the south and uppermost Pannonian on the north).

According to Jaskó [1959], the variation of water level within the main karst aquifer depends on the possibility of losses from the amount of water stored. Recharged from northern and southern directions and tapped by abundant springs on the west and east, the aquifer has its underground divide in the line of Urkut-Bakonybél, where the hydrostatic level of the water is at 280 to 225 m above sea level with a northward decrease in elevation. Both to the west and east of this watershed, the water level gradually sinks to 100 to 110 m above sea level at the west end of lake Balaton and in Budapest on the east. The slope of the water level is uneven, the steepest part being immediately west of the underground divide, where the level sinks 100 m within

10 km distance; the most gentle part is between the Bakony and Budapest, where there is a drop of 50 m within 100 km.

This natural karst hydrorelief brought about by infiltration and tapping by springs is responsible for the natural flow of the main karst waters, which is, however, a weather-dependent process, as both the main karst water level and the yield of the springs show marked fluctuations. Within the main karst aquifer the amplitude of water level fluctuations attains, without artificial interference, 8 m in the Hauptdolomit, but is smaller in the Dachsteinkalk. The highest yield of the springs in extreme cases may be as much as 100 times the lowest yield, while the average yields of particular years may show differences with a factor of ten. From these data it is evident that flow and water level of karst waters fluctuate continuously even under natural conditions.

According to the calculations of KESSLER [1959], the rate of recharge of the Bakony main karst waters from precipitation over a 666 km^2 area is 318 m^3/min, a figure consistent with the annual yields of the springs. These data readily illustrate the amount of water involved in the circulation of the main karst waters. Accordingly, in the Transdanubian Central Mountains the rate of recharge would be 1,200 m^3/min.

Let us develop further this train of thought: if the thickness of the water-bearing karstified zone is assumed to be as low as 100 m—dispensing with the deeper caverns and channels—and if the buried areas of the main karst aquifer within the waterstoring body of the Transdanubian Central Mountains are also taken into consideration, and if a 2% void ratio is assumed, the amount of stored water will be $8 \cdot 10^9$ m^3.

The yields of the springs are high, a typical feature of karst regions; exceeding 10 m^3/min for a good many of them. The highest figures so far obtained were 75 m^3/min on the west border of the Bakony, 35 m^3/min on the northwest foreland of the Vértes, and 30 m^3/min in a group of springs in Budapest.

In the coal mines of Dorog the heaviest water entry observed at a pressure of 10 atm was 80 m^3/min, but values of 50–60 m^3/min were common. Under the influence of water inrushes and connected continuous drainage between 1948 and 1960, the hydrostatic level of the main karst aquifer at a distance of 14 km from the mines in the northwestern Vértes dropped 10 m.

The amount of water entering the mines from the main karst water-bearing rocks has practically remained unchanged for decades. The springs issuing from Middle Cretaceous and Eocene limestones are of low yield, but the initial rates of water entries in mines may be very high. So in the Balinka coal mine, at 20 atm pressure, a rate of 20 m^3/min was observed and the water inrush from the Eocene limestone hanging wall of the coal-bearing sequence at Ajka proceeded at a rate of 50 m^3/min. These water entries, however, tend to decline rapidly and drop to a fracture of the original values within a month or year. Their area of depression of the water level is so large that large-scale drainage will completely eliminate any water hazard to mining.

The waters differ in chemical composition. As shown by DARÁNYI [1966a], although waters of equal composition may occur at various places, knowledge of the chemistry of the waters occurring in given places may provide valuable information on the source of the particular water occurrences. In practice, this can be used in case of water inrushes, and as, if the source of water is known, the expected course of an inrush and the most useful measures of protection can be anticipated.

The chemical constituents of the water of the main karst, in general consists chiefly of Ca- and Mg-bicarbonate and a slight sulphate content; its hardness attains 20–30 German degrees. The water of the completely isolated Middle Cretaceous limestones is very soft (1 or 2 German degrees) and it is characterized by the presence of bicarbonate. Although the water of the Middle Cretaceous and Eocene limestones, recharged directly from precipitation, shows a wide range of changes in concentration as a function of the alternation of humid and drier seasons, it is characterized by 16–22 German degrees of hardness, as well as by the association of Na-ions with Mg-ions in a comparatively high concentration.

In Budapest, where the Danube flows along a large step-fault, on the left bank the Upper Triassic surface of the main karst aquifer lies 1,000 m deep, whereas on the right bank it is exposed, emerging 530 m above sea level. Along this fault, parallel to the course of the Danube, on the right bank there are numerous subthermal and hot springs which, if hit by drilling on the left bank, surge above the ground, a reason why Budapest has become a city of spas. The total yield of the subthermal springs of Budapest is 30,000 m^3/day, that of the hot springs attains 44,000 m^3/day.

Northeast Range

The Northeast Range covers some 1,500 km^2 of area, of which at least 900 km^2 are Palaeozoic and Mesozoic areas; the area of karstified rocks being more than 300 km^2. Their virtual extent is, however, greater, as the Aggtelek karst is divided by the Hungaro-Czechoslovakian frontier and the contiguous karst region extends far beyond the frontier. This circumstance can by no means be neglected, particularly from the hydrogeologic point of view. The annual precipitation amounts to some 700 mm.

The Bükk Mountains are surrounded by Tertiary rocks which do not restrict the flow of the waters of the karstified rocks to such an extent as was shown for the Bakony Mountains, as these rocks are more reduced in thickness and as the accumulation of water here is due to the impervious layers underlying the karstified formation. The springs and karst channels in the Northeast Range lie above the base level, so that their water reserves and dynamism differ from that of the Bakony. The karstified rocks occurring here are mainly of an emergent—subordinately stratum-like—karst type, a fact proved by the high, 300–630 m, level of the spring outlets. The deeper-dropped portions at the edges are but a small fraction of the whole karst massif and in these cases some subsurface losses through the unconsolidated Tertiary rocks can be admitted, especially in north- and eastward directions.

As shown by SCHMIDT et al. [1962], of the 148 million m^3 of annual precipitation, 52 million m^3 infiltrate into the karst, of which some 40 million m^3 are discharged through the springs and some 12 million m³ is lost through the unconsolidated rocks to the basins. The annual precipitation of the Aggtelek karst attains 60 million m³, the amount infiltrated is estimated at 20 million m³ which corresponds to the yields of the springs. Thus possible losses due to filtration through the unconsolidated Tertiary rocks surrounding the karst are improbable.

The chemical composition of the waters is characterized by bicarbonate with subordinate amounts of Mg. Where Mg is absent, the water must have been in contact with karstified limestones only. The waters coming from the more remote parts of the catchment basin and having been in contact with dolomites always show a higher Mg content. Total hardness is 16–19 German degrees, carbonate-induced hardness being by 1–3 German degrees lower. Water with a chemical composition that differs from the above characteristics, shows that it was mixed with waters of a different type occurring in the basin.

The springs are very numerous, nearly 400 in number, with yields lower than 500 l/min in the majority of the cases. There are 22 springs whose average yield is higher. The most important spring is near Miskolc, on the east border of the Bükk, with 20 m³/min; within the Aggtelek karst, the Jósva Spring has a yield of 15 m³/min.

The temperature of the springs usually corresponds to the mean annual temperature, ranging from 7–10° C. The ascending waters along faults are subthermal to warm, varying between 14 and 36° C.

Mining is practised only in the basins surrounding the borders of the Northeast Range. Therefore, the immediate influence of the karst waters is not felt in the mines, where no experience in karst hydrology can therefore be gained.

A phenomenon observed for many years is that the yields of the springs change with a regular periodicity. This was noticed particularly in the Aggtelek karst. After a very thorough analysis, MAUCHA [1967], relying on PAPP [1941], explained it by a lunisolar earth tide influence. According to this theory, the tidal changes in karst water level are an indirect result of the tide-induced deformation observable in the solid crust. The vertical walls of the lithoclase network of the karstified rocks are tightened by 2.5 μ with a 12-h periodicity. Tightening thus induces an indirect tidal flood which lifts the karst water level. Consequently, the aclimatic water level fluctuations of the underground watercourses feeding the karst springs within the Aggtelek karst may cause 700 l/min differences in yield. The "eruptions" of the siphoned karst springs are controlled by the tidal waves with a 33% probability.

Mecsek Mountains

The Mecsek's area is but a small fraction of that of the Transdanubian Central Mountains and the Northeast Range. Its karstified catchment area is 32 km², a figure to which can be added the 12 km² area of the non-karstified surface drained by the karst. In addition, the 18 km² area of the isolated blocks of the Villány Mountains cannot be ignored. The annual precipitation amounts to 800 mm.

The total annual amount of precipitation falling in the catchment area of the

Mecsek is 34 million m³, of which 12.5 million m³ infiltrate into the karst caverns and channels. The total annual yield of the springs is 7 million m³; thus 5.5 million m³ of water is lost through infiltration into the surrounding Tertiary basin deposits.

The recharge to the karst of the Villány Mountains is 4.7 million m³ of water a year, and the total annual yield of the springs is 1.3 million m³ indicating an annual loss of 3.4 million m³ to the Tertiary basins surrounding the mountain blocks.

According to SCHMIDT et al. [1962], two types of karst springs can be distinguished on the south flank of the Mecsek and in the central areas, there are karst springs of stratigraphic or tectonic control, whereas the northwest area is characterized by karst springs issuing at the base level. Since the impervious underlying beds are not deep, the springs are of shallow-karst type.

In the south foreland of the Villány Mountains the karst is tapped by hot springs. This phenomenon can be ascribed to the uniform behaviour of the Triassic, Jurassic and Cretaceous limestones, for they are all fractured, massive, subject to karstification and dropped down and buried by younger sediments beyond the edges.

Some of the karst springs issue directly from karstified rocks, whilst some flow upward through the Tertiary deposits. The large springs have yields varying from 2 to 8 m³/min. In the Villány Mountains a spring has yielded 3 m³ of water a minute at 62°C.

The waters show chemical compositions typical of karst waters: comparatively high in Ca-bicarbonate, with smaller amounts of Mg-bicarbonate and a hardness of about 20 German degrees. It is, however, noticeable that if the water moves upward to the surface through the Upper Tertiary of Pleistocene deposits, the Na content, otherwise very low, will be relatively high. The subsurface thermal waters also contain some sodium chloride.

PRACTICAL PROBLEMS AND THEIR APPROACH

Some of the karst waters can be used for domestic water supply, irrigation and industrial purposes. On the other hand, protection against flooding by groundwater is necessary for the mining industry.

The utilization of water from a spring or a well does not face any particular difficulty, if the hydrostatic level and the rate or recharge are known, and both the need for water and the costs are taken into consideration.

Considerable troubles may, however, be caused by the karst waters in the mining industry, especially so in the Transdanubian Central Mountains, where browncoal, bauxites and manganese-ore bodies are underlain in the majority of the cases by the main karst aquifer and its hydrostatic level.

The great number of water inrushes showed that in the places where the thickness in metres of the impervious layer protecting the coal seam against water in its footwall was one and a half more times the absolute value (expressed in atm) of the water pressure applied to the bottom surface of the coal seam, the water inrushes diminished markedly in number or vanished altogether. Consequently, a specific protective layer thickness of 1.5 m/atm provides relative safety for mining operations. This value, however, is arbitrary, as the protecting effect of the layer is a function of petrophysical parameters. In deep-seated underground workings within karst water hazard areas with 30 atm footwall water pressure, it is conceivable that on a large open (exposed area) or on the surface of free contact between rock pressure and water pressure, the protective footwall will react by plastic deformation or fracture, if it is not thick enough or has unfavourable petrophysical characteristics.

As regards the hanging water hazard, practical experience is similar to the case of the specific protective layer. According to our knowledge the underground workings in a coal seam can be considered safe from water inrushes from the roof, if after mining the protective hanging wall remains from 8 to 18 times thicker than the coal itself, these values depending on petrophysical characteristics.

The protection offered by a protective layer is always to be understood as being efficient within particular tectonic blocks in undisturbed tectonic settings. Large faults always threaten with flooding from the footwall along the lower strike and from the roof along the upper strike, as they use up the protective layer in dependence on the ratio of fault thrown to the thickness of the protective layer. Moreover, we have to take into consideration the faults which cannot be recognized by exploratory drilling. The lower limit of detectability varies as a function of the depth of prospecting, but may attain 10–15 m even in holes 200–400 m deep.

Attempts at finding a clear-cut theory for explaining the hydrogeologic function of faults have also been made, as also in the case with protective strata. They have started from the protective layer thickness/throw ratio, but some difficulties were encountered since one of the prerequisites of a perfect solution to the aforementioned problem is to express in absolute values, the degree of protection offered by the protective layers.

Beside the hydrogeologic role of faults, the hydrogeologic survey of an area has to study the structure of the Mesozoic bedrock and overlying younger sedimentary cover separately. This is a fundamental and very complex task. Since the throw of a fault may also not be the same over its full length and since the thickness of single strata may vary as a result of differences in structure and subsequent erosion, it is obvious that the hydrology of an area cannot be evaluated in terms of average values but only by considering the continual changes in geologic environment.

Recently it has been attempted to find relationships between the degree of fracturing of a coalfield, i.e., the orientation, number and total length of faults per unit area on the one hand, and the frequency of water inrushes on the other. This is undoubtedly a good approach, though its solution is hindered by many factors including the aforementioned fact that the bedrock is strikingly different in structure from its Tertiary and Quaternary cover. This implies an a priori error in practice, as mining operators plot the structure on the basis of the coal seams which according to their age, are less "faulted" than, e.g., the underlying Upper Triassic basement, whose intensively tectonized structure is unexplored. From this it is evident that in one and the same field two different degrees of faulting are encountered, both being equally significant from the hydrogeologic point of view. Of the two—bedrock and cover—the common, younger structure offers the more visible direct connection, but the degree of water hazard is defined by the interrelation of both the buried structure of the old surface and that of the superimposed stratum. This allows us to decipher the puzzle of water inrushes which at first glance seem to be enigmatic, but which are due to faulting of the basement.

Problems in interpretation may also occur when hydrogeologic relationships are considered on the basis of the younger structure as obtained for the coal seams alone, since a fault does not produce the same effect throughout the whole field. Large faults are, as a rule, boundary faults, along which water-control piers are left. They are not approached by mining operations, thus being explored the least of all. Therefore, they can be taken into consideration only in as much as information on them is available. Consequently, they cannot always be relied upon in drawing final conclusions. The accuracy of the results also depends on the technology of mining and on whether a fault was reached from the direction of its upper or from its lower strike. In addition, errors may be due to intersections of faults—particularly when hidden in the basement—which may provoke unforeseen water inrushes. A question to be answered in each case is the distance of water entries from the particular faults, i.e., to decide the reliability of the relationships between individual water inrushes and faults. We could continue this list of problems, but these examples seem to be enough to illustrate their manifold nature and the uncertainty of the methods.

Although the Hungarian karst morphologically corresponds to the general notion of karst, its mode of occurrence gives us a special status. This holds true particularly for the Transdanubian Central Mountains, whose Upper Triassic main karst aquifer lies for the most part below the hydrostatic level and is covered by impervious sediments. This is what defines its peculiar hydrogeologic character. The Northeast Range belongs chiefly to the emergent type of karst, and the marked karstification of the Aggtelek karst has brought about one of the largest cave systems of the world with significant stalactites and stalagmites. The Mecsek belongs to the third type: the shallow karst. Each of the three has a different structure, the fault/fold ratio being different in each of them. They also differ in hydrogeologic significance, the most peculiar hydrogeology being characteristic of the Transdanubian Central Mountains, with the most intensive faulting and most varied mining features.

REFERENCES

BALOGH, K., 1964. A Bükkhegység földtani képződményei. (Geologic relations of the Bükk.) *Magy. Állami Földt. Int. Évkönyve*, 48(2):245–553.
BENDEFY, L., 1964. Geokinetic and crustal structure conditions of Hungary asrecorded by repeated precision levellings. *Acta Geol. Acad. Sci. Hung.*, 8(1/4): 395–411.

BERTALAN, K., 1962. A magyar karszt és barlangdokumentáció problémai. (Problems of Hungarian karst and cave documentation.) *Magy. Állami Földt. Int. Évi Jelentése*, 1962:555-561.
DARÁNYI, F., 1957. Adatok az Ajka környéki kréta kifejlödéséhez. (Cretaceous sediments of Ajka.) *Bányász. Lapok*, 90:253-256.
DARÁNYI, F., 1966a. A Bakonyhegység karszthidrológiai kérdései. (Karst hydrology of the Bakony.) *Hidrol. Közl.*,46(5):211-218.
DARÁNYI, F., 1966b. Adatok a Bakony hegység szerkezetéhez. (The structure of the Bakony.) *Földt. Közl.*, 96:280-291.
FÖLDVÁRY, A., 1952. A szabadbattyáni ólomérc és kövületes karbonelöfordulás. (Lead-bearing Carboniferous sediments of Transdanubia.) *Magy. Tud. Akad. Müszaki Tud. Oszt. Közlemen.*, 5:25-53.
FÜLÖP, J., 1961. Magyarország krétaidószaki képzödményei. (Cretaceous sediments of Hungary.) *Magy. Állami Földt. Int. Évkönyve*,45(3):577-587.
JASKÓ, S., 1959. A földtani fölépités és a karsztviz elterjedésének kapcsolata a Dunántuli Középhegységben. (Geology and karst hydrology of Transdanubia.) *Hidrol. Közl.*,39(4):289-298.
KESSLER, H., 1955. A karsztból tartósan kitermelhetö vizmennyiség és a beszivárgási százalék számitása. (Infiltration percentage and water production from karstified rocks.) *Hidrol. Közl.*,35(5/6): 213-222.
KESSLER, H., 1959. Az országos forrásnyilvántartás. (Well register of Hungary.) *Vizgazdálkodási Tud. Kut. Int. Tanulmányok Kut. Eredmények*, 7:1-122.
KISS, J., 1951. A szabadbattyáni Szárhegy földtani és genetikai adatai. (Geologic and genetic data of Szárhegy.) *Földt. Közl.*, 81:264-274.
LÓCZY, L., BOCKH, J., BITTNER, A., HALAVÁTS, GY., LACZKÖ, D., PAPP, K., VADASZ, E. and VITALIS,J., 1913. A Balaton környékének geológiai képzödményei és ezeknek vidékek szerinti telepedése. (Balaton monography.) *Balaton Tud. Tanulmányozásának Eredményei, Budapest*, 1:5-356.
MAUCHA, L., 1967. A karsztvizszint árápály jelenségének kimutatása. (Hydrostatic surface and tidal phenomena.) *Bányász. Kut. Int. Közlemen., Elökészületben.*
NOSZKY, J., 1934. Adatok az északi Bakony krétaképzödményeinek ismeretéhez. (Cretaceous sediments of the Bakony.) *Földt. Közl.*,64:99-136.
NOSZKY, J., 1943. Elözetes jelentés a Szentgál környéki földtani felvételröl. (Brief geology of Szentgál.) *Magy. Állami Földt. Int. Évi Jelentése*, 1:3-6.
ORAVECZ, J., 1964. Szilur képzödmények Magyarországon. (Silurian sediments in Hungary.) *Földt. Közl.*,94:3-9.
PAPP, F., 1941. Dunántul karsztvizei és a feltárás lehetóségei Budapesten. (Karst waters of Transdanubia.) *Hidrol. Közl.*, 21(7/12):257-306.
SCHMIDT, E.R., 1957. *Geomechanika*. Akad. Kiadó, Budapest, 170 pp.
SCHMIDT, E. R., ALMÁSSY, E., BERTALAN, K., EMBER, K., ERHARDT, GY., FERENCE, K., LÁNG, G., NÉMETH, L., OZONAI, GY. and RAVASZ, C., 1962. *Vázlatok és Tanulmányok Magyarország Vizföldtani Atlaszához. (Studies on the Hydrogeologic Atlas of Hungary.)* Müszaki Könyvkiado, Budapest, 30 pp.
SCHRÉTER, Z., 1959. A Bükk-hegység tengeri eredetü permi képzödményei. (Marine Permian sediments in the Bükk.) *Földt. Közl.*,89(4):364-373.
SZABÓNÉ-DRUBINA, M., 1956. Az eplényi mangánércelófordulás közettani viszonyai. (Petrographical conditions of the maganese ores of Eplény.) *Magy. Állami Földt. Int. Évi Jelentése*, 1955/1956: 331-338.
SZALAI, T., 1951. Adatok a Dunántul hegyszerkezetéhez. (Structural conditions of Transdanubia.) *Bányász. Lapok*, 84(6): 543-550.
SZALAI, T., 1958. Geotektonische Synthese der Karpaten. *Geofiz. Közlemen.*, 7:111-145.
TELEGDI-ROTH, K., 1935. Adatok az Északi Bakonyból a Magyar Középsó Tömeg fiatal mezozóos fejlódéstörténetéhez. (Genetic data of the development of the Hungarian internide.) *Magy. Tud. Akad. Mat. Term. Tud. Ért.*, 52:205-247.
VADÁSZ, E., 1960. *Magyarország Földtana. (Geology of Hungary.)* Akad. Kiadó, Budapest, 320 pp.
WEIN, Gy., 1952. A Mecsekhegység hidrogeológiája. *Földrajzi Ért.*, 32:237-250.

Chapter 9

Karst of Czechoslovakia

J. BYSTRICKÝ, E. MAZÚR AND J. JAKÁL

Slovakian Academy, Bratislava (Czechoslovakia)

INTRODUCTION

The karst districts in Czechoslovakia extend over a relatively large area, i.e., more than 3,000 km². The distribution of the karst regions is not centred, but is much dispersed over areas of sizes varying from small islands up to relatively large karst regions of some hundreds of km². At the same time the karst phenomena are also very unequally developed in the individual karst districts. This fact depends on a whole range of factors about which we shall speak further.

Notwithstanding considerable scattering of the karst districts and the variety of the karst development, we can distinguish in Czechoslovakia two regions with a different areal distribution of karst, different size of karst districts and abundance of karst phenomena; these are the Bohemian upland and the Carpathians. In the Bohemian upland, the karst regions are found only sporadically, in relatively small areas and largely with a limited number of forms. The exception here consists of the Moravian karst with relatively well developed features. This region has been the one most studied in Czechoslovakia and it is well-known even in world literature.

In contrast to this, in the Carpathian region in Slovakia, more than 90% of the karst districts in Czechoslovakia are to be found, i.e., more than 2,700 km². Many of the karst districts here extend over some hundreds of km² and the degree of the karst phenomena is also more intense. For these reasons and also in the interests of filling up considerable gaps in international knowledge of the karst districts in Slovakia, this region is the object of this study.

The history of the investigation of the karst phenomena and karst regions of Slovakia is not particularly rich, even though we can date it from the beginning of the 20th century according to the works of SAWICKI [1908, 1909]. The Slovakian karst area was studied later also by VITÁSEK [1930], ROTH [1939], JANÁČEK [1941] and others. Some problems of the genesis of the Slovakian karst in relation to the surrounding region were solved by LÁNG [1949]. Among the more important works are later studies by SENEŠ [1957], KEMÉNY [1961] and LUKNIŠ [1962].

Among the other karst regions, the Low Tatras have been examined, chiefly from the speleological aspect, by DROPPA [1957]. KUBÍNY [1959] devoted himself to the study of surface karst in the Low and Western Tatras. LUKNIŠ prepared a monograph on the karst of the Slovenský Raj [1945] and he devoted himself also to certain karst regions of the Little Carpathians [1946] and of the Tisovec karst [1948]. The karst regions of the Malá Fatra have been included in the larger monograph by MAZÚR [1963].

The karst regions in the high mountains were studied by SEKYRA [1954] for the Belanské Tatras; by LOUČEK [1956] for the Low Tatras; and MAZÚR [1962] for the Western Tatras. In several studies, DROPPA [1957,1966,1967] devoted his attention mainly to the underground karst phenomena in Slovakia, but also partly to other problems.

Special geologic investigations concerning the karst regions are relatively numerous also. Amongst these studies are those of HOMOLA [1951], MAHEL' [1957], ANDRUSOV [1964–1965], BYSTRICKÝ [1964] and others. Some problems of the karst hydrology have been studied by KULLMANN [1964]. Fossil soils and crustal weathering have been examined by SMOLÍKOVÁ [1962], BORZA and MARTÍNY [1964] and others.

Among the journals which devote their attention especially to the karst problems in Czechoslovakia are *Československý kras* and *Slovenský kras*. Reports on karst studies often appear in *Geografický časopis, Sborník čs. spol. zeměpisné*, sporadically in geological journals and in *Archeologické rozhledy* as also in the journal *Slovenská archeológia*, etc.

CONDITIONS OF DEVELOPMENT OF THE KARST PHENOMENA[1]

As already stated in the introduction, the karst phenomena in the various karst regions are not very uniformly developed and this depends on a whole range of factors starting with the lithologic-structural properties of rocks, through the relief, weathering products, the climate, water, up to the vegetation, and other factors.

For an easier understanding of the classification of Czechoslovakian karst regions, which are the object of further study in these pages, it is necessary to analyse concisely the above-mentioned elements and their time–space relationships.

Geologic conditions

The Western Carpathians are a zonal mountain chain of the Alpine–Carpathian Mountain system with a marked division into two longitudinal zones which differ in their morphological character and overall development and structure. They are the Outer Carpathians in the north and the Inner Carpathians in the south. The Outer Carpathians are composed mainly of flysch sediments of Cretaceous and Palaeogene age (hence they are also called the "Flysch Carpathians"); the older formations (Jurassic) are represented only sporadically in their western section. The chief structural element of the Inner Carpathians is composed of crystalline, Palaeozoic and Mesozoic rocks. A common characteristic of both zones is the complex nappe (over-

---

[1] Written by J. Bystrický.

thrust) structure, but each of them has its own particularities resulting from the different character of the sediments and the period of its structural uplift. The nappe structure of the Outer Carpathians resulted from the folding Carpathian geosyncline in the Miocene (following the Oligocene and before the Burdigal–Savian phase, after the Lower and before the Upper Tortonian—the Styrian phase), while in the Inner Carpathians it was already at the end of the Lower Cretaceous (after the Cenomanian and Turonian and before the Senonian—the sub-Hercynian phase). The Miocene orogenetic movements which gave rise to the Alpine-type structure of the Outer Carpathians were manifested in the Inner Carpathians only in the last formation of the pre-Senonian tectonic units or they gave rise to folds of large amplitudes—mega-anticlines and mega-synclines—and to large thrusts with a displacement mainly towards the south which cut off their interior flanks. As a result of vertical movements during these and younger phases, older formations rose to the surface: the present crystalline cores of the so-called "core mountain chains". The "Klippen zone" which extends between the Outer and Inner Carpathians along the whole of the Western Carpathians as a 5–50 km broad belt has a special position and structure. The complex, of chaotic structure, of the Klippen zone is due chiefly to the fact that the sedimentation in the Klippen zone during the Mesozoic was in the area of the northern border of the Inner Carpathians and it underwent Alpine-type folding in the Cretaceous. But from the Upper Cretaceous it was a part of the area of the sedimentation of the Outer Carpathians. The Miocene orogenic movements (Savian and Styrian) almost entirely veiled the nappe structure.

In view of the various developments and the structure of the different zones of the Western Carpathians, primary conditions were created for the origin of karst in very different ways. In the Outer Carpathians, the karst phenomena are only in some bioherm limestones of the Jurassic (Pavlovské Hills, Štramberk). On account of their small thickness, but especially on account of the small extent of these chemically very pure limestones only surface karst phenomena were formed here. In the Klippen zone the limestones are found in much larger quantity, but their jutting out in the form of small blocks projecting from the flysch and flyschoid sediments of the Upper Cretaceous and Palaeogene, as well as the varied facies of the Jurassic limestones and of the Lower Cretaceous, did not favour the development of the karst. The karst phenomena are most extensively and most magnificently developed only in the Inner Carpathians. The limestone–dolomite complex of the Triassic is present in all the tectonic units and in many places it forms substantial parts of mountains or even whole chains.

The structure of the Inner Carpathians, as we have already stated, is the result mainly of pre-Senonian orogenetic movements. Through the folding of the Mesozoic geosyncline there arose a complex system of tectonic units superimposed on one another from south to north, of which the most characteristic for the Inner Carpathians are the nappes. From north to south these are the main tectonic units: Tatrides, Ultratatrides, Veporides, Ultraveporides and Gemerides [BIELY et al., 1968]. Each of the given units is composed of crystalline rocks and of a "sedimentary cover" (Late

Palaeozoic, but particularly Mesozoic). The sedimentary cover of some of the units is, against the underlying crystalline rocks, more or less autochthonous (Tatrides, Veporides). In other tectonic units, the sedimentary cover was entirely separated from the underlying crystalline rocks and was shifted to the north to the outer tectonic units in the form of nappes (Krížná nappe, Choč nappe etc.). Each of the tectonic units mentioned has a peculiar litho-stratigraphic content, especially the Mesozoic, which is the reflection of the facies division of the Carpathian geosyncline into more or less parallel longitudinal zones and of the withdrawal of the geosyncline axis during its development towards the north (Fig. 1).

The *Tatrides* represent the northernmost and thereby also the lowest main tectonic unit. The original sedimentary cover (in places Carboniferous and Permian, Lower Triassic up to Turonian) in relation to the crystalline rocks (crystalline schists and granitoids) is mostly autochthonous and it has a simple structure. Only in some mountain ranges (e.g., in the Low Tatras and in the High Tatras) it forms inclined, horizontal folds and slices. The Mesozoic of the sedimentary mantle is characterized by generally shallow-sea facies and a frequent hiatus in the Rhaetian. Some facies of the Liassic (stained marls) and Dogger–Malm (radiolarites), however, point out that in the anticlinal zone of the Tatrides in the Jurassic there was a region with sedimentation in the deeper sea which is characteristic for the more southerly Ultratatride zone.

The next higher tectonic unit is formed by the *Ultratatrides*. The crystalline rocks of this unit are known from the surface of only a small district near Staré Hory. It is assumed that its chief mass is hidden by the Veporides on the scar dividing the Tatrides from the Veporides. The sedimentary cover (Late Palaeozoic and Early Triassic up to Cenomanian) was thrust over the Tatrides and is known as the Vysoká and Krížná nappes. The latter is the largest and the most extensive of all the nappes in the Inner Carpathians. It represents a simple bed resting on the Mesozoic of the Tatrides (Tribeč, Vel'ká Fatra mountains), or it forms digitations and slices (Vysoké Tatry, Malá Fatra). The *Veporides* form a tectonic unit with the diaphtorized crystalline rocks and also with a more or less metamorphosed sedimentary cover, including the Carboniferous up to Neocomian strata. Of the tectonic unit of the *Ultraveporides*, we know only that the sedimentary cover moved to the Tatride zone as the Choč nappe. It is assumed that its crystalline rocks were left on the scar between the Veporides and the Gemerides. The Choč nappe has a stratigraphic expanse of the Carboniferous up to the Neocomian, but its entire succession of strata is very rare. At the frontal parts its only representatives are limestones and dolomites of the Triassic resting in the form of nappe dross (lambeaux de recouvrement) on the Neocomian–Albian–Cenomanian Krížná nappe. In the more interior parts, besides these, Lower Triassic, Permian and Carboniferous rocks are also represented. The different development of the Triassic in individual "digitations" and slices leads towards the assumption that the Choč nappe is a group of some Ultraveporide nappes [ANDRUSOV, 1964–1965].

KARST OF CZECHOSLOVAKIA

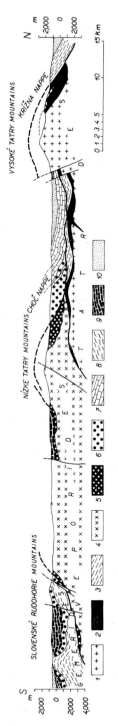

Fig.1. Profile of the West Carpathians. *1* = Tatride crystalline; *2* = Tatrides Mesozoic; *3* = Ultratatrides Mesozoic (Krížná nappe); *4* = Veporide crystalline; *5* = Veporides Mesozoic; *6* = Upper Palaeozoic (in the area of the Nízke Tatry Mountains: Ultraveporide—in the Slovenské Mountains: Gemeride); *7* = Ultraveporides Mesozoic (Choč nappe); *8* = Gemeride crystalline; *9* = Gemerides Mesozoic; *10* = Tertiary (in depressions). (After BIELY et al., 1968.)

The *Gemerides* form the innermost main tectonic unit. In contrast to the preceding ones, the crystalline layer does not consist of crystalline schists and Hercynian granitoids, but of a bulky complex of weakly metamorphosed sediments and vulcanites of Early Palaeozoic, Carboniferous and Permian rocks. The sedimentary mantle is Mesozoic, of which at the present we know only the Triassic and, from some small remnants, the Jurassic also. The nappe position of the Gemeride Mesozoic is evident only at the frontal parts of the Spiš nappe, which rests on the Mesozoic of the Veporides (Muraň Platform, Stratená upland) or on the Choč nappe (Drienok nappe in the Strážov upland, the nappes of the Minor Carpathians and of the Vel'ká Fatra). In the more southerly regions of its occurrence, in the Slovakian karst district, it rests on its original Palaeozoic base. The shifting of nappes, particularly the Križná and Choč, is accompanied by secondary shiftings conditioned by the jutting out of plastic series of strata (Carpathian Keuper, Rhaetian, Liassic and also to a certain extent Lunz Beds and Rhinegraben schists) in the masses of rigid limestones and dolomites, the consequence of which is the occurrence of further nappes and digitations, reduction to the part of slipping out of series of entire strata, or, on the contrary, their accumulation. The most diverse complications arise, such as a superposition of stratigraphically distant layers of one and the same unit, or even of various tectonic units (e.g., granites, Carpathian Keuper, and the Neocomian of the Križná nappe), of a superposition of strata of the same age and various tectonic units (e.g., on the Middle Triassic dolomites of the Križná nappe there are Middle Triassic dolomites of the Choč nappe); or there arise successions with an apparently normal superposition (e.g., on the Lower Triassic of the Tatrides, there immediately follow the Middle Triassic of the Križná nappe and the like).

The nappes of the Inner Carpathians are overthrust nappes on an approximately horizontal surface. Normally the separation of the nappes occurred at the base of the limestone–dolomite complex of the Middle Triassic, hence on the border with the plastic strata of the Lower Triassic. The lower members of the nappes—the Late Palaeozoic and the Lower Triassic rocks—jut out only in the neighbourhood of the root zones ["troncature basale"; ANDRUSOV, 1965].

The pre-Senonian nappe structure is followed by the transgression of the Upper Cretaceous in the "Gosau" development, and mostly, however, by the transgression of the Eocene. The orogenetic movements following the Palaeogene, as we have already stated above, gave rise to mega-anticlines whose alinement is diagonal to the system of folds and pre-Senonian nappes. Due to their erosion and vertical movements after the faults cut off most of the southern flanks of the mega-anticlines, the crystalline rocks were exposed at the surface. Thereby the base of the present core of the mountain ranges and their asymmetrical structure and internal basins arose. In addition to the subjacent faults and reverse faults, there arose also during the Neogene phase the system of cross faults on which occurred the immense volcanic activity in the Tortonian. At the same time the system of intermontane basins was developing and it was substantially in accordance with the present-day system.

In the Mesozoic of the Inner Carpathians, the carbonate rocks (limestones, dolomites and marls) are predominant but, with rare exceptions, the karst phenomena are only in Triassic limestones. These consist chiefly of a series of strata of dark limestones known as the Guttenstein Limestones and the strata of light massive limestones of the Wetterstein type. The Guttenstein Limestones generally form the base of the Middle Triassic of all the tectonic units; the latter form an essential element of the Middle and Upper Triassic of the Gemeride nappes.

Guttenstein Limestones are chiefly thick bedded and massive, but in some places they consist of thin beds of limestones of dark colours which in places contain a network of white veins of calcite of different thickness. Some of these strata are represented by black-grey "wormy limestones" (calcaires vermiculés), thinly bedded grey dolomites and lenses of crinoidal and brecciated limestones. The Guttenstein Limestones are strata with relatively little change of facies. The differences in their development among the individual tectonic units consist of those mentioned above and in the extent and thickness of the strata. Dolomite zones as well as various manifestations of dolomitization of limestone beds are mostly represented in the Tatrides. For this reason also the lithologic division of the Middle Triassic into two parts—the lower consisting of Guttenstein Limestones, the upper of beds of dolomites—is much less marked in the Tatrides than in the other tectonic units, although even in these the Guttenstein Limestones contain some dolomites. In the Krížna nappe the Guttenstein Limestones belong to the Anisian, and the dolomites in their overlying rocks to the Ladinian. The Choč nappe where the Guttenstein Limestones form the lower part of the Anisian contrasts to this; the dolomites in the overlying rocks belong to the Upper Anisian or (in the Čierny Váh development) they form the whole Ladinian which (in the Biely Váh development) is represented by the Reifling Limestones. In the Triassic of the Gemeride nappes, the immediately overlying rocks of Lower Anisian Guttenstein Limestones are either the light massive limestones of the Wetterstein type (Murán Plateau, Vernar Mesozoic, Stratená Mountains) or the well differentiated strata of the dark-grey dolomites and it is on these that follow the strata of the light massive limestones (the Slovakian karst).

The Guttenstein Limestones are chemically very pure (see Table I for composition), but in the Tatride Series, because of their small thickness and considerable share of dolomites and dolomitic limestones, very weak karst phenomena occur in them. In the Guttenstein Limestones of the Krížna nappe, the karst is more extensive; there are numerous smaller caves and also one of the largest caves in Slovakia—the Demänovská Cave on the northern slopes of the Nízke Tatry (Lower Tatra). In the Guttenstein Limestones of the Choč nappe, the karst phenomena developed only in the tectonic outliers of a greater areal extent (for example, the western slopes of the Vel'ká Fatra, in the Strážov Mountains) or in the mountain chains of this nappe, built up almost exclusively of Triassic (Choč Mountains).

Even though the lithologic conditions for karstification in the Guttenstein Limestones of the Krížna and Choč nappes are substantially more favourable than in

TABLE I

COMPOSITION OF THE GUTTENSTEIN LIMESTONES

Oxides	Percentage ranges
CO_2	43.04–43.93
Al_2O	0.08– 0.66
SiO_2	0.06– 1.10
Fe_2O_3	0.09– 0.58
CaO	51.91–55.70
MgO	0.52– 2.82

the Tatrides, its development in the extent observed would hardly be possible without a further lithologic factor, namely the strata of dolomites and of their overlying rock, of the impermeable strata of clayey and marly shales of the Lower Triassic, or the marls of the Lower Cretaceous in their substratum. The dolomites are mostly of a coarse-bedded structure, coarse-grained texture and microgranular, with locally brecciated areas with lenses of crinoidal dolomites, or local areas of dark, thin-bedded limestones. Among the dolomites of various tectonic units there are no other essential lithologic differences. These are only in a different stratigraphic extent. In the Tatrides they belong mostly to the Ladinian; in the Krížná nappe to the Upper Triassic; in the Choč nappe (in the Biely Váh development) they are of Upper Anisian age; in the Čierny Váh they also include the whole Ladinian. Generally they are strongly cracked and disintegrated into granular material ranging in size from small gravel to sand.

In addition to the above-mentioned dolomites of the Middle Triassic, there occur in the Choč nappe strata of dolomites of the Upper Triassic (Upper Carnian–Norian), which are separated from the Middle Triassic by the Lunz Beds (Lower Carnian). In the Krížná nappe and in the Čierny Váh development of the Choč nappe, where the Lunz Beds are only occasionally developed (primarily and tectonically), they form Middle Triassic dolomites together with the Upper Triassic dolomites, a uniform, lithologically indivisible mass that forms the greater part of the mountains. The significance of this immense mass of dolomites in the overlying rocks of the Guttenstein Limestones lays mainly in the fact that, as the strata are very permeable they underlie a region in which karst waters form the streams in the Guttenstein Limestones. The subterranean karst in the Guttenstein Limestones resulting from local penetration of surface waters is very rare in the Carpathians (for example the Demänovské caves).

The light massive limestones (those of the Wetterstein type) are the most characteristic and at the same time the thickest strata of the Gemeride nappes (see Table II for the chemical composition of these limestones). Their most substantial part consists of the organo-detrital limestones of light colours in which occur granular dolomites like lenses, or limestones with different lithological composition. They consist of nodular dark-grey, rose and red limestones (Schreyerlam, Reifling, Hallstatt), which renders it possible for us to divide lithologically this otherwise very monotonous series of strata. The thickness and stratigraphic extent of light limestones

TABLE II

COMPOSITION OF THE WETTERSTEIN LIMESTONES (in %)

Oxides	Včeláre deposit	Gombasek deposit	Drienovec deposit
SiO_2	0.22– 1.52	0.31– 1.26	0.53– 1.48
Al_2O_3	0.22– 1.06	0.12– 0.49	0.41– 0.83
Fe_2O_3	0.10– 0.45	0.11– 0.31	0.18– 0.34
CaO	53.63–55.95	54.17–55.12	54.15–55.20
MgO	0.33– 0.97	0.36– 1.07	0.19– 0.73

is very variable. In the tectonic outliers of the Strážov nappe, the Drienka and Vernár nappes are found, mainly of Upper Anisian and Ladinian age in the Spiš nappe (Muráň Plateau, Stratená Mountains) they stretch from the Upper Anisian to the Norian.

On account of the considerable thickness of the light massive limestones (generally ca. 1,200 m), the horizons of dolomites, of nodular limestones and limestones with chert have no influence either on the development of surface or underground karst phenomena.

The distribution of limestones and dolomites of the Triassic and hence also of karst phenomena within the framework of the Mesozoic is determined by the tectonic conditions. These are, as we have already mentioned, very complicated in the Inner Carpathians and often very different even in one and the same mountain chain. The tectonics of the Inner Western Carpathians is, namely, the result of several orogenetic Alpine phases of foldings: during Cretaceous phases a nappe structure arose and post-Palaeogenic phases gave rise to macro-folds accompanied by longitudinal and transverse faults.

The nappe structures consist of units having the character of immense folds with a displacement towards the north, composed of a crystalline core and its Mesozoic cover (Tatrides, Veporides, Gemerides) and of tectonic units represented mainly by the Mesozoic removed from its original substratum and thrust as extensive nappes towards the north. These are the Ultratatride nappes, lying over the Mesozoic of the Veporides and Ultratatrides, and finally the uppermost are the Gemeride nappes thrust across the Mesozoic of the Veporides and Ultraveporides up to the zone of the Tatrides. The nappe structure is the most distinctive in the northern group of core mountains, where on the Tatride Mesozoic most often two nappes lie (Krížna and Choč); in places, however, even three (the two mentioned above and the Strážov nappe). The least marked is the nappe structure in the south, in the Spiš-Gemer Ore Mountains, where in the rock overlying the crystalline core of the Gemerides, only one Mesozoic series, namely the Mesozoic of the Gemerides, is exposed.

While the differences in the tectonic predisposition for the occurrence of karst between the outer and the inner parts of the Western Carpathians follow from the general movement of pre-Senonian tectonic units towards the north, the differences in the longitudinal direction of the mountain chain are rather the reflection of complicated interior tectonics, as much of autochthon as of shifted Mesozoic series. The nappes, as thrust masses with only developed normal flanks do not have the shape of a simple mildly folded board, but are often according to the shear surface of the second order, disintegrate into a system of slices and digitations of different sizes, whose sedimentary series of the same nappe is repeated several times above itself. Alternatively there comes a reduction in the individual series of strata or even an entire exposure of individual nappes. Not infrequently a lower nappe was ground down by an upper one, and thereby also the immediate tectonic superposition of the limestone–dolomite complexes of the Triassic of various tectonic units. The tectonic deformations (reduction, boudinage) affected most of the plastic series of strata of the Lower Triassic, Jurassic and Lower Cretaceous, and even the limestone–dolomite complexes of the Triassic were affected by them. Hence there is a considerable range in distribution and thickness: in the Tatrides and Veporides from 0 m to 400 m, in the Krížna nappe from 0 to 700 m, in the Choč nappe from 0 to 1,200 m, and in the Gemerides from some hundreds of metres up to 2,000 m.

Even though the pre-Senonian nappe tectonics of the Inner Carpathians is very marked, it was not until the post-Palaeogene orogenic phases that a base for their division into mountain chains and inner basins was formed. The mountain chains of the Inner Carpathians have their foundation in the system of post-Palaeogene mega-anticlines, from whose arc they were gradually removed by erosion and denudation, starting with the Neogene, the sediments of the Palaeogene, the Mesozoic of the shifted nappes and the Mesozoic of the crystalline cores themselves. During the formation of the mega-anticlines, longitudinal faults also occurred. The most marked among them are faults in the southern boundary of the external row of mega-anticlines which conditioned the asymmetric structure of the core mountains.

Through the rise of mega-folds and their raising on longitudinal faults, the slope was substantially enlarged in the originally subhorizontal shear surface areas of individual nappes and individual strata of the Mesozoic which were shaped in detail by the erosion of rivers and the denudation processes.

In the northern part of the Inner Carpathians, in the so-called zone of the core mountains, in which the system of mega-anticlines is the most marked and the most divided, the strata of the Triassic—the Choč and the Strážov nappes—are lithologically and tectonically the most favourable for the development of karst. These were so much affected by erosion that only isolated nappe ruins were preserved. In the southern part, the so-called zone of the Slovakian Ore Mountains, which represents a single immense anticline, the pre-Senonian structure is not broken into a range of blocks and hence the Mesozoic of the Gemerides lying on its original substratum (the Palaeozoic of the Spiš-Gemer Ore Mountains), or as the nappe on the Mesozoic of the Veporides (Muráň Plateau) takes up extensive connected areas bordering on the crystalline core of the mega-anticline from the north (Slovenský raj upland) and from the south (the Slovakian karst). The fundamental feature of the tectonics of the Gemeride

Mesozoic, represented chiefly by non-permeable strata of the Lower Triassic and by a thick complex of limestones of the Wetterstein type, is a peneplain system of parallel synclines and anticlines of various breadth, having a E–W course with transverse wavy axes. This altogether simple structure is complicated by a system of longitudinal tectonic areas of a thrust character and a system of transverse faults.

Relief and its reflection in the differentiation of the karst phenomena

The relief of the Western Carpathians as one of the fundamental elements conditioning the differentiated development of karst, is very varied, whether we take into account the vertical division (in absolute or relative values), slopes, the dissection, the age, the origin or dynamics of the landscape development. The fact that the karst regions are bound to different physiognomic-genetic types of relief means, therefore, a considerable spatially differentiated development of the karst phenomena.

Speaking of the geological aspects of the karst rocks in the Western Carpathians, we already mentioned above all the aspect of their origin and tectonic development in the course of the Mesozoic and Palaeogene, as the constituents of the individual tectonic zones and only concisely did we indicate the further development in the Neogene. During this geological period the Western Carpathians came already permanently under the influence of the subaerial destruction and at that time the fundamental aspects of the present-day relief were already laid. The geomorphological development of the highlands went on by stages in the Neogene [MAZÚR, 1965]. The lowering and the levelling of the mountain system was interrupted by several phases of intensive tectonic movements. These movements, however, are characterized, in contrast to the older tectonic periods, by the predominance of a vertical trend against a tangential one. The nappe zonal mesostructure had been disturbed, especially in the Inner Western Carpathians, by younger faults in various directions and by non-uniform vertical movements of individual crustal blocks along them. Thus, in the course of the Neogene there arose an irregular mosaic of variously raised individualized crustal mountain blocks of various heights, and between them subsiding depressions or basins [the so-called neostructure; MAZÚR, 1965]. Hence we are justified in speaking about the Western Carpathians as a region of basins. The subaerial destruction and principally the non-uniform tectonic movements have thus spread the originally connected nappe zones and, within their framework, even have isolated the carbonate complexes one from another into various large areas, with considerable vertical differences and angular shapes, according to how they belonged to individual orographic units.

Dependent on the erosion–denudation processes and on tectonics, the relief of the Western Carpathians acquired, form the viewpoint of development of the karst phenomenon, very diverse properties.

In the sub-Carpathian basins the pre-Neogene complexes have sunk deeply and are covered by a thick mantle of the Neogene, so that the karst phenomenon is entirely missing there.

In intramontane basins the pre-Neogene formations also have sunk relatively, and they are mostly covered by Palaeogene or Neogene up to Quaternary sediments. In spite of this, in the basins there are at least small areas of carbonate rocks exposed which partially developed karst; or the karst phenomena are developing under a loose, mostly Quaternary cover (most frequently under sheets of glaciofluvial or periglacial cones); occasionally exhumed karst occurs, e.g., in the Hron Valley, in the Liptov Basin, in the Zvolen Basin etc. [LENČO, 1956; SEKYRA, 1956; DROPPA, 1967].

The most extensive range of conditions from extremely unfavourable ones up to the most favourable for the development of karst can be observed in the mountain macroforms of the Inner Carpathians.

From the climatic–morphological point of view it is necessary to mention the extremely elevated high mountains (High Tatras, Low Tatras, Malá Fatra) with high mountain karst contrary to the other mountainous regions.

A much greater differentiation of karst regions may be observed in dependence on the morphologic–structural properties of the landscape.

One of the most typical mountain structures is the monoclinal horst with a crystalline core asymmetrically laid as a result of one-sided thrust, so that the carbonate complexes jut out in the shape of inclined structures only on one side of the mountain chain (e.g., the High Tatras, Low Tatras, Malá Fatra and the like).

Another structural type, the Appalachian structure, is classically developed in the Strážov Mountains. It is represented by the open folds of Mesozoic complexes without the crystalline core, or the crystalline core outcropping excentrically.

At other places the Mesozoic complexes outcrop in the form of horsts without a crystalline core (e.g., the northern part of the Little Carpathians, the Choč Mountains), or as residual blocks preserved in the form of small plateaus overlying the core (e.g., Žiar).

A particular structural type in the Inner Western Carpathians, the Ore Mountains structure, is the most favourable for the development of karst. The Mesozoic complex in detail, is a complicated fold structure essentially in a horizontal to sub-horizontal position on the underlying non-karstified rocks. Besides this, these rocks are relatively well preserved in large areas in the form of plateaus (e.g., the Slovakian karst, Muráň Plateau, Slovenský raj).

The least favourable structure for the development of karst is the Klippen zone, where the karst rocks outcrop in the shape of isolated morphologically strongly exposed monadnocks (Klippen).

In summary, the morphological conditions of the Western Carpathians are mostly not very favourable for the full development of karst—mainly as a consequence of tectonic–erosive destruction of the originally more continuous Mesozoic nappe-folded structures into mosaic faulted–folded structures, whereby the karst rocks became extremely exposed to the course of non-karst morphological processes. The extreme slopes, the dissected relief, and the influence of the morphological processes of the neighbouring non-karst districts allowed, for this reason, even in regions with favourable lithological properties of the rocks, only an incomplete and one-sided development of the karst phenomenon. For the development of a typical surface and underground karst, the most suitable regions have shown to be, above all, the Neogene level plateaus of the Slovakian Ore Mountains, and also locally the plateaus of the horst mountains (Little Carpathians, Strážov Mountains, Žiar, etc.).

The structural–tectonic and morphological character of the Western Carpathians indicated above, caused very different conditions for the development of the karst phenomena from those in the Alps or Dinarides or Hellenides. In the latter, there have been preserved in essence connected zones of folded Mesozoic and in them extensive areas of carbonate complexes. The karst regions, therefore, cover a considerable area, and have a greater wealth of forms and a more fully developed karst hydrography. Even the influence of the non-karst environment is here far smaller than in the Carpathians. The vertical range in the Alps and Dinarides is also greater than in the Carpathians, with a more marked climatic differentiation of the karst. For these reasons, the Carpathian karst appears typologically quite characteristic. It seems that it is closer to the central European karst of the Variscan mountains than to the Alpine–Dinaric mountains.

The other physical-geographical conditions

In the Western Carpathians, the karst regions are for the most part denuded, in places up to the rock substratum; or they are covered in varying degree by the preserved soil cover (continuous in the forested districts, sporadic in the unforested ones), mostly of the rendzina type [MIČIAN, 1966]. In the region of the Ore Mountains, thanks to a considerable level surface, remainders of red earth have been preserved, mainly in the fissures and at the bottom of the karst holes [BORZA and MARTÍNY, 1964]. Crusts of weathering of this type are the most wide spread in the Slovakian karst region and in the Muráň karst, but also in the Považský Inovec, in the central valley of the Hron River and elsewhere [SMOLÍKOVÁ, 1962]. In places small remnants of the sedimentary mantle rock have also been preserved on the plateaus. In the higher mountainous districts and the more dissected karst regions (Nízke Tatry, Malá and Vel'ká Fatra etc.), they are not found on account of the strong erosion. In the Slovakian Ore Mountains there occur karst districts covered by a volcanic mantle (environs of Španie polje). In the high mountainous karst territories there is only a discontinuous soil cover in the form of gravel soils and only in places do Alpine rendzinas occur [TARÁBEK, 1958].

The karst surface in the tectonically subsided karst regions of basins is often covered by Neogene sediments and even more often by gravel mantles. These are the regions of covered karst.

From the climatic viewpoint, the karst of the Western Carpathians belongs mostly to the karst of the temperate zone of the central European climatic region. Only small islands of the highest mountain chains rise above the tree line and belong to the Alpine karst.

Even the climatic conditions for the development of karst in the Western Carpathians are, to a considerable extent, variable and whether it is also a question of rainfall, temperature, evaporation, etc. and taking into account rather considerable vertical differences of elevation in the mountainous region (250–1,400 m above sea level), still it is not possible, on the basis of present knowledge, to follow the differentiation of the development of karst, conditioned climatically. Table III shows some climatic factors from several of the karst regions according to *The Atlas of the Climate of Czechoslovakia* [ANONYMOUS, 1958].

The climatic elements in the Western Carpathians are on the whole favourable to the development of karst. The karst regions (according to Köppen's classification) are found mostly in the climatic zones Dfb and Dfc [KONČEK, 1964].

When comparing the total amount of rainfall for a year in the individual karst regions, we do not find such differences as would influence more distinctly the degree of the karst processes. In most of the territories, the yearly amount of rainfall is 700–1,000 mm, with average yearly temperatures of 6°–8°C (the Slovakian karst 700 mm, 7°–8°C, Little Carpathians 800 mm, 8°–9°C, Nízke Tatry 800–1,000 mm, 5°C). In some mountainous districts the amount of rainfall is markedly influenced by differences in altitude of the relief. In such regions, the rainfall in the lower altitudes amounts to 800 mm and in the higher altitudes to 1,200 mm, e.g., the Choč Mountains [ANONYMOUS, 1958]. Generally the amount of rainfall increases towards the west.

Greater climatic changes influencing the development of karst may be noticed in a vertical direction in the regions above the tree line. These belong to the ETG climate. Low temperatures which annually amount to an average of 0°–2°C and the large amount of yearly rainfall (in places more than 1,800 mm) accelerate to a great extent the process of karst formation and affect the development of high mountainous karst types. Weathering of the limestone as a whole by frost is considerable.

In comparison with the classical karst regions such as the Mediterranean or the tropical karst, the Czechoslovakian karst has less favourable climatic conditions. In contrast to the tropics, the rainfall is smaller, the temperature lower; contrary to the Mediterranean, where there is a marked maximum rainfall, in Czechoslovakia the rainfall is distributed relatively uniformly throughout the whole year.

The differentiation of the karst in the Western Carpathians is also conditioned by palaeo-climatic conditions, essentially different from the tropical regions and up to a certain extent even from the Mediterranean regions. In the course of the Neogene, the territory passed alternately through wetter and drier periods of a sub-tropical climate, while in the Pleistocene again, several cold and warm periods alternated which is necessarily also reflected in the morphology of the karst regions.

From the above-mentioned properties of the relief and climate of the Western Carpathians, there follow the particular hydrographic or hydrologic conditions of the karst territories.

The regions relatively subsided, i.e., the intramontane basins, are abundantly supplied with water on the one hand from local rainfall, but mainly by mountain

TABLE III

SOME CLIMATIC FACTORS FOR SEVERAL KARST REGIONS

Region	Height (a.s.l.)	Average yearly temperature (°C)	Maximum temperature (°C)	Minimum temperature (°C)	Average yearly rainfall (mm)	Maximum rainfall (mm)	Minimum rainfall (mm)
Rožňava	289	8.0	18.5(VII)	−3.8(I)	668	89(VI)	30(I)
Dobšina-Čuntava	1,106	4.1	14.0(VII)	−5.7(I)	982	122(VI)	48(I)
Liptovský Hrádok	648	5.9	16.0(VII)	−5.3(I)	744	95(VII)	36(II)
Motyčky	681	5.7	15.7(VII)	−5.2(I)	1,085	109(VI)	62(II)
Pružina	381	7.7	17.9(VII)	−3.1(I)	860	103(VII)	42(II)
Banská Bystrica	348	7.0	18.7(VII)	−4.2(I)	853	88(V)	51(II)
Dumbier-chata	1,740	0.8	9.3(VII)	−7.3(II)	1,328	147(VII)	78(I)
Červené vrchy	1,800	0.0	10.0(VII)	−8.0(I)	1,800	200(VII)	80(II)

I = January; II = February; V = May; VI = June; VII = July.

rivers from the adjacent mountain ranges and they are characterized by a constant level in the groundwaters. The development of the karst phenomena is here, however, limited by the fact that the karst complexes are mostly covered with impermeable non-karstified rocks or with loose formations of the Quaternary. Only in few places do the islands of karst rise to the surface with karst hydrography.

The elevated mountainous karst areas have, on the contrary, substantially less favourable hydrographic conditions. The considerable relative heights, great slopes and the relatively small size of the karst areas condition the very fast circulation of the groundwater, or the quick surface outflow by canyon into the adjacent basins. This goes on in spite of the vegetation cover and relatively considerable rainfall. A constant level of groundwater is here entirely absent (especially in the monoclinal steep karst belts and horsts) or it is of an extremely fluctuating nature (in the folded structures). The development of karst forms, in consequence of the above facts, is strongly influenced by erosive processes of allochthonous rivers, modeling of the slopes, or by periglacial processes above the tree line.

The most favourable conditions for the development of karst hydrography from the aspect of structure, lithology and morphology are in the karst plateaus of the Ore Mountains region. There are even plateaus of considerable extent. Owing to the depressed position of the karst complex against the non-karstified substratum and the flatness and sufficient rainfall, these regions are characterized by signs of karst hydrography.

The vegetation cover of the karst regions of the Western Carpathians has been mostly changed and removed by anthropogenous activity. In particular the karst plateaus of the Ore Mountains region are often deforested and changed to small pastures and meadows with unproductive shrub growth, and xerophilous and thermophilous undergrowth. Considerable areas are covered with coniferous forests.

The other karst areas are mostly forested, in the higher altitudes by coniferous forests and in the lower regions by deciduous forests (oak and hornbeam). In places the smaller karst plateaus of the horst mountains are deforested and changed into grasslands, meadows and fields. In high mountain regions dwarf pine and sub-Alpine flora cover the ground.

TYPOLOGICAL DIVISION OF THE KARST IN THE WEST CARPATHIANS[1]

Dependent upon a very differentiated character of physical elements and their interaction in the Carpathian area, as we have already stated above, the karst phenomena have developed in a different way in the various Carpathian regions.

With regard to the fact that CVIJIĆ's [1960] classical division of the karst [MICHOVSKA, 1957; DROPPA, 1966] cannot entirely express the diversity of the karst

---

[1] Written by E. Mazúr and J. Jakál

regions, we shall try, on the ground of several criteria following from the foregoing analysis, to introduce a new typological division of the karst in Czechoslovakia which would more fully express its phenomenal and spatial differentiation.

As the fundamental criteria in the classification can be cited the following: (*1*) the representation and state of development of the surface karst forms; (*2*) the representation and state of development of underground karst phenomena; (*3*) the character of the karst hydrography; (*4*) the climate of the karst region; (*5*) the morphological aspects of the karst region; (*6*) the structural–lithological character of the karst region; and (*7*) the extent of the karst region, etc.

Within the framework of the temperate climatic zone, we can include the karst of Czechoslovakia in the central European karst, in contrast to the Mediterranean karst. This division was originally made by SAWICKI [1909], and is later used in Soviet literature in the classification of karst regions of the U.S.S.R. by MAKSIMOVIČ and GORBUNOVA [1958] and GVOZDETSKIY [1965]. PANOŠ [1965] also recommends a division into a special central European type of karst.

In the framework of the central European karst, some islands of Alpine karst above the tree line are dispersed in the Western Carpathians (Fig. 2).

The central European karst of the Western Carpathians is divided into three

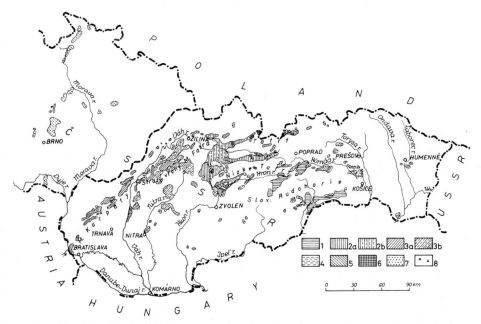

Fig. 2. Types of karst areas in the West Carpathians. *1* = Plateau karst areas (complete karst); *2* = karst of the monoclinal ridges (imperfect karst): *2a* = in a typical development; *2b* = in a less typical development; *3* = karst of the horsts and combined structures (imperfect karst): *3a* = in a typical development; *3b* = in a less typical development; *4* = basin karst (covered karst); *5* = sporadic karst of the Klippen structure; *6* = Alpine karst; *7* = karst of the Bohemian Massif; *8* = accessible caves.

typologically different groups which again may be divided into sub-types, as the following classification shows.

(*A*) Central European karst of the temperate zone.
    (*1*) The plateau karst of the semi-massive structures—complete karst.
    (*2*) Articulated karst of the fault-folded mosaic structure—imperfect karst.
        (*a*) Karst of the monoclinal ridges.
        (*b*) Karst of the horsts and of the combined fault-folded structures.
        (*c*) Karst of the basins—covered karst.
    (*3*) Sporadic karst of the Klippen structure.
(*B*) Alpine karst.

Central European karst of the temperate zone

Plateau karst—complete karst

In contrast to tradition [KUNSKÝ, 1950; SEKYRA, 1954; MICHOVSKÁ, 1957; JANÁČIK, 1961; DROPPA, 1966, etc.] based on CVIJIĆ's [1960] division of karst, we do not class plateau karst as intermediate karst, but we look upon it as complete karst. We are led to this fact, because it has a development with the whole richness of surface karst forms, including poljes and underground forms and a characteristic hydrographic régime. For example, the representative karst phenomena include lapies, karst holes, valley sinks, poljes, dry valleys, blind and half-blind valleys, canyons, sinks, springs, extensive cave systems, etc. Very likely only the classical development of cones is lacking, which is conditioned by climate.

The plateau karst is the most characteristic type in the Western Carpathians and territorially it takes up the largest connected areas. The plateau karst is bound only to a single morphological-structural type of upland region, and that is the semi-massive block of the Slovenské rudohorie (Slovakian Ore Mountains, Fig. 3). The most suitable conditions existed there for the development of karst, i.e., a considerable thickness and extent, as well as purity of karst rocks, in addition to depressed position, relatively flat initial relief, etc. The plateau karst is represented by three regions, viz.: Slovenský kras, Muránska planina and Slovenský raj, i.e., a total area of more than 1,000 km². Their characteristic features are high-level plateaus, relatively sharply limited by the neighbouring non-karst regions. A partial obstacle to the process of karst formation is formed by recent and fossilized weathering products, which block the outflowing fissures. The vegetation cover takes up a considerable area, especially on the Muráň Plateau and Slovenský raj, but to a lesser extent in the Slovenský kras (Fig. 4–6).

A common feature of the development of these regions was the levelling of the surface by the corrosive and erosive activity of rivers in the period of alternate humid and dry hot sub-tropical climates during the Pannonian [MAZÚR, 1965]. It is very likely that the levelling occurred by way of recession of the slopes and the rise of border plains which had been inundated by sediments transported by rivers from

Fig. 3. View on the Slovakian karst from the north. On the foreground is the Rožňava Basin. (Photo E. Mazúr.)

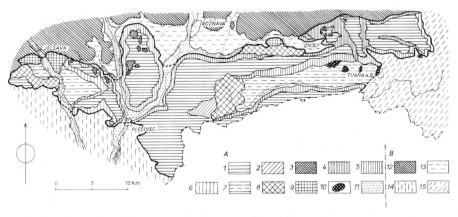

Fig. 4. Slovakian karst region. *A*. Slovakian karst area: *1* = karst plateau; *2* = border-karst plateau; *3* = residual hills; *4* = karst narrows; *5* = steep slopes of the karst plateau (escarpments); *6* = gentle slopes of the karst plateaus; *7* = karst basins (poljes and semi-poljes); *8* = dissected fluviokarst areas; *9* = fluviokarst structural ridges; *10* = strong residuals on the bottom of the karst basins; *11* = sediment filling of the canyon bottoms. *B*. Non-karst areas: *12* = Slovakian Ore Mountains; *13* = Rožňava Basin; *14* = Rimava Basin; *15* = Košice Basin.

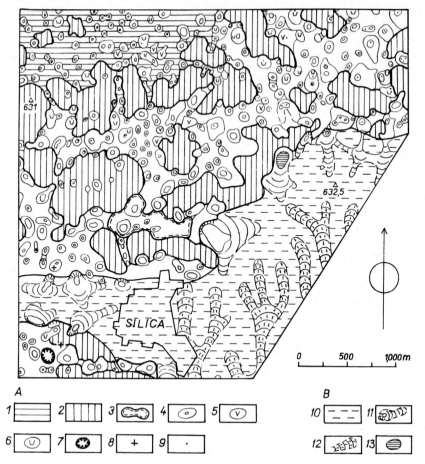

Fig. 5. Detailed map of the northern part of the Silica karst plateau (Slovakian karst). *A*. Karst relief: *1* = young karst depressions (furrows); *2* = convex surface of the karst plateau (remnants of the Neogene level); *3* = uvala; *4* = bowl dolines; *5* = funnel dolines; *6* = intermediate (compound) type of dolines; *7* = collapse dolines; *8* = abysses (aven type); *9* = ponors. *B*. Relief on non-karst rocks: *10* = leveled relief; *11* = blind dellen; *12* = periglacial valley; *13* = lake.

the neighbouring non-karstified areas. The Slovenský kras (Slovakian karst) at this period could have had, at least partially, the character of a typical cone karst [LUKNIŠ, 1962]. At the period of the Rhodanian tectonic movements, there occurred a raising up of the originally levelled surface and there followed a vertical underground drainage, as well as an intensive karst formation which goes on even at present. The water courses cut deep beds, formed canyons and divided the original uniform surface into a system of several plateaus [SAWICKI, 1909; LUKNIŠ, 1962; MAZÚR, 1965; Fig. 7].

The canyons in the Slovakian karst attain a relatively considerable width and they are filled in places to depths of up to 100 m with a thick Pliocene–Quaternary cover. From this we conclude that there were fluctuating movements, which condi-

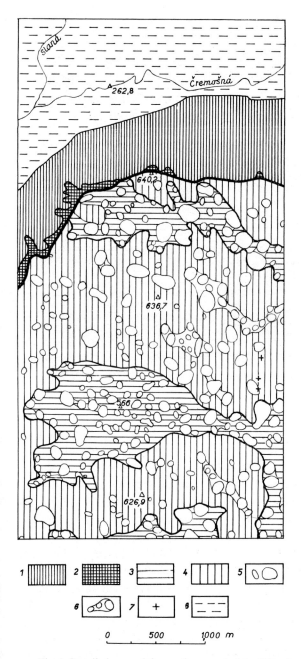

Fig. 6. Detailed map of the southern part of the Silica karst plateau (Slovakian karst). *1* = Steep slopes of plateau with talus; *2* = escarpments; *3* = young karst depressions; *4* = convex surface of the karst plateau (remnants of the Neogene level); *5* = dolines; *6* = composite dolines; *7* = abysses (aven type); *8* = bottom of the Rožňava Basin.

Fig. 7. Turňa karst basin with strong residuals on the bottom. On both sides karst plateaus of the Slovakian karst are developed. (Photo P. Plesnik.)

tioned the very complicated development of the karst phenomena. The plateaus are raised above the bottoms of the canyons by 300–400 m steep rocky escarpments. In the region of the plateau karst, various forms of lapies, karst holes, valley sinks and blind valleys can be seen as well as chasms of the aven type [SAWICKI, 1909; ROTH, 1939; KEMÉNY 1961; Fig. 8]. The character of the lapies points to the fact that there is the question of originally covered lapies which were uncovered by deforestation and the destruction of soil cover. On the plateaus it is possible to observe also the features of the older relief which is manifested by a net of concave forms and depressions. The conical hills rise above the plateaus to relative heights of 80–100 m, in places even more, and they can be considered as witnesses of the tropical cone karst from the Neogene [LUKNIŠ, 1962]. The present relics of the cones have been considerably eroded and their slopes have been rendered more moderate especially by the action of the periglacial modeling. Among the best known caves are the Domica caves in the Slovakian karst which attain, together with Aggtelek in Hungary, a length of 21 km, the Gombasek Cave, the Jasovská Cave, the Bear Cave, the Dobšina Ice Cave etc.

Articulated karst—imperfect karst
This is bound with mountain horsts, open folds, or grabens – basins, of the Tatra–

Fig. 8. Detailed relief of the Silica karst plateau (Slovakian karst). (Photo E. Mazúr.)

Fatra region in the Western Carpathians. It corresponds in substance to CVIJIĆ's [1960] intermediate karst. There is question here of karst regions with less favourable conditions for the development of the karst phenomena than in the preceding type. This is conditioned partly by a smaller area, and partly by a smaller thickness, frequent soil coverage of karst rocks and morphologically unfavourable position. For this reason only some karst phenomena have developed in the different types of the divided karst.

Karst of the monoclinal ridges. This is bound with carbonate rocks asymmetrically situated at the so-called core mountains in the shape of the monoclinal structures, or rising as independent monoclinal structures without a core. As a consequence of the action of allogene rivers and exposed morphological positions, the karst areas are divided by deep valleys into elongated monoclinal crests. Such a divided surface is not favourable for the development of surface karst phenomena. The slopes and crests are only weakly covered with weathered material, often denuded and relics of weathering in the caves are rare. In spite of a relatively weak weathering cover, the vegetation mantle is almost continuous, in higher altitudes with coniferous and in the lower altitudes by deciduous vegetation. Lapies and dolines occur only sporadically. More often there is underground karst. On the one hand there are chasms of the aven type [DROPPA, 1957; KUBÍNY, 1959], on the other hand there are extensive underground caves with a classical development, often with several floors above the lowest level and with a rich stalactite and stalagmite ornamentation. Among the hydrologic phenomena are numerous sinks and springs. In places there are dry valleys, and very frequently typical forms in these regions are narrow as fluvial-karst forms. This type of karst is classically developed, especially in the northern slopes of Nízke Tatry—the well-known Demänovské jaskyne (caves) and others; a smaller region in the southern part of Nízke Tatry; the monoclinal ridges of the Tatras— the Balenská Cave (up to a height of about 1,400 m above sea level); the Vel'ka Fatra—the Harmanec Cave, Malá Fatra, Choč Mountains and the like.

Karst horsts and combined fold-fault structures. These are bound with carbonate rocks preserved in various positions in the mountain horsts, or in the fold-fault structures. The carbonate rocks are preserved in relatively small areas, many of which are covered by soil, on various slopes and positions and therefore in this sub-type of karst very different varieties of development of the karst phenomena can be observed within relatively small distances. On the one hand there are small karst plateaus with a relatively rich presentation of surface forms and sub-surface cavities, on the other hand again a type of karst similar to that of the preceding monoclinal structure, with a predominance of sub-surface formations, intensely dissected. Among the most significant karst regions of this type are the Žiar Mountains, Čachtice karst (the Little Carpathians) and especially

Fig.9. Basin karst (partly exhumed) in the Zvolen Basin. (Photo E. Mazúr.)

the Mojtín and Slatina karst in the Strážov Mountains. Naturally, with regard to the relatively small extent, especially of the karst plateaus, there are sub-surface cavities, small joint caves and the karst hydrography is strongly influenced by allogene water courses. A frequent sign of this sub-type of karst territory are fluvial-karst narrows (canyons).

Basin karst—covered karst. This is bound to the tectonically depressed complexes of the carbonate rocks, and occurs most often in the intra-montane basins. In places this karst rises to the surface, but more frequently the complexes of karst rocks are covered, chiefly with loose Neogene and Quaternary formations (periglacial, and glacio-fluvial cones, terraces, etc., see Fig. 9). The surface karst features are chiefly dolines, but in places there are lapies and sub-surface cavities, often bound to the system of fluvial terraces [SEKYRA, 1954; DROPPA, 1967]. Among the most significant regions of the basin karst are the following: Važec karst, karst of the Štrba upheaval, the karst regions of the Hron Valley and of the Zvolen Basin (Šumiacky, Ponický karst, and the like). The supply of water in these regions is very favourable on the one hand from the local rainfall, and on the other, from the allogene waters from the mountains surrounding the basins.

Fig. 10. Narrow in the Peniny Mountains (sporadic karst type). (Photo P. Plesník.)

Sporadic karst of the Klippen structure

The karst of this type is bound to the small islands of carbonate rocks, appearing in the form of strong residuals of the Klippen zone. The karst rocks represent here most often Jurassic limestones, strongly folded, sometimes pure but mostly covered with soil. Even though the thickness of the Klippen limestones is not small (in certain Klippen it attains the order of some hundreds of metres in thickness), nevertheless their too small area and extreme morphological position do not allow here of an entire development of the karst phenomena. The Klippen residuals appear in essence as tectonic lenses, that rise up strikingly to some hundreds of metres above the surrounding area. In places they are rocky, but mostly they are covered with continuous forest (Fig. 10). The surface karst forms occur here only sporadically in the form of lapies; still more rarely there are dolines; here and there dry valleys occur. Among the phenomena of the karst hydrography are frequent karst springs. Among the forms of sub-surface karst are joint chimneys, as a rule filled in with material from crusts of weathering of the Neogene, further small caves and abris. Allogene rivers often cut the Klippen strong residuals with wild narrows (gorges). The best-known narrows are the Manín in the Váh Valley, and in Spiš in the region of the Klippen group Peniny.

Fig. 11. Monoclinal step karst in the group of Červené vrchy, Tatra Mountains (Alpine karst type). (Photo E. Mazúr.)

Alpine karst

Alpine karst in the Western Carpathians represents a particular type, above all from the viewpoint of climatic conditions. Regions rising above the tree line, i.e., about 1,400 m above sea level, are classified as Alpine karst (Fig. 11).

The climatic conditions as a consequence of a high rainfall (up to about 1,800 mm), of low temperatures (average yearly temperature about 0 °C or below 0 °C) and cold water with a high content of CO_2, speed up the solution of limestones. Low temperatures and frequent fluctuations in daily temperatures favour the mechanical weathering, especially frost weathering. A lack of soil cover and of the forestation makes direct contact of water with the karst rock possible and this indirectly promotes the speeding up of the karst process. The karst rocks consist chiefly of intensely folded limestones and lime dolomites.

The karst penomenon here is very well developed. Among the surface forms are especially the various classically developed types of lapies, further there are dolines and chasms (aven type). In places steppe karst as described by Bögli [1960] occurs. Among sub-surface forms there are found extensive cave spaces with a dominant vertical tendency. Besides these numerous sinks and springs are found.

Spacially this type of karst is most extensive in Belanské Tatry [Sekyra, 1954], in the Western Tatras [in the classical form especially in the group of Červené vrchy–Mazúr, 1962], in the Nízke Tatry [Louček, 1956], and also locally in the Vel'ká Malá Fatra Mountains.

REFERENCES

Andrusov, D., 1964–1965. *Geologie der Tschechoslowakischen Karpaten.* Verlag der Slowakischen Akademie der Wissenschaften, Bratislava, 1:263 pp.; 2:443 pp.
Anonymous, 1958. *Atlas of the Climate of Czechoslovakia.* Ustredna Správa Geodézie a Kartografie, Prague, 99 maps.
Anonymous, 1964. *Vysvetlivky k Prehladnej Geologickej Mape C.S.S.R. 1:200,000.* Geofond, Bratislava.
Biely, A., Bystrický, J. and Fusan, O., 1968. Zur Problematik der "Subtatrischen Decken" in den Westkarpaten. *Geol. Sb.,* 1:295–296.
Bögli, A., 1960. Kalklösung und Karrenbildung. *Z. Geomorphol.,* 2:4–21.
Borza, K. and Martíny, E., 1964. Verwitterungsrinden, Bauxitlagerstätten und "terra rossa" in den slowakischen Karpaten. *Geol. Sb.,* 1:9–26.
Büdel, J., 1951. Fossiler Tropenkarst in den Schwäbischen Alb und den Ostalpen. *Erdkunde,* 5:168–170.
Bystrický, J., 1964. *Stratigraphie und Dasycladaceen des Gebirges Slovensky Kras.* Ústredny Ústav Geologicky, Bratislava, 203 pp.
Corbel, J., 1959. Érosion en terrain calcaire. *Ann. Géograph.,* 366:97–120.
Cvijić, J., 1893. Das Karstphänomen. *Geograph. Abhandl. (Penck),* 5:217–330.
Cvijić, J., 1960. La géographie des terrains calcaires. *Acad. Serbe Sci. Arts,* 341(26):1–212.
Droppa, A., 1957. *Die Höhlen Demänovské Jaskyne.* Slovakian Academy of Science, Bratislava, 289 pp.
Droppa, A., 1966. Characteristics of the karst region in the Carpathians. *Probl. Speleol. Res.,* 2:23–30.

DROPPA, A., 1967. Karsterscheinungen im Tale des Flusses Biely Váh. *Geograf. Časopis*, 2:141–153.
GVOZDETSKIY, N. A., 1965. Types of karst in the U.S.S.R. *Probl. Speleol. Res.*, 1:47–54.
HOMOLA, V., 1951. The stratigraphy and paleogeography of the Mesozoic of the south Slovakian Karst. *Sb. Ústredny Ústav Geol.*, 28:153–200.
JANÁČEK, J., 1941. Morphological–hydrological observations in the Turna River Basin. *Rozpravy Cesk. Akad., Ser. 2*, 2(5):1–17.
JANÁČIK, P., 1961. Beitrag zur Kenntnis des Karstgebietes im Stražov-Bergland mit besonderer Berücksichtigung des Mojtiner Karstgebietes. *Slovensky Kras*, 4:3–34.
KAYSER, K., 1955. Karstrandebene und Poljeboden. *Erdkunde*, 9(1):60–64.
KEMÉNY, A., 1961. Geomorphological conditions of the Koniar Plateau. *Geograf. Časopis*, 2:104–139.
KONČEK, M., 1964. Sketch of climatic conditions in Slovakia. *Geograf. Časopis*, 2:160–184.
KUBÍNY, D., 1959. Karstsysteme in Hüllenserien der Liptauer und der Niederen Tatra. *Slovensky Kras*, 3:3–20.
KULLMAN, E., 1964. Karstgewässer in der Slowakei und deren hydrogeologische Durchforschung. *Geol. Prace, Zpravy*, 32:9–28.
KUNSKÝ, J., 1950. *Kras a Jeskyné*. Prírodovědecké Nakladatel'ství, Prague, 200 pp.
LÁNG, S., 1949. Geomorfológiai és hidrológiai tanulmányok Gömörben. *Hidrol. Kozl.*, 29(1/2):5–6; 29(8/9):2–289.
LEHMANN, H., 1936. Morphologische Studien auf Java. *Geograph. Abhandl.*, 3(9):114.
LEHMANN, H., 1959. Studien über Poljen in den Venezianischen Voralpen und im Hochapennin. *Erdkunde*, 13(4):258–289.
LEHMANN, H. and SUNARTADIRDJA, M., 1960. Der tropischen Karst von Maros und Nord-Bone in SW-Celebes (Sulwesi). *Z. Geomorphol., Suppl.*, 2:49–65.
LENČO, V., 1956. Karst Phenomena of Poniky, Môlča and Lehotka. *Ochrana Prírody*, 1:21–29.
LOUČEK, D., 1956. The Alpine karst in Dumbier (Low Tatra). *Rozpravy Česk. Akad. Nauk*, 3:45.
LOUIS, H., 1956. Das Problem der Karstniveaus. *Rept. Comm. on Karst Phenomena—Intern. Geograph. Union, New York*, pp.24–30.
LOUIS, H., 1956. Die Entstehung der Poljen und ihre Stellung in der Karstabtragung. *Erdkunde*, 10(1):33–53.
LUKNIŠ, M., 1945. Contribution à la géomorphologie du karst superficiel dans la Ztratenská hornatina. *Sb. Prirody Slovenske Univ.*, Bratislava, 1945:46 pp.
LUKNIŠ, M., 1946. Quelques remarques sur la géomorphologie de la Porte de Beckov et des régions avoisinantes. *Práce Štát. Geol. Ústavu*, 15:46.
LUKNIŠ, M., 1948. Le karst à l'ouest de Tisovec en Slovaquie. *Československy Kras*, 1:85–94.
LUKNIŠ, M., 1962. Geomorfologicky prehl'ad. Vysvetlivky k prehl'ad. geologickej mape Č.S.S.R. 1:200,000. *List Rimavská Sobota*, 1:81–88.
MAHEL, M., 1957. Geológia Stratenskej hornatiny. *Geol. Práce (Bratislava)*, 48a:201.
MAKSIMOVIČ, G. A. and GORBUNOVA, K. A., 1958. *Karst Permskoj Oblasti*. Permskoje Knižnoje, Perm, 184 pp.
MAZÚR, E., 1955. Beitrag zur Morphologie des Wassergebietes des Studeny-Baches in der Liptauer Tatra. *Geograf. Časopis*, 1:15–45.
MAZÚR, E., 1962. Die Formen des Hochgebirgskarst in den Červené vrchy (Westliche Tatra). *Geograf. Časopis*, 2:87–104.
MAZÚR, E., 1963. *The Žilina Basin and the Adjacent Mountains*. Slovakian Academy of Science, Bratislava, 185 pp.
MAZÚR, E., 1964. Zu den Grundsätzen der geomorphologischen Rayonierung der Westkarpaten. *Geograf. Časopis*, 3:281–288.
MAZÚR, E., 1965. Major features of the West Carpathians as a result of young tectonic movements. *Probl. West Carpath. Geomorphol.*, pp.9–53.
MICHOVSKA, J., 1957. The types of the Czechoslovakian karst. *Československy Kras*, 2:60–68.
MIČÍAN, L., 1966. Übersichtliche bodengeographische Gliederung der Slowakei. *Geograf. Časopis*, 4:296–311.
PANOŠ, V., 1965. Genetic features of a specific type of the karst in the central European climate morphogenetic area. *Probl. Speleol. Res.*, 1:11–23.
PANOŠ, V. and ŠTELCL, O., 1966. Development of isolated limestone hills in Cuba. *Československy Kras*, 18:7–22.

ROGLIĆ, J., 1954. Korrosive Ebenen im Dinarischen Karst. *Erdkunde*, 8(2):113–114.
ROGLIĆ, J., 1957. Zaravni na vapnencima. *Geograf. Glasnik*, 19:103–134.
ROGLIĆ, J., 1960. Das Verhältnis der Flusserosion zum Karstprozess. *Z. Geomorphol.*, 4(2):116–128.
ROTH, Z., 1939. Nekolik geomorfologickych poznámek o Jihoslov. krasu a Silické Lednici. *Rozpravy Česk Akad.*, *Ser. 2*, 49(8):1–24.
SAWICKI, L., 1908. *Szkis Krasu Slowackiego z Pogladem na Gykl. Geograficzny w Ogóle.* Kosmos, Lwów, 48 pp.
SAWICKI, L., 1909. Ein Beitrag zum geographischen Zyklus im Karst. *Geograph. Z.*, 15:187–204; 259–281.
SEKYRA, J., 1954. *The Alpine Karst in the Bělské Tatry.* Slovakian Academy of Science, Prague, 141 pp.
SEKYRA, J., 1956. The geomorphology of the south of Kralova Hola (1943)—Šumiacky kras. *Sb. Česk. Akad. Nauk*, 41:193–209.
SENEŠ, J., 1957. Ergebnisse der speleologischen und geomorphologischen Erforschung der Höhle. Hačavská jaskyna im Slowakischen Karst. *Geograf. Časopis*, 1:27–39.
SMOLÍKOVÁ, L., 1962. Different forms of occurrence of terrae calcis in karst areas of Slovakia. *Československy Kras*, 14:93–100.
TARÁBEK, K., 1958. Bericht über die Bodenerforschung in der Belaer Tatra. *Sb. Prác Tatra Nár. Parku.*, 2:11–19.
VITÁSEK, F., 1930. Le karst de Silice et sa grotte de glace. *Sb. Cesk. Akad. Nauk*, 36:200–212.

Chapter 10

Karst of Poland

J. GLAZEK[1], T. DABROWSKI[1] AND R. GRADZIŃSKI[2]

[1] Geological Institute, Warsaw (Poland)
[2] Department of Geology, Jagellonian University, Cracow (Poland)

LITHOSTRATIGRAPHIC AND SEDIMENTOLOGIC ASPECTS

Karst in Poland is known primarily in carbonate rocks, and more rarely in gypsum and rock salt (Fig.1). The main occurrences of these rocks are noted here in stratigraphic order*.

Among the Upper Proterozoic and Lower Palaeozoic low-grade metamorphosed geosynclinal deposits of the Sudetes, karst phenomena are connected with small areas built up of thick-bedded marbles with a total surface of about 20 km².

The Upper Palaeozoic carbonate rocks are exposed in the Holy Cross Mountains and in the Silesia–Cracow upland. In the Holy Cross Mountains, the Middle and Upper Devonian carbonate formation is exposed over a wide area of about 500 km². It attains over 500 m in thickness. Karst phenomena are best developed in light-coloured, thick-bedded Givetian and Frasnian, mainly organogenic limestones in the southern part of this area. Northwest of Cracow, karst phenomena occur in shallow water carbonate rocks of the Middle and Upper Devonian and Lower Carboniferous Ages. Some karst phenomena are known in the Zechstein carbonate and gypsum and salt deposits occurring in the wide area of western Poland.

The Triassic (Röt and Muschelkalk) carbonate rocks, about 200 m thick, occur in the Silesia–Cracow upland, border zone of the Holy Cross Mountains and in the north Sudetic Basin. In these rocks there are usually numerous but small karst forms due to many marly intercalations and fine bedding. The Triassic carbonate rocks of the High Tatra Mountains are about 400–600 m thick. Because of the smaller quantity of marly intercalations, the longest and deepest caves in Poland are developed there.

In the Upper Jurassic deposits the karst phenomena are connected mainly with the platy and biohermal butte-limestones (Felsenkalk) of the Oxfordian and Lower Kimmeridgian. The Upper Jurassic calcareous rocks form the most extensive karst areas in Poland. On the northwestern slope of the Silesia–Cracow upland, the Upper Jurassic limestones (over 300 m thick) form a belt 170 km long and 20 km broad. The Upper Jurassic limestones in the border zone of the Holy Cross Mountains are

---

* The basic textbook and source of bibliography on the geology of Poland is that by KSIAZKIEWICZ et al. [1968].

even thicker (up to 1,000 m). In the Tatra Mountains a 250 m thick calcareous formation is exposed. This formation is mainly composed of thick-bedded, pelagic limestones, with detrital limestones of the Middle Jurassic in the bottom part and the Urgonian facies at the top. Particularly large caves occur in these limestones.

In the Upper Cretaceous deposits, there are soft, chalky limestones (chalk in the eastern part of Poland), marly, siliceous limestones and gaizes, in some places attaining 1,000 m in thickness. The Upper Cretaceous calcareous deposits are exposed primarily in the Lublin upland, where karst phenomena are known over an area of about 2,000 km^2 [MARUSZCZAK, 1966].

In the Carpathian Foredeep, particularly on its northern border, the Tortonian gypsum (attaining about 50 m in thickness) is widely distributed and is strongly karstificated.

TECTONIC SETTING OF THE AREA

The northeastern part of Poland belongs to the east European Platform, the southern part to the Carpathian Belt. The rest of Poland (about 2/3 of the whole area) is a part of the young epi-Variscan, middle European Platform (Fig. 1).

The epi-Variscan Platform is the most interesting area, because its southern margin was subjected to uplift, which formed a belt of uplands parallel to the Carpathian Arc. Karst forms of various ages occur in these uplands. They are connected with subsequent phases of emersion (Fig. 2,3).

In the Sudetes, which comprise a Pliocene horst [DYJOR, 1968] of very complicated structure, Upper Proterozoic and Lower Palaeozoic marbles occur within the epimetamorphic Caledonian Series, in the form of isolated lenses which dip steeply and are cut by faults. The Zechstein and Middle Triassic carbonate and gypsum deposits occur in the flat and faulted epi-Variscan north Sudetic Basin.

In the Variscan basement of the Silesia–Cracow upland, the Upper Palaeozoic carbonate rocks occur in some elevation of northwest–southeast direction among horizontal younger deposits. Triassic carbonate rocks in the centre of the Upper Silesian Coal Basin, are preserved only in grabens or in the form of residual hills, but in the northern and northeastern margin of this upland, the Triassic and Jurassic deposits form a vast monocline, dipping east-northeast at a low angle. The southern part of this monocline is more elevated and transected by numerous Miocene faults.

The Holy Cross Mountains form a Variscan anticlinorium, which was lifted up during the Palaeogene. The trend of these mountains is from west-northwest towards east-northeast. The synclines within this anticlinorium are built mainly of Devonian carbonate rocks. The axes of the folds plunge towards the west-northwest. The Variscan massif of the Holy Cross Mountains is bordered by a less deformed Mesozoic Series. The whole area is cut by numerous faults of Variscan and Tertiary age.

Northwest of the Holy Cross Mountains, a swell extends beneath Cenozoic deposits. It is built of the Mesozoic rocks, which are pierced by diapirs of Zechstein salt.

Fig. 1. Sketch-map showing the occurrences of karst rocks in Poland. *1* = Lower Paleozoic and Upper Proterozoic marbles; *2* = Devonian and partly Carboniferous limestones and dolomites; *3* = Permian carbonate and gypsum-saline deposits; *4* = Röt and Muschelkalk carbonate deposits; *5* = Upper Jurassic limestones; *6* = Upper Cretaceous limey deposits; *7* = Triassic, Jurassic and Lower Cretaceous carbonate deposits in the Carpathians; *8* = Upper Miocene gypsum and limey deposits; *9* = faults; *10* = Carpathian overthrust; *11* = ice-transported blocks of Senonian limestones; *12* = southern limit of Mindel Glacial; *N.S.B.* = north Sudetic Basin; *U.S.C.B.* = Upper Silesian Coal Basin; *P.K.B.* = Pieniny Klippen Belt.

East of the Holy Cross Mountains in the southern, uplifted, part of the pericratonic basin, which is called the Lublin upland, almost undisturbed Upper Cretaceous deposits are exposed.

South of the Middle Polish uplands lies the Carpathian Foredeep, filled up with Miocene sediments. Near the margin of the Carpathians, the lower parts of these deposits, including the Lower Tortonian, are folded, but the south slopes of the uplands are covered by horizontal Miocene sediments.

In the High Tatra Mountains, complicated tectonics of scale type gave rise to narrow belts of carbonate rocks directed approximately west–east and dipping towards

Fig. 2. Map showing the distribution of karst in Poland. *1* = Site of uppermost Carboniferous–Middle Triassic karst (*S.G.* = Stare Gliny); *2* = site of uppermost Triassic–Middle Jurassic karst; *3* = site of Lower and Middle Cretaceous karst; *4-11* = young post-Cretaceous karst phenomena; *4* = carbonate karst in lowland and uplands; *5* = fragments of Palaeogene peneplain developed on the carbonate rocks; *6* = gypsum-saline karst; *7* = gypsum karst; *8* = site of Pliocene–Lower Pleistocene vertebrate faunas in karst forms (*R.K.* = Rębielice Królewskie ; *K.* = Kamyk; *P.* = Podlesice); *9* = karst forms in the ice-transported blocks; *10* = high-mountainous karst; *11* = larger caves (*R.C.* = Raj Cave; *N.C.* = Niedźwiedzia Cave; *S.C.* = Śniezna Cave; *C.C.* = Czarna Cave).

the north. These structures are complicated by transversal elevations and depressions. Since the Miocene, the High Tatra Mountains have been subjected to isostatic uplift, which has finally caused the development of high-mountainous karst during the Quaternary.

PHASES OF KARSTIFICATION

Numerous localities of fossil karst are known in Poland [GILEWSKA, 1964; GRADZIŃSKI and WÓJCIK, 1966; GLAZEK, 1971].

The first well-documented karstification phase began during the Carboniferous

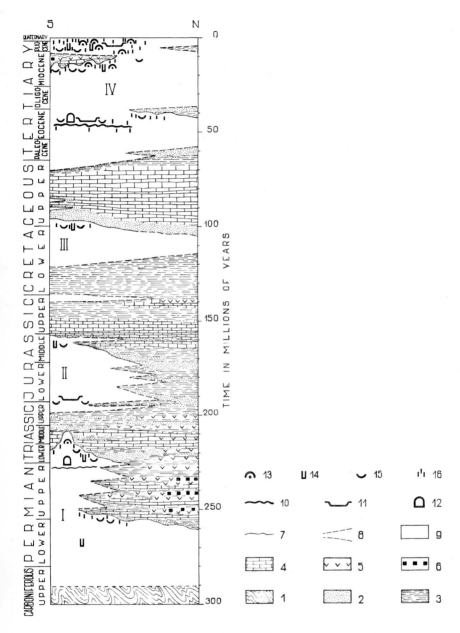

Fig.3. Time distribution of karst phenomena and marine sediments in Central Poland. *1* = Folded Variscan or Caledonian substratum; *2–6* = marine and partly lacustrine sediments: *2* = psammitic and psephitic; *3* = argillaceous; *4* = carbonate; *5* = gypsum and anhydrite; *6* = saline; *7* = stated erosional surface; *8* = supposed limits of sedimentation; *9* = land environments; *10* = karst corrosional surface; *11* = uvalas and poljes; *12* = cones (mogotes); *13* = caves; *14* = shafts; *15* = dolines; *16* = lapies; *I, II, III, IV* are phases of karstification.

(before the Stephanian) after the Variscan movements and ended in the Middle Triassic (Fig.3,4). The karst developed in conditions of a warm climate, with semiarid and humid phases. A corrosional surface was then formed on the carbonate rocks, over which isolated, steep-walled hills (mogotes?), often more than 100 m high, dominated. This surface was gradually inundated during the Triassic. A cave-fragment with a bone breccia of Middle Triassic age in a fossil mogote(?) wall at Stare Gliny (Fig. 4) has been described [TARLO, 1959; LIS and WÓJCIK, 1960].

The next phase of karstification has been established only in the Silesia–Cracow upland, where the karst processes have developed in the Triassic carbonate rocks. Dolines, uvalas and even small poljes were then formed which were then filled with variegated clays, siltstones and sands with pollen grains of the lowermost Jurassic age [GÓRZYŃSKI, 1963]. These karst processes were developed there under humid and warm climatic conditions (subtropical?). This phase was shorter than the preceding one and its extent was much smaller.

Again in the Silesia–Cracow upland, karst processes developed during the Lower Cretaceous. Dolines and shafts were noted there filled up with clays, sandstones, conglomerates and marls of the Upper Cretaceous [BUKOWY, 1956]. The climate was probably moderately warm, subtropical with humid and semiarid phases.

The last, long phase of karstification in the belt of uplands began with the emersion at the end of the Cretaceous or in the Palaeocene and lasted to Recent time. On the northern side of the Middle Polish uplands, marshes, lakes and even seas—for a short period of time—existed in which clays, sands and brown coals were laid down during the Tertiary. In some places, these deposits were noted on the karstificated slopes of these uplands. During the Tortonian, there were some bays penetrating the Middle Polish uplands from the south, where the Tortonian deposits rest on the karstificated surface of the Mesozoic carbonate rocks. In the Carpathians the karst development began after the Miocene emersion.

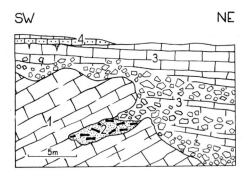

Fig.4. Schematic section of a Middle-Triassic cave at Stare Gliny. *1* = *Amphipora* Dolomites (Givetian); *2* = cave bone breccia (Middle Triassic); *3* = *Diplopora* Dolomites including debris of Givetian dolomites (Middle Triassic); *4* = sandstone passing into limestones in the top (Callovian). These sediments filled up the small karst crevices.

Fillings of karst forms from the Pliocene and Lower Pleistocene contain numerous remains of vertebrates. Older, Pliocene faunas have been encountered in the bone breccias (Weze I, Podlesice) and in red residual clays (Rebielice Królewskie, Weze II); younger—Early Pleistocene—faunas were found in red-brownish clays (Kadzielnia Hill in Kielce, Kamyk). The famous site Weze I is the most interesting one [SAMSONOWICZ, 1934], as the richest vertebrate fauna [see, e.g., KOWALSKI, 1964; SULIMSKI, 1964] was found there. The Pleistocene Glacials restricted karst development in Poland, especially during the maximal—Mindel Glacial— when the glacier from Scandinavia reached the Sudetes and the Carpathians.

The last karstification phase was characterized by a variable climate [TYCZYŃSKA, 1957]. The Upper Cretaceous regression took place under subtropical climatic conditions. A similar climate prevailed until the Tortonian [SZAFER, 1961; BALUK and RADWAŃSKI, 1967]. During the Pliocene, Mediterranean and moderate phases alternate with steppe semiarid climatic conditions [SZAFER, 1954; KOWALSKI, 1964; SULIMSKI, 1964]. Later in the Quaternary, a moderate and cool climate existed during the Lower Pleistocene, Interglacials and Holocene, whereas an arctic climate existed during the Glacials [RÓŻYCKI, 1965, 1967]. The development of karst phenomena during the Quaternary was of lesser importance, because of unfavourable climatic conditions, both arid arctic in the Glacials and semiarid in the Interglacials [MARKOWICZ-LOHINOWICZ, 1969]. Young, Late Pleistocene and Holocene karst forms play a major role only in the Tatra Mountains [RUDNICKI, 1967; GLAZEK, 1968].

The intensity of present-day karstification in Poland is characterized by calculations of the ablation (karst corrosion) coefficient, which ranges within the limits of 10–20 m³/km² per year) in the zone of the uplands [MARUSZCZAK, 1966; MARKOWICZ-LOHINOWICZ, 1968; LISZKOWSKI, 1971]. A larger intensity of karstification (30–50 m³/km² per year is noted in the mountains in the south of Poland [PULINA, 1965; RUDNICKI, 1967].

MORPHOLOGIC CONSEQUENCES

The morphology of the karst areas in Poland was developed during the last polycyclic phase of karstification. Where the calcareous and saline–gypsum rocks occur near the surface, small karst forms in them are to be found, in the form of soil lapies and dolines. Such phenomena were observed even in the ice-transported blocks of limestones in moraines. Usually these forms are filled with the Tertiary and Quaternary sediments, which mark their upper limit of age; but in the case of loose pervious sediments, such karst forms may be younger, as they could have developed under a thin overburden (up to about 10 m in thickness).

Karst morphology in Poland predominates only in some parts of the Middle Polish uplands, where the morphology was not deformed by the glacier of the maximal—Mindel Glacial. The only wide area of this kind is that of the Upper Jurassic limestone belt of the Silesia–Cracow upland, which had comprised a nunatak

[RÓŻYCKI, 1965,1967]. The Palaeogene peneplain is preserved in that area, above which there dominate the residual hills considered as remnants of mogotes (Fig. 5) formed under tropical and subtropical climatic conditions [KLIMASZEWSKI, 1958; POKORNY, 1963; RÓŻYCKI, 1967]. These forms exhibit, however, a distinct lithological character as they are built of more resistant bioherms of "butte-limestones" (Felsenkalk), situated within layered limestones. Several dozen caves of shaft character (up to 35 m deep) are situated within these hills and are usually filled up with calcite. Probably these are fragments of an older generation of karst conduits, the development of which began during the Palaeogene. There are numerous undrained depressions filled with residual sediments and sands within the peneplain. Deep valleys, frequently dry, are cut into the peneplain. These developed during the Pliocene and the beginning of the Pleistocene, prior to the Mindel Glacial [DŽULYŃSKI et al., 1966]. The origin of the numerous caves (about 500) is connected with the subsequent phases of valley development [GRADZIŃSKI, 1962]. They are entirely dry and have been filled with deposits since the Riss Glacial [MADEYSKA-NIKLEWSKA, 1969].

The morphology of some parts of the Sudetes is of similar character, with fragments of peneplain and also remnants of mogotes and numerous caves [PULINA, 1965]. Among them, the newly discovered Niedźwiedzia Cave which is about 600 m long, has the richest dripstone formation in Poland [M. Pulina, personal communication, 1968]. A similar situation occurs in the Holy Cross Mountains, where about 50 caves

Fig.5. Residual hill above Palaeogene corrosional surface southwest of Częstochowa. (Photo by R. Gradziński.)

are known in the Devonian limestones. The Raj Cave near Kielce, about 270 m long, has very rich dripstone formations [GRADZIŃSKI and WRÓBLEWSKI, 1968].

In the Lublin upland, numerous undrained depressions occur in the Upper Cretaceous rocks, which are shallow and comparatively broad and variable in size (dolinas, uvalas and poljes). Also karst corrosional plains occur there [MARUSZCZAK, 1966]. Occasionally in the morphological culminations, there are small bare dolinas, but the majority of these forms are covered with sands and gravels, with deluvial silts, peats or lakes at the bottom. There is a lack of underground karst forms. This assemblage of forms is regarded by MARUSZCZAK [1966] as being typical for the karst of the chalk.

In the High Tatra Mountains numerous surface karst forms show a definite zonal arrangement. Lapies (typical karren) associated with less frequent bare dolinas predominate at above 1,500 m a.s.l. Dolinas reproduced in moraines which cover calcareous rocks, are predominant below this level. These zones correspond to those differentiated in the Alps, but their boundaries occur about 200 m lower in the Tatra Mountains than in the Alps (about 1,700 m) because of the slightly colder climate of the former [GLAZEK, 1968]. All these karst forms are younger than the Würm morphology as the latter was deformed by them. To older forms dome-shaped peaks of some limestone ridges, which rise over fragments of the elevated Neogene gradational surface belong. They are probably remnants of initial domes (mogotes), formed under subtropical conditions during the Upper Miocene and Lower Pliocene [GLAZEK, 1968].

Over 200 caves are known in the Tatra Mountains. Among them there are the longest caves in Poland, e.g., the Czarna Cave, 6 km long, the Mietusia Cave, 5 km long, and the Zimna Cave, 4.2 km long [RUDNICKI, 1967]. Deep caves of aven and gouffre type occur in ridges. Among them, the Śnieżna cave system reaches about 770 m in depth [KOISAR, 1969]. Exsurgence cave systems and resurgence cave systems of various age have been encountered in the Tatra Mountains. The former are usually developed perpendicularly to the valleys and are connected with gouffres, while the latter ones are parallel to the valleys and show numerous allochthonous gravel deposits [WÓJCIK, 1966; RUDNICKI, 1967]. A storied arrangement of horizontal conduits and scanty dripstone are characteristic features of the Tatra Mountains caves. The highest conduit levels, situated over 200 m above the present valleys have developed at the boundary of the Miocene and Pliocene. The Pliocene conduits are situated more than 100 m above the valleys. The Pleistocene ones are situated below, and the avens and gouffres are of the same age. They result from infiltration of cold waters during the recession periods of mountain glaciers [RUDNICKI, 1967]. Little caves hanging high over the valleys also occur there. They are independent of the karst drainage systems and are of superficial origin (grottes cutanées); their age is Late Pleistocene [GLAZEK, 1969].

The karst morphology of the Tortonian gypsum in the area south of the Holy Cross Mountains is of particular interest. Numerous solution cavities and dolines, as well as less common karst shafts, caves, uvalas, valley sinks and small poljes were noted there [FLIS, 1954]. Solution of gypsum takes place mainly along interstratal surfaces and this leads to the formation of caves and collapse dolines. Flatly lying gypsum beds form erosion scarps at the valley slopes. In these scarps, interstratal suberosion has caused local dips towards the valleys. Less frequent are dolinas caused by surface corrosion, and these are reproduced in the sandy overburden of uppermost Miocene and Pleistocene age. This karst is of Holocene age. The karst forms are typical, but smaller than in areas of typical limestone karst. The gypsum karst develops quickly and the karst forms also quickly disappear. In the salt deposits of the southern margin of the Carpathian Foredeep, natural caverns occur, with beautiful halite crystals and dripstones. The latter develop also in the old galleries of the Wieliczka salt mine.

Solution cavities and caves occur in the Zechstein gypsum deposits in the north Sudetic Basin [SACHANBIŃSKI, 1967]. Numerous suberosional caves are actually formed in the peak parts of the salt domes in northern Poland [POBORSKI, 1957].

HYDROGEOLOGIC CONSEQUENCES

The occurrences of karst waters are of importance in some regions of the Middle Polish uplands and in the Tatra Mountains. Recharge of karst waters takes place mainly by the infiltration of precipitation and, in some places, of river waters. Average annual precipitation ranges from 500–800 mm in the belt of uplands, whereas in the Tatra Mountains it is 1,100–1,700 mm. The rainfalls are unevenly distributed during the year, nevertheless the amount of infiltrating waters is reduced during the winter and rapidly increased in the early spring as a consequence of the thawing of snow. The increased supply of karst waters in the spring is accompanied by minimum hardness, increased corrosion and maximum yield of the springs, but during the winter the maximum hardness of karst waters occurs with minimum yield of springs [DABROWSKI, 1967a; RUDNICKI, 1967; MARKOWICZ-LOHINOWICZ, 1968]. However, in July, when the rainfall in Poland is greatest, in the belt of uplands the yield of the karst springs decreases because the evaporation there is so great that only a small part of the rain waters infiltrates to the depth of the karst massif [MARKOWICZ-LOHINOWICZ, 1968]. On the other hand, in the Tatra Mountains, where the evaporation is comparatively lower, an increase of the yield of the karst springs occurs. Overall mineralization of the karst waters ranges within the limits 200–600 mg/l [KOTLICKA, 1962; MOWSZOWICZ and OLASZEK, 1965; MARKOWICZ-LOHINOWICZ, 1968], only in the Tatra Mountains is it less than 200 mg/l [DABROWSKI, 1967a]. Among the karst waters of Poland the $HCO_3:SO_4:Ca:Mg$ type dominates while $HCO_3:SO_4:Ca$ and $HCO_3:Ca$ types are less frequent. High contents of sulfate are characteristic of karst waters in the Silesia–Cracow upland. It is primarily caused by enrichment with SO_4^{2-} of infiltration of waters by industrial wastes [MARKOWICZ-LOHINOWICZ, 1968] and partly by the presence of sulphide ore deposits in the karstificated Triassic rocks.

Local geological and morphological conditions control the outflow direction of the karst waters, the velocity of outflow and reserves of these waters. In the Triassic limestones and dolomites of the Silesia–Cracow upland, several reservoirs of karst waters occur. But, in the Upper Jurassic limestones on the northeast slope of this upland, the karst waters form the water table of a "nappe en reseau" type at a depth of up to 100 m [MARKOWICZ-LOHINOWICZ, 1968]. The karst springs along the Upper Jurassic scarp and in the gap of the Warta River drained these waters. The total flow of karst waters within the gap section of the Warta River was estimated at 400–500 l/sec before the exploitation.

The karst waters of the central part of the Holy Cross Mountains occur in the Devonian rocks and show an outflow to the west-northwest, along the dip of the synclinal axis. In this direction, westward of Kielce, some karst springs had a yield of up to 82 l/sec before exploitation of the water. Karst waters of the Mesozoic border zone of the Holy Cross Mountains occur in the Middle Triassic and Upper Jurassic rocks. The biggest springs near Tomaszów Mazowiecki had a yield of 220 l/sec in 1933; this had dropped to 80 l/sec in 1962/1963 due to water demand [MOWSZOWICZ and OLASZEK, 1965].

In the Tatra Mountains, the paths of the karst-water circulation are better recognized as a result of numerous water colorations. The karst regions are drained by copious exsurgences, some of which reach a yield as high as 3,500 l/sec. The paths of karst-water circulation are developed in stepped levels, so there is no uniform water table. The exsurgence systems are independent of the drainage system of the surface waters. The trends of flows in these systems are directed along the structural belts toward the transversal depressions. Flow velocity in exsurgence systems was under 100 m/h, but under the bottoms of the valleys across the structural belts, resurgence systems are developed, in which flow velocities range from 100 to 3,300 m/h [DABROWSKI, 1967a, b; RUDNICKI, 1967].

A deep-seated circulation of karst waters in the Zechstein sediments of the north Sudetic Basin has been encountered at a depth of 1,000 m (i.e., about 750 m below sea level), whereas the piezometric water table lies at 5–30 m below ground level [KRASOŃ and WÓJCIK, 1965].

PRACTICAL PROBLEMS AND THEIR APPROACH

Although the karst regions in Poland occupy the surface of only about 8,000 km^2 (i.e., 2.5% of the surface area of the country), karst waters constitute the basis for the water supply of both population and industry in some regions.

In the Silesia–Cracow upland, the karst waters from the Triassic deposits constitute the basis for the water supply of the Upper Silesian industrial district. Waters from the Upper Jurassic limestones of this upland are exploited for the larger towns, in the south for Cracow, and in the north for Czestochowa. In this latter region, the maximum yield of a drilled well reached 500 m³/h with depressions less than 1 m. Waters from the Devonian rocks of the Holy Cross Mountains are exploited for the water supply of Kielce. Similarly karst waters from the Triassic and Jurassic limestones of the Mesozoic border zone of the Holy Cross Mountains are exploited for the water supply systems of some smaller towns.

Within the north Sudetic Basin, the karst waters cause serious obstacles to mining. Between 1944 and 1961 there were a number of mines flooded, with violent water irruptions reaching 1,000 m³/h [KRASOŃ and WÓJCIK, 1965]. Similarly, karst waters in the Middle Triassic deposits of the Silesia–Cracow upland cause difficulties in the exploitation of the zinc–lead ores.

In the karst regions of Poland, karst processes have caused little damage to buildings [LISZKOWSKI, 1971]. During the construction of a metallurgical combine near Czestochowa, in the zone of the Upper Jurassic outcrops, on the northeast slope of the Silesia–Cracow upland the karst caused considerable difficulty, which was controlled by different geological methods, drillings and resistivity measurements followed by cementation [BAZYŃSKI, 1960]. Some devastation of buildings occurred during the last century at Inowroclaw, the town sited on the salt diapire, where suberosion of gypsum and salt caused serious danger [POBORSKI, 1957].

In the fillings of karst depressions, which are mainly of Tertiary age, concentrations of several raw materials occur in the Silesia–Cracow upland and in the Holy Cross Mountains, where secondary concentrations of galena (Holy Cross Mountains), calamine (Silesia–Cracow upland), limonite, wad, fireclays and sands were periodically exploited. Recently the Palaeogene moulding sands from the dolinas and uvalas on the Upper Jurassic limestones in the northeast slope of the Silesia–Cracow upland have been exploited on a larger scale. Calcite (from dripstones) and bats' guano was exploited from the caves of this area.

The most interesting karst forms in Poland are officially preserved as inanimate natural monuments, where they occur in national parks (Tatra National Park and Ojcowski National Park near Cracow) or when they constitute separate reserve. Amongst the karst caves of Poland, only three (of minor size) have been equipped with electric lighting and are accessible to tourists (Mroźna Cave in the Tatra Mountains, Smocza Cave in Cracow and Raj Cave near Kielce).

ACKNOWLEDGEMENTS

The authors are indebted to Dr. J. Kutek, Dr. A Radwański, Dr. J. Rudnicki and Mgr. M. Markowicz-Lohinowicz for helpful discussion; to Dr. J. Lefeld, Dr. F. Simpson and Dr. A. Radwański for help in the preparation of the English text; and to Mr. A. Kozlowski for preparation of the drawings.

REFERENCES

Baluk, W. and Radwański, A., 1967. Miocene cirripeds domiciled in corals. *Acta Palaeontol. Polon.*, 12:457–521.
Bazyński, J., 1960. Engineering geological characteristic of karst in the vicinity of Częstochowa. *Przeglad Geol.*, 8:430–432. (In Polish, with English summary.)
Bukowy, S., 1956. Geology of the area between Cracow and Korzkwia. *Biul. Inst. Geol.*, 108:17–82. (In Polish, with English summary.)
Dabrowski, T., 1967a. Karst hydrogeology of the Polish Tatra Mountains obtained by colouring methods. *Steirische Beitr. Hydrogeol.*, 1966/1967:219–226.
Dabrowski, T., 1967b. Underground karst flows of streams in the western Tatras. *Acta Geol. Polon.*, 17:593–622. (In Polish, with English summary.)
Dyjor, S., 1968. Marine horizons within Poznan clays. *Kwart. Geol.*, 12:941–957. (In Polish, with English summary.)
Džulyński, S., Henkjel, A., Klimek, K. and Pokorny, J., 1966. The development of valleys in the southern part of the Cracow upland. *Rocznik Polsk. Towarz. Geol.*, 36:329–343. (In Polish, with English summary.)
Flis, J., 1954. Gypsum karst of the Nida Trough. *Prace Geograf., Polska Akad. Nauk Inst. Geograf.*, 1:1–73. (In Polish, with English summary.)
Gilewska, S., 1964. Fossil karst in Poland. *Erdkunde*, 18:124–135.
Glazek, J., 1968. Karst phenomena in the eastern part of the Polish Tatra Mountains, south Poland. *Congr. Intern. Spéléologie Yougoslavie, 4me, 1965, Actes*, 3:445–449.
Glazek, J., 1969. The superficial origin of the Woloszynskie caves (Tatra Mountains) and their age. *Speleologia*, 4:53–64. (In Polish, with English summary.)
Glazek, J., 1971. Phases of karstification in the epi-Variscan Platform of Poland. *Bull. Acad. Polon. Sci., Sér. Sci., Géol. Géograph.*, 19, in press.
Górzyński, Z., 1963. Methods and preliminary results of search works for clayey raw materials in the Upper Silesia area. *Przegląd Geol.*, 11:483–486. (In Polish, with English summary.)
Gradziński, R., 1962. Origin and development of subterranean karst in the southern part of the Cracow upland. *Rocznik Polsk. Towarz. Geol.*, 32:429–492. (In Polish, with English summary.)
Gradziński, R. and Wójcik, Z., 1966. Fossil karst in Poland. *Prace Muzeum Ziemi*, 9:151–222. (In Polish, with English summary.)
Gradziński, R. and Wróblewski, T., 1968. Speleothems in the Raj Cave. *Ochrona Przyrody*, 33:281–307. (In Polish, with English summary.)
Klimaszewski, M., 1958. The geomorphological development of Poland's territory in the pre-Quaternary Period. *Przegląd Geograf.*, 30:3–43. (In Polish, with English summary.)
Koisar, B., 1969. New discoveries in the Sniezna Cave. *Speleologia*, 4:33–37. (In Polish, with English summary.)
Kotlicka, G., 1962. Remarks on the mineralization of subterranean waters of the Silesia–Cracow region. *Kwart. Geol.*, 6:372–382. (In Polish, with English summary.)
Kowalski, K., 1951–1954. Les cavernes de la Pologne. *Panstwowe Muzeum Archeol., Warszawa*, 1:1–466; 2:1–186; 3:1–192. (In Polish, with French summary.)
Kowalski, K., 1964. Palaeoecology of mammals from the Pliocene and Early Pleistocene of Poland. *Acta Theriologica*, 8:73–88. (In Polish, with English summary.)
Krasoń, J. and Wójcik, Z., 1965. The deep karst of the Boleslawiec syncline in the Sudetes Mountains. *Acta Geol. Polon.*, 15:179–212. (In Polish, with English summary.)
Ksiazkiewicz, M., Samsonowicz, J. and Rühle, E., 1968. *An Outline of the Geology of Poland*. Sci. Publ. Foreign Coop. Center, Warsaw, 414 pp.
Lis, J. and Wójcik, Z., 1960. Triassic bone breccia and karst forms in Stare Gliny Quarry near Olkusz (Cracow region). *Kwart. Geol.*, 4:55–74. (In Polish, with English summary.)
Liszkowski, J., 1971. Engineering-geological character of the karst development on the base of the lithology of the Upper Jurassic deposits in the northeastern border zone of the Holy Cross Mountains. *Biul. Inst. Geol.*, in press. (In Polish, with English summary.)
Madeyska-Niklewska, T., 1969. Upper Pleistocene deposits in the caves of the Cracow upland. *Acta Geol. Polon.*, 19:341–392. (In Polish, with English summary.)

Markowicz-Lohinowicz, M., 1968. Corrosion karstique contemporaine dans le massif calcaire du Jura de Częstochowa. *Speleologia*, 3:55–75. (In Polish, with French summary.)

Markowicz-Lohinowicz, M., 1969. Essai d'évaluation de l'intensité de corrosion karstique pendant le Quaternaire dans le massif calcaire du Jura de Częstochowa. *Speleologia*, 4:19–26. (In Polish, with French summary.)

Maruszczak, H., 1966. Phénomènes karstiques dans les roches du Crétacé supérieur entre la Vistule et le Bug (type de karst crayeux). *Przegląd Geograf.*, 38:339–370. (In Polish, with French summary.)

Mowszowicz, J. and Olaszek, R., 1965. *The Blue Springs*. Łódzkie Towarzystwo Naukowe, Łódz, 78 pp. (In Polish, with English summary.)

Poborski, J., 1957. Ausbildung des Gipshutes und Entwicklung der Karstphänomene auf dem Salzstock von Inowrocław. *Arch. Górnictwa*, 2:225–248. (In Polish, with German summary.)

Pokorny, J., 1963. The development of mogotes in the southern part of the Cracow upland. *Bull. Acad. Polon. Sci., Sér. Sci., Géol. Géograph.*, 11:169–175.

Pulina, M., 1965. *Zjawiska Krasowe w Sudetach (The Karst Phenomena in the Sudetes)*. Thesis, Inst. Geograf. Polska Akad. Nauk, Warszawa, 237 pp. (In Polish.)

Różycki, S. Z., 1965. Traits principaux de la stratigraphie et de la paléomorphologie de la Pologne pendant le Quaternaire. *Intern. Congr. Quaternary, 6th, Warsaw, 1961, Rept.*, 1:123–142.

Różycki, S. Z., 1967. *The Pleistocene of Middle Poland*. Panstwowe Wydawnictwo Naukowe, Warszawa, 251 pp. (In Polish, with English summary.)

Rudnicki, J., 1967. Origin and age of the western Tatra caverns. *Acta Geol. Polon.*, 17:521–591. (In Polish, with English summary.)

Sachanbiński, M., 1967. Gypsum karst in the vicinities of Niwnice. *Przegląd Geol.*, 15:337–339. (In Polish, with English summary.)

Samsonowicz, J., 1934. Sur les phénomènes karstiques et la brèche osseuse de Węze près de Dzialoszyn sur la Warta. *Zabytki Przyrody Nieozywionej Ziem Rzeczypospolitej Polsk.*, 3:151–162. (In Polish.)

Sulimski, A., 1964. Pliocene Lagomorpha and Rodentia from Węze I (Poland). *Acta Palaeontol. Polon.*, 9:149–261.

Szafer, W., 1954. Pliocene flora from the vicinity of Czorsztyn (West Carpathians) and its relationship to the Pleistocene. *Prace Inst. Geol.*, 11:1–238. (In Polish, with English summary.)

Szafer, W., 1961. Miocene flora from Stare Gliwice in Upper Silesia. *Prace Inst. Geol.*, 33:1–205. (In Polish, with English summary.)

Tarlo, L. B., 1959. A new Middle Triassic reptile fauna from fissures in the Middle Devonian limestones of Poland. *Proc. Geol. Soc. London*, 1538:63–64.

Tyczyńska, M., 1957. Climat de Pologne au Tertiaire et au Quaternaire. *Czasopismo Geograf.*, 28:131–170. (In Polish, with French summary.)

Wójcik, Z., 1966. On the origin and age of clastic deposits in the Tatra caves. *Prace Muzeum Ziemi*, 9:3–130.

Chapter 11

Karst of Rumania

M. D. BLEAHU

Geological Institute, Bucharest (Rumania)

INTRODUCTION

The karst of Rumania has acquired a certain renown due to the works of Racovitza and his co-worker Jeannel [JEANNEL and RACOVITZA, 1929], who established the bases of biospeleology as a result of their investigations, particularly in this country. This accounts for the fact that some karst zones and some Rumanian caves have become familiar, especially to biologists. Notwithstanding this renown, the karst only covers a limited area of 4,400 km² of Rumania, only 1.4% of the total surface. This low percentage is due to Pliocene and Quaternary deposits which overlie the old formations, including the karst forming rocks, in broad regions.

In Rumania the karst is developed mainly on limestones, dolomites, salt and gypsum. Owing to the limited extent of the salt and gypsum, karst in these areas will be briefly presented at the end of this paper, which will in the main deal with carbonate rocks.

LITHOSTRATIGRAPHIC AND SEDIMENTOLOGIC ASPECTS

From the petrographic standpoint, a first distinction to be made among karst rocks involves the metamorphic and sedimentary dolomites and limestones. The former amount to 8%, mainly constituted of epizonal, old, pre-Hercynian-age reef formations, seldom mesozonally metamorphosed. In spite of metamorphism, the palaeogeographic aspect can be reconstituted, i.e., the extension of the bioherm and biostroms in the Poiana Rusca Mountains [PAPIU et al., 1964], the former massive, the others bedded. In other metamorphosed limestones, remnants of crinoids or spores can be identified, thus proving their sedimentary origin; but for those intercalated in deposits prior to the Upper Proterozoic, it may be assumed that they generated following chemical precipitation.

Metamorphosed limestones and dolomites form 82% of the total of carbonate rocks, i.e., an average surface of 3,600 km². These figures do not include the underlying formations of Palaeozoic and Mesozoic age in the basement of the Rumanian Plain. Limestones and dolomites outcrops are of Triassic (17.8%), of Jurassic and Cretaceous (47.3%), and of Neozoic (16.8%) age. A percentage separation between limestones and dolomites could not be established on account of both their frequent alternation and the dolomitization phenomena by which the limestones gradually pass into dolomites. It may be estimated that the dolomites, widespread particularly during the Middle Triassic, do not represent more than 1/8 of the limestones.

In general, limestones in Rumania display Alpine facies, thus justifying the use of Alpine denominations (Hallstatt, Gutenstein, Gosau, etc.). Dobrogea is the sole district in which Jurassic limestones show fauna of the Swabian facies.

Genetically, limestones can be divided into two groups: (*1*) open-sea carbonate associations; and (*2*) shallow-water marine and marginal marine associations.

Open-sea carbonate associations comprise: (*a*) pelagic limestones; and (*b*) mixed deposits of carbonate and terrigenous rocks.

Open-sea carbonate associations

The pelagic limestones
These limestones are exclusively of Mesozoic age and sometimes cover relatively large areas determining the formation of karst. Such are the Gutenstein and Hallstatt limestones in the Apuseni Mountains (Fig. 1, 26, 27, 28), well layered and exceeding 100 m in thickness, and the Jurassic and Cretaceous limestones in west Banat (Fig. 1, 22) which occur at several levels (Oxfordian, Kimmeridgian, Tithonian, Berriasian, Valanginian), on which exo- and endokarst-rich platforms developed. The Malm in the Almaj Mountains (Fig. 1, 21), of ammonitico rosso type, is less important from the standpoint of karstification, as well as limestones with Aptychus in Pieniny facies, in the east Carpathians (Fig. 1, 2). The pelagic deposits include south Dobrogea chalk.

The mixed deposits of terrigenous and carbonate rocks
These deposits of open-sea environment are less important from the point of view of karst, since they comprise thin beds of limestones interlayered with siltstones, marls and/or shales. This category comprises the limestones in the Aptychus Beds in the Trascau Mountains (Fig. 1, 25) and in Sinaia Beds of Valanginian–Hauterivian age, as well as other limestones from flysch sequences where the detrital component part is a turbidite.

Shallow-water marine and marginal marine associations

The greatest part of the Rumanian limestones are of reef detrital origin, laid in shallow sea environment. As usual, bioconstructed limestones with colonial organisms in growth position are rare. This applies to the limestones in the Middle Kimmeridgian in Dobrogea (Fig. 1, 31)—a bioherm— some limestones in Gosau facies within the Apuseni Mountains (Fig. 1, 24, 28)—bioherms—and a few reef cores, in reef limestone masses of the Pădurea Craiului (Fig. 1, 28) and the Dîmbovicioara Col (Fig. 1, 13) With these exceptions, the reef limestones are built up of limy material derived from

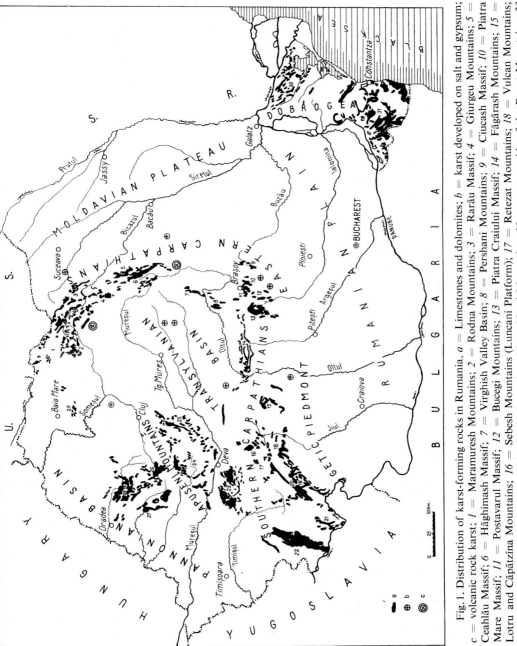

Fig.1. Distribution of karst-forming rocks in Rumania. a = Limestones and dolomites; b = karst developed on salt and gypsum; c = volcanic rock karst; 1 = Hăghimash Massif; 2 = Rodna Mountains; 3 = Rarău Massif; 4 = Giurgeu Mountains; 5 = Ceahlău Massif; 6 = Maramuresh Mountains; 7 = Virghish Valley Basin; 8 = Pershani Mountains; 9 = Ciucash Massif; 10 = Piatra Mare Massif; 11 = Postavarul Massif; 12 = Bucegi Mountains; 13 = Piatra Craiului Massif; 14 = Făgărash Mountains; 15 = Lotru and Căpătzîna Mountains; 16 = Sebesh Mountains (Luncani Platform); 17 = Retezat Mountains; 18 = Vulcan Mountains; 19 = Mehedintzi Plateau; 20 = Mehedintzi Mountains; 21 = Almaj Mountains; 22 = western side of the Banat Mountains; 23 = Poiana Rusca Mountains; 24 = Metaliferi Mountains; 25 = Trascău Mountains; 26 = Bihor Mountains; 27 = Codru-Moma Mountains; 28 = Pădurea Craiului Mountains; 29 = Preluca Massif; 30 = northern Dobrogea; 31 = southern Dobrogea.

bioherms. The place and mode of consolidation, as well as the presence or absence of terrigenous material results in a very large range of types of limestone. Since the stratonomic characters play the major role in the formation of karst, carbonate rocks, in this category, will be grouped with regard to bedding.

Massive limestones or dolomites

In these deposits bedding is uncommon—the dolomites are either imperfectly bedded or have beds exceeding 10 m in thickness. Taking into account the age and occurrence of carbonate rocks, those mainly involved in karst formation are: the Anisian (massive dolomites in the east Carpathians), the Ladinian (limestones containing Dasyclad algae in the east Carpathians and corals and brachiopods in the Apuseni Mountains), the Norian (limestones with corals, brachiopods and lamellibranches lumachelles in the east Carpathians and limestones with *Megalodontes* in the Apuseni Mountains), the Tithonian (limestones with *Nerinea* throughout the Carpathians and Dobrogea), and the Urgonian (limestones with *Pachiodontes* and corals throughout the Carpathians).

The thickness of these limestones is always notable (more than 100 m), and the superposition of differently aged bed complexes often leads to a considerable total thickness (for instance the Urgonian limestones directly overlying the Tithonian limestones).

Bedded limestones

This type of limestones is more frequently encountered in the sedimentary sequences, which do not attain the thickness recorded for the massive limestone and, therefore, are less important for the karst. They are deposits on the flank of reefs or in some places they are detrital with intraformational reworkings, in other places they are consolidated loam from reefs. Among the bedded limestones are those of the Campilian, Anisian, Ladinian and Carnian (in the east Carpathians and Apuseni Mountains), Rhaetian and Middle Jurassic (Apuseni Mountains), Bathonian–Kimmeridgian and Barremian (Dobrogea) Aptian–Albian (Apuseni Mountains), Eocene (nummulitic limestones in Maramuresh and Dobrogea, and the detrital limestones in the Transylvanian Basin), Sarmatian (Dobrogea).

Littoral deposits comprise either oolitic limestones in the Sarmatian of the Moldavian Plateau, or detrital rocks where limestones constitute the predominant element (e.g., calcarenites with well-rounded quartz granules in the Aptian within the Metaliferi Mountains, see Fig. 1, *25*), or abrupt cliff breccias in the Albian, within the Bucegi Massif (Fig. 1, *12*).

TECTONIC SETTING

In Rumania limestones occur in three different geological settings: (*1*) in zones of the pre-Alpine shields; (*2*) in zones of Alpine folding in the Carpathians; and (*3*) in the post-tectonic Neozoic depressions.

Zones of pre-Alpine shields

The Moesian Shield, located in the southern part of the country, is overlain in the Rumanian Plain by Neogene and Quaternary deposits and outcrops only in the south and central Dobrogea. The Moesian Shield was definitively consolidated during the Late Assyntian foldings, which are overlain by a Palaeozoic cover (4,000 m in thickness) and Mesozoic (5,000 m in thickness), about half the thickness of which are limestones and dolomites. The drilling data show that the sedimentary cover is slightly folded, but appears strongly faulted and divided into compartments. In Dobrogea, Upper Jurassic and Barremian limestones and dolomites, Senonian chalk and Sarmatian limestones occur. Excepting the Jurassic, which forms a syncline in the centre of Dobrogea, the other formations are not folded, the Sarmatian limestone flag which covers the whole south Dobrogea playing the main part as regards the relief.

Within the Moldavian Plateau, a fragment of the Russian Shield, limestones occur in the Tortonian and Sarmatian, showing reduced thickness, and being in fact a part of a monocline dipping southwards. The limestones form very conspicuous cuestas.

Zones of Alpine folding in the Carpathians

In the Carpathians limestones and dolomites form the sedimentary cover of some metamorphosed pre-Alpine massifs (of Hercynian age or even older), being folded together with the latter in the Alpine orogenic phases during the Cretaceous Period. These limestones which belong, in the main, to three sedimentary cycles, namely Triassic, Jurassic and Cretaceous, display—from the structural point of view—three types, as they have been affected differently by the tectonic movements, and hence have generated specific karstic forms.

Limy plateaus

In the Rumanian Carpathians there are but a few limy plateaus of great extent. The most important are to be found in the Sebesh Mountains (Fig. 1, *16*), in the west Banat Mountains (Fig. 1, *22*), Bihor Mountains (Fig.1, *26*), Codru-Moma Mountains (Fig. 1, *27*) and Pădurea Craiului Mountains (Fig.1, *28*). These limestones have relatively great thicknesses, alternating with complexes of non-carbonate rocks. They form wide folds with gentle dips which sometimes form synclinoria. These folded structures are faulted into block compartments. Thus, horst and graben structures divide a synclinorium into compartments (Fig. 2A). In case of a general vergence of the faults (in the autochthonous of the overthrusts), step faults hading against the dip, separate the compartments where the limestone levels are repeated. Their total thickness is great, e.g., in the Pădurea Craiului Mountains.

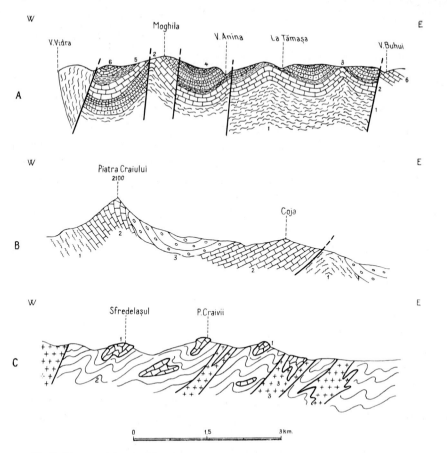

Fig.2. Structural types of karst.
A. Limy plateau. Geological section on the western side of the Banat Mountains. *1* = Sandstones and marls (Lias–Dogger); *2* = limestones (Callovian); *3* = marls (Oxfordian); *4* = limestones (Lusitanian–Tithonian); *5* = marls (Hauterivian–Lower Barremian); *6* = limestones (Urgonian).
B. Limy ridges. Geological section through the Piatra Craiului Massif. *1* = Crystalline schists; *2* = limestones (Callovian–Berriasian); *3* = conglomerates (Vraconian–Cenomanian).
C. Isolated calcareous massifs. Geological section in the Metaliferi Mountains. *1* = Olistoliths of Jurassic limestones; *2* = Lower Cretaceous flysch; *3* = diabases.

Limy ridges

These are derived from limy bed complexes whose thickness is not very great and they are found between non-karst-forming rocks. When such limestones form steep-limbed folds, they undergo erosion, appearing as very prominent limy ridges. Such are, for instance, the Haghimash Massif (Fig. 1, *6*), the Piatra Craiului Massif (Fig. 1, *13*) and the Trascau Massif (Fig. 1, *25*). These ridges are less affected by longitudinal faults and more frequently are bordered by such faults, taking on a horst aspect. However, they are cut by transverse faults, some of which have horizontal displacement.

Isolated calcareous massifs

These massifs occur amidst the non-karst deposits of various size (ranging from a few or tens of metres to a few kilometres.) In some places, they represent reefs or deposits bordering reef in autochthonous position, as for example, Mount Tesla in Cîucash (Fig.1,*9*), part of the limestones in the Ceahlău Mountains (Fig.1,*5*) or the Dimbau Mountain in the Metaliferi Mountains (Fig.1,*25*). In other cases they represent elements of megabreccia, sedimented in flysch deposits, i.e., olistoliths or more precisely, calcolistoliths, for example, the limestones in the Piatra Mare (Fig.1, *10*) and Bucegi Mountains (Fig.1,*12*) and particularly, the multitude of klippen in the Metaliferi Mountains (Fig.1,*25*, Fig.2C). Finally such limestone massifs may also be tectonic outliers, pushed out and truncated especially in the thrust phenomena as, e.g., in the southern end of the Mehedintzi Mountains (Fig.1,*19*).

In all of the above cases, due to the difference in competence, the limestones have been subjected to pressures that have led to their strong jointing during their folding.

Post-tectonic Neozoic depressions

The main folding phases in the Carpathians date from Cretaceous time, after which the geosynclinal flysch arose in the external zone of the ridge, accompanied inside by the formation of sedimentation basins. In the latter the sedimentation has a marine, neritic, shore or lacustrine character. These limestones are not very thick; they were affected very little by tectonic movements, showing a quasi-horizontal position and a slightly fissured or faulted structure. They may be compared—from the standpoint of the degree and mode of tectonization—to limestones found in the pre-Alpine shields.

PHASES OF KARSTIFICATION AND EVOLUTION OF THE KARST AREAS

In the geological evolution of Rumania, several karstification periods might be distinguished: during the Upper Triassic, the Cretaceous (Neocomian, Albian and Senonian), and in the course of the Neozoic. The first periods are represented by fossil karst showing a corrosion and erosion relief covered by transgressive or residual materials; the present karst relief is due to the last period.

In the Apuseni Mountains, within the Bihor autochthonous (Fig.1,*28*), after the deposition of the Middle Triassic limestones, there followed a period of emergence during which a karst relief formed on the surface. This relief underwent a filling with detrital formations of Rhaetian and Liassic age. Likewise, in the mass of Ladinian limestones, karstic solution openings are encountered, filled with breccias which probably correspond to the same period.

The same process took place subsequently to the deposition of the Tithonian limestones, their karstification occurring in the course of the Berriasian and Valanginian. The filling was made up of residual deposits, nowadays transformed into bauxites, transgressively overlain by the Hauterivian lacustrine limestones (Fig.3).

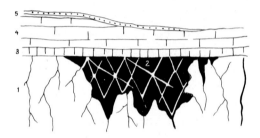

Fig.3. Type of fossil karst. Geological section in the Pădurea Craiului Mountains. *1* = Jurassic limestones; *2* = bauxites (Valanginian); *3* = lacustrine limestones (Hauterivian); *4* = reef limestones (Barremian); *5* = marls (Aptian). (After D. Patrulius, personal communication.)

Bauxites of this age occur in the Bihor (Fig.1,*26*) and Pădurea Craiului (Fig.1,*28*) Mountains. In the Sebesh Mountains (Fig.1,*16*) there exist, though only locally, Albian bauxites, whereas in the Metaliferi Mountains (Fig.1,24) their age is assigned to the Senonian.

The present karst forms spring up at the beginning of Neozoic, when owing to Laramian movements, the greatest part of the central zone in the Carpathians underwent uplift with emergence. For this reason, the history of the Tertiary of the Carpathians is particularly that of a subaerial evolution, whose predominant character is the polycyclic evolution. DE MARTONNE [1922] was the scientist who established the existence of three important block uplifts of the Carpathian ridge, each uplift being followed by a modelling of peneplains. Concomitantly with the modelling surfaces, karstification processes also took place in the zones where denudation led to the outcropping of the carbonate formations.

During the Eocene, subsequent to the uplift at the end of the Cretaceous, the upper peneplain named Borascu was formed; however, no important karst relief developed, the limestones still being overlain by their stratigraphic or tectonic cover (thrust nappes). As a result of the Savian movements, a second block uplift of the Carpathians took place; this was followed in the Miocene by the modelling of the second erosion surface named Riu Ses, which did not reach the degree of a peneplain. In this phase limestones denuded by erosion were strongly karstified on the surface. Circulation of water draining with a base level, determined by the Miocene sea surrounding the Carpathians, led to the formation of a series of cavities in the phreatic zone in the limestone. The last uplift of the Carpathians occurred in the Villafranchian, due to the Vallachian movements. Concomitantly with the uplift of the karstplains which were formed at the level of the Rîu Ses erosion, the caves corresponding to the latter, were exposed in the vadose zone as the phreatic zone was lowered. At the same time, a new erosion level took place, the third one which has been named the Gornovitza, and together with it, a new generation of karstplains was born. New karst cavities, related to the latter, developed in the phreatic zone, through which the underground drainage of the surface became possible. Subsequently, the deepening of the hydrographic network during the present erosion cycle determined the lowering of the local base levels, the lower caves changing over to a vadose zone. These caves which adapted themselves to the above levels by free-surface flow and overdeepening, allow a correlation with the fluviatile terraces, although they display phreatic patterns. Their recent change-over to a vadose regime seems to explain the failure of any deposits older than the Middle Pleistocene within these caves.

During the Upper Pleistocene, the periglacial climate changed the Tertiary morphology of the karst, especially through a strong congelifraction, and the karstification slowed down because of the permafrost. However, the limestone districts were not touched by the Carpathian glaciation.

Under the present temperature conditions, soil formation and the covering up of karst zones, took place as a function of the local climate, determined by altitude.

MORPHOLOGIC FEATURES

The lithologic nature, the structural position, the degree of tectonics and the stage of evolution are the factors responsible for the present morphological aspect of the karst, each one having a predominating influence at a certain scale.

At the regional scale, four karst types can be distinguished, corresponding to the above-mentioned structural positions:

(1) *Karst of non-folded zones* (pre-Alpine shields or post-tectonic depressions), where the weak tectonization of the limestones obviously limited the development of the entire range of forms. Small-sized dolines, shallow gorges and short horizontal caves, particularly bedding caves are found here; but seepage-water caves are, however, highly predominant. No potholes are encountered. Due to the reduced thickness of the limy layer, the normal hydrographic network is not dismembered. Except for abrupt cliffs on the borders of the valleys cut in the limestones, the relief appears rather uniform. A single exception is south Dobrogea (Fig.1,*31*), where the extended horizontal plate of the Sarmatian limestones, partially covered by loess, generates a large zone without surface outflow; however, it does not reveal the structure of a polje, since no flat alluvial bottom appears, neither is it bordered by faults nor flooded by water.

(2) *Karst plateaus* (karstplains), exhibiting the whole rich range of the karst forms, whose development obviously was favoured by the great thickness of limestones. These plateaus are bordered by scarps, at least on one side, and water from below the whole plateau is drained at the foot of the scarp. The surface of the plateaus is undulating, with a dismembered hydrographic network, blind valleys, closed depressions, doline and lapies areas. The slight dip of the limestones contributes to the development of an extended underground horizontal hydrographic network on some levels connected by shafts, or with the surface by potholes. The few typical poljes existing in Rumania are found within such plateaus.

Alternating limestones and impermeable rocks provide step relief, because at their contact, the water accumulated on the impermeable rocks disappears in the limestones, forming a positive step relief. Similarly, at the outflow from limestones, a line of karst springs located in blind valleys causes a negative step.

The overthrust or antithetic faulted structures, although providing a very thick limestone mass, do not generate a holokarst with drainage below the thalwegs in the aquifer nappes within the detrital rocks in the plains bordering karstic massifs.

(3) *The ridge karsts* are characterized by the prevailingly vertical streams of water, favoured by the strongly dipping position of the limestones. Such streams cause erosion of the vertical walls, sharp edges and pillars. The allochthonous valleys which cross the thin limestone ridges, generate savage gorges with fossil caves carved within their walls. The narrowness of these limestones obviously impedes large underground streams, the caves being developed in particular by seepage water. The vertical bedding favours the development of the potholes. The development of small plateaus on the ridge is dependent on the width of the limestone outcrop. Such plateaus exist for instance in the Trascau Mountains, with zones comprising dolines and small clint areas, but in the Piatra Craiului Mountains, where the limestone ridge ends with a crest a few metres wide, such forms are absent.

(4) *The karsts formed on isolated massifs* are but slightly developed. When limestones are found on ridges, they are carved by differential erosion, appearing sharply cut into the landscape. When located at the bottom of the valleys they mostly appear epigenetically cut by the allochthonous valleys. In general, such massifs display restricted underground water streams and, consequently, only small-sized caves due to the seepage water are to be found. The surface of the limestones is weakly modelled, with few dolines or lapies.

If on a regional scale the structural position prevails for the relief, the individual karst forms are due mainly to the lithology and degree of tectonic deformation. From this point of view, the following points may be established.

In comparison with the massive limestones, there are more dolines and lapies in bedded limestone. Although more numerous they are not very deep. As a result, the bedded limestones generate a chaotically disturbed relief, but this is only its surface aspect. The massive limestones display large, sharply outlined forms, yielding to a greater degree of deviation from the normal relief, and karstification appears deeper (e.g., the total disorganization of the hydrographic networks). The above two types of karstification may be illustrated by the forms they have generated, on the one hand the Middle Triassic layered limestones, and on the other hand the Tithonian–Urgonian massive limestones in the Bihor Mountains (Fig. 1,26).

The stratigraphic and tectonic conditions are best reflected within the endokarstic morphology.

Rumania possesses more than a thousand cavities, which might be divided into caves (slopes up to 30°), and potholes (slopes ranging from 70° to 90°). Like in most karst areas all over the world, the intermediate slopes are exceptions, a fact that accounts for hydrodynamic laws, but is due, also, to the position of the limestones, either almost horizontal or strongly dipping. As our data on the influence of stratigraphy and tectonics upon the small-sized underground cavities correspond to those obtained all over the world, this question will not be discussed in the present paper.

The main and decisive part in the endokarstic relief is played by the evolution factor acting upon karst zones. From this point of view, the polycyclic evolution of the Carpathians, with alternate levelling and uplift, appears the most important. Out of the total number of cavities 72% are fossil and only the remaining 28% are active ones; this may be explained by the short time interval elapsed from the last uplift of the mountains. Although there are high karst platforms, the total number of potholes is reduced (20%) in comparison with the number of caves (80%), which also may account for a rhythmic uplift of the massif in which the underground networks always display horizontal drainage. But the most characteristic fact is that caves exceeding 100 m in length exhibit forms of phreatic flow, even if they are afterwards carved by free-surface flows. Thus phreatic patterns—pendants, sponge-works, anastomosis, wall pockets, flutes, ceiling channels and blades—are very frequent.

KARST DEVELOPED ON SALT AND GYPSUM

Salt occurs relatively frequently under the form of salt rocks, in the Miocene formations. These massifs sometimes outcrop, but mostly are overlain by non-karst-forming rocks.

When salt outcrops, typical dissolution forms develop on it, i.e., lapies, dolines, natural bridges or even blind valleys. The most spectacular are the lapies which may take the aspect of flutes or saw teeth.

Where the salt is in the basement rocks, collapse sinks or suffusion phenomena may be produced at the surface by underground solution, thus provoking dolines or karstic collapse sinks and valleys [SENCU, 1965]. This category also comprises forms resulting from the activity of man, e.g., collapsed salt mines with sinkhole ponds and potholes. The growth process is active in such cavities as long as salt comes into contact with fresh water. The process ceases when water is saturated.

Gypsum, which occurs in the same formations as salt is less widespread. Lapies, dolines and small-sized caves are to be found on gypsum or even an underground capture of some surface river, but owing to the rapidity of the dissolution, these forms display only an ephemeral existence.

VOLCANIC ROCK KARST

In the volcanic formations of the east Carpathians, a few caves are known; they are regarded as karst features. As a matter of fact some are lava caverns with opal, while other ones are formed by the underground erosion of tuffs among the layers of cemented agglomerates [NAUM and BUTNARIU, 1967]. The term karst cannot be applied here as it is not a matter of dissolution, but at the most might be employed only for lapies and dolines formed on andesites, by dissolution of feldspars.

HYDROGEOLOGIC CONDITIONS

Streams which cross the limestone district are almost without exception allogene, being the main water source of the karst. However, we must add to this the seepage water and that originating from endokarstic condensation.

The water of the major allogene streams crosses the limestone zones, being only partially lost, while small streams completely disappear—once exposed to the limestones in caves or swallow holes. As in all karstic districts elsewhere, for a great number of places where water becomes lost, only a moderate number of places of rising water could be found, this fact indicating a systematic reorganization of the water network.

Since in the Carpathians the limestones have a more elevated position as compared to the surrounding plains, and do not continue below them, draining occurs at the foot of the limy massifs, the whole amount of accumulated water being discharged through a line of karst springs and vauclusian springs.

We have no proofs available so far about the presence of ground water in the limestones within the Carpathians, but the endokarstic morphology pleads for its existence in the past, so that Davis' and Bretz's two-cycles theory might be envisaged here. However, the Carpathians present a special case which may be assigned to the polycyclic evolution of this mountainous ridge.

The situation is more particular for the pre-Alpine shields. In the limestones of Dobrogea, such ground water was discovered by drillings. The limestones, as much as 500 m below sea level, have karst cavities, although they have never been emerged or affected by a vadose regime. They comprise a fresh-water nappe which drains at a rate of flow amounting to 2 m/sec into a lake near the Black Sea. In some places the water is artesian and is slightly mineralized. The origin of the water is not yet known, but it can be tentatively assumed that part of it is from the Danube, part from seepage, and that the favourable geological structure (dip of beds) determines a flow at a great depth. The most important conclusion we can draw from the above is the fact that karstification may also occur under the water table.

PRACTICAL PROBLEMS AND THEIR APPROACH

The karst areas in the Carpathians are mountainous, either uninhabited or with a low population density. Excepting some villages lacking water, the problem does not represent any difficulties.

It is only in zones exeeding 1,800 m in altitude that—due to strong congelifraction and high aggressiveness of the water—the karst appears bare; elsewhere it is covered, the soil formation being active (mountain rendzina). Therefore, no problems in connection with soil improvement are to be raised. The massive exploitation of forests, however, generates the future threat of soil erosion, without possible replacement, since its formation required other hydrologic conditions which are no longer available if we take into account the block uplift and the deepening of the underground karstic networks.

The periodic flooding of poljes is to be considered only for restricted areas, owing to their rare occurrence and reduced size. Poljes occur in inhabited zones only within the Mehedintzi Plateau (Fig. 1,*19*), where agriculture is practised, and crops are frequently endangered [BLEAHU et al., 1963].

The most important are the karst waters, which—owing to their considerable flow on the border of the depression—constitute a perfect water source. Since sanitary conditions cannot be assured, the reception basins being frequently unknown, these waters are submitted to purifying treatments. Important water catchments are to be found on the southern slopes of the Vulcan Mountains (Fig. 1,*18*), west Banat Mountains (Fig. 1,*22*) and Pădurea Craiului Mountains (Fig. 1,*28*), whereas in Dobrogea water is pumped from the phreatic karst nappes.

Karst thermo-mineral waters represent a particular case, exhibiting a different genetical process involving only karstic circulation, as observed in the Băile Herculane Springs (Fig. 1,*19*), and in the Mangalia (Fig. 1,*31*) and Geoagiu (Fig. 1,*24*) areas. The natural and artificial salt lakes in karstic depressions seem no less important.

However, the most important economic role of the Rumanian karst is tourism. The picturesque scenery, the wild forest covering several karst zones, gorges and particularly the caves (the longest reaching 11,000 m and still developing) are first class attractions which have so far not been opened to tourists.

REFERENCES

BLEAHU, M. and RUSU, T., 1964. The karst of Rumania—brief outlook. *Rev. Roumaine Géol. Géophys. Géograph., Sér. Géograph.*, 12(1):193–202.

BLEAHU, M., DECU, A. and DECU, V., 1963. Das hydrographische System von Zaton-Ponoare. *Rev. Géol. Géograph.*, 7(1):147–156.

DE MARTONNE, E., 1922. Excursions géographiques de l'Institut de Géographie de l'Université de Cluj en 1921. *Résultats Sci.*, 1:1–211.

JEANNEL, R. and RACOVITZA, E.G., 1929. Énumération des grottes visitées. septième série. *Arch. Zool. Exp. Gén.*, 68:293–608.

NAUM, T. and BUTNARIU, E., 1967. Le vulkano-karst de Calimani (Carpathes roumaines). *Ann. Spéléol.*, 22:727–755.

PAPIU, V. C., POPESCU, A. and SERAFIMOVICI, V., 1964. Importanta litogenetica a rocilor carbonatate din epizona masivului Poiana Rusca. *Inst. Geol., Dari Seama Sedintelor*, 49:53–66.

SENCU, V. 1965. Carstul din sarea gema in România. *Studii Geol. Geofiz. Geograph., Ser. Geograph.*, 12(1):47–63.

Chapter 12

Karst of the U.S.S.R.*

I. V. POPOV[1], N. A. GVOZDETSKIY[2], A. G. CHIKISHEV[3] and B. I. KUDELIN[1]

[1] Faculty of Geology, Moscow State University, Moscow (U.S.S.R.)
[2] Faculty of Geography, Moscow State University, Moscow (U.S.S.R.)
[3] Institute of Geography, Academy of Sciences, Moscow (U.S.S.R.)

INTRODUCTION**

Specific studies of karst in the former Russian Empire have been well known since the beginning of the 18th century. In 1702 Semen Remezov described the Kungur Cavern at the direction of Peter I. In the works of V. N. Tatishchev on the Urals, in the course of description of the karst found there its origin is correctly explained as a result of the activity of subsurface and surface waters and of weathering by frost. Scientific studies of karst were carried out in passing during the famous travels of academicians of the Academy of Sciences of St. Petersburg during the second half of the 18th century: I. I. Lepekhin, P. P. Rychkov, V. M. Severgin, and others. The karst phenomena they found (caverns, sinkholes, pits, small disappearing streams, and others) were described and measured in connection with general geographic and geologic description of various localities in Russia. Information on the karst of the Urals can be found in a series of geologic works on mineral deposits published in "*Gornyy Zhurnal*" in various years during the 19th century. Karst was investigated professionally at the end of that century in connection with railway construction (D. L. Ivanov, F. N. Chernyshev, N. I. Karakash, A. A. Shtukenberg, N. P. Barbot de Marni, and others). During efforts to improve agriculture and to combat soil erosion and gullies, light was thrown on the karst of the central regions of the European part of Russia by the investigations of V. V. Dokuchayev, A. S. Kuzmenko, and others. KRUBER's [1915] works on karst in the Crimea gave systematic descriptions of karst forms and of the hydrography of the karst, and analysis of the conditions of formation and development of karst in the mountainous region of the Crimea (Gornyy Krym). A monograph among this material summed up the scientific ideas concerning karst which were held at that time, largely following the concepts of DAVIS [1899] on geographical cycles of development of the topography of the earth's surface, which are based in very small degree on the analysis of concrete geologic conditions.

The approach to the study of karst through historical geology can be noted in the

---

* Manuscript translated from Russian by F. W. Trainer, U.S. Geological Survey, Washington, D.C. (U.S.A.).
** Written by I. V. Popov.

work of VARSONOF'YEVA [1915] on the Ufimsk Plateau, and of NOINSKIY [1913] on karst of the Samarskaya Luka on the Volga, which were published in the late prerevolutionary years.

The rapid and extensive development of the national economy of the U.S.S.R. in the post-revolutionary years required comprehensive, deep, and concrete study of the natural conditions governing the construction of hydropower, irrigation, highway, and other projects. Numerous expeditions carried out geographic and geologic investigations in all parts of the country, and particularly along the valleys of the large rivers—the Volga, Kama, Angara, and others. As a result, in addition to the practical results which had been required, a great deal of information was obtained for interpretation of the regional conditions of the development of karst and for elaboration of the theory of karst processes under diverse regional geologic and zonal geographic conditions. The number of printed materials on karst in the U.S.S.R. has grown enormously. Enumeration of only the large and important works would be difficult. We shall indicate here only some of the principal authors of generalized theoretical monographs: BARKOV [1932,1933]; SOKOLOV [1937,1960]; ZAYTSEV [1940]; DZENS-LITOVSKIY [1940,1946]; YAKUSHOVA [1949]; GVOZDETSKIY [1954]; SOKOLOV [1962]; MAKSIMOVICH [1963]; RODIONOV [1963]; STUPISHIN [1967].

A great deal of work has been devoted to study of the solvent capacity of natural waters of diverse composition, to clarify the conditions under which karst develops. SOKOLOV [1962] summarizes data from an examination of this problem. The work of LAPTEV [1939] and his conclusions about the carbonate capacity of aqueous solutions are very significant for practical appraisal of solution phenomena in the processes of karst formation. The characteristics of solution of limestone–dolomite rocks are considered in the works of RODIONOV [1948,1949,1963].

Concepts which Soviet investigators have established concerning the conditions and forms of development of karst, as a result of its study in the U.S.S.R., can be summarized briefly as follows.

Geographers and geologists in the U.S.S.R. apply the term karst to an aggregate of geologic processes and of phenomena created by them in the earth's crust and on its surface—phenomena produced by solution of the rocks and expressed through the formation of openings, through the destruction and alteration of the structure and condition of the rocks, and through the creation of a particular type of groundwater circulation and regime and of a characteristic regional topography and a characteristic regime of the drainage network [POPOV, 1956]. This definition of karst was adopted by the Conference for the Study of Karst held in Moscow in 1956.

The fundamental and general conditions controlling the development of karst were formulated by F. P. Savarenskiy [here quoted according to GORDEYEV, 1948]: "The presence of soluble rocks, accessible to the infiltration of water along systems of fractures, is necessary for the development of karst ... Not only the presence of water in the fractured rocks, but also conditions which provide for the movement of this water, are necessary for the development of karst ..."

The following may be considered general laws of the development of karst: (*1*) decrease of karstification with depth; (*2*) intensification of the karst process near valleys, with correspondingly greater karstification near valleys than near divides in massifs; (*3*) the development of karst is very closely related to climate and, for karst in carbonate rocks, to vegetation (which is a product of climate); and (*4*) the nature of karst development is related to geology and geologic history [SOKOLOV, 1951, 1962].

Dependence on the character of groundwater drainage may be considered [LYKOSHIN, 1968] a fifth law of karst development. Thus, drainage may occur under the influence of: (*1*) nearby erosional cuts; (*2*) distant erosional cuts or other centers of groundwater discharge; and (*3*) zones of tectonic disturbance. The last of these may give rise to downward movement of groundwater in some instances and to upward movement in others. These types of karst may be combined with one another to a certain extent.

The distribution of karst is related most closely to the characteristics of fracturing of the rocks and of movement of water through them. Several hydrodynamic zones can be distinguished according to the conditions of circulation of water in a karst massif: a surficial zone of descending flow, with a subzone of suspended watercourses in local confining rocks or in rocks less affected by karst processes, and a transitional zone. Other zones can be distinguished below the transitional zone: a horizontal zone of free circulation, a confined zone, and a zone beneath a river channel toward the master stream [MAKSIMOVICH, 1962].

Regionalization according to the character of the karst should be carried out through consideration of the following criteria: (*1*) petrographic character of the rocks in the geologic section; (*2*) tectonic structure; (*3*) geomorphology; and (*4*) hydrologic, hydrogeologic and climatic conditions.

A basic prerequisite for the development of karst is the presence of rocks susceptible to solution by water. Such rocks occur in the carbonate formations (limestones, dolomitized limestones, chalk) and in the lagoonal or sulfate-dolomite formations (dolomites, gypsum, anhydrite, and salt rocks).

Another equally necessary condition is the circulation in these rocks of water which attacks them. For this to be possible the rocks must somehow be brought to the earth's surface or to a position near it, by tectonic processes, and must be exposed by denudation or must be connected with aquifers. Because rocks which are susceptible to solution by water are in themselves impermeable or very slightly permeable (as, for example, chalk), the circulation of water through them is possible only where they have been disrupted (as by tectonic, lithogenetic, or exogenetic fractures; by fractures along stratification; or by other fractures), or where they have lithogenetic porosity (as, for example, some dolomites).

From the foregoing it may be concluded that study of the history of karst formation during development of the geology and geomorphology of a region (including consideration of palaeoclimatological conditions) is of exceptional value in appraisal of karst phenomena. Failure to consider this conclusion adequately is a principal cause of differences of opinion expressed in well-known discussions of karst hydrography among west European investigators [SAVARENSKIY, 1933]. The decisive influence of the tectonic development of the earth's crust begins at the time of deposition of formations composed of soluble rocks, inasmuch as each formation originates under a definite tectonic regime. The disposition in the earth's crust and at its surface of rocks subject to karst processes, in both ancient and recent geologic epochs, is also important. The most recent tectonic movements, which control the development of local geomorphology, are subject in greater or lesser degree to the structural-tectonic plan of the earth's crust formed during earlier periods. Tectonics and geomorphology also determine hydrodynamic conditions in the earth's crust.

This manifold and deep influence of tectonics on the conditions under which karst develops forces us to take the structural-tectonic criterion as the principal one in regionalization (in the case of the U.S.S.R.) in accordance with the conditions of development of the karst and with its character. For this purpose the territory of the U.S.S.R. may be divided into the following regions of first rank:

(*A*) The Russian or east European Platform (with its pre-Caledonian cover).

(*B*) The Urals orogenic belt of the Hercynian geosyncline.

(*C*) The Alpine folded zone (in the southern framework of the Russian Platform).

(*D*) The Alpine folded zone of the Kopet-Dag and the Pamirs.

(*E*) The Turansk Platform.

(*F*) The orogenic belt of the eastern part of central Asia on the folded Palaeozoic basement.

(*G*) The Palaeozoic folded country of Kazakhstan.

(*H*) The south Siberia folded zone with a Palaeozoic folded basement.

(*I*) The Baikal orogenic belt (in the framework of the Siberian Platform).

(*J*) The Siberian Platform (with its pre-Caledonian cover).

(*K*) The Zeysk–Bureinsk Platform (with its Palaeozoic cover).

(*L*) The Mesozoic folded zone of the Far East.

(*M*) The Cenozoic folded zone of the Far East.

(*N*) The Caledonian–Hercynian folded zone of the Taymyr.

Structural-tectonic regions in which karst is unknown or has not been studied have not been placed in the summary. Karst undoubtedly exists in regions (*J*), (*K*), (*L*) and (*M*), and there are several single summaries of it, but general studies of the karst of these regions have not been made. Data for these regions are therefore not included in the description of karst in the U.S.S.R.

THE RUSSIAN PLATFORM[1]

The Russian Platform is the eastern part of the east European Platform. Karst is described for that part of this region which extends from the boundary of the U.S.S.R..

---

[1] Written by I.V. Popov.

on the west, to and including the Cisuralian (Predural'skiy) Trough on the east. Several structural-tectonic stages are distinguished: the Precambrian, made up of metamorphic rocks; and the Caledonian, Hercynian and Alpine, which form the sedimentary cover of the platform. Rocks of differing degrees of solubility occur at various levels and over diverse areas in the sedimentary cover. The conditions of karst development are determined by the petrographic character of the rocks and by their position and attitude, which in turn determine hydrogeologic factors in development of the karst. Regularities in the distribution of karst in the Russian Platform are determined by its tectonic pattern, and principally by this pattern in the sedimentary cover. A brief description of the karst of the Russian Platform, according to its structure, is given below.

Karst of the crystalline basement of the Russian Platform

Limestones, dolomites, and marbles are found among the Proterozoic metamorphic rocks in small areas of the Baltic Shield. Limestones are well known in the region of Lake Yelm and on the coast of Oleniy Island. Dense fractured dolomites are exposed near the Arctic Circle, at Medvezh'ya Gora and at other places. Pure and dolomitized limestones are developed in the vicinity of the city of Sortavala, where their thickness attains 300–400 m. A belt of marbles extends along the coast of Povenets Bay. Information concerning the karst has been assembled along the way during the work of various assignments. Cavities and caverns in fractured limestones in the curved part of a fold are exposed on Oleniy Island. One cavern, 20–40 m long and 2 m high, is oriented along the trends of the principal tectonic dislocations. Cavities 0.5–1.5 m high have been revealed at various depths by boreholes.

Near the city of Petrozavodsk karst occurs in marbleized pure dolomites as pores and caverns whose dimensions range from a fraction of a millimeter to 5–6 cm. Many of the caverns are filled with secondary dolomite, calcite, and quartz, but there are also unfilled openings.

There is no information on the age of the karst of the Baltic Shield. Probably it is ancient and was formed during Palaeozoic and Mesozoic time.

In rocks of the Ukraine crystalline shield karst has been described in the Krivorozhsk and Kremenchugsk regions, in dolomites and carbonatized quartzites. Large cavities and the destruction of carbonatized rocks have been discovered at depths of 200–600 m in ancient valleys of the Zheltaya, Saksagana, and Ingul rivers. For example, a meridional opening in the Zheltaya River Valley is more than 1,500 m long, 19–25 m wide, and about 350 m deep. It is full of sand–clay rocks and water. According to SHCHERBAKOV [1961] its formation is related to the pre-Palaeogene valley of the Zheltaya River.

Carbonate and carbonatized rocks have been revealed by many wells in the region of the Voronezh crystalline massif, but no information is available concerning karst in the Precambrian rocks.

Karst in rocks of the southern buried slope of the Baltic Shield

This region extends westward from a point south of the Ladoga Lake to the shore of the Bay of Finland, and terminates in the Estonian Archipelago. It is an uplifted and dissected plain formed of Ordovician and Silurian carbonate rocks. Different parts of the plain bear different names: in the Leningrad oblast'[1], the Silurian or Ordovician plateau with the Izhora upland; in the E.S.S.R., the Estonian Plateau with the Pandivera upland. Its lowest parts are in the Saarema and Mukhu islands and in the lower course of the Volkhov River.

The carbonate rock formations in the Caledonian structural stage are limestones, many of which have been dolomitized; some are clayey and some contain interbeds of oil shale. The strata dip gently (3–4 m/km) to the south. The total thickness of the carbonate rock formations is as great as 100 m in the north, and several hundred metres in the south. Intensive uplifts since Quaternary time have had substantial influence on the conditions of karst development. Tectonic disruption of the rocks by fractures of two trends, with azimuths 20–60° NE and 300°–320° NW, has been no less substantial a factor. The fractures are steeply dipping, nearly vertical.

The surface and subsurface forms of the karst are expressed most sharply in the central and western parts of the region. Numerous sinkholes are grouped along the trends of the principal fractures. Larger sinkholes have formed in low places, where the Quaternary cover is thinner (less than 5–6 m). The sinkholes are from 2 to 50 m in diameter, and 1–2 to 10–15 m deep. Ditch-like surface forms of the karst, which are widely developed, have been given the local name "kurisu" in Estonia. They are the principal routes for the absorption of surface water and for its transfer into the groundwater body. Subsurface karst openings which are accessible for observation are longer than 150–200 m. Horizontal karst channels 0.5–2 m wide and as high as 2 m are observed at depths from 1–6 to 30 m. The karst of this region has great significance as an accumulator of groundwater and as a factor which complicates the exploitation of oil-shale deposits.

Karst of the Latvian saddle, and karst of the Belorusskaya Arch

The karst of this region has developed in sulphate-dolomite formations of the Early Palaeozoic structural stage (Upper Devonian) and of the Late Palaeozoic Stage (Upper Permian). Thus the prerequisites are present here for the development of both sulfate and carbonate (dolomite) karst.

Gypsum karst has formed in the northern part of the Lithuanian S.S.R. in Shelonskikh strata of the Upper Devonian. There are two patches of gypsiferous rocks here—dolomites, marls, and clays, with bands of gypsum as thick as 13–16 m. On divides the soluble rocks are covered by Quaternary deposits which contain moraine

---

[1] The Russian word "oblast'" refers to an administrative subdivision of territory.

at one or two horizons. Abundant rainfall, intensive local uplift, and relatively deep erosion have promoted intensive development of gypsum karst. For example, 4,000 surficial karst features have been counted in an area of 160 km^2 in northern Lithuania. They are sinkholes (formed by subsidence and by infiltration), depressions, and karst lakes. The degree of mineralization of the water from several springs—2,000 mg/l and more—is evidence of the intensity of solution of the gypsum. The waters of the local spas (Baldone, Kemeri, Druskeniki, and others) are related to the karst.

Carbonate (dolomite) karst appears in two basic forms: (*1*) small cavernous openings, of regional distribution, in the Upper Devonian dolomites; and (*2*) large karst forms (sinkholes and pits). Observations in the region of the Vitebsk rapids on the Zapadnaya Dvina River and elsewhere show that the regional development of caverns bears no relation in character or distribution to the local stream network. It is related to distant base levels of drainage (Rizhskiy Bay, in the Baltic Sea, and earlier basins there).

The large karst forms are related to the present and ancestral stream networks. Pre-Quaternary and Early Quaternary valleys, incised to depths of 60–80 m, were filled with Quaternary deposits. The modern stream network has in part inherited former valleys having the same orientation. The old overdeepened valleys were and are of substantial importance in development of the karst.

Dolomites in the valley of the Zapadnaya Dvina River have been modified by karst processes, with the formation of caverns, dolomitic meal, and pits. Below the city of Plyavinyas, pits in Upper Devonian rocks are covered by Quaternary deposits which in some places have been disturbed and in others have not. Because of this burial of the pits, and because of their location in the ancient, pre-Quaternary valley of the Zapadnaya Dvina, the pits are considered to be of pre-Quaternary age. The ancient valley in its time exposed an aquifer in Podsnetogorsk sands, which led to undermining, collapse of the sands which covered the dolomite, and the formation of karst in the dolomite. Thus these pits have a complex origin by solution and undermining [SKUODIS, 1959].

In a reach of 50 km along the former valley, 250 such pits have been counted.

Dolomite karst, in the form of extensive cavern development and of collapse of the Upper Devonian rocks, is also found in the upper part of the Velikaya River Basin, in the region of the city of Pskov, and elsewhere.

Karst of the southern limb of the Moscow tectonic basin

Another place where karst has developed in Upper Devonian rocks is on the south limb of the Moscow Depression. Here the Devonian rocks are limestones and dolomites and interbedded marls and clays, with gypsum beds and lenses of considerable thickness at their base. Dipping northward, the Devonian rocks are succeeded by Carboniferous deposits assigned to the carbonate formations of the Hercynian structural stage. The Devonian rocks have been strongly karstified where they outcrop. Groundwater has played a principal role in formation of the karst, discharging into the modern and ancient valleys of the Oka, Upa, and other rivers along systems of tectonic fractures. The zone of active water exchange (within 5–8 km of the river), with differences of altitude of 20–25 m in the belt along the streams, is the zone of greatest development of the modern karst. The ancient karst developed by stages: (a) in Famennian and Frasnian time; (b) in pre-Variscan time; (c) in pre-Upper Jurassic time, beginning at the end of the Carboniferous and including the greater part of the Jurassic; (d) during the Cretaceous and Palaeogene; and (e) during the Early Quaternary and continuing to the present. Manifestation of the karst is very rare at places where it is covered by Jurassic clay or by moraine more than 5–7 m thick. Pumping for water supply is a significant factor in stimulating present-day karst processes in the southern limb of the Moscow Depression.

Modern karst has developed at a number of places where Quaternary and Recent erosion has exposed karst which is older than Upper Jurassic, in Palaeozoic rocks. This considerably complicates the course and forms of development of the karst, which in turn considerably complicates the exploitation of some coal deposits (through instability of roofs, eruptions of quicksand, etc.).

Karst of the western and northwestern limb of the Moscow tectonic basin

This karst region extends as a wide zone from the headwaters of the Dnieper and the Zapadnaya Dvina to the White Sea. The essential features of the topography were established in pre-Quaternary time, and its final formation was influenced by Quaternary glaciers, glacial streams, and karst erosional processes. The region is dissected by deep (70–200 m) pre-Quaternary valleys, many of which were inherited by the present stream system. The conditions favourable for development of karst are a thin cover of Quaternary deposits resting on soluble Permian and Carboniferous rocks; deep incision by the ancient streams; and several features of the modern relief, especially a high ledge of Carboniferous limestones (the "karbonovyy glint").

Karst has formed in Lower Carboniferous limestones in the western part of this region, in the upper reaches of the Zapadnaya Dvina and Dnieper rivers. The rocks are cavernous; karst has been discovered at depths as great as 60 m. In the Msta River Basin karst is widely developed in the rocks of the Carboniferous ledge and the ancient plateau (Lower Carboniferous limestones and dolomites). The porosity of the rocks (openings with dimensions of 10–15 cm) is very irregular, being greater in the dolomites and lesser in the limestones. Subsurface and surface karst forms are closely related to the nature of the fracturing of the rocks. Cavities and small caves in dolomites in the bank of the Msta River are related to horizontal bedding fractures; small chains of sinkholes extend parallel to the tectonic fractures, which trend northwest, east–west and north–south; karst gullies are oriented along the courses of vertical and steeply dipping fractures of northeast strike. Considerable cavernous porosity and collapse of the rocks, and numerous karst springs, coincide with the rims of the ancient valleys.

The surface forms are sinkholes, small troughs, karst gullies, dry valleys, karst lakes, and streams which go underground. Distribution of the sinkholes is irregular—from 1 to 5/km². Subsurface karst features revealed by boreholes, mine workings, and quarries are conduits of various lengths; dolomitic meal; and buried sinkholes filled with varicolored clays of Carboniferous age, or covered by Variscan clays.

There are large sinkholes and troughs within 50 km west of the city of Borovichi, in the Uglovka region. Caverns exposed in a quarry here have been studied to a depth of 50 m (A. G. Chikishev, personal observation, 1968).

Karst in the Tikhvinsk rayon[1], formed in Carboniferous limestones overlain by thin Quaternary deposits, consists of caverns, karst lakes, and sinkholes. In part these forms are of pre-glacial origin, as is shown by a sinkhole 45 m deep, filled with glacial moraine, which has been revealed by a well.

Karst lakes are abundant on the plateau south of Lake Onezhskoye, in the vicinity of Lake Beloye, in rocks of Middle and Upper Carboniferous age. These karst lakes disappear periodically. There is also a group of karst lakes (Shim, Kusht, and others), situated between Lake Onezhskoye and Lake Beloye, which are joined by subsurface karst conduits. Sinkholes with diameters of 100 m, and small karst lakes, have been noted near the city of Kargopol' and near Lake Lacha. There have been instances of karst collapse here. Five to seven sinkholes per hectare have formed in the karst area.

The karst near Lake Onezhskoye has developed in several stages; the oldest occurred during the time between the Early and Middle Carboniferous. The rocks of these two ages are separated by an old weathered zone.

In the interfluve between the Onega and the Severnaya Dvina karst has formed in Upper Carboniferous limestones and dolomites, in gypsum-dolomite rocks of Lower Permian age, and in limestones of the Kazanian Stage of the Upper Permian. Sinkholes more than 15 m deep are widespread, singly and in groups. Some karst lakes disappear periodically. Grist mills have been built in very large karst springs. Water-bearing limestones about 40 m thick, which have been extensively karstified, hinder exploitation of economic mineral deposits.

Karst phenomena are well known in Middle Carboniferous rocks on the Onezhsk Peninsula. Dolomites are nearly everywhere cavernous, porous, and of cellular structure. Cavities attain 20–40 cm, and wells have revealed openings as high as 3–4 m. Sinkholes are commonly dry; near some lakes they swallow up part of the lake water.

On the Pinega River and the Kuloy River karst has developed in Middle Carboniferous limestones and in gypsum-dolomite rocks of Lower Permian age. A cavern in the watershed of the Pinega and the Kuloy is named Kulogorsk [A. G. Chikishev, personal communication, 1965]. Sinkholes 10 m deep and 20–30 m in diameter are found in the upper course of the Kuloy River, in the lower course of the Mezen' River, and on the Zimniy coast of the White Sea, in areas underlain by Lower Permian

---

[1] The Russian word "rayon" refers to an administrative subdivision of territory.

gypsum-dolomite rocks. The sinkholes are so closely spaced that it is difficult to walk between them. The gypsum occurs in complex sequences, not interbedded with dolomite, and the karstification has been more intense and has gone to greater depth in this rock.

The northern part of this region has been uplifted and subjected to intense denudation, and carbonate and sulphate karst developed here until the onset of the glacial periods.

Karst of the southeastern part of the Moscow tectonic basin

This part of the basin is complicated tectonically by the Oksko–Tsninsk Arch, in the west, and by the Alatyrsk-Gor'kovsk uplift in the east. The Oksko–Tsninsk Arch is made up of Carboniferous limestones and dolomites, bordered by Permian gypsum-dolomite rocks which have been intensely karstified both at depth and at the surface. Boreholes have revealed subsurface openings everywhere, partly filled with calcite and, in places, with breccias and dolomite meal. Mealy dolomites and carbonate meal are widespread in the zone through which groundwater levels fluctuate. They have been traced over enormous areas (120×25–30 km), with layers 1–3 m thick and locally as thick as 10 m. These deposits originated during formation of the pre-Jurassic weathering crust, which began to form at the end of Early Permian time; in the Carboniferous rocks it may have begun even sooner. From the Late Cretaceous until the end of the Neogene karst formed beneath a cover of thick terrigenous deposits of the Tartarian stage of the Upper Permian. The Dneprovsk (Riss) glacier levelled the pre-Quaternary surficial karst features, and they were buried beneath a moraine and beneath fluvioglacial deposits. At present karst is developing along the modern stream valleys and in valleys buried beneath water-bearing fluvioglacial sands. At these places karst cave-ins are associated with the removal of large quantities of sand into subsurface karst openings; as a result, very large sinkholes form, and they can cause great damage to roads and in populated places.

Karst is extensively developed in gypsum-anhydrite rocks in the basins of tributaries of the Oka and Volga—the Tesha, the Serezh, and the P'yana. There are rather large caverns (the Balakhoninsk, and others).

Karst of the Timan Ridge

The Timan Ridge is a large and complex uplift whose formation began at the end of the Riphean; the last phases of the orogenesis are dated at the end of the Permian Period. Crystalline rocks outcrop in arched parts of the structure; the Palaeozoic rocks are crumpled into linear and domal folds. Karst phenomena are restricted to the carbonate, sulphate-carbonate, and sulphate rocks of Devonian, Carboniferous, and Permian age. Modern karst is widespread and locally is sharply expressed where gypsum and gypsum-carbonate rocks are at shallow depth. The density of sinkholes at some places is more than 10/km². their diameters are commonly 5–10 m, and at some places 15–60 m, and their depths 8–60 m. A considerable part of the area is covered by fields of sinkholes, karst lakes, and blind gullies.

Karst has been revealed in the modern and ancient valley of the Vychegda River by drilling beneath the modern and old alluvium. The presence of fresh cave-ins shows the high activity of the modern karst processes. The most widespread karst is in the Vym' River Basin, in a sequence of Lower Permian gypsum and carbonate rocks.

In the central Timan Ridge karst is well known in limestones and dolomites of the Middle and Upper Carboniferous and in marbleized limestones and dolomites of Proterozoic–Cambrian age. There are caverns in the Carboniferous limestones and dolomites.

Very little is known about the karst of the northern Timan Ridge. Karst has been mentioned with reference to the Middle Carboniferous and Lower Permian limestones. Caverns occur in the Silurian limestones.

Karst of the Volga-Ural Arch and the Vyatsk swell

Karst in the region of the Volga-Ural'sk Arch is widespread in Permian rocks (sulphate-dolomite rocks of the Zechstein type, and gypsum-anhydrite and salt-bearing formations) along the Vyatka River in the Prikamsk Povolzh'ye, along the Kama River and its tributaries, and in the Samarskaya Luka on the Volga.

Complex fluctuations of base level relative to the position of the soluble Palaeozoic rocks caused the formation of several cycles of karst in these rocks. Such cycles formed at the boundary of the Early and Late Permian, at the beginning of the Early Jurassic, and in the Neogene. Several additional small fluctuations of base level occurred during Quaternary time. Deep incision of the ancient streams favoured karst development. The Neogene valleys of the Kama and its tributaries, and of the ancestral Volga (ancestral Kama), were 150 m and more, deeper than those of the present. The influence of the karst preserved along the old valleys since that time is shown now by the formation of large and deep cave-ins. For example, a case is known in which a tractor working in a field sank deeply into the ground. This cave-in originated at a depth of 90 m in the ancient buried valley of the ancestral Zay River.

Construction of reservoirs on the Volga and Kama has changed hydrodynamic conditions in their banks, which are made up of rocks already karstified, and the karst processes are dying out. However, it is possible that they will become active under certain conditions, through undermining related to annual water-level fluctuations in the reservoirs. Several such phenomena have been observed on the Kama River.

Karst has developed extensively in the crest of the Ufimsk Arch in limestones and dolomites of the Artinskian and Kungurian stages of the Lower Permian. In the western limb of the Ufimsk Arch karst is well developed in gypsum-anhydrite and limestone-dolomite rocks of these stages along the Irena River. There are numerous sinkholes, blind valleys, and dry ravines. Near the Birsk River they are as much as 100–150 m in diameter and 40–50 deep. They were formed by subsidence.

Strata of the Ufimsk and Kungurian stages are nearly horizontal in the Ufa River region. The valley walls of the Belaya and Ufa rivers are more karstified; 30–50 sinkholes/km² occur in individual areas. Most of them are 30–50 m in diameter and as deep as 5 m. One sinkhole in the city of Ufa is about 400 m wide and 60 m deep. One to three sinkholes form every year on hillsides in the Ufa River Valley, many of them by sudden subsidence.

There is a series of large caverns in the part of this region near the Urals. The well-known ice cave of Kungur is located near the city of Kungur, on the right bank of the Sylva River. There are 58 grottoes with interconnecting passages in the part of the cave which has been studied. The total extent of the cave is 4.6 km. The cave contains 36 lakes; their water levels fluctuate with that of the Sylva River. Various forms of ice have accumulated in the cave: crystals, stalagmites, stalactites, columns, crusts of ice on the walls and ceiling, etc. The forward part of the cave is cold and ice is preserved in it the year around; the deeper part is warm, with a constant temperature of 5°. Recent investigations have shown that the cave has four levels which are in diverse stages of development or have even reached old age.

Karst of the southwestern borderland of the Russian Platform and the outer zone of the Ciscarpathian Trough

A belt of gypsum-anhydrite rocks, in the outer zone of the Ciscarpathian (Predkarpatskiy) Trough within the U.S.S.R., extends southwest from the national boundary at the city of Nemirov, through the headwaters regions of the Dnestr and Prut rivers, to the city of Chernovtsy. The sequence of sulphate rocks is underlain by Lower Tevtonsk beds (sandstones, limestones, marls) and overlapped by Upper Tortonian limestones. The thickness of these sulphate rocks increases southwestward from 10 to 50 m. Abundant springs, karst cavities, and collapse sinkholes are common near bedding surfaces in the rocks. Deposits of native sulphur in the northern part of the belt of sulphate rocks are related to the karst history. The total result of karst activity, expressed as the volume of sinkholes and other cavities, amounts to 23% of the total volume of gypsum, while that of the subsurface forms is 8%. Old sinkholes are commonly filled with clay.

The Upper Tortonian limestones have been strongly karstified and as a result they are very porous (porosity is as great as 36–37%). In the southwestern borderland of the Russian Platform karst has developed in Upper Cretaceous limestones (chalk and chalky marls), and in Upper Tortonian gypsum–anhydrite rocks.

In northern Podol'sk the Tortonian gypsum is as thick as 30 m. Sinkholes are from 2 to 15 m in diameter and 3–5 m deep. Karst occurs as sinkholes, shafts, and caverns in the slopes of the Dnestr River Valley, in two outcrop belts of reef limestones (Tortonian to the west and Middle Sarmatian to the east), and in Tortonian limestones as much as 250 m thick. A wide belt of Pontian limestone crag has been affirmed in the right-bank region of the Reyt River (Faleshty–Orggeyev), from the dropping down of the drilling tools in boreholes and from the loss of drilling mud. Further evidence of the karst has not been observed.

Several large caverns are known to occur in gypsum in separate Tortonian limestones in Podol'sk. They are the following: Ozernaya (more than 21.6 km long), Kristal'naya and Optimisticheskaya (each about 19 km), Mlynka (about 15 km), and Verteba (about 9 km). In the Dnestr Valley karst occurs in Silurian dolomites and organic limestones [TATARINOV, 1965; A. G. Chikishev, personal communication, 1965].

Chalk karst in rocks of the Russian Platform

Deposits of chalk and chalky marl are widely distributed west and north of the Dnieper–Donets Basin, in the Saratov–Ul'yanovsk Trough, and in the continuation of the latter southward from Saratov. The development of karst processes in these rocks has several distinctive features related to their comparatively friable structure and to the weak cementation of their very minute particles of calcite and shells of microorganisms. Water penetrating the pores in the rock dissolves the carbonate cement which binds the particles, releases them, carries part of them away through fractures, and dissolves part of them. Thus the formation of openings in the chalk is a result of solution (that is, of the karst process), and of washing out and mechanical transport through fractures. Softened chalk is formed in situ and redeposited in fractures and cavities. In some cases it is somewhat fluid, it flows, and it can be set in motion by pumping. Sinkholes appear at the surface. Subsurface channels 20–25 cm wide, visible in the walls of quarries, have been widened by the solution of fractures and have had iron deposited on their walls. A quarry exposed a cavern 2 m high in the Starooskol'sk Chalk. The greatest development of karst in the chalk occurs near rivers, where the chalk contains more water than near divides. Pre-Palaeogene, Neogene, and Quaternary chalk karst can be distinguished. Karst processes undoubtedly are active in the chalk at the present time.

THE ALPINE FOLDED ZONE IN THE SOUTHERN AND SOUTHWESTERN FRAMEWORK OF THE RUSSIAN PLATFORM[1]

The Crimea

One of the classic karst regions of the Soviet Union is the mountainous part of the Crimea (Gornyy Krym), which belongs to the zone of Alpine folding and represents the individual Crimean anticlinorium.

---

[1] Written by N.A. Gvozdetskiy.

The karst of the mountainous part of the Crimea, principally in Upper Jurassic limestones, serves as a clear example of denuded karst of the Mediterranean type. At many places the limestone is devoid of soil cover and forms karren fields. Vast, enclosed, undrained regions contain every possible type of enclosed karst forms of diverse ages: basins, among them features which from their hydrographic character remind one of poljes, swallow holes, and sinkholes which have at their bottoms shafts and fractures. The basins in the loftiest parts of the Yayla—a monoclinal and folding limestone plateau in the Crimea, dissected by karst valleys—were formed during the Glacial Epoch through the solution of limestones by meltwater from snowfields and snow patches. Meltwater from snow is still a very important agent in karst formation [GVOZDETSKIY, 1954; DUBLYANSKIY, 1963].

There are many karst shafts in the Yayla, with depths as great as 261 m (the Molodezhnaya shaft on the Karabi-Yayla), and caverns with stalactites. The system of Krasniye caverns, in the Dolgorukovskaya Yayla, has been investigated over an extent of 12.5 km.

The karst of the mountainous part of the Crimea has been described in a monograph by KRUBER [1915], in a series of later papers [CHURINOV, 1961; IVANOV and IL'INA, 1965; DUBLYANSKIY, 1968] and also in sections in generalized monographs on karst [ZAYTSEV, 1940; GVOZDETSKIY, 1954; SOKOLOV, 1962; RODIONOV, 1963]. In the Yayla, karst water is used for the water supply of the health resorts of the southern coast of the Crimea.

Cuesta foothills are present on the north side of the Crimean Yayla. Karst phenomena have been noted in Palaeogene and Neogene limestones here.

The plain of the steppe part of the Crimea, which occupies the northern and central parts of the Crimean Peninsula, is part of the epi-Hercynian Skifsk Platform. Subsurface and surface karst forms have developed in Palaeogene and Neogene limestones which have been scarcely or slightly dislocated.

The Caucasus

Karst in the Caucasus occurs principally in the mountain system of the Greater Caucasus (Bol'shoy Kavkaz). It is considerably less developed in the several regions of the Ciscaucasus (Predkavkaz'ye), the Lesser Caucasus (Malyy Kavkaz), and the Armyansk Highland.

In the distribution of karst regions, which coincide for the most part with the limestones of the range, ridge, and plateau that border the glaciated upland, the Greater Caucasus closely resembles the Alps [GVOZDETSKIY, 1958]. GVOZDETSKIY [1952a, 1954; see also RODIONOV, 1963] has distinguished eleven karst regions in a belt of this borderland. This can now be supplemented on the basis of new work, particularly for the western, eastern, and southeastern parts of the Greater Caucasus.

The karst of the Caucasus has developed in rocks (travertines and calcareous conglomerates) of very diverse geologic age—from Palaeozoic to Quaternary. However, karst is more widespread in Upper Jurassic and Cretaceous rocks—limestones and some dolomites and gypsum. Diversity of tectonic structures (the north Caucasus monocline; systems of anticlines and synclines, some narrow, some wider) creates diverse conditions for the circulation of groundwater and is expressed in the nature of the surface and subsurface karst forms. The intensive tectonic fracturing which is characteristic of the karstified rocks in the Caucasus is exceptionally important in the morphology of the karst.

Karst formation had already begun, at a number of places, during Cretaceous and Early Palaeogene time. For example, in the western part of the south slope of the Greater Caucasus Eocene clays lie transgressively on Jurassic limestones, filling deep basins and sinkholes in them. These are buried or fossil karst forms. The karst has undergone a long history of formation which is expressed in the present morphology. It began to develop at the end of the Pliocene, when a mature relief had formed in the Caucasus, with wide valleys and extensive planation of the surface. Early Quaternary uplift led to the incision of a dense network of river valleys and to the relative subsidence of the groundwater level. Development of the karst continued against this background.

The onset of glaciation partly preserved the limestone surfaces with a cover of firn and of glaciers in the highlands. Below, on the other hand, where intensive thawing of ice, firn, and snow occurred, the karst processes continued in action. In some regions sinkholes were filled with glacial drift. Development of the karst was intensified during the waning of glaciation and the melting of firn and ice in most parts of the leading ranges. A considerable part of the surficial karst, especially in areas freed of ice and firn, was formed during Postglacial time.

Thus the age of the karst in the Causasus is quite varied. In many regions karst is being formed today; the karst processes are particularly active in the relatively warm and humid regions of the western half of the south slope of the Greater Caucasus, where abundance of water (the solvent) and the activity of biologic agents increase their aggressiveness. In the eastern regions, which are less humid, development of the karst is limited by lack of moisture, but even so modern karst is developing there. Its form clearly reflects the high-altitude landscape of the Caucasus. For example, variants of the denuded karst in the high mountains, in mountains of intermediate height, and in the low mountains clearly differ in their morphology.

Karst is locally present on the north slope of the westernmost section of the Greater Caucasus (the Black Sea Caucasus). However, moving from there eastward, we meet typical karst regions—the Chernogor'ye Plateau in the Pshekha River Basin and the wide plateau of the Fisht-Oshten-Lagonaki Massif, bounded to the east by the headwaters regions of the Belaya River.

Karst in the Fisht-Oshten-Lagonaki Massif has formed in Upper Jurassic dolomitized limestones. In the southern part of the massif the limestone beds have been folded, while to the north they dip gently northwest. A high-mountain variant of denuded karst is typically expressed in the south, in the most uplifted part of the massif, with karst shafts in which the snow has not melted, even by the end of the summer, and with karren fields characterized by deep dissection [GVOZDETSKIY, 1965]. There are karst basins in the massif, the largest of them occupying nearly the whole of the Lagonaki Plateau. Many sinkholes are strewn everywhere; in the southern part of the massif they have been clogged with glacial drift, a fact which indicates their pre-Glacial age. However, many of the sinkholes were formed during Postglacial time.

The karst belt of the Peredovyy Range, where Palaeozoic (Devonian) and Triassic limestones are subject to leaching, extends southeast from the Fisht-Oshten-Lagonaki Massif. There are many sinkholes (formed by surface leaching and by subsidence), basins, and shafts; a few karren, and many springs. The karst has been "superimposed" on the old glacial landscape, and most of the surficial karst forms are Postglacial.

A belt of karstified rocks extends from the Belaya River gorge, on the east-southeast, along the entire north slope of the Greater Caucasus. The karst regions of the north Caucasus cuesta, the transitional region of cuesta structure and folds, and the limestone terrain of Dagestan are part of this belt.

The Skalistyy Range, the southern and highest cuesta in the western region of the north Caucasus cuestas, is underlain by Upper Jurassic limestone strata which dip $5°-12°$ northeast and are broken by tectonic fractures, with which the trends of the canyons coincide. Hydrogeologically, the monocline is an artesian system with regions of recharge in surface parts of the cuesta, which form a plateau sloping gently north-northeast. The water flows north-northeast from here, down the dip of the strata. The Upper Jurassic limestones which have been karstified are covered by a varicolored Tithonian sequence with gypsum and anhydrite that have been extensively dissolved (it forms an elongated depression north of the crest of the Skalistyy Range). The Jurassic limestones extend beneath the strongly karstified sequence of Lower Cretaceous (Valangian) limestones which make up the north slope of the range.

The various karst forms—sinkholes and pits, grouped in bunches and chains; complex sinkholes, shafts, niches, and caverns; in places, karren; residual karst forms; gullies and blind ravines, formed by erosion due to karst and other processes—have diverse morphology depending on geologic conditions. The karst which is on the structural surface of the Valangian limestone sequence is covered by thin deposits of sand and clay and by unconsolidated Hauterivian deposits, or which in places is denuded, is represented by small karren fields [GVOZDETSKIY, 1965]. The southwestern ridges of the Skalistyy Range, which were more uplifted in the east (Mount Gud, Bermamyt), were subject to glaciation. The karst forms are of Postglacial age. Thin-bedded Upper Cretaceous limestones which have locally been karstified form the ridge of the next cuesta to the north (the Borgustan Range, Dzhinal, and others).

The eastern region of the north Caucasus cuestas, the immediate continuation of the foregoing region, extends from Kich-Malka to the Fiagdon River. The Upper Jurassic and, in particular, the Lower Cretaceous (Valangian) limestones have been karstified. As in the preceding region, the limestones have monoclinal structure; to the east, however, the angle of dip of the strata increases (because of which the Skalistyy Range also takes on a more "bristling" character), and against the background of monoclinal dip appear folds, faults, and (to the east) thrusts.

In the western part of the region [GVOZDETSKIY, 1964,1967], at an altitude of about 1,800–2,000 m, the characteristic features are large basins with sloping walls, and poorly defined depressions. Lower on the slope sinkholes, shallow shafts, and particularly large subsidence sinkholes are found. Formation of the basins and depressions seems to be related to the surficial solutional activity of meltwater from large patches of snow which last through most of the summer at levels below those of the firn fields and ice of the Last Glacial Epoch. The sinkholes due to subsidence formed as a result of subsurface solution by meltwater moving north-northeast down the slope of the strata. Water probably was considerably more abundant in the Valangian limestones then than it is now.

Still larger subsidence sinkholes are characteristic of the Cherek River Basin in the central part of the region. According to data of geologist KUZNETSOV [1928], the Kel'-Ketchkhen sink is 405×213 m in plan and 177 m deep. The very large dimensions of this form can be explained by the great height of the Skalistyy Range and the resulting greater glaciation. There are karst lakes in the region. Tserik-Kyel Lake, in the Cherek River Valley, occupies a karst shaft which was formed by rising artesian water. It is 258 m deep and 235×125 m in plan [KUZNETSOV, 1928].

The tectonics are more complex in the transitional "cuesta-fold" belt and the Andiysk Range; faults, thrusts, folds, and steep inclination of strata in monoclinal ridges are to be observed. The karst has formed in Upper Jurassic and Lower Cretaceous (Valangian) limestones. Solution forms and springs are present.

LILIENBERG [1959] distinguished several regions in mountainous calcareous Dagestan which differ in degree of karstification of the rocks and in the character of the karst features. The karst is in Upper Jurassic and Lower Cretaceous limestones and dolomites and in Upper Cretaceous sequences of limestone and marl, which occur as steeply-dipping beds in monoclines or as layers of diverse dip in folds. There are also both subsurface karst forms—caverns and natural shafts—and surface forms such as basins, karren, depressions, sinkholes (formed by solution and by subsidence), karst wells, and residual forms. There are karst lakes and springs. The mountainous calcareous Dagestan region extends a little way into Azerbaydzhan, where there are caverns.

In the western part of the south slope of the Greater Caucasus karst has formed in the Upper Cretaceous flysch formations, which contain considerable amounts of carbonate rock and which have been folded. The most notable karst phenomena are limited to areas of tectonic disturbance [KOLODYAZHNAYA, 1965].

In the more easterly regions of the south slope of the western half of the Greater Caucasus (from the vicinity of Sochi and farther east) several limestone formations of Upper Jurasic and Lower and Upper Cretaceous age, and in part Palaeogene Limestones and Neogene and Lower Quaternary calcareous conglomerates, have been karstified. Numerous sinkholes of irregular form (which depends on the fracturing) occur in the arches of anticlines in Upper Cretaceous limestones near Sochi, and there are karst wells and caverns, partly described as long ago as the work of MARTEL [1909]. The Vorondovskiy cavern system, in the headwaters region of the Kudepsta and Vostochnaya Khosta rivers, has a total length of about 5 km, according to SOKOLOV [1959]. The karst shaft Nazarovsk, in the same region as the headwaters of the Khosta, is the deepest karst opening in the U.S.S.R. (about 500 m deep); the Velichest-Vennaya shaft is also found here.

Upper Jurassic and Cretaceous limestones form a belt of karst massifs, ranges, and plateaus in western Georgia, extending 320 km from the Psou River to the divide between the Black Sea and the Caspian Sea, encompassing parts of the ranges of Abkhaziya, Megreliya, Imeretiya, Lechkhumi, and Racha. Fold structures are wider in the eastern part of this belt than in the west, where they form a plateau in which karstification has been intense (Askhi, Nakeral'sk in the Rachinsk Range, and at other places). Karren, depressions, sinkholes, and basins are characteristic of the surface of the limestones of the plateau and massifs. At many places rivers and creeks sink into the limestone. Subsurface streams appear suddenly from caverns and karst channels. There are notable caverns with stalactites. The Anakopiysk chasm near Novyy Afon is among the largest caves of karst in the world. It includes an enormous room (about 950,000 m^3). Abrskil Cavern, near the city of Ochamchira, has a total length of about 3 km. Karst shafts, 70–160 m deep, have been described in the massifs of Arabik [KIKNADZE, 1965] and Kvir. There is relict karst at Lake Ertso, in the eastern part of the region. In high-mountain localities the karst phenomena have been formed against a background of Late Glacial events, but the formation of many chasms and karst wells preceded them [MARUASHVILI and TINTILOZOV, 1963].

Individual karst forms are known in the Ciscaucasus, near the Mineralovodsk group of laccoliths [GVOZDETSKIY, 1954, 1964]. Karst has formed in Sarmatian limestones and calcareous sandstones and conglomerates in the Stavropolsk upland.

Karst has developed in Upper Jurassic and Upper Cretaceous limestones of some regions in the Lesser Caucasus and in spurs of the Zangezursk Range. Several karst regions are distinguished here: in northwestern Azerbaydzhan, on the north slope of the Shakhdagsk Range, in the Karabakhsk Range, and in the eastern spurs of the southern part of the Zangezursk Range. The principal forms examined and described in these regions are caverns. The Inek Magarasy Cavern in the Shakhdagsk Range has a subterranean stream. The Azykhsk Cavern, in the Karabakhsk Range, which consists of several grottoes separated by narrow passages, contains formations of sinter and dripstone.

Study of karst in the Caucasus is related to problems of water supply, damming of mineral springs, and hydro-engineering construction, and to other practical tasks.

KARST OF THE EASTERN CARPATHIANS [1]

In the Soviet Union the east or forested Carpathians are a mountain chain as much as 90 km wide, extending southeastward for 250 km. Karst occurs in the south slope of the Carpathians and in its foothills.

The Soviet Carpathians, which are the karst region of the eastern Carpathians, are subdivided into three karst provinces—the Ciscarpathian (Predkarpatskaya), central Carpathian, and trans-Carpathian (Zakarpatskaya) provinces—which differ from one another in the nature and intensity of their karst processes.

The *Ciscarpathian karst province* coincides with the Ciscarpathian border trough, a region of Neogene submergence. The trough was filled with a thick sequence of sandy–clayey molasse and of evaporite deposits of Miocene age which overlie the Upper Cretaceous, Jurassic, and Palaeozoic rocks (principally carbonate rocks) that constitute the basement of the trough.

The Lower Miocene evaporite formations are at great depth (deeper than 1,000 m), but because of great dislocation of the Miocene deposits they are near the surface at some places and are accessible to water from the atmosphere. Sulphate deposits in the evaporites form the so-called caps, which are hypergene products of leaching.

Karst in the Ciscarpathian Trough has formed principally in an outer zone where gypsum, anhydrite, and partly limestone have been affected. Tectonic movements have had a significant influence on development of karst processes here, which began prior to the Miocene and have extended into more recent time [BUROV, 1963]. Sinkholes, swallow holes, and ponors are widespread. Large subsurface cavities occur where confined sulphate–carbonate water is in contact with the Russian Plain. This has been found in the Nemirovsk and Yzovsk deposits, in the Ratinsk Limestones, at depths of 50–70 m. Evaporite deposits—sodium salt (rock salt), potassium salt, and partly gypsum—have been karstified in an inner zone of the trough. Karst has formed particularly widely in the Dombarovsk part of the Kalushsk potash deposit, where as a result of tectonic movements the salt body is partly in the zone of active groundwater circulation. Modern processes have not produced many karst features. In the past, however, during the period of ascent of the salt massif, karst formation progressed very intensively, as is shown by the presence of gypsum–clay caprock as thick as 18 m. According to KOROTKEVICH [1961, p.189], "...a layer of potash salts more than 110 m thick must have been dissolved to form it."

The *central Carpathian karst province* encompasses the mountainous Carpathians. Karst has been noted only in the southern part of the province, in a belt of Mesozoic deposits which extends northwestward from Rakhov to Perechin. Triassic and Jurassic carbonate rocks have been exposed by erosion only in the southeastern part of this belt—in the Chivchinsk Mountains, in the Rakhov Massif of crystalline rocks,

---

[1] Written by A.G. Chikishev.

and in the Ugol'ka River Basin; northwest of this region they are covered by younger deposits. The belt of outcrops is very narrow (10–20 km). A second belt, of Jurassic limestones, has been traced parallel to the basic belt, through the zone of central Carpathian structures, between the upper valleys of the Chernaya Tissa and Belaya Tissa rivers. Exposures of Jurassic rocks in this zone have been established as far as the Chernaya Cheremosh River.

The karst is more extensive developed in the region of the Dragovskiy Menchul Mountain. Twenty-three caverns, richly ornamented with sinter, have been discovered here [CHERNYSH, 1966] in high cliffs of dense gray marbleized Jurassic limestones [SLAVIN, 1953], in the upper reaches of the Malaya and Velikaya Ugol'ka rivers. A vertical shaft called the Druzhba Cavern is of particular interest. It begins as a cave-in 21 m deep which leads into a complex system of vertical and horizontal passages. The total length of horizontal galleries in the cavern exceeds 200 m, and it is 46 m deep. The Belaya Stena Cave is 101 m long, Molochnyy Kamen' Cave is 90 m long, and Greben' Cave 71 m. On the right bank of the Malaya Ugol'ka River, 200 m west of the cliff of Chur', is a beautiful karst bridge up to 3 m high and more than 10 m long.

Karst has been recorded in the *trans-Carpathian karst province* only within the upper Tissa Basin, which extends for 65 km along the Tissa, between its tributaries the Rika and the Shopurka, as a belt as much as 25 km wide. The depression is filled

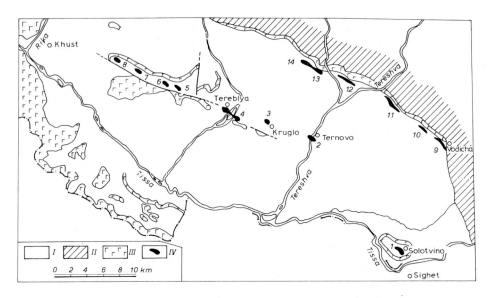

Fig.1. Diagrammatic map of the distribution of salt structures in the upper Tissa Basin. *I* = Miocene, Pliocene and Quaternary deposits which fill the basin; *II* = Palaeogene and Jurassic; *III* = volcanic and volcanogenic rocks; *IV* = salt structures: *1* = Solotvinsk; *2* = Ternovsk; *3* = Kruglo; *4* = Tereblinsk; *5,6* = Aleksandrovsk; *7* = Danilovsk; *8* = Baranovsk; *9* = Voditsk; *10* = Baskheu; *11* = Solyana; *12* = Nizhne Novoselitsk; *13* = Verkhne Novoselitsk; *14* = Vulikovets. (After KORENEVSKIY, 1961.)

with a thick sequence of Neogene rocks—sand–clay, evaporite, and tufogene rocks—which unconformably overlie an intensely dislocated basement of Cretaceous and Palaeogene rocks that is buried to depths as great as 5,000–6,000 m. Salt-diapir structures are typical of the depression; they coincide with the cores of brachyanticlines and with large fractures where rock salt breaks through the rocks (up to 800 m thick) which cover it and rises to higher horizons as far as the surface.

Karst has formed in massifs of rock salt (Fig. 1) which occur along the northeast edge of the basin and in its axial region near the surface, but in the Solotvina region are exposed as capricious salt crags [IVANOV and LEVITSKIY, 1960]. The surface of the salt cupolas, covered by a caprock of gypsum and clay, has been strongly karstified. Diverse karst features have formed here, their origin being related to the activity of salty and nearly salty water which has come from the infiltration of precipitation and by underflow from neighbouring regions. Mining of salt by the open method, the construction of shallow shafts, and particularly the subsurface leaching of salt have promoted intensification of the karst processes and the formation of saucer-shaped and cone-shaped sinkholes of subsidence and collapse, hollows, shafts, sinkholes formed by leaching, swallow holes, and caverns [KORENEVSKIY, 1961].

Karst is most widespread in the Novoselitsk–Voditsk, Boron'yevsk–Ternovsk, and Solotvinsk regions.

The Novoselitsk–Voditsk karst region encompasses salt structures which extend as a narrow belt along the north edge of the basin from Ol'khuchik to Voditsa. Rock salt lies near the surface at a number of places. The presence of numerous salt springs shows considerable development of the karst [KORENEVSKIY, 1959]. Karst cave-ins at many places are related to leaching of salt at old workings.

Karst in the Boron'yevsk–Ternovsk region, in the axial part of the upper Tissa Basin, has developed particularly northwest of the Tereblya River. Here large sinkholes, many of them filled with water and thus transformed into lakes, occur at the sites of old prospect pits and salt shafts. Cave-ins, sites of subsidence, and lakes related to karst formation and to the collapse of old salt shafts are also known in other places. Subsurface openings have also been recorded, along with the surface features [KORENEVSKIY, 1959,1961].

The karst has been best studied in the Solotvinsk karst region, where a salt dome cuts through a sequence of Miocene deposits and comes out upon the present surface. In the arched part of the salt massif, sinkholes formed by subsidence, basins and small depressions, and shafts have developed in the mantling deposits of clay and gypsum and are related to leaching at the surface of the salt dome by water from above the salt. Karst has formed particularly intensively at the sites of old salt mines and at places where rock salt outcrops at the surface. Depressions occur everywhere on the slopes of salt crags; they are in the form of branching furrows, trenches, and small channels separated from one another by sharp ridges and pinnacles. Salt "mushrooms" are also found here, in various stages of development. Subsurface features are found along with the surficial ones. Special studies have shown that the rock salt has been affected by karst processes to a depth of about 100 m at the contact of the salt with the surrounding rocks [DZENS-LITOVSKIY, 1966].

THE URAL OROGENIC BELT OF THE HERCYNIAN SYNCLINE[1]

The modern Urals and their northward extension, the Pay-Khoy Range, are a narrow western zone of the vast Hercynian geosyncline which extended far to the east during the Palaeozoic and now, as the west Siberian Platform, is concealed beneath a thick cover of Mesozoic and Cenozoic terrigenous deposits. At the end of the Palaeozoic the geosynclinal environment changed to that of a platform, and the Ural Mountains were formed by arching and block-faulting orogenic movements; during Mesozoic and Palaeogene time the region underwent denudation and was transformed into a peneplain. Late Cenozoic orogenesis was subordinate to the inherited Hercynian structural-tectonic plan, whose basic elements were large, deep fractures and grabens of meridional trend. These elements had been established during Precambrian time and were further complicated by younger faulting, but the meridional orientation of the deep fractures remained unchanged and it determined the plan of the arching and block-faulting movements which formed the present Urals.

Characteristics of the tectonic structure controlled the meridional trend of the mountain ranges and of the lowlands which separate them and with which the river valleys coincide. At places where the rocks have been more weakened by tectonic fracturing the rivers have cut through the short east–west distance from one meridional valley to the next. In accordance with this history the valleys are of different ages in their different parts: meridional reaches are older and east–west reaches younger. (The younger reaches of valley have been more strongly karstified.) In the Pay-Khoy and Polyarnyy ranges, and locally in the south Urals, the trends of orographic and hydrographic elements depart from this plan.

Carbonate rocks assume an essential part in the structure of the synclines, as well as in the structure of the buried Urals to the west and east, which form the Cisural and Turgaysk troughs. At such places many of the carbonate rocks are now deeply buried; their ancient karst features are revealed by boreholes and mine workings at great depth.

Carbonate rocks in the anticlinoria, anticlinal zones, and fault-block uplifts are principally exposed at higher places in the erosional sections. They are Lower and

---

[1] Written by I. V. Popov and A. G. Chikishev.

Fig. 2. Diagram of the regionalization of karst in the Urals (prepared by A.G. Chikishev).
Karst regions: I = the Paykhoysk region; II = the Urals north of the Arctic Circle; III = the Arctic Urals; IV = the northern Urals; V = the central Urals; VI = the southern Urals.
Karst provinces: A = west slope of the Urals; B = central Urals; C = east slope of the Urals; D = trans-Urals region.
Karst districts: 1 = Yangereysk; 2 = Sredne Karaysk; 3 = upper Usinsk; 4 = Lemvinsk–Kozhimsk; 5 = Un'insk–Podcheremsk; 6 = Kolvo-Vishersk; 7 = Verkhnevishersk; 8 = Sos'vo-Loz'vinsk; 9 = Yayvo-Kos'vinsk; 10 = Pashiysk–Chusovsk; 11 = Ufimsk–Serginsk; 12 = Tagilo–Chusovsk; 13 = Neyvo-Bagaryaksk; 14 = Verkhneaysk; 15 = Sredneyuryuzansk; 19 = Inzero–Satkinsk; 20 = Verkhneural'sk; 21 = Sugomaksk–Miassk; 22 = Gumbeysk–Suunduksk; 23 = Techensko–Uvel'sk.

KARST OF THE U.S.S.R.

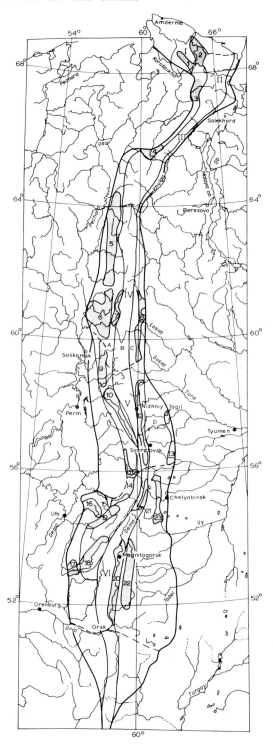

Middle Palaeozoic rocks. The considerable metamorphism of these rocks is not conducive to the development of karst in them.

The Ural karst country is subdivided into six karst regions—the Pay-Khoy, trans-Polyarno (Zapolyarno)-Ural, Polyarno-Ural, north Ural, central Ural, and south Ural regions—which, being situated in diverse bioclimatic environments, are characterized by specific assemblages and diverse intensities of exogenetic karst processes [CHIKISHEV, 1967, 1968]. At the same time morphostructural characteristics permit our distinguishing four karst provinces of meridional trend (provided the Cisural border trough is provisionally assigned to the Russian Plain). These provinces are the west slope of the Urals, the central Urals, the east slope of the Urals, and the trans-Ural region. They can be subdivided into 23 karst districts (Fig. 2) on the bases of geologic structure, relief, climate, and intensity of karst processes.

The karst province of the west slope of the Urals

A number of tectonic structures can be distinguished in this province. The most significant are the Polyudov Kamen' faults, the Bel'skiy Trough, and the Simskaya Trough, the southwestern part of the Zilairsk synclinorium, and others. Part of the carbonate rocks are of Silurian age and of platform type, and belong to the Caledonian structural stage. They are bioclastic limestones and brecciated dolomites, exposed in the inner valleys of the trans-Polyarnyy and Polyarnyy regions, the north Urals, and the south Urals. Part of the carbonate rocks belong to the Hercynian structural stage. They are limestones, dolomitized limestones, and unfossiliferous dolomites of the Zhivetsk Stage of the Middle and Upper Devonian. Carbonate rocks are represented in each of the three divisions of the Carboniferous System along the entire extent of the province. Permian deposits, compressed into synclinal structures, are carbonate dolomite–sulphate rocks; they are, however, of limited extent.

The essential prerequisites for the development of karst, both in this province and in the whole of the Urals, are brought about by tectonic disturbance of the rocks. The most recent tectonic movements, locally with a rate of 20 cm/100 years, renew the open fracturing of the rocks; this facilitates the circulation of water through them, and the formation of karst. Considerable karst formation has occurred where rivers cut through youthful arching uplifts. Karst features can be observed beneath the modern river levels in regions of dominantly negative movements and in areas which have been relatively stable during the general arching uplift of the Urals.

Karst has been noted in the Pay-Khoy, trans-Polyarno-Ural, Polyarno-Ural, north Ural, central Ural, and south Ural provinces. The karst has formed in the presence of permafrost in the Pay-Khoy and the trans-Polyarnyy Urals. Here the rocks affected are Silurian, Devonian, and Carboniferous carbonate rocks. Small caverns have been recorded in the headwaters regions of the Bol'shaya and Malaya Usa rivers. A borehole 20 km northeast of Vorkuta revealed karst openings as much as 6 m across, up to a depth of 75 m.

Three karst districts are distinguished on the west slope of the north Urals: the Un'insk-Podcheremsk, the Kolvo-Vishersk, and the Verkhnevishersk. The karst has formed in Devonian, Carboniferous, and Lower Permian limestones. Caverns are extensively developed along with surficial features. Seventy caverns and shafts have been studied along the banks of the Un'a and Pechora rivers; the largest are the Medvezh'ya (480 m long), the Un'insk (390 m long), and the Shezhimsk cave-shaft (20 m deep). Many caverns are full of glacial sediment washed into them through ponors. Sinkholes and swallow holes are widespread in the Kolvo–Vishersk region; some of them are as much as 200 m in diameter and 15–20 m deep. One of the largest caverns in the Urals, the Div'ya, is also found here; it has been studied and mapped over a length of 3,240 m [CHIKISHEV, 1967]. It is in Lower Permian limestone.

Karst has developed extensively on the west slope of the central Urals, along the Chusovaya (Fig. 3), Kos'va, Kizel, Us'va, and Vil'va rivers, and other streams. Their valleys have the character of canyons. Streams diminish sharply, or disappear, at some places. The largest karst openings are at the levels of the river terraces. However, active karst extends to a depth of 150–160 m along zones of strong tectonic disruption. There are subsurface openings as much as 3–4 m across, many of them containing deposits of sand and clay and of brecciated carbonate rocks.

The greatest development of karst is in the Yayvo–Kos'vinsk, Pashiysko–Chusovsk, and Ufimsko–Serginsk karst districts.

The most extensive karst in the Yayvo–Kos'vinsk district is in Visean limestones

Fig.3. Karstified limestones in the middle course of the Chusovaya River; west slope of the central Urals. (Photograph by A.G. Chikishev.)

and dolomites. Of the surficial karst features, sinkholes are particularly abundant; their mean density is 20–30/km². They are along lines of tectonic disruption, along the strike of the rocks, and along contacts with rocks of coal-bearing formations. The largest sinkholes reach 80–100 m, and even 300 m, in diameter. Many small streams disappear into ponors and fractures. Thus, four small streams which flow from the west slope of the Belyy Spoy Range escape into sinkholes and flow 7 km in subsurface channels. Many valleys have been transformed into blind creeks. Numerous karst openings were recorded during well-drilling near the city of Kizel, where karst features extend to a depth of 1,000 m. There are many caves. One of them, the Kizelovsk Cavern, has several levels with a total length of grottoes and galleries of about 800 m [Maksimovich and Gorbunova, 1958]. The coefficient of karst development in chemically pure Visean limestones amounts to 29%, while on the average it is 7–8% [Bunina, 1947]. The inflow of water to shafts in the Kizelovsk Coal Basin amounts to 2,700 m³/h [Pecherkin and Karzenkov, 1964].

The most karstified rocks in the Pashiysk–Chusovsk district are Upper Devonian dolomites, and dolomites and the purest limestones of the Visean and of the Middle Carboniferous. Of the sinkholes, those seen most frequently are 6–10 m in diameter and as deep as 2 m. East of the city of Chusovoy there are abundant sinkholes 50–80 m across and 10–15 m deep. At some places the number of sinkholes is more than 80/km² [Milikhiker, 1962]. Complex large depressions have been formed by the merging of several sinkholes. On a tributary of the Shaytanka River there is a karst basin 80 m long, 20 m wide, and 18 m deep. Blind valleys are the typical karst features. Ascending karst springs are abundant. The velocity of flow in subsurface karst streams, as determined by the use of dyes, ranges from 1.5 to 11 cm/sec.

There are about 50 caverns in the Pashiysk–Chusovsk district. They are generally narrow, low corridors which gradually constrict with depth and pass into fissures that go along the trend of the fractures. The caverns are small. As a rule their length does not exceed 10–15 m, and their height 2–4 m; only some of the largest attain lengths of 30 m and heights of 8 m [Chikishev, 1958, 1964]. The Pashiysk Cavern occupies a special place, being situated on the left bank of the Vizhay River, 2.5 km from the city of Pashiya. It consists of fourteen grottoes, and the total length of the cavern is about 400 m. Drilling work here, at a depth of 100 m, has revealed karst openings as wide as 3.5 m, mostly filled with loam and clay.

Karst in this province on the west slope in the south Urals has formed in Devonian and Carboniferous limestones along the Ay, Yuryuzan', Sim, and Belaya rivers. Four karst districts are distinguished: the Verkhneaysk, Sredneyuryuzansk, Verkhnesimsk, and Srednebel'sk districts.

Along the upper course of the Ay River Devonian and Carboniferous limestones in the east limb of the large Suleymanovsk brachyanticline, which has been complicated by deformation, have been karstified. Of the karst features, caverns are of particular interest. The largest one, the Laklinsk, is about 350 m long and has grottoes as high as 18 m. A 47-m deep karst well is situated on the right side of the Kurgazak gorge. Subsurface openings, located during well drilling, are widespread in the region of the Uluirsk syncline.

Karst in the Sredneyuryuzansk and Verkhnesimsk districts has formed in Devonian and Carboniferous limestones and dolomites in synclines on the slopes of the Simsk Trough. The most diverse karst features are present; caverns, which attain lengths of 1,150 m (the Sukhoatinsk Cavern), are particularly abundant. A karst stream near Min'yar station is of interest; it escapes from a cavern and descends as a cascade 5 m high.

Karst is widely developed in Devonian and Carboniferous limestones in the middle course of the Belaya River, where the direction of the stream changes from latitudinal to meridional. This karst is of diverse ages. Youthful features, basically limited to the bottoms of the valleys, are sinks and subsidence pits with depths to 3 m. Youthful features are also found on high erosional terraces. A distinctive microrelief of sinks can be seen on the valley walls and on the high-terrace surfaces. Relicts of old sinkholes are also found locally on the valley floors.

The karst province of the central Urals

This province is situated between two deep faults, the Zapadnyy and the Glavnyy. In general this is the most uplifted part of the Urals. It consists of a series of tectonic structures, principally anticlinoria. Owing to this structure, ancient rocks which underwent deep metamorphism have been exposed at the surface. Carbonate rocks are of very limited occurrence in Pay-Khoy and in the central and south Urals, and the karst is dispersed.

Karst in Pay-Khoy, in the Srednekaraysk district, has developed in a permafrost environment.

In the central Urals karst has been noted along the right tributaries of the Chusovaya River in the Koyvo–Serebryansk region [CHIKISHEV, 1964]. This region extends as a narrow belt along the west limb of the central Urals anticlinorium. Small sinkholes have formed in Ordovician, Silurian, and partly Devonian carbonate rocks. The depth of the sinks commonly does not exceed 1–1.5 m, and their diameter 5–7 m. Some of them hold water and are karst lakes. Some are very deep; thus, Bezdonnoye Lake, not far from the village of Visim, attains a depth of 55 m. On the bottom of the lake is the trunk of a downed tree; its presence indicates the recent origin of the lake as a result of subsidence. Large youthful karst features seem not to be characteristic of this region, however, as a result of the shallow dissection of the river valleys and of the consequent high groundwater level.

In the south Urals karst has developed in the Riphean and Lower Palaeozoic limestones and dolomites in the Inzerskiy synclinorium and the Bakalo–Satkinsk uplift, and also in Devonian and Carboniferous limestones and dolomites in the headwaters region of the Belaya River. Because strongly metamorphosed rocks are widespread the karst is rather weakly developed; large features are rarely found.

Caverns and other subsurface cavities revealed by wells are of small diameter and commonly are full of sand and clay. The celebrated Kapova Cavern, which owes its origin to the activity of the Shuul'gan River, is situated here. The total length of the Kapova Cavern exceeds 1,600 m. The cavern has two levels, which characteristically contain large grottoes and wide passages. Drawings (horse, mammoth, rhinoceros, bear) found in the cave date from the Upper Palaeolithic (20,000 years ago). The multilevel Sumgan cavern system, also in the Verkhnebel'sk district, begins in a vertical shaft 70 m deep and has a total extent of 3,670 m (Fig. 4).

The karst province of the east slope of the Urals

This province is east of the Glavnyy deep fracture, between the central Urals anticlinorium and the Uralo–Tobol'sk anticlinorium. Tectonically the region is a series of synclinoria whose cores are rocks of the Hercynian structural stage. In the northern part of the Urals these rocks are principally graywackes and, to a lesser degree, carbonate rocks. They are exposed only along the west limb of a synclinorium, inasmuch as the axial part and east limb have been strongly depressed and have been covered by Mesozoic–Cenozoic terrigenous rocks from west Siberia. Farther south, in the central part of the Urals, the Tagil'sk synclinorium in the Caledonian structural stage is made up of igneous and metamorphic rocks. The Hercynian structural stage, in the axial part of the synclinorium, is represented by terrigenous flysch of Middle and Upper Devonian age, and by a thick sequence of carbonate rocks of all three subdivisions of the Carboniferous System except for the Visean Stage, which consists of terrigenous and coal-bearing formations. The Magnitogorsk synclinorium, in the south Urals, is made up of volcanic-sedimentary and intrusive rocks. Marbles occur in metamorphosed rocks of the Wenlock and Ludlow. A thick sequence of limestones belongs to the Frasnian and Famennian stages of the Upper Devonian.

Tectonic deformation and the composition of the Palaeozoic carbonate rocks created conditions which, during the whole time of continental regime (since the Jurassic and, probably, earlier), favoured the development of karst in Silurian limestones and marbles and Devonian limestones on the east slope of the Urals.

On the east slope of the north Urals karst has been recorded in Upper Silurian and Devonian limestones in the Petropavlovsk uplift, which is an overturned morphostructure. Sinkholes as wide as 100 m and as deep as 12 m are the dominant surficial karst features. The very large Svetlaya Cavern is about 110 m long. Boreholes near Ivdel' revealed karst cavities at a depth of 40 m [CHIKISHEV, 1967]. Various manifestations of karst are also known near the city of Severoural'sk, where Lower Devonian rocks have been karstified to considerable depth. The age of the karst is pre-Eifelian, but the karst activity also continued into the Mesozoic. Solution processes have developed widely in connection with recent pumping from shafts, especially along the valleys of the Vagran, Kal'ya, and Sarayna rivers. Karst has also been noted in the Sos'vo–Loz'vinsk district, in the Volchansk and Veselovsk coal fields, where it has

Fig.4. Entrance to the Sumgan Cavern, which begins with a vertical shaft, 70 m deep; southern Urals. (Photograph by A.G. Chikishev.)

formed in Devonian limestones which had been transformed into tectonic breccia.

Karst development has been relatively slight in the Tagilo–Verkhnechusovsk district in the central Urals. The karst is restricted to Upper Silurian and Lower Devonian limestones, dolomites, and marbles. Sinkholes are small—0.5–5 m deep and 2–10 m in diameter—and they commonly are dry. In some sections the river valleys are blind valleys. Subsurface karst cavities as much as 3–4 m across, usually filled with sand and clay, are found to a depth of 50 m. Old buried karst features are of rather considerable dimensions: 400–500 m long, 200 m wide, and 100 m deep. They are characteristically restricted to the contacts of limestones and volcanic-sedimentary rocks. Small caverns are found in natural exposures on the Tura River.

Two karst districts are distinguished in the Magnitogorsk synclinorium in the south Urals: the Sugomaksk–Miassk and the Verkhneural'sk districts. The karst is related to isolated insular outcrops of Devonian and Lower Carboniferous limestones and marbles. Old karst features are especially widespread as large sinks on whose sides karren have formed; they are now in various degrees of degradation. Modern karst occurs only in the floors of the river valleys. The intensity of modern karst formation is significantly less than that of Neogene and Lower Quaternary time. The Sugomaksk Cavern, more than 80 m long, is on the slope of Mount Sugomak. Karst cavities with diameters up to 3 m have been found by drilling, at a depth of 50–70 m, on the left bank of the Yangel'ka River (a tributary of the Ural River).

The trans-Ural karst province

Tectonically, this province is in the east Ural synclinorium, which outcrops from under a Mesozoic–Cenozoic cover south of 59°N latitude. The region is an upland plain with gentle hills; only a small part of it, between the Tagil and Tura rivers, is a lowland. The altitude of the land surface, other than on individual summits, ranges from 400 to 200 m, becoming lower to the east in the direction of west Siberia.

Considerable development of karst has occurred in the Neyvo–Bagaryaksk district, in the central Urals, in carbonate rocks of Carboniferous, Devonian, and (in part) Silurian age. There are some tens of large and small belts of limestone, separated by terrigenous and volcanic rocks; the surface is marked by sinkholes, ponors, blind valleys, and ravines. More than 70 caverns have been described in the district; they are commonly 10–20 m long, and rarely 40 m or longer, and are restricted to erosional horizons. Karst springs with flows of 0.2–29 l/sec discharge along the river banks. Buried karst features include sinkholes, swallow holes, and relicts.

In the southern trans-Ural region karst has been recorded in Lower Carboniferous limestones in the Gumbeysk–Suunduksk and Techensk–Uvel'sk districts. Caverns and little chains of saucers and sinkholes occur along the contacts of carbonate rocks with volcanic and terrigenous deposits. There are karst lakes, many due to subsidence. Old karst features—large depressions and sinkholes—have been filled with Mesozoic and Cenozoic deposits. Groundwater drains into the karst limestones from the adjacent unkarstified areas. The flow of karst water is typically toward the rivers and creeks which drain the region.

Summing up, it should be noted that karst has formed in a very irregular manner in the Urals. It is most widespread on the west slope, where it is limited to Middle and Upper Palaeozoic limestones. It is poorly developed on the east slope and in the trans-Ural region, where it is in strongly metamorphosed Palaeozoic limestones which extend as rather narrow meridional belts between volcanic and terrigenous rocks. Its distribution is rather restricted in the central Urals, where the karst has much greater surface expression than in the aforementioned regions. Differences in the intensity of karst formation are observed not only while going from west to east but also from north to south. The most highly developed karst is in the central Urals, where large karst cavities have been followed to a depth of 1,000 m. Karst is poorly developed to the south of the central Urals and, particularly, to the north. This shows the important role of bioclimatic factors in the development of karst.

THE TURANSK PLATFORM[1]

The Turansk Platform is an epi-Hercynian platform. Basically it corresponds to the wide desert plain called the Turansk Lowland. The Hercynian folds which make up the platform are covered by Mesozoic–Cenozoic calcareous and gypsum-bearing rocks. The Hercynian folds reach the surface in local uplifts, forming small mountain ridges, as for example in the central part of the Kyzylkum Desert. They contain Palaeozoic limestones. The rocks have been attacked by solution, both in the Mesozoic–Cenozoic cover and in elevated parts of the basement, and karst phenomena are observed in both. In large part these phenomena originated in another palaeogeographic setting in the past, under a climate more humid than that of the present. The formation of karst under the present desert conditions is limited by insufficient moisture, although it has not completely ceased. Important problems in the study of karst in the deserts of central Asia are as follows: (*1*) delimitation of features due to modern karst processes from ancient features which formed under another geographic setting in the past; and (*2*) establishment of the specific processes and factors which permit the formation of karst under present desert conditions with low precipitation (less than 200 mm/year, and locally less than 100 mm).

In order to explain the degrees of intensity of modern karst processes in the deserts of central Asia we must study a general map of the groundwater flow; the local characteristics of the circulation; the chemistry and chemical activity of groundwater in the desert karst regions of central Asia; the role of non-uniformity of streamflow in space and time; the significance of unconsolidated cover deposits; and, finally, the conditions and possibility of condensation of water, and the role of water of condensation in karst formation.

---

[1] Written by N.A. Gvozdetskiy.

BOL'SHAKOV [1962] referring to N. I. Plotnikov's data, indicated the significance of deep regional groundwater flow (below the local base level of erosion) from the direction of the interconnected geologic structures (which are exposed along the southwest slope of the Karatau Range, in the Kyzylkum Desert, in the Karakum, and in other places) with which the accumulation of the modern groundwater reserves is related. In the view of Bol'shakov, deep regional groundwater flow in the deserts and semi-deserts of central Asia promotes modern karst and solution processes. Karst should develop in soluble rocks which lie in the path of the deep regional flow of fresh groundwater.

FEDOROVICH [1949, 1962] has indicated the significance of non-uniform streamflow in space. It is important to emphasize the seasonal non-uniformity of flow; the significance for karst development of soaking of the ground in spring; and the increase in streamflow and in infiltration during the melting of snow, during spring rains, and during winter thaws.

POSLAVSKAYA [1954] has noted the role of unconsolidated cover deposits in the concentration of water above rocks which are being karstified. In his opinion the surficial deposits facilitate the accumulation and prevent the run-off of moisture, promoting leaching of the soluble rocks—limestone and gypsum—which lie beneath. Considering this circumstance, however, it is impossible to leave without attention the circulation of groundwater in these soluble rocks. But where the subsurface cavities are old (as, for example, in Ustyurt) water which accumulates in the unconsolidated cover deposits and percolates from them into the limestone and gypsum may play a fundamental role in weakening the roofs of the cavities and in collapse at the surface.

The question of the role of water of condensation in forming karst in the deserts of central Asia is very important. BOL'SHAKOV [1962] summed up many facts which indicate the possibility that water of condensation has an essential role in formation of the karst. The undersaturation of water of condensation with respect to dissolved substances, and hence its aggressiveness, is particularly important. In BOL'SHAKOV's opinion [1962, p. 143], there is reason "...to assume the significant role of the condensation of water vapor from the air in the formation of groundwater, both in deserts generally and in the deserts and semi-deserts of central Asia in particular".

Where the cover deposits consist of eolian sand it is necessary to consider not only the role of thermal condensation but also that of the absorption of water vapor by sorption, within the soil. According to KOLOSKOV's [1937] data, hygroscopic absorption, or the sorption process, in sand may give not only the so-called "hygroscopic water" but also film and capillary water. These processes do not yield gravitational water which could reach soluble rocks; but because they occur, greater quantities of meteoric water and of water of thermal condensation can percolate through the soil and the upper layers of the rocks. Moisture which otherwise would form film and capillary water in dry rock can thus take part in the formation of karst.

The Mangyshlak–Ustyurtsk region is of greatest interest among the karst regions of the Turansk Platform. The karst is in Neogene limestones and gypsum (the limestones are chiefly Sarmatian and the gypsum Tortonian) in the upper structural stage of the Ustyurt Plateau and of southern Mangyshlak. The Neogene sedimentary cover which forms this structural stage is gently folded.

KUZNETSOV [1963, 1965] distinguishes nineteen karst regions in Mangyshlak–Ustyurtsk; they are restricted to the limbs of large structures, and in a number of instances to zones of large flexures characterized by increased fracturing of the Miocene carbonate rocks. Diverse karst forms are found: small depressions and swallow holes, ponors, sinkholes, large cave-ins and vertical and overhanging rims ("any"), natural wells, caverns, basins and valley-like depressions. In the opinion of KUZNETSOV [1963, 1965], GOLTS and EPIFANOV [1965], and other investigators, large karst relief features in Ustyurtsk are fundamentally relicts related to the more humid climatic conditions of the past. According to Kuznetsov, karst formation was vigorous during the Lower Pliocene, when karst depressions and large subsurface openings were formed; their roofs were subsequently destroyed (in part by solution along fractures), and this led to formation of large cave-ins at the surface. Small depressions formed by solution of the rocks at the surface are modern features formed entirely by karst processes [KUZNETSOV, 1965].

Caverns, widespread in the limestone cliffs of Mangyshlak, are due to sea waves and karst processes. Beds of marl and the less-tough limestones interbedded with the Middle Sarmatian shelly limestones were subject, first of all, to abrasion [KLEYNER, 1959]. Features of limestone karst which formed through solution by sea water are well known at the Cape of Izenty-Aral on the Aral Sea. LUPPOV [1948] is inclined to attach great significance to karst processes in the formation of the Sarykamyshsk Basin, adjoining the Ustyurtsk Plateau. GERASIMOV [1940] had earlier indicated the significance of solution and karst in formation of the Sarykamyshsk Basin.

Karst phenomena have been noted in residual mountains and heights of the Kyzylkum Desert, where there are small caverns, grottoes, niches, overhangs, karst springs in Palaeozoic limestones, and features formed by solution and undermining in salt-bearing deposits.

The separate karst regions and areas in the Mangyshlak–Ustyurtsk region, in the Kyzylkum Desert, and in several other places, taken together, give an interesting picture of the distribution of ancient and modern karst forms and processes in the Turansk Platform.

THE SIBERIAN PLATFORM[1]

Central Siberia and part of east Siberia are in the Siberian Platform. It is bounded on the west by the valley of the Yenisei River; on the east by the Lena River to the mouth of the Aldan River, and then by the Aldan River and the Stanovoy Range; on the south by the fold region near Lake Baikal; and on the north by the Taymyrsk folded

---

[1] Written by I.V. Popov (after data from RODIONOV, 1948, 1949, 1963).

zone. The soluble rocks are of Proterozoic and Palaeozoic age. The karst of the Siberian Platform has developed under permafrost conditions—that is, in rocks which are "eternally frozen". The thickness of the permafrost ranges from 20–50 m in the south to 300–400 m in the north. Summer thawing in the north is from several decimeters to 1–1.2 m, and in the south to 1.5–2.5 m.

The Siberian Platform is, over its vast extent, of non-uniform structure. Because of this, and because of diversity in conditions of physical geography, the development of karst and the forms it assumes are different in different parts of the platform. Generally, the karst has formed in Sinian and Lower and Middle Palaeozoic rocks, and it does not occur in rocks of the Mesozoic–Cenozoic structural stage.

In the region of the Khantaysk–Rybninsk uplift, on the northwest edge of the platform, karst is developing in Sinian, Cambrian, Ordovician, Silurian, and Lower Carboniferous rocks.

The total thickness of carbonate rocks is more than 3,000 m, while that of a gypsum–anhydrite sequence in the northern part of the region is 150 m. Karst springs discharge as much as 1–5 l/sec. There is no information about large karst features in the literature.

The Turukhansk karst region is one of block faulting. It is underlain by rocks of Sinian, Lower Cambrian, Ordovician, and Silurian age, which are fractured, cavernous, and porous limestones and dolomites with interbedded sandstone and shale. The total thickness of these rocks is as much as 4,900 m. Devonian rocks, as thick as 300 m, are principally terrigenous deposits with limestone interbeds up to 17 m thick. Many sinkholes are found on the divides of the rivers; they include both youthful sinks with steep sides which are nearly overhanging, and old sinks which have become marshy. Many karst lakes, some of them elongated meridionally and submeridionally and some round in plan, occur on a high terrace of the Yenisei River. Sinkholes in the basin of the Nizhnaya Tunguska River are cup-shaped, 1–2 m in diameter and 2–3 m deep; there are also steep-walled karst basins, 200–300 m in diameter and as deep as 40 m. Numerous taliks in the permafrost permit the development of karst, and they now store karst water.

The Yenisei Ridge is a mega-anticlinorium with a core of Precambrian rocks. Its topography is characterized by low mountains with deeply dissected valleys and flat summits. The limestones and dolomites are more than 4,000 m thick. Karst features, which are everywhere, formed in connection with Palaeogene, Neogene, and Quaternary uplifts of the Yenisei Ridge. They are sinkholes, small depressions, and basins. The Lower Cambrian rocks have been intensely karstified. Boreholes in the Ust'–Angara synclinorium have found subsurface karst, in the form of small cavities, at depths of 150–200 m. In the marginal trough near the Yenisei (Priyenisey) the surficial karst features are large and small sinkholes, poljes, and blind valleys. Gypsum and salt karst have formed, in addition to that in the carbonate rocks, with the formation of depressions and abundant mineralized water. Neogene karst has been recorded—sinkholes filled with pre-Quaternary deposits.

In the Silurian rocks on the northeast slope of the Yenisei uplift there are characteristically parts of the geologic section in which interbeds and lenses of gypsum, from 1–5 cm to 6 m thick, occur in the limestones and dolomites. Openings in the carbonate rocks have dimensions from 1–2 to 15 cm. Karst poljes, 5–6 km long and 2 km wide, occur on the divides. The karst features have been filled with Palaeogene lake and slope deposits. Sinkholes and cave-ins 0.5–3 m deep and 2–10 m in diameter lend the relief a pitted character along the valleys of the Podkamennaya Tunguska and Stolovaya rivers and other streams and in their watersheds. The diameters of large sinkholes in the Bichel'ya River Valley attain 100 m, and sinks with diameters as large as 70 m occur on the divides. Fresh sinkholes, and springs whose water contains as much as 2.5 g/l of mineral matter, show that the karst is now being formed.

The divide between the Podkamennaya Tunguska and Angara rivers, which is a large asymmetrical anticline several kilometres across, is made up of Sinian and Cambrian rocks. The Sinian consists of terrigenous deposits with interbedded fractured limestones and dolomites. The Cambrian rocks are fractured and cavernous limestones and dolomites with interbedded sandstones and shales and small lenses of gypsum. Modern and ancient karst features are most widespread in a ring-like zone of Cambrian carbonate rocks. Wide bowl-shaped depressions occur in Sinian rocks in the central part of the Chadobetsk uplift; there are also old karst features, now filled with unconsolidated deposits. Several large sinkholes with dimensions of 80–100 m, which remind one of poljes, are situated at the contacts of the Cambrian terrigenous rocks and the Sinian carbonate rocks. Modern sinkholes formed by solution at the surface and by infiltration occur on flat divides and their gentle slopes. Subsidence sinkholes, 50–80 m across and with walls as much as 10 m high, are known in wide, flat-bottomed valleys. Many old sinkholes are found, generally together with modern ones.

The topography in the Rybinsk Depression, in the southwestern part of the Siberian Platform, is characterized by low mountains and by plains. Devonian limestones outcrop at its edges. Small sinkholes find little expression in the topography. In the Taseyeva River Valley and on the left bank of the Angara, at places underlain by gypsiferous rocks of Upper Cambrian age, there are small single sinkholes and springs with highly mineralized water, which indicate the activity of salt-karst processes at depth.

The Prisayan–Kansk region, in the middle course of the Biryusa River, has well developed and rather deep valleys (100–200 m deep, relative to the divides). Limestones and dolomites occur in Devonian terrigenous-carbonate rock formations, and limestones and dolomites with interbeds of gypsum–anhydrite rock in the Lower Cambrian. Middle and Upper Devonian limestones and dolomites have been the most karstified. There are sinkholes, ponors, dry gullies, and karst springs. In Upper Devonian calcareous sandstones the karst features combine with those due to undermining, forming swallow holes, basins, trenches due to settling, and, more rarely, sinkholes, which are restricted to divides and to the gentle slopes of valleys. Karst features in the

Lower Cambrian rocks are sinkholes, caverns, various other cavities, breccias due to solution, and karst springs. Salt-bearing and gypsiferous dolomites and limestones on the right bank of the Biryusa River have been opened up by solution and contain small grottoes. The dolomites have been intensely broken down and locally have been transformed into dolomitic meal.

The Angara–Katangsk uplift, in the central part of the Irkutsk oblast', encompasses the regions adjacent to the Angara River north of the city of Bratsk. This is a gently rolling plateau dissected by the deep valleys of the Angara River system. Karst, slightly developed, is found in restricted areas where Upper Ordovician gypsiferous siltstones, gypsum, and dolomite outcrop in thin (up to 0.6 m) interbeds. Brooks disappear into the karst. The high dissolved load (up to 2.4 g/l) of the river water is evidence that chemical denudation of the rocks is in progress.

The Irkutsk piedmont trough, ancient and of erosional and tectonic origin, is filled with Jurassic continental deposits.

In the Prisayansk region it is slightly dissected, with low hills and relief as great as 400 m. Soluble Lower Cambrian rocks are widespread; they are limestones, dolomites, and dolomitized limestones. Terrigenous rocks are found only as thin strata and lenses.

Traces of ancient karst are preserved in the highest positions in the topography as buried sinkholes filled with kaolin and varicolored clays, as zones of brecciated rocks, and as individual karst cavities. The sinkholes are 5–10 to 30–50 m deep and 15–100 m wide. A zone of carbonate breccias, found on the divides, is more than 10 m thick and extends across the divides more than 300 m. Another manifestation of ancient karst is a rather thick (up to 4 m), red weathering crust on Jurassic rocks above the Cambrian dolomites; farther from the contact it is replaced by dolomitic meal and strongly fractured rocks with a total thickness of 100–150 m. Sinkholes above the present level of erosion are filled with red clay.

Considerable elevation (as much as 200–300 m) of a locality above the chief drainage, dissected relief, and monoclinal structure of the carbonate rocks have created conditions favourable for the development of Quaternary and Recent karst. Ancient development of karst plays a large role in formation of the modern karst. Quaternary karst occurs along valleys and ravines, and on their slopes and divides, as sinkholes, caverns, ponors, dry creeks, springs, and other features. The sinks are usually small—2–8 m deep for widths of 15–50 m. Most of them formed by subsidence. The small chains of sinks which are observed are restricted to the dominant systems of fractures (30°–40° NE and 300°–310° NW). Many small caverns are known in the Irkutsk River Valley. Sinkholes, and a karst shaft 63 m deep which ends in a room 16 m long, 12 m wide, and 16 m high, occur on the left bank of the Kitoy River. Large caverns are unknown in the Prisayansk region.

Karst processes have played an important role in the formation and widening of trenches due to settlement which are widespread on the slopes.

As a rule, strongly karstified interstream areas lack surface run-off; the precipitation is almost completely absorbed by the numerous ponors, and goes to depth. Not only small flows of water but also intermediate ones (up to 300 l/sec) go beneath the surface, and this results in substantial loss of water from large rivers.

The slope of the platform in the Irkutsk piedmont trough, in the southern part of the region near the Angara River between the cities of Balagansk and Usol'ye, is a flat upland strongly dissected by river valleys 140–370 m deep. Lower Cambrian carbonate formations are covered by terrigenous deposits of the upper Lena Series, of the Upper Cambrian, and of the Jurassic. The Lower Cambrian rocks consist of various dolomites, dolomitized limestones, and limestones with beds of gypsum and anhydrite. Thin strata and lenses of sandstone and marl are interbedded with the dolomites. Dolomite breccias are widespread in individual horizons. The thickness of the sequence of carbonate rocks ranges from 0–100 to 450 m.

The upper Lena rocks are gypsiferous, with beds and lenses of gypsum particularly in their lowest parts. Three stages in development of the karst have been established: Middle Cambrian; pre-Jurassic (Early Mesozoic); and post-Jurassic. The first of these is known from small buried sinkholes and older horizons of carbonate breccias, and from local diminution of the upper horizons of the Lower Cambrian (Angara Series). The pre-Jurassic karst is more widespread in the southern half of the region, where the red terrigenous sequence of the Upper Cambrian was eroded long before Jurassic time. There are vast fields of buried sinkholes filled with kaolin, and karst cavities filled with breccias that have been cemented with carbonate minerals. The formation of much of the brecciated dolomite and breccia, which is of regional distribution, and also silicification of rocks near the top of the Angara Series, are related to the pre-Jurassic karst. Formation of the karst led to differences in the lithology and thickness of the Angara Series: in the southern part of the area gypsum has in large part been leached from the rocks, while in the north the gypsum has been preserved and is of wide occurrence. The breccia-like dolomites in the south correspond to gypsiferous rocks of the north, although the thickness of these dolomites is 2–3 times less.

Quaternary and Recent karst is observed everywhere in the form of sinkholes, natural wells, swallow holes, gullies, dry creeks, karst lakes and springs, and travertine deposits. The subsurface features are caverns, karst cavities (empty and filled), various breccias, and porous rock. Karst has formed most strongly in gypsiferous rocks of the upper Angara sub-series, in lesser degree in carbonate rocks of the lower Angara sub-series, and very slightly in the terrigenous rocks of the upper Lena Series. Carbonate karst occurs in rocks of the Angara Series, and sulphate karst in rocks of the Angara Series and in gypsiferous rocks of the upper Lena Series. Salt karst has been discovered in boreholes in rocks of the Usol'sk Series, at depths of about 1,000 m; it does not appear at the surface. The sinkholes here are of subsidence, undermining-subsidence, and polygenetic origin. They are from several metres to 24 m deep and from 5–10 to 20 m wide. They are in irregular groups and in lines which trend along the dominant systems of tectonic fractures.

Large caverns are known: the Balagansk, Khudugunsk, and Orgaleysk caverns developed in gypsiferous rocks of the upper Angara sub-series. The total length of passages in the Balagansk Cavern is 1,200 m, and their volume exceeds 5,000 m^3. The Khudugunsk Cavern is somewhat smaller, the length of its passages being about 1,000 m. It is a classic example of a cavern developed along tectonic fractures. The Orgaleysk Cavern is of the type formed by collapse. Local overdeepening of river valleys at several places is also related to the karst.

Karst affects the mining of local coal deposits, causing disruption and deformation of coal beds and decrease in coal reserves. In the area of gypsiferous rocks of the upper Angara sub-series, active karst creates hazards to construction work. The regeneration of karst processes and the formation of cave-ins was observed during construction of the reservoir of the Bratsk hydroelectric station on the Angara River.

The Prisayan–Baikal extension of the Siberian Platform borders the Irkutsk amphitheater on the southwest and southeast. The carbonate rocks here are black bituminous limestones of the Lower Cambrian Oselochnaya series and rose-colored dolomites of the Lower Cambrian Karagassk Series; marbles and calcitophyres of the Archean Kitoysk Series; and crystalline limestones and dolomites of the Proterozoic Kamchadal'sk and Goloustensk Series. There are single sinkholes and groups of them, caverns, karst cavities, breccias, dry creeks, and karst springs. The caverns are restricted to limestones of the Ulumtuysk and Goloustensk Series, of Proterozoic age, at heights of 60–250 m above the level of Lake Baikal. Their lengths range from several metres to 30 m.

The Angara–Lena marginal trough occupies the eastern and southeastern margin of the Siberian Platform. The topography is made up of ridges, large depressions, low plateaus, and a vast region of flat uplands, deeply dissected by the valleys of tributaries of the Angara and Lena Rivers. The rocks in which karst has formed are limestones, dolomites, dolomitized limestones, and dolomite breccias of the Lower Cambrian Angara Series. In their upper part, at their contact with the upper Lena Series, the rocks contain lenses of gypsum. The total thickness of the carbonate rocks is about 1,000 m. Tertiary, Early Quaternary, and Recent stages can be distinguished in the karst development in the region. The Tertiary and Quaternary stages led to the origin of the brecciated zones, to dolomitization, and to the formation of dolomite meal, various kinds of cavities, and other features. The carbonate breccias are most widespread in the Kirenga River Valley. The most karstified rocks are in the arches of anticlines. Ancient karst is represented by depressions on the interfluves of the Lena and Malaya Chuya rivers; the depressions are as long as 5 km, and 1.5–2 km wide. They are filled with brown and red clays and interbedded sand. Where ancient karst developed, modern karst is also of wide occurrence in the form of sinkholes, ponors, dry creeks, caverns, karst springs, and other features. Sinkholes occur particularly on old surfaces of planation on regional divides of large rivers. Lakes are situated in many depressions and subsidence sinkholes. For example, there is a karst lake 14 m deep and 36–45 m wide on the left bank of the Kirenga River. A spring with a discharge of 100 l/sec flows from the lake. Many small caverns formed during development of the Quaternary karst. The formation of areas of steppe among the taiga, at an altitude of 1,150 m, was related to the karst, as was drying of marshes and local degradation of permafrost. The frozen ground is also degraded near large karst springs.

In the Angara–Lena region karst has formed principally in rocks of Ordovician carbonate-terrigenous formations, but it is slightly developed because the strata of carbonate rock are thin. There are elongated sinkholes, trenches, and swallow holes, which are related to the brows of large and high walls of valleys. The depth of the sinkholes and trenches locally attains 10–15 m at places where the trenches are 0.5–2 km long. Several caverns, in the form of elongated shafts as deep as 20 m or of chink-like cavities 1–8 m (more rarely up to 15 m) wide and 150–200 m long, have been noted in the valleys of the Angara and Ilim rivers. Formation of these caverns was related to settlement of rocks on slopes.

Karst has developed rather intensively at places where Lower Cambrian carbonate or sulphate rocks outcrop in the cores of anticlines. This is related to the considerable fracturing of the rocks and to the discharge, here, of saline groundwater. The rocks have been transformed into carbonate breccia, dolomitic meal, and highly porous dolomites.

In the eastern part of the Angara–Lena region Lower Cambrian sequences of limestone, dolomite, and marl are locally gypsiferous, while at depth there are thick sequences of rock salt. In consequence, gypsum karst and deep salt karst are also typically present. A special study of the karst has not been made in the Anabarsk Massif, in the Vilyuysk and Tunguska basins, in the region of the Kempendyaysk salt-dome structures, and on the east slopes of the Aldansk Shield. However, adequate data have already been collected concerning the general distribution of karst in these regions, not only in unfrozen rocks but also in permafrost. (The character of karst in permafrost differs significantly from that in unfrozen rocks; it is therefore distinguished as an independent geographical type of frozen-ground karst according to GVOZ-DETSKIY, 1954.]

The thickness of the frozen ground here ranges from 50 to 200 m in the southern and southwestern regions, and to as much as 300–600 m in the northern regions. Climate and the frozen ground are unfavourable for the development of karst; on the other hand, geomorphologic and geologic conditions favour it. In the Aldansk Shield and the Anabarsk Massif, which are underlain by crystalline schists, gneisses, and granites, the slopes of plateaus and uplands are overlapped by Lower Palaeozoic carbonate rocks with karst. At the junction of the Tungussk and Vilyuysk basin there are Cambrian, Ordovician, and Silurian rocks in which karst has formed, and numerous trap-rock intrusives. Except for the Kempendyaysk salt-dome structures the interior regions of the syneclise are underlain by terrigenous rocks. This determines the nature of the karst and the areal distribution of its various types. The total thickness of the karstified rocks is 1,000–2,000 m; in the greater part of the region they are bedded nearly horizontally or with very gentle monoclinal dip. The greatest fracturing of the rocks is observed in regions adjacent to zones of folding and areas of trap intrusions. Water beneath and between layers of permafrost finds flow paths through them. Karst is found more rarely in the north than in the south. The principal types of karst here are carbonate (the most widespread), gypsum, and salt karst. The features found more frequently are sinkholes, small depressions, pits, poljes, karst wells, caverns, niches, overhangs, corridors, fissures, small karst streams and brooks, springs, karst lakes and cave-ins, karren, and deposits of calcareous travertine.

The modern surficial karst is related to water above the permafrost. It includes both covered and denuded karst, the latter being found more rarely. Cone-shaped sinks are predominant; oval, asymmetric, and saucer-shaped sinks are uncommon. Their diameters range from 5–10 to 100–150 m. Sinkholes formed by collapse are restricted to valleys, to dissected parts of watersheds and socle terraces near valleys, and to river channels. Features of the surficial karst also include fresh and saline lakes, which are related to carbonate karst and to salt karst, respectively. Deep karst has developed very widely but has been inadequately studied. There are disappearing streams, springs, caverns, and subsurface cavities which have been revealed by deep drilling. The deep karst is related to waters which occur between and beneath layers of permafrost.

The karst processes are active now, as a result of the widespread occurrence of very recent tectonic movements.

The eastern slope of the Aldansk Shield is a plateau which has been deeply dissected (300–350 m). Here the karst is in Sinian rocks (limestones, dolomites, dolomitic limestones) 600–1,100 m thick; and in Cambrian rocks (limestones, calcareous dolomites, calcareous shales, calcareous argillites, and others) 900–1,420 m thick. The rocks have been strongly fractured, and in a number of formations they are highly porous. Sinkholes, karren, and karst basins are widespread. The basins are probably related to sagging of the roofs of subsurface karst cavities. Sinkholes are found on the divides and along the channels of several rivers; some of them may be above karst cavities. Their maximum depth is 10 m, and their diameters exceed 50 m. Karst processes also contributed to the formation of ancient surfaces of planation at altitudes of 1,000–1,100 m. Traces of the karst include sinkholes, hollows, and other features. There is a series of large karst basins which contain lakes and marshes. Innumerable sinkholes, filled with water and thus transformed into lakes, are on the watershed plateaus, which are covered by stunted forest and bogs. Many of the brooks which flow out of them end in closed ("blind") valleys. The sudden appearance of rivers from valleys which have the form of closed sinks with nearly perpendicular walls is no less characteristic.

Ice caverns in the Sinian dolomites are of great interest. They have two levels, with two, three, or more rooms joined by narrow passages in each. The orientation of the rooms and passages conforms to the two strikes of the tectonic fractures. The rooms in the upper level are partly clogged by blue-white ice; they also have ice stalagmites. It is essential to note that an accumulation of gravel underlies the continuous ice floor in the first room of the Abogy-Dzhe Cavern. Intensive karst formation began in the Oligocene; it may be assumed to have weakened in Middle and Late Quaternary time under the intense chilling of this region, which was not subjected to glaciation.

The Taymyr folded zone is a Late Palaeozoic arched uplift in the northern part of the Taymyr Peninsula, separated from the Siberian Platform by the north Siberian Lowland. The Byrrang Mountains occupy the main area. The valleys which separate the ranges are glacial troughs cut to depths of 500–700 m. Permafrost extends to depths of 400–600 m everywhere. Interconnected taliks (thawed zones) are possibly the largest water reservoirs.

Active manifestations of the karst are known in Upper Silurian and in Devonian deposits. The former are in synclines. Gray and dark gray limestones and dolomites, both thick-bedded and interbedded with shales, are predominant; their total thickness is 1,500–2,000 m. Caverns with paintings, niches, and sinkholes are found in cliffs along the river banks. Some of the caverns are 6–8 m long and 3–4 m wide and high.

Organogenic limestones in the lower part of the Lower Devonian have a total thickness of 600–700 m. The upper part contains gypsum and anhydrite, which constitute 50% of the rocks. The gypsum and anhydrite are strongly fractured, and have been karstified. They have caverns, niches, and sinkholes.

There are limestones and dolomites in Cambrian, Ordovician, Devonian, and Carboniferous deposits, but we have no data on karst in them.

THE OROGENIC BELT OF THE EASTERN PART OF CENTRAL ASIA, WITH A FOLDED PALAEOZOIC BASEMENT[1]

Here we find the Tarbagatay Range and the Dzhungarsk Alatau Mountain system, with Hercynian folds; the Tyan'-Shan', with Caledonian and Hercynian folding in the northern part and the Hercynian structural stage in the southern and the northern part of the Pamir–Alaya–Gissaro–Alaysk Mountain system, with Hercynian structures.

Karst phenomena are known to have formed in Palaeozoic limestones in the Tarbagatay Range. The presence of karst is assumed, though seemingly in rather restricted degree, in the Dzhungarsk Alatau system, which has been distinguished as a special region in regionalization of the karst of central Asia [GVOZDETSKIY, 1957]. Karst in the Tyan'-Shan' and in the northern chains of the Pamirs and Alaya is much more widespread and better studied. Here the karst regions of the Karatau, of the western Tyan'-Shan', of the inner and central Tyan'-Shan', of the Hercynian chains of the Pamirs and Alaya, and elsewhere have been distinguished.

A distinctive central Asiatic variant of denuded limestone karst is characteristic of the low-mountain desert regions of the Tyan'-Shan' and of the northern chains of the Pamirs and Alaya. It has formed in Palaeozoic rocks, with karren which have been

---

[1] Written by N.A. Gvozdetskiy.

destroyed by physical weathering, caverns, and numerous niches in canyon walls, but with slight development of sinkholes and other closed surficial karst forms. But in regions of mountains of intermediate height which receive greater precipitation, sinkholes appear in the limestones; while in the high parts of these mountains the pitted karst topography sometimes takes on an unusually typical expression, as, for example, in the Kyrktau Plateau at the western tip of the Zeravshansk Range. In some regions karst wells and shafts occur in Palaeozoic limestones.

Karst phenomena are strongly developed in Upper Devonian and Lower Carboniferous limestones and dolomites in the Karatau Range. There are caverns, the larger of which were formed owing to sulphuric acid, or the so-called "ore" karst [Gvozdetskiy, 1954, pp. 146–147]; and abundant sinkholes, niches, and karren [Olli, 1958]. The edges and overhangs of the karren are often separated from the main limestone and dolomite slabs as a result of intense physical weathering under the sharply continental climate. Just such crumbling karren fields are characteristic of the central Asiatic variant of denuded karst [Gvozdetskiy, 1954, 1960]. There are siphon springs of karst origin in the Karatau.

The role of calcareous rocks in the ranges of the western Tyan'-Shan' region is important. For example, flat divides of part of the Ugamsk Range, with abundant sinkholes and "blind" valleys (closed at the lower end) are formed in Palaeozoic limestones. Limestones form the foremost, moderately high range of the Kazy-Kyrt (in the southern Chimkentsk region), in which karst features are also widespread. The narrow and deep gorge of the lower course of the Ugam River, through which the Ugam flows into the Chirchik River, is in limestones. The carbonate karst of the Chirchik Valley receives special notice because the Charvaksk dam and hydro-electric station were built here, near Khodzhikent. Various solutional features in the limestones (cave passages, niches, etc.) can also be seen at Khodzhikent.

The karst of the north slope of the Tersk–Alatau Range belongs to the northern Tyan'-Shan' region. At the natural boundary of the Uch-Kashk, between the upper reaches of the Kyr-Kyr and Tyup rivers, sinkholes and karst wells formed by collapse have been described which are at an altitude of 3,000–3,200 m. There is a small karst cavern in the Palaeozoic limestones of the gorge of the Chon-Kyzylsu River.

Surficial features due to solution—niches in limestone cliffs, and other karst phenomena—are rather widespread in the interior and central Tyan'-Shan'. In pre-revolutionary time the reports of the mining engineer K. I. Argentov described periodic karst lakes with communicating subsurface channels—swallow holes—in Kyl'su Canyon (in the Kokkiy River Basin); the karst springs at Biryuk-Bulak, in limestones of the north slope of the Atbashi Range and Bogoshta Pass; canyons in Lower Carboniferous limestones with caverns, grottoes, niches, and various weathering features in the vicinity of Sonkel' Lake and in the Narynsk Range and the Aksaya Basin (Tekelik River canyon). Gvozdetskiy [1954] observed niches and cavern openings in Palaeozoic limestones of the interior Tyan'-Shan' in the Naryn River Valley above the city of Naryn; in the canyon along the route from Kochkorka to

Naryn, south of Dolon Pass; and in the central Tyan'-Shan', in the Sary-Chat River Valley. A. P. Gorbunov explored a cavern in the Sarydzhaz River Valley above the mouth of the Ottuk River, on the south slope of the Tersk–Alatau Range. Numerous sinkholes along the Naryn River below the mouth of the Alabug, in the interior of the Tyan'-Shan', and farther south in the southwest part of the inner Tyan'-Shan', have been formed by solution of lenses of rock salt in Tertiary deposits.

D. L. Armand examined a cavern with large calcite crystals, opened in Devonian limestones in the northern part of the Fergansk Range, in the Baubashat Massif.

Caverns with large calcite crystals are known also in the Pamir–Alaya Mountain system, as, for example, in Upper Silurian limestones in the headwater region of the Magian River, a left tributary of the Zeravshan River. These are not ordinary karst caverns, in the opinion of LEVEN [1940], but were opened up by hydrothermal solutions circulating through tectonic fractures.

In Palaeozoic limestones in the foremost chains and massifs of the Alaya Range (in southeast Fergana), as in the Zeravshan Basin and in the Fergansk Range, there are caverns and cavities of the vertical-channel type whose walls are coated with crusts of large crystals of calcite and of barite. The Bol'shoy Baritovoy Cavern, near the Aravan River, which was described by FERSMAN [1952], and other investigators, is indeed a wonderful natural phenomenon. The Chil'-Ustunsk Cavern, which is of considerable dimensions and has stalagtitic columns, stalactites, and stalagmites, is of particular interest among the numerous caverns situated near the Oshskikh Mountains. These caverns contain barite, as well as the stalactitic crusts of calcite.

In general the caverns of southern Fergana are of considerable interest mineralogically. Celestite (strontium sulphate) and other minerals are found in them, as well as calcite and barite. FERSMAN [1952], for example, notes the strontium caverns of Lyakan (between Sokh and Isfara), whose walls are covered with crusts of celestite.

Karst in the Aravan River region has developed in several stages. The dissected topography of pre-Pliocene time was covered almost completely by the sequence of continental deposits, N_2+Q_1 (Neogene$_2$+Quaternary$_1$, or Pliocene and Lower Quaternary). Formation of the present-day topography began with uplift of the region. Burial of ancient limestone ridges beneath the N_2+Q_1 deposits was governed by the degree of prior incision and widening of the valleys. Some karst cavities (caverns, natural shafts) were already being formed before burial of the ancient topography by the N_2+Q_1 deposits; this is shown by traces of their fillings. Some of the cavities were formed during early stages of development of the modern topography (that is, at the beginning of uplift and dissection of the initial surface). The Dikobraz Cavern, west of the Tange gorge not far from the Bol'shoy Baritovoy Cavern, is an example. Youthful karst forms of the present stage and of the preceding stage are related to the present topography, which had already formed, and to the rims of youthful epigenetic gorges. The large number of small niche-caves in the edges of gorges and canyons is very typical of the karst in this region. The original conduits of earlier channels of the Aravan, situated not far below the channel at various levels, are also youthful karst forms [GVOZDETSKIY, 1960].

The Akburinsk karst region, east of the Aravan karst region, has a history of similar topographic development. Karst features are also widespread to the west, in the advance ranges and massifs of the Alaya and Turkestan ranges. Caverns, to say nothing of shallow niches, occur almost everywhere—in Abshirsa, in Shakhimardan, in the Khaydarkan and Sokh regions, down to the boundary of Shadymir, west of Isfara, where the Kan-i-Gut Cavern is situated. In all parts of the southern mountain framework of the Fergana Valley, from Akbura and Aravan and farther west, solution has played a large role in development of limestone canyons along tectonic fractures. There are vauclusian springs (karst springs, of large flow, whose discharge does not cease during low-water periods) in Abshir-Bulak, in the Abshirsa Canyon, and elsewhere. Surface forms which may be related to the sinkholes are very exceptional. The surfaces of the limestone ranges are bare and rocky. They are distinctive karren areas with corroded blocks and fragments separated from the main massif by weathering.

Bare corroded limestone crags are also characteristic of the locality where the celebrated Kan-i-Gut Cavern ("the mine of disappearance", or "the mine of death") occurs. This cavern, which is related to the so-called "ore karst" [GVOZDETSKIY, 1954, pp. 146–147), is distinguished for the grandiose nature of its rooms, which reach heights of 40–50 m.

The Kyrktau Plateau, in the Chakyl–Kalyan Mountains in the western part of the Zeravshansk Range, is distinguished by very strongly developed karst which is forming in Devonian and Silurian limestones. The Kyrktau Plateau, with numerous surface forms—sinkholes, basins, half-disintegrated and disintegrated karren, with chains of sinks in hollows near the ridge on the north slope—is a continental analogue of the *yaila* (a monoclinal and folding limestone) in the mountainous part of the Crimea. To the west, nearer the pass of Takhtakarach, in the headwaters region of the Bul'bul'zarsa and Maydansa, there are also large sinkholes, while on the plateau-like ridge there are destroyed karren fields. One sinkhole passes with depth into a natural shaft with a total depth of 81 m (in the Aravan region there is an even deeper shaft, which is more than 140 m deep). The grotto of Amankutan, nearby in this same region, is famous as a Palaeolithic camping place. The well-known Kattaopa and Yettykyz caverns are in the southern spurs of the western part of the Zeravshansk Range.

West of Takhtakarach Pass, in the Karatyube Massif, is the original Amir-Temir Cavern which was formed through solution of limestone beds that lie between granites. The cave is an opening in the granite, and only remnants of the limestone strata show its karst origin.

There are also relict features of tropical karst in the northern ranges of the Pamirs and the Alaya. SEKYRA [1964] considers rocky limestone remnants near the city of Osh, and also limestone summits at Taldyk Pass in the Alaysk Range at an altitude greater than 4,500 m, to be such relicts. On the Iskanderdar'ye River, near Lake Iskanderkul', there are groups of limestone outliers which SHCHUKIN [1964, p. 103] compares with the towers and cones of tropical karst.

Karst features in the folded Palaeozoic rocks of the Tyan'-Shan' and the northern ranges of the Pamirs and the Alaya are limited to Mesozoic and Cenozoic rocks deformed during Alpine tectonic movements of the Palaeozoic fold basement. Such, for example, are karst forms in synclinal plateaus in the western foothills of the southern tip of the Fergana Range (where it joins the Alaysk Range), in the Tarana Basin on the divide between the Laysu and Buyg rivers.

Movements along faults in the Palaeozoic folded basement explain the origin of folds in the Mesozoic–Cenozoic cover of the southwestern tip of the Gissarsk Range which are related to the epi-Hercynian platform. In the Shurobsaya Basin region, in the southwestern spurs of the Gissarsk Range, the karst has formed principally in Upper Jurassic rocks; limestones are weakly karstified, and gypsum strongly so. There are caverns, sinkholes (some grouped in chains), karst wells, basins, disappearing streams, and streams in caverns. A large cavern, opened in Upper Jurassic limestones in the Baysuntau Mountains near the village of Machaya, is of interest for its fossils. Karst in the region of the Kugitangtau Range is restricted to Jurassic and Cretaceous sequences containing limestones, gypsum, and rock salt. The Karlyuksk Cavern is about 3 km in total length. The karst of all these regions, in terms of the lithologic, tectonic, and even climatic conditions of its development, closely resembles that already described in the eastern part of the Alpine folded zone of the U.S.S.R.

KARST OF THE TRANS-BAIKAL REGION[1]

OSOKIN [1965] described karst in the trans-Baikal region (Zabaykal'ye). Data on karst here were available during the first half of the 18th century [GMELIN, 1751–1752], but descriptions of it were fortuitous until recently. Carbonate rocks are found in the southeast in metamorphic sequences of Archean, Proterozoic, and Lower and Middle Palaeozoic age. The thickest carbonate sequences (about 1,000 m) of the Lower Palaeozoic are known in the Klikchinsk Range; near the city of Verkhnyaya Borzya they are stratigraphically younger, but the same Lower Palaeozoic carbonate sequence is about 700 m thick. This is the thickest and most widespread occurrence of carbonate rocks in the trans-Baikal region. Judging from the course of tectonic development of the eastern trans-Baikal region, conditions may have been suitable for karst formation during the Middle and Late Carboniferous. Uplifts or processes of erosion and possibly of karst formation occurred during Late Permian and Early and Middle Triassic times. With recession of the sea in the Late Jurassic, establishment of the centers of present-day karst formation could begin. Tectonic movements at the boundary between the Late Jurassic and Early Cretaceous caused fracturing of the carbonate rocks and created conditions for the development of karst. The structure of the earth's crust assumed a platform character. Local structures, topography, climate, and regional hydrology

---

[1] Written by I.V. Popov.

determined the sites and intensities at which karst formed. The well-known role of recent volcanism as a factor in karst formation is feasible. Intensification of karst processes was possible during the Cretaceous Period when the climate in the eastern trans-Baikal region was warm and humid. The epoch of most intense karst formation seemingly was related to neotectonic movements at a time before permafrost had developed here. The modern conditions of karst formation are characterized by the combined activity of neotectonic movements, of fractures related to it, and of a severe continental climate. A large part (90%) of the precipitation falls during the non-freezing period of the year, which makes possible the seasonal formation of karst. However, the small quantity of rainfall (200–400 mm) renders this possibility small. Thus, karst phenomena here are principally old.

Sinkholes and caverns are well known in the Kotuy-Nuku Range. The sinkholes are small; their large axes measure 10–20 m, their small axes 2–13 m, and their depths 1–5 m. In the Kharonorsk brown-coal region sinkholes are known which are pear-shaped, and about 60 m long and 30 m wide. The formation of frost on the edges of some sinkholes in winter shows the possible presence of subsurface cavities in such places. There are two sinkholes on the divide of the Omon River and its tributary, the Turga, which are connected by a cavern about 60 m long. There is a famous cavern named Uzornaya. It is essentially two caverns which extend horizontally, beginning in two sinkholes situated on a slope. With distance from the entrance the cavern narrows and becomes lower. Its total length is 50 m. There is yellow-red clay on the floor in the upper cavern, and ice in the lower. Surficial karst features are also known in two places on the Vitimsk Plateau. Surface karst features are generally rather rare in the trans-Baikal region. Deep karst is common in the form of caverns; porous rock and disappearing surface streams are also common. The Indoga River frequently disappears, leaving a dry channel. The circulation of water in carbonate rocks in the Indoga River Valley, beneath permafrost, leads to the formation of deep karst. Springs with highly mineralized water (to 0.42%) discharge at the bases of slopes underlain by limestones. The depth of karstification has not been established; small caverns have been found to a depth of 84 m.

Deep karst has also been recorded in limestones near the city of Nerchinsk. Widened karst fractures, which are from 1 cm to 1 m wide, are filled with ocher-yellow clay deposits. These deposits are observed to a depth of 10–11 m in wells. Karst phenomena in the Nerchinsk region are restricted to marbles, dolomitized marbles, and their contacts with marbleized limestones. Karstification of the rocks amounts to 1%, and rarely to 2.5%. Karst fractures, caverns, and cavities are revealed by boreholes in limestone deposits at Ust'-Borzinsk. The caverns are up to 5 cm in diameter, and the cavities to 7 m. The amount of karstification ranges from 0.9 to 34.17%, averaging 8.29%.

Karst appears to be widespread also in other parts of the trans-Baikal region, but as yet there is no substantiation of this from drilling or from hydrogeologic and geophysical studies.

KARST OF PALAEOZOIC FOLD STRUCTURES OF SOUTHERN SIBERIA[1]

This karst region is tectonically an east–west extension of a series of Late Precambrian, early Caledonian, and Middle and Late Palaeozoic fold structures. Carbonate rocks of Proterozoic (Sinian), Cambrian, Silurian, Devonian, and Carboniferous age occur in these structures.

The Altay karst province

The Altay karst province can be subdivided into a number of karst regions. In the *Aleysk karst region* karst is developed in marbleized limestones of Lower Silurian (Llandovery and Wenlock) age which are as thick as 300 m. In the Charysh River Valley oval and circular karst basins, 200–300 m across and from 2–3 to 5 m deep, and caverns, springs, and small streams which disappear into karst openings, are well known. The Upper Silurian is composed almost entirely of limestones, marbleized limestones, and marbles which have a total thickness of 400–1,200 m. Karren and sinkholes are the surface expression of the karst. The sinkholes were formed by surficial solution, by infiltration, and by collapse. Those formed by infiltration are not more than 6 m deep and 6–15 m in diameter; those due to collapse are about 2 m in diameter and 8–9 m deep. Many sinkholes are known along the Ina River; they are 30×22 m in plan and 3–4 m deep, and there are as many as 30/km^2. Dry ravines, and streams which flow into sinkholes, are widespread. Along the banks of the Charysh and Ina rivers and other streams are caverns, as long as 20–30 m, which consist of narrow passages and wide openings. The greatest length explored in the Nebo Cavern is 125 m [ROZEN, 1954]. Formation of the caverns dates from the end of the Neogene and the beginning of the Quaternary. The presence of fresh collapse sinkholes and large karst springs shows that karst processes are continuing at the present time.

Carbonate rocks in the *Anuysk–Chuysk karst region* occur in Lower and Upper Silurian and in Middle Devonian deposits. The thickness of the entire Lower Silurian sequence is 150–160 m, of which more than half is limestone. Karst has been noted along the valleys of tributaries of the Anuy and Charysh rivers. The Bol'shaya Karakol'sk and Yabagansk caverns are famous. Abundant karst springs show the activity of the karst processes. Data on karst in the Upper Silurian limestones, which are 750–1,500 m thick, are very limited. Caverns on the Anuy River, and karst springs, are well known.

Karst is forming in Sinian and Lower and Middle Cambrian rocks in the *Katunsk karst region*. The Sinian rocks are marbleized limestones, marbles, and dolomites with interbeds of quartzite and slate. The series is 2,000–3,000 m thick. Old sinkholes in the Sinian rocks are filled with the redeposited products of weathering; they are of pre-Cretaceous age. Modern sinkholes occur along the Ashiyakhta, Kurutkan, and

---
[1] Written by I.V. Popov.

Chiya rivers; among them are some formed by collapse. Many rivers (the Cheposh, Bulukha, Aksu, and others) flow into the ground in the karst terrain. The Baragal River, for example, flows 500–600 m beneath the ground, and the length of its dry channel is 8–9 km. Among a number of well-known caverns, the Arydunsk Cavern is 150 m long. Little is known of the formation of karst in the Cambrian marbleized limestones. Rivers and brooks which disappear beneath the earth surface and discharge into Lake Teletskoye are mentioned in the literature, and there are sinkholes east of the city of Gorno-Altaysk. The Kinderlinsk Cavern, which discharges into a left tributary of the Katuna River 10 km from the health resort of Chemal, has a wide mouth which changes, farther along, into fan-shaped diverging branches. The oldest stage in development of the karst was in a continental period during pre-Lower Cambrian time, when it occurred in Sinian rocks. The second stage is dated as Late Jurassic and Early Cretaceous, on the basis of Early Cretaceous deposits which fill the sinkholes. The last stage, which began at the end of the Palaeogene during general uplift of the mountainous part of the Altay (Gornyy Altay), has continued to the present.

The *Salair karst region* is on an asymmetric plateau. Karst is developing in rocks of Sinian, Lower and Upper Cambrian, Lower Silurian, Upper Silurian (Ludlow), and Lower and Middle Devonian age. Karst in the Sinian marbleized limestones (1,500–2,000 m thick) has been little studied. Sinkholes are known in only one area. Karst springs issue from these limestones at a number of localities. The most intensely karstified rocks are the Lower Cambrian limestones. They are dense, massive, and marbleized, with interbeds of sandstone and shale; their thickness is 2,000–3,000 m. Sinkholes of the most diverse forms are the most widespread karst phenomena. They attain 200 m and more in diameter and 5–20 m in depth. Among them are well-like features, with ponors at the bottom, which probably originated through collapse. Old sinkholes are filled with the redeposited products of weathering. The oldest deposits in the sinkholes are of Albian age. A few caverns are known, some of them with stalactites. Caverns, and also cavities revealed by drilling, are 2–5 m in size. Sinkholes in the Upper Cambrian limestones are filled with varicolored clays.

Karst features in the Silurian limestones are sinkholes and blind gullies, but they are restricted to the headwaters region of the Berda and Ina rivers and to other localities.

The Lower Devonian and the Eifelian Stage of the Middle Devonian in Salair consist entirely of limestones (up to 1,350 m thick). Sinkholes and blind valleys have developed along the trends of fractures. Mine workings expose cavities, some of which are filled by clays with limestone rubble and blocks. Well-like collapse sinkholes show that karst processes are continuing their destruction of the rocks. There are caverns along the Kara-Chumysh River and in other places.

Karst developed here in Late Cambrian, Early Eifelian, pre-Late Jurassic, Late Jurassic, and Early Cretaceous time. The karst processes became active again in the Late Oligocene, but formation of the modern karst features began in Holocene time.

The Tomsk–Kolyvansk karst province

Carbonate rocks in this province are of very restricted occurrence, in lenses and interbeds among non-karstified rocks of the Middle Devonian, of the Frasnian Stage, and of the Lower Carboniferous. In the Devonian rocks the interbeds are from 20–30 to 70–80 m thick, and in the Frasnian Stage up to 400–500 m. Sinkholes in Frasnian rocks in the northeastern part of the province are filled with red-brown ochers and fire clays of Lower Cretaceous age. There are no sinkholes with topographic expression. Somewhat to the south, where Frasnian rocks outcrop, there are modern sinkholes. Old and modern sinks are known in the southwestern part of the province, in the headwaters part of the Ina River Basin.

The Kuznetsk–Alatausk karst province

This province is in the Kuznetsk Alatau Range. There are carbonate rocks among the Proterozoic and Lower and Middle Cambrian rocks. The Proterozoic carbonate rocks are marbles about 2,000 m thick. They contain small chains of depressions and sinkholes in the northern margin of the region, which are filled with Lower Cretaceous deposits. Modern karst is found in the upper basin of the Tada River as sinkholes which are 2–40 m across and to 20 m deep. Near the Taskyl and Tserkovnaya mountains the karst features are complex because of undermining of the diluvial loams which overlap the karstified limestones. Sinkholes are 5–40 m across and 2–10 m deep. Karst basins of low relief are closed depressions as long as 100–150 m; their slopes are covered and overgrown with conifer forest. The bottoms of many sinkholes are silted, and the sinkholes are full of water. Subsurface features are karstified fractures, channels, and caverns. The largest of these openings attains a volume of 300 m^3.

The Sinian rocks are crystalline limestones, marbles, dolomitized limestones, and dolomites; their total thickness is 2,000–3,000 m. There are karst depressions as large as 40–50 hectares in area and from 0.5–65 m deep. Sinkholes reach sizes of 150–190 × 200–260 m and depths as great as 144 m. There are as many as 35/km^2. The depth within which these karst features occur is about 160 m. They are of Mesozoic and Cenozoic age. The sinkholes of the modern karst, and disappearing streams, occur in Sinian rocks on the northwest margin of the region. In the Lower Cambrian the karstified rocks are dolomitized limestones, dolomites, limestones, and marbles, with a thickness of 1,600–2,500 m. Depressions filled with redeposited weathering products are old karst features; they are as deep as 100 m, as long as 1,000 m, and from 10 to 32 m wide. They were formed in Mesozoic time or perhaps earlier. Modern sinkholes on the right bank of the Kiya River reach depths of 3 m and diameters of 20 m. There are many sinks in the Abakansk Range; they are 1–5 m in diameter and 2–3 m deep. In ravines the diameters of sinkholes attain 100 m, and there are disappearing brooks. Many caverns are found in the Kiya and Syi River basins. The Krest Cavern is nearly vertical and 10 m deep. Its walls are covered with ice. There are a number of other caverns with interesting solution features, with ice stalactites, with pockets as much as 42 m deep, and other rare phenomena. There are caverns which contain 1,500 m³ of ice. The gigantic stalactite "Pagoda" is in the Borodinsk Cavern, in Middle Cambrian rocks. The karstification extends to a depth of 45 m.

The karst province of western and eastern Sayan

Karst phenomena are rare in west Sayan; they are irregularly distributed in Lower Cambrian rocks. Sinkholes are restricted to small chains along tectonic contacts. They are as large as 10 m in diameter and 5–7 m deep. Some sinkholes swallow brooks. The age of the karst seems to be Quaternary. Subsurface openings along large fractures are old karst features; the openings are filled with Mesozoic–Palaeogene clays. Many of the sinkholes are situated 800–900 m above the local base level of erosion; this shows that here karst can developed at great depths.

There are several areas in the karst region of east Sayan where karst is being formed in Lower and Upper Cambrian rocks. The Lower Cambrian carbonate rocks occur as interbeds alternating with terrigenous rocks and marls. In the Upper Cambrian the carbonate rocks make up sequences tens and hundreds of metres thick. The greatest development of karst is on the watersheds of the Yenisei and its tributaries, the Biryusa, Bazaikha, and other streams, and on other watersheds. There are sinkholes, karst springs, and numerous caverns, large and small. Most caverns were formed through the combination of vertical shafts and of erratic low passages which meet at angles. Several caverns—Divnaya, Babiseysk, Kubinsk, and others—have been thoroughly investigated. Quaint sinter formations are observed in many: the gigantic "Lyustra", "Os'minog", and others. In the southern part of east Sayan there are Archean, Proterozoic, and Middle Cambrian rocks which are frozen. The temperature in karst openings in the frozen rocks is characteristically above freezing [SALONENKO, 1960].

The Divnaya Cavern is on the left bank of the Yenisei, 1.5 km southeast of the mouth of the Biryusa River and 150 m from the shore. It is in dark-gray, massive limestones. The cavern is on two levels, and horizontal. The total length of passages is 500 m, and their depth 50 m. Seven grottoes have been counted. There are many deposits of white sinter, stalactites, and stalagmites, some of which are transparent; cave pearls are frequently found on the floors of the grottoes. There are lakes in the grottoes. Still more caverns are found nearby, some of them of smaller dimensions.

The Kubinsk Cavern is on the left bank of the Biryusa River at an altitude of about 260 m. Its total extent is about 2,000 m, and its depth 250 m. Its entrance is at almost the very summit of a sheer cliff. A 20-m shaft reaches a gallery with three branches. Two of the passages are very low and narrow, and 10–15 m long; the third passage, gradually widening, meets another shaft 20 m deep after a distance of 15 m. A small sloping passage leads to a third shaft, which in turn leads out to the Fidelya Grotto. The walls of the passage are covered with white sinter, and on the floor are many cave pearls. There are many white stalactites and stalagmites in the Fidelya Grotto. A system of inclined passages, the "Bolshebnyy Labyrinth", leads into the Grandioznyy, Nadezhd, and Ozernyy grottoes. The walls and half the floor of the Ozernyy Grotto are covered with streams of "rock milk"; a lake occupies the other half of the floor. There are also the Zaozernyy and Skrytyy grottoes.

In the southern folded framework of the Siberian Platform karst is well known in the Tuvinsk Trough, in the Baikal folded zone, and in the Mongolo–Okhotsk folded region. East of the Siberian Platform karst is found in the Verkhoyano–Chukotsk region in the Setta–Dabansk anticlinorium and Yudomo–Maysk miogeosyncline, and in the Kalymsk and Molonsk massifs. Karst is well known in the Sikhote–Alin' folded region, in the Khantaysk medial massif, and in Sakhalin Island.

UNDERGROUND FLOW IN KARST REGIONS OF THE U.S.S.R.[1]

As is well known, karst develops under the influence of the solutional and mechanical activity of flowing water. Groundwater circulating in massifs of fractured and karstified rocks on a continent is finally discharged into rivers, through karst springs or directly through the stream channels. Determination of underground flow into rivers (base run-off) in karst regions is therefore of outstanding scientific and practical interest, first of all as an indicator of the intensity of karst processes. A no less important aspect of the study of underground flow lies in determination of natural groundwater resources through the magnitude of underground flow. Detailed investigations according to karst regions have been made in the U.S.S.R. as part of the more general task of determining underground flow for the entire country [KUDELIN, 1966], the data for which were used in establishing the overall water balance. Underground flow into rivers, among them those in whose basins karst phenomena have developed, was determined by the combined hydrologic-hydrogeologic method of genetic separation of the river hydrograph, which we have described in other works [KUDELIN, 1958, 1960]. The use of other methods of calculating underground flow—for example, the hydrodynamic methods—encounters difficulties in determination of the calculated parameters for fractured karstified rocks, especially in large regions.

The basic results of the investigation of underground flow in several regions of the U.S.S.R. are given very briefly below.

Underground flow into rivers occurs under the influence of three fundamental factors: climate, topography, and the structural-hydrogeologic factor.

Climate imparts a clear latitudinal zonation to underground flow. The modulus of underground flow in the European part of the U.S.S.R. decreases regularly from 6 to 4 l/sec per km^2 in the north and northwest, to a fraction of 1 l/sec per km^2 in the south and southeast. This range corresponds to a decrease in precipitation from

---

[1] Written by B.I. Kudelin.

700–600 to 400–300 mm/year and an increase in evaporation in the same direction.

Topography, which is closely related to the structure of the earth's crust, imparts vertical zoning to underground flow that depends on vertical zonation in the distribution of precipitation. On platforms the largest magnitudes of underground flow are in uplands; in fold-mountain structures, other conditions being equal, it increases with height of the region (to well-known limits) along with increase in the quantity of precipitation and increase in the depth and areal density of erosional dissection.

The hydrogeologic factor interrupts the regular zonal character of the areal distribution of underground flow. This is displayed most sharply in regions where water occurs in fractured and karstified rocks. As a rule, the presence of karst sharply increases the magnitude of the modulus of underground flow in comparison with its background values.

The Urals serve as a typical example of the influence of karst on underground flow. The considerable flooding in the region and the high mean annual moduli of underground flow (as high as 10–12 l/sec per km^2) are related to the formation of karst in the Palaeozoic rocks on the west slope of the Urals. Here the streams, in their middle courses, have cut into a sequence of karstified rocks, and they drain these aquifers. At the same time, in a series of basins which have no karst, the modulus of underground flow does not exceed 3–4 l/sec per km^2. On the east slope of the Urals karst waters in separate areas in the Sos'va and Vagran River basins have mean annual moduli of underground flow of 3–3.5 l/sec per km^2, while the values are 2 l/sec per km^2 and less in basins which lack karst.

In the Ufimsk Plateau, which is underlain by strongly karstified carbonate and gypsiferous rocks of Lower Permian age that easily absorb precipitation, the modulus of underground flow characteristically increases to 4 l/sec per km^2 in the basin of the Ufa River and its tributaries the Ay, Yuryuzana, and other streams, although in regions adjacent to the plateau this modulus is 2–3 l/sec per km^2.

The decisive influence of the karst on underground flow is observed in the Crimea. In the mountainous part of the Crimea, under the same climatic conditions, underground flow develops only in the region of strongly karstified Jurassic carbonate rocks which undergo flooding. The modulus reaches 5–15 l/sec per km^2 and higher in these rocks, in the midst of impervious Triassic, Jurassic, and Cretaceous rocks which are practically dry.

The effect of karst on the distribution of underground flow has also been recorded in other places: in karst regions in marl–chalk rocks in the middle course of the Volga, in the basins of the Serezha, Tesha, and P'yana rivers, the values of the modulus are anomalous.

In an area in the north, in the Yemtsa and Vaymuga River basins in the Onego–Severo–Dvinsk interfluve, the mean annual modulus of underground flow amounts to 3–5 l/sec per km^2 in intensely karstified Carboniferous limestones and dolomites. The background value of the modulus is 2.0–2.5 l/sec per km^2. In the karstified basin of the Kulaya River, underlain by Carboniferous limestones and dolomites and by Lower

Permian gypsiferous deposits, the modulus ranges from 5.0 to 6.0 l/sec per km².

In Timan the modulus is observed to increase from the periphery of the region toward its center, attaining a value of 4 l/sec per km². This is explained by the development of karst and fracture-karst waters in the Middle and Upper Palaeozoic carbonate rocks.

Increase in the modulus of underground flow can also be observed in the region near the Black Sea (Prichernomorskaya zone) on the south slope of the Greater Caucasus (Sochi-Sukhumi), where subsurface flow attains 35–40 l/sec per km².

There are analogous examples of the influence of the karst on underground flow in the Asiatic part of the U.S.S.R. Considerable resources of fresh groundwater are concentrated in the southern part of the Siberian Platform, in the basin of the upper course of the Lena River (on the west slope of the mountains near Lake Baikal), where Cambrian carbonate rocks have mean annual moduli of underground flow of 2–5 l/sec per km².

The maximum values of the mean annual modulus in the northern and highest part of the Yenisei Ridge, in the Podkammenaya Tunguska River Basin, are restricted to Precambrian carbonate rocks. These values are 4.5–5 l/sec per km².

In the Sayan–Altaysk fold-mountain region, on the western slopes which consist of Palaeozoic carbonate rocks, the mean annual modulus of underground flow attains maximum values of 5–9 l/sec per km², whereas the background values are from 1 to 2 l/sec per km².

In the Verkhoyan–Chukotsk fold-mountain region the maximum value of the mean annual modulus is 2.0 l/sec per km²; this is related to a series of block uplifts of the Kolymsk medial massif which is composed of karstified carbonate rock sequences.

Water-bearing Proterozoic and Cambrian carbonate rocks contribute to the streamflow in the basin of the middle course of the Amur River, and in the Malyy Khingan Range in the upper course of the Biraya River. Underground flow in these deposits increases to 3 l/sec per km².

Not only the mean annual moduli of underground flow but also the minimum values of the modulus show appreciable increase in karst regions. Thus, during the period of stable low water the mean annual modulus for the Vishera River Basin amounts to 4.5 l/sec per km², while for karstified basins of rivers on the east slope of the Urals it is 1.5 l/sec per km²; in the Ufimsk and Onego–Severo–Dvinsk plateaus it is 3–3.5 l/sec per km², and on the Kuloysk Plateau it is as high as 4.0 l/sec per km² and higher. At the same time, in non-karstified river basins in these regions, the minimum moduli are in the range from 1 to 2 l/sec per km². A considerable increase in minimum moduli, relative to their values for that zone, is observed in karst areas of the south slope of the Greater Caucasus, of the Crimea, of the middle Volga Basin, and of other regions. This shows the regulating influence of karst on river flow, which has been established by many investigators of the hydrology and hydrogeology of karst regions.

The regulating role of the karst is evident not only in the redistribution of flow within the year—decrease in the volume of the spring flood and increase in low-water discharge of rivers—but also in the relative proportions of underground and surface flow. The fraction of underground flow in the total annual river discharge (the so-called coefficient of underground feeding of rivers) in karst regions is invariably increased. In most of the European part of the U.S.S.R. underground flow amounts to 20–30% of the total streamflow, and in the south to somewhat less; and in the northeastern part of the country, in regions where the rocks are perennially frozen, to less than 10%. On the other hand, in karst regions of the Urals, the Ufimsk and Onego–Severo–Dvinsk plateaus, and the central Devonian area, the coefficient of underground feeding of rivers increases to 40–50%, while in the Kuloysk Plateau, in the Greater Caucasus, and in the Crimea it attains 50% and more.

In a number of regions in the Asiatic part of the U.S.S.R.—the upper course of the Lena River, the northern part of the Yenisei Ridge, the west slopes of Vostochyy Sayan—the coefficient of underground feeding of rivers attains maximum values of 40–50% in karstified basins; in karstified rocks of the Altay and of Salair it amounts to 30–40%.

A third parameter, the coefficient of underground flow (the ratio of the quantity of underground flow to that of precipitation), is also sharply increased in karst regions. According to mean annual data it equals 30–40% and more in karstified regions in the Kuloysk Plateau, the west slope of the Urals, the Greater Caucasus, the southern Siberian Platform, and the west slopes of the Sayan–Altay fold-mountain region, while its value is 10–20% in non-karstified regions. In the Crimean yayla the coefficient of underground flow reaches 70%.

In the Silurian plateau, on the shore of the Gulf of Finland, in an area of about 3,000 km² where modern carbonate karst is developing intensively, the karst is expressed at the surface as single sinkholes and groups of sinkholes. Surface flow is generally absent. All the precipitation except the part of it lost to evaporation and transpiration is absorbed by the karst formations and recharges the karst water, contributing to underground flow. According to water-balance studies by KOLOTIL' SHCHIKOV [1962], 41–61% of the precipitation was wasted to evaporation during the period 1948–1954; the remaining 39–59% went to inflow and infiltration into the limestones. For this 7-year period, on the average, 38.7% of the total precipitation became spring discharge, but 2.5–23.2% was transformed into groundwater reserves and entered deeper underground flow.

Effective absorption of precipitation by karst formations, and strong underground flow in these rocks, lead to the absence of surface flow or to its strong diminution. This phenomenon is well known, and is observed in separate areas of the Crimean Yayla, on the Ufimsk Plateau, and in other regions of the European and Asiatic parts of the U.S.S.R. Similar phenomena are observed in many karst regions of Italy, Morocco, Australia, France, the Balkan Peninsula, etc.

The "coefficient of absorption" of surface flow in the southern Urals, in the

Kamenka, Blinovka, Termenevka, Pokrovka, and other river basins, was observed by MOLITVIN [1962] to be 4–5 l/sec per km^2 in 1958 and 6–7 l/sec per km^2 in 1954; in the Uluir River Basin it amounted to 4–6 l/sec per km^2 in 1954. In karstified river basins in the Onego–Severo–Dvinsk watershed and in the Pyarga River Basin the coefficient of absorption, according to Molitvin's data, amounted to 6.5 l/sec per km^2 in 1949; in the Sukhaya Sheleksna Basin it was about 7.5 l/sec per km^2. These streams have constant flow only in the reaches near their mouths, and this as a result of the discharge of karst water. Streamflow is seen along the entire extent of a river during the spring flood; at this time considerable losses in flow occur to karst formations in the bottom of the channel and in the valley.

However, losses of surface flow and of precipitation are of largely local character and lead to redistribution of subsurface flow in relatively restricted areas. In this regard the results of studies by MOLITVIN [1962] in karst regions of the northern and southern Urals and in the Onego–Severo–Dvinsk watershed are very significant. A water balance for the Vagran and Sos'va rivers for 1947, computed by Molitvin, showed that practically all the water absorbed in the upper parts of these river basins returns to the rivers "...when the rivers flow from a limestone belt into non-karstified rocks, not affecting the volume at the principal downstream gaging station in the long run" [MOLITVIN, 1962]. The same phenomenon was observed in the southern Urals by Molitvin. Water lost from surface and channel flow in the Kamenka River returned by subsurface routes to the Ay River, which is the principal drainage course in the region. The mean annual coefficient of streamflow is stable in the Ay River Basin (area 5,384 km^2). Channel flow in the Yemtsa River—the principal stream in the Onego–Severo–Dvinsk watershed—also undergoes great variation along its length which is related to the diverse features of carbonate karst in its basin. However, the coefficient of flow levels off for the area of the basin (1,467 km^2) and approaches the "climatic norm". Flow which is lost "...in the basins of the Pyarga and Sukhaya Sheleksna rivers returns to the Yemtsa River by subsurface routes, forming large springs in its valley" [MOLITVIN, 1962].

This penomenon is very significant in the study of underground flow in karst regions. It is necessary, in regional appraisal of the magnitude of underground flow and of natural groundwater resources in karstified areas by the method of separation of river hydrographs, to study basins which are sufficiently large. Moreover, these basins should belong to the type of so-called "neutral rivers", according to the classification of MOLITVIN [1962] and MAKSIMOVICH [1963b], in which the exchange of water between surface and subsurface is of "closed" character; that is, the absorption of surface and channel flows in some parts of a basin is compensated by intensive groundwater discharge in others. Other types of rivers—those with dominant loss of flow into karst formations or with partial loss of flow into karst—cannot be used for appraisal and mapping of underground flow and of natural groundwater resources because in such instances understated values for the moduli of underground flow are obtained. On the other hand, karst streams with predominant groundwater drainage give exaggerated values for this modulus because of groundwater movement from neighbouring basins and the non-coincidence of the surface and subsurface drainage areas. These rivers can be used for this type of appraisal where it is possible to determine with adequate precision the subsurface drainage areas.

The characteristics of underground flow through karst regions which are brought out in the preceding paragraphs were obtained by separation of hydrographs of the total streamflow in sufficiently large basins—from 1,000–3,000 to 10,000 km² and, rarely, to 30,000 km², which can be taken as basins of "neutral rivers". However, the location of gaging stations on the rivers did not always correspond precisely with the boundaries of the karstified rocks. In the sense brought out above, the data are somewhat provisional, although they also reflect fundamental characteristics of the formation of underground flow in karst regions. These characteristics would be expressed still more clearly if the gaging stations were specially located.

On the basis of available material, obtained to the present time by many investigators, the following conclusions can be drawn concerning the influence of karst on underground flow:

(*1*) Karst leads to increase in underground flow. Mean annual, annual, maximum, and minimum moduli and coefficients of underground flow, and coefficients of underground feeding of rivers, are sharply increased in karst regions, relative to their background values.

(*2*) Karst disrupts the smooth zonal character of the areal distribution of values of underground flow which is determined by latitudinal climatic zoning or by vertical zoning in mountain regions.

(*3*) Sharp fluctuations in underground and surface flow are observed within karst regions. Areas with abundant water alternate with areas in which there is relatively much less water; this is related to the nature of the karst and to the hydrography of the karst regions.

(*4*) Karst leads to a redistribution of underground flow into a river within relatively restricted areas. In study of the groundwater resources of karst regions through genetic separation of hydrographs of total streamflow it is necessary to take sufficiently large "neutral" basins within which the complete cycle of water exchange between surface and subsurface is carried out.

(*5*) The large magnitudes of the modulus of underground flow and of the coefficients of underground flow and underground feeding of rivers in karstified regions are explained not only by conditions especially favourable for the infiltration of precipitation and of underground flow into the karst formations but also by unusual features of the regime of karst water, which is distinguished by the great dynamicity, turbulence, and high flow velocity of the groundwater; and by its relatively short flow paths and well-developed areas of recharge and discharge. The mean coefficients of dynamicity of the underground flow under conditions which are otherwise similar are significantly higher for karst water than for other types of groundwater (except for fracture waters in fold-mountain structures).

(6) In many karstified massifs the type of underground-flow regime (which depends on the type of karst) reminds one, in its general features, of the regime of streamflow, sometimes differing from the latter only in lag in the flow maxima and minima.

(7) Storage of water in karstified rocks is relatively restricted, as compared with that in unconsolidated sedimentary deposits. Where the average porosity of large blocks of rock in karstified massifs is a few percent, that of unconsolidated sedimentary deposits amounts to tens of percent. The magnitudes of the moduli of underground flow (base run-off) of the two types of deposits bear the reverse relationship. These circumstances lead to the fact that the period of renewal of karst water is shorter than that of water in sedimentary deposits.

REFERENCES

ANONYMOUS, 1961. *Tectonic Map of the U.S.S.R. Scale 1:25,000,000.* (Under general editorship of A.A. Bogdanov.)
ARGENTOV, K. I., 1911. Preliminary report on the geological investigations of the Przhevalsk district of the Semirechensk region in 1910. *Mountain J.*, 2(37). (In Russian.)
BALKOV, V. A., 1964. Vliyaniye karsta na vodnyy balans i stok. (The influence of karst on the water budget and on flow.) *Uch. Zap., Permsk. Gos. Univ.*, 112.
BARKOV, A.S., 1932. Karst Samarskoy Luki. (Karst of the Samarskaya Luka.) *Zemlevedeniye*, 34(1–2).
BARKOV, A.S., 1933. Karst Vostochno-evropeyskoy ravniny. (Karst of the east European Plain.) *Tr. Vses. Geograf. S"yezda*, 3.
BASKOV, K.A. and KORNUTOVA, YE. I., 1959. Karstovyye yavleniya v Yuzhnoy Yakutii. (Karst phenomena in southern Yakutsk.) *Materialy Vses. Geol. Inst., Novaya Ser., Obshch. Ser.*, 24.
BOGAT'KO, N. M., 1963, *Nekotoryye Novyye Dannyye o Karste i Karstovykh Vodakh Khabarovskogo Kraya i Amurskoy Oblasti. (Some New Data on Karst and Karst Waters of the Khabarovsk Kray and the Amursk Oblast'.)* Nauka, Khabarovsk.
BOL'SHAKOV, N. M., 1944. Novyye danyye po karstu Severo-vostochnogo Salaira. (New data on karst of northeastern Salair.) *Izv. Tomsk. Industr. Inst.*, 62(1).
BOL'SHAKOV, P. M., 1962. Zamechaniya po dokladu B. A. Fedorovicha "Osobennosti migratsii rastvorov i obrazovaniya kor i karsta v pustynyakh." (Comment on B. A. Fedorovich's report "Characteristics of the migration of solutions and of the formation of crusts and karstin deserts.") In: *Obshchiye Voprosy Karstovedeniya. (General Problems of the Study of Karst.)* Akad. Nauk S.S.S.R., Moscow.
BUNINA, M.V., 1947. Karstovyye yavleniya v otlozheniyakh vizeyskogo yarusa v predelakh poley shakht 6 i 15 Kizelovskogo kamennougol'nogo basseyna. (Karst phenomena in deposits of the Visean Stage in the areas of shafts 6 and 15 of the Kizelovsk Coal Basin.) In: *Tezisy Permskoy Karstovoy Konferentsii 26–31 Yanvarya 1947. (Theses of the Perm Karst Conference, 26–31 January 1947.)* Nauka, Perm.
BUROV, V.P., 1960. Karstovyye formy rel'yefa v yugo-zapadnoy chasti Kuznetskogo Alatau. (Karst relief features in the southwestern part of the Kuznetsk Alatau.) *Uch. Zap. Tomsk. Univ.*, 36.
BUROV, V. S., 1963. Ocherk tektoniki vneshney zony Predkarpatskogo progiba. (Outline of the tectonics of the outer zone of the Carpathian Foredeep.) *Sov. Geol.*, 11.
CHERNYSH, L. V., 1966. Po stalaktitovym peshcheram Zakarpat'ya. (Through the stalactite caverns of the trans-Carpathian region.) In: *Karpatskiye Zapovedniki. (Carpathian Preserves.)* Karpaty, Uzhgorod.
CHIKISHEV, A.G., 1958. Karst v basseyne reki Chusovoy na zapadnom sklone Srednego Urala. (Karst in the Chusovaya River Basin on the west slope of the central Urals.) In: *Regional'noye Karstovedeniye. (The Regional Study of Karst.)* Akad Nauk S.S.S.R., Moscow.

CHIKISHEV, A.G., 1964. Karst Srednego Urala i ego narodnokhozyaystvennoye znacheniye. (Karst of the central Urals and its significance for the national economy.) In: *Karst i ego Narodnokhozyaystvennoye Znacheniye. (Karst and its Significance for the National Economy.)* Nauka, Moscow.

CHIKISHEV, A.G., 1967. Rayonirovaniye podzemnykh karstovykh form Urala. (Regionalization of subsurface karst features of the Urals.) *Zemlevedeniye*, 7(47).

CHIKISHEV, A. G., 1968. Osobennosti glubinnogo karsta i speleologicheskoye rayonirovaniye Urala. (Characteristics of deep karst and speleological regionalization of the Urals.) *Proc. Intern. Congr. Speleology, 4th, Ljubljana, 1965*, 3.

CHURINOV, M.V., 1961. O karste Gornogo Kryma. (Karst of the mountainous part of the Crimea.) In: *Regional'noye Karstovedeniye. (The Regional Study of Karst.)* Akad. Nauk S.S.S.R., Moscow.

DAVIS, W. M., 1899. The geographical cycle. *Geograph. J.*, 14.

DOBROVOL'SKIY, M. N., 1965. Kratkiye svedeniya o krupneyshikh peshcherakh Sredney Sibiri. (Brief information on the largest caverns of central Siberia.) In: *Peshchery (Caverns.)—Inst. Karstovedeniya Speleologii Permsk. Univ., Permsk. Otd. Geograf. Obshchestva S.S.S.R.*, 5(6).

DOLGUSHIN, I. YU., 1961. Karstovyye yavleniya v predelakh Aldano–Timptonskogo vodorazdela. (Karst phenomena in the Aldan–Timptonsk watershed.) In: *Regional'noye Karstovedeniye. (The Regional Study of Karst.)* Akad. Nauk S.S.S.R., Moscow.

DUBLYANSKIY, V. N., 1963. O roli snega v zakarstovyvanii i pitanii karstovykh vod. (The role of snow in karstification and in the recharge of karst waters.) *Izv. Akad. Nauk S.S.S.R., Ser. Geograf.*, 2.

DUBLYANSKIY, V. N., 1968. Nekotoryye voprosy gidrogeologii karsta Gornogo Kryma. (Some problems of the hydrogeology of karst in the mountainous part of the Crimea.) *Proc. Intern. Congr. Speleology, 4th, Ljubljana, 1965*,3.

DZENS-LITOVSKIY, A. I., 1940. Karst solyanykh mestorozhdeniy S.S.S.R. (Karst of salt deposits of the U.S.S.R.) *Izv. Vses. Geograf. Obshchestva*, 72(6).

DZENS-LITOVSKIY, A. I., 1966, *Solyanoy Karst S.S.S.R. (Salt Karst of the U.S.S.R.)* Nedra, Leningrad.

FEDOROVICH, B. A., 1949. O roli karsta v rel'yefe pustyn'. (The role of karst in desert topography.) *Tr. Inst. Geograf. Akad. Nauk S.S.S.R.*, 43.

FEDOROVICH, B. A., 1962. Osobennosti migratsii rastvorov i obrazovaniya kor i karsta v pustynyakh. (Characteristics of the migration of solutions and of the formation of crusts and karst in deserts.) In: *Obshchiye Voprosy Karstovedeniya. (General Problems of the Study of Karst.)* Akad. Nauk S.S.S.R., Moscow.

FERSMAN, A. YE., 1952. Geokhimiya peshcher. (The geochemistry of caverns.) *Priroda*, 3.

GAVRILOV, A. M., 1960. O vliyanii karsta na stok malykh rek. (On the influence of karst on the flow of small rivers.) *Izv. Vses. Geograf. Obshchestva*, 92(3).

GERASIMOV, I. P., 1940. Fiziko-geograficheskiy ocherk Sarykamysha. (Outline of the physical geography of Sarykamysh.) *Tr. Inst. Geograf. Akad. Nauk S.S.S.R.*, 35.

GLUSHKO, V.V., 1958. Osnovnyye cherty tektoniki Predkarpatskogo progiba i prilegayushchey chasti Russkoy platformy. (Fundamental features of the tectonics of the Carpathian Foredeep and the adjacent part of the Russian Platform.) *Geol. Sb. L'vovsk. Geol. Obshchestva*, 5–6.

GMELIN, I.G., 1751–1752. *Die Reise in Sibirien von 1733 bis1743.* Göttingen.

GOLTS, S.I. and EPIFANOV, M.I., 1965. Karst depressions in central and east Ustyurt. In: *Karst Types of the U.S.S.R.,15*. Nauka, Moscow. (In Russian.)

GORDEYEV, D.I., 1948. Voprosy karsta v rabotakh akademika F. P. Savarenskogo. (Karst problems in the works of academician F. P. Savarenskiy.) *Tr. Permsk. Karstovoy Konf., Perm, 1946*,1.

GVOZDETSKIY, N. A., 1952a. Notes on the division into districts of the karst in the Greater Caucasus. In: *Geographical Collections, 1. Geomorphology and Palaeogeography*. Geograph. Soc. U.S.S.R., Moscow. (In Russian.)

GVOZDETSKIY, N. A., 1952b. Karstovyye yavleniya v Priangar'ye. (Karst phenomena in the region near the Angara River.) *Uch. Zap., Mosk. Gos. Univ.*, 160(5).

GVOZDETSKIY, N. A., 1954. *Karst. (Karst.)* 2nd ed., Geografizdat, Moscow.

GVOZDETSKIY, N. A., 1957. O rasprostranenii karstovykh yavleniy v pustynyakh i gorakh Sredney Azii. (The distribution of karst phenomena in the deserts and mountains of central Asia.) *Vopr. Geograf., Sb.*,40.

GVOZDETSKIY, N. A., 1953. Kras Velkého Kavkazu. (Karst of the Greater Caucasus.) *Česk. Kras*, 11.

GVOZDETSKIY, N. A., 1960. K voprosu o rasprostranenii i osobennostyakh karsta v gorakh Sredney Azii. (The problem of the distribution and characteristics of karst in the mountains of central Asia.) *Zemlevedeniye, Novaya Ser.*, 5(45).
GVOZDETSKIY, N. A., 1964. Karst rayona Kavkazskikh Mineral'nykh Vod. (Karst of the region of the Caucasus mineral waters.) In: *Karst i ego Narodnokhozyaystvennoye Znacheniye. (Karst and its Significance for the National Economy.)* Nauka, Moscow.
GVOZDETSKIY, N. A., 1965. Tipy karsta Severnogo Kavkaza. (Types of karst of the northern Caucasus.) In: *Tipy Karsta v S.S.S.R. (Types of Karst in the U.S.S.R.)* Nauka, Moscow.
GVOZDETSKIY, N. A., 1967. Karst mezhdurech'ya Malki i Gundelena (Severnyy Kavkaz). (Karst of the Malka–Gundelen interfluve, northern Caucasus.) *Zemlevedeniye, Novaya Ser.*, 7(47).
IVANOV, A. A. and LEVITSKIY, YU. F., 1960. Geologiya galogennykh otlozheniy (formatsiy) S.S.S.R. (Geology of halide deposits of the U.S.S.R.) *Tr. Vses. Nauchn. Issled. Geol. Inst.*, 35.
IVANOV, B. N. and IL'INA, S. M., 1965. Otkrytyy (golyy) karst Gornogo Kryma. (Denuded karst of the mountainous part of the Crimea.) In: *Tipy Karsta v S.S.S.R. (Types of Karst in the U.S.S.R.)* Nauka, Moscow.
KHOROSHIKH, P. P., 1949a. Peshchery Altaya. (Caverns of the Altay.) *Priroda*, 4.
KHOROSHIKH, P. P., 1949b. Issledovaniye peshchery Salaira. (Investigations of the caverns of Salair.) *Priroda*, 8.
KHOROSHIKH, P. P., 1949c. Peshchery Zabaykal'ya. (Caverns of the trans-Baikal region.) *Tr. Kyakhtinsk. Krayevedchesk. Muzeya Kyakhtinsk. Otd. Vses. Geograf. Obshchestva*, 16(1).
KHOROSHIKH, P. P., 1955. *Po Peshcheram Pribaykal'ya. (Through the Caverns of the Region near Lake Baikal.)* Nauka, Irkutsk.
KIKNADZE, T. Z., 1965. Some conditions affecting the karst development in the Arabika Massif (Caucasus, western Georgia). *Probl. Speleol. Res.*, Prague.
KIRILOV, M. V., 1957. K voprosu o karstovykh protsessakh na territorii Krasnoyarskogo kraya. (On the problem of karst processes in the Krasnoyarsk Kray.) *Uch. Zap. Krasnoyark. Ped. Inst.*, 8.
KLEYNER, YU. M., 1959. Morskiye peshchery i karst vostochnogo poberezh'ya Kaspiya. (Marine caves and karst of the east coast of the Caspian Sea.) In: *Speleologiya i Karstovedeniye. (Speleology and the Study of Karst).* Mosk. Obshch. Isp. Prirody, Moscow.
KOLODYAZHNAYA, A. A., 1965. K ^{v}rst Flishevoy Formatsii Yugo-Zapadnogo Sklona Kavkaza. (Karst of the Flysch of the Southwest Slope of the Caucasus.) Nauka, Moscow.
KOLOSKOV, P. I., 1937. Natural conditions of condensation of atmospheric vapors. *Probl. Phys. Geograph.*, 4. (In Russian.)
KOLOSKOV, P. I. 1938. Sorption as one of the origins of well- and groundwater. *Probl. Phys. Geograph.*, 6. (In Russian.)
KOLOTIL'SHCHIKOV, V. K., 1962. Rezhim karstovykh vod "siluriyskogo plato." (Regime of karst waters in the "Silurian plateau".) In: *Spetsial'nyye Voprosy Karstovedeniya. (Special Problems of the Study of Karst.)* Akad. Nauk S.S.S.R., Moscow.
KORENEVSKIY, S. M., 1959. Geologicheskaya kharakteristika solyanykh struktur Verkhnetissenskoy vpadiny. (Geologic character of salt structures of the upper Tissensk Basin.) *Tr. Vses. Nauchn. Issled. Inst. Gallurg.*, 35.
KORENEVSKIY, S. M., 1961. Solyanoy karst Verkhnetissenskoy vpadiny. (Salt karst of the upper Tissensk Basin.) In: *Regional'noye Karstovedeniye. (The Regional Study of Karst.)* Akad. Nauk S.S.S.R., Moscow.
KOROTKEVICH, G. V., 1961. O nekotorykh osobennostyakh razvitiya solyanogo karsta. (Some peculiarities of the development of salt karst.) *Dokl. Akad. Nauk S.S.S.R.*, 136(1).
KOROZHUYEV, S.S., 1965. Karst Yakutii. (The karst of Yakutsk.) In: *Tipy Karsta v S.S.S.R. (Types of Karst in the U.S.S.R.)* Nauka, Moscow.
KRUBER, A. A., 1900. O karstovykh yavleniyakh v Rossii. (Karst phenomena in Russia.) *Zemlevedeniye*, 7(4).
KRUBER, A. A., 1915. *Karstovaya Oblast' Gornogo Kryma. (The Karst Region of the Mountainous Part of the Crimea.)* Nauka, Moscow.
KUDELIN, B. I., 1958. The principles of regional estimation of underground water natural resources and the water balance problem. *Intern. Assoc. Sci. Hydrol., Gen. Assembly Toronto, 1957*, 44(2): 150–167.

KUDELIN, B. I., 1960. *Printsipy Regional'noy Otsenki Yestestvennykh Resursov Podzemnykh Vod. (Principles of Regional Appraisal of Natural Groundwater Resources.)* Izd. Mosk. Univ., Moscow.

KUDELIN, B. I. (Editor), 1966. *Podzemnyy Stok na Territorii S.S.S.R. (Underground Flow in the U.S.S.R.)* Mosk. Gos. Univ., Moscow.

KUZNETSOV, I. G., 1928. Ozero Tserik-Kel' i drugiye formy karsta v izvestnyakakh Skalistogo khrebta na Severnom Kavkaze. (Lake Tserik-Kel' and other karst features in the limestones of the Skalistyy Range in the northern Caucasus.) *Izv. Gos. Russk. Geograf. Obshchestva*, 60(2).

KUZNETSOV, YU. YA., 1963. Karst Ustyurta. (The karst of Ustyurt.) *Zemlevedeniye, Novaya Ser.*, 6(46).

KUZNETSOV, YU. YA., 1965. Plato Ustyurt kak primer karstovoy oblasti pustyni. (The Ustyurt Plateau as an example of a desert karst region.) In: *Tipy Karsta v S.S.S.R. (Types of Karst in the U.S.S.R.)* Nauka, Moscow.

LAPTEV, F. F., 1939. Agressivnoye deystviye vody na karbonatnyye porody, gipsy, i beton. (The aggressive effects of water on carbonate rocks, gypsum, and concrete.) *Tr. Spetsgeo*, 1.

LEVEN, YA. A., 1940. Caves at the upper Magian River. *News All Union Geograph. Soc.*, 72(2). (In Russian.)

LILIENBERG, D. A., 1959. Karstovyye rayony i peshchery Dagestana. (Karst regions and caverns of Dagestan.) In: *Speleologiya i Karstovedeniye. (Speleology and the Study of Karst.)* Mosk. Obshch. Isp. Prirody, Moscow.

LUPPOV, N. P., 1948. O genezise Sarykamyshskoy vpadiny. (On the origin of the Sarykamyshsk Depression.) *Izv. Vses. Geograf. Obshchestva*, 80(2).

LYKOSHIN, A. G., 1960. Nekotoryye gidrodinamicheskiye zakonomernosti razvitiya karsta v platformennykh usloviyakh. (Some hydrodynamic regularities in the development of karst under platform conditions.) *Zemlevedeniye*, 5.

MAKHAYEV, V. N., 1939. Ledyanaya peshchera Abogi-Dzhe. (The ice cavern of Aboga-Dzhe.) *Izv. Gos. Geograf. Obshchestva*, 71(6).

MAKSIMOVICH, G. A., 1962. Gidrodinamicheskiye zony karstovykh vod i osnovyye tipy podzemnogo stoka. (Hydrodynamic zones of karst waters and basic types of underground flow.) In: *Spetsial'nyye Voprosy Karstovedeniya. (Special Problems of the Study of Karst.)* Akad. Nauk S.S.S.R., Moscow.

MAKSIMOVICH, G. A., 1963a. *Osnovy Karstovedeniya. Voprosy Morfologii, Speleologii i Gidrogeologii Karsta, 1. (Principles of the Study of Karst. Problems of the Morphology, Speleology, and Hydrogeology of Karst, 1.)* Nauka, Perm.

MAKSIMOVICH, G. A., 1963b. Nekotoryye voprosy gidrologii karstovykh oblastey. (Some problems of the hydrology of karst regions.) In: *Metodika Izucheniya Karsta, 8. (Methodology of the Study of Karst, 8.* Nauka, Perm.

MAKSIMOVICH, G.A. and GORBUNOVA, K. A., 1958. *Karst Permskoy Oblasti. (Karst of the Perm Oblast'.)* Nauka, Perm.

MAKSIMOVICH, G. A. and KOSTAREV, V.P., 1964. Karstovaya oblast' Kuznetskogo Alatau. (The Kuznetsk Alatau karst region.) *Nauchn. Tr. Permsk. Politekhn. Inst., Sb.*, 12(2).

MARTEL, E. A., 1909. *La Côte d'Azur Russe (Riviera du Caucase)*. Gauthier-Villars, Paris.

MARUASHVILI, L.I. and TINTILOZOV, Z. K., 1963. Results of new speleological investigations in the karstbelt of western Gruzia (1957–1960). *Zemlevedeniye*, 6(16). (In Russian.)

MASLOV, V. P., 1934. Balaganskaya peshchera. (The Balagansk Cavern.) *Byul. Mosk. Obshchestva Ispytateley Prirody, Novaya Ser., Otd. Geol.*, 12(1).

MILIKHIKER, SH. G., 1962. Issledovaniya karsta v rayone srednego techeniya reki Chusovoy v svyazi s gidrotekhnicheskim stroitel'stvom. (Investigations of karst in the region of the middle course of the Chusovaya River in connection with hydraulic structures.) In: *Spetsial'nyy Voprosy Karstovedeniya. (Special Problems of the Study of Karst.)* Akad. Nauk S.S.S.R., Moscow.

MIROSHNIKOV, L. D., 1962. Karst arkticheskoy chasti Sibirskoy platformy. (Karst of the Arctic part of the Siberian Platform.) *Sov. Geol.*, 7.

MOLITVIN, P.V., 1962. Metodika gidrologicheskikh issledovaniy v karstovykh rayonakh Severnogo i Yuzhnogo Urala i Onego–Severo–Dvinskogo vodorazdela. (Methodology of hydrologic investigations in karst regions of the northern and central Urals and of the Onego–Severo–Dvinsk watershed.) In: *Spetsial'nyye Voprosy Karstovedeniya. (Special Problems of the Study of Karst.)* Akad. Nauk S.S.S.R., Moscow.

NOINSKIY, M. E., 1913. Samarskaya Luka. Geologicheskoye issledovaniye. (The Samarskaya Luka. Geological investigation.) *Tr. Obshchestva Yestestvoisp. Kazansk. Univ.*, Kazan, 14(4–6).
OLLI, YE. I., 1958. Karst khrebta Bol'shoy Karatau. (Karst of the Bol'shoy Karatau Range.) *Byul. Mosk. Obshchestva Ispytateley Prirody, Otd. Geol.*, 33(3).
OSOKIN, I. M., 1965. O karste Vostochnogo Zabaykal'ya. (On karst of the eastern trans-Baikal region.) In: *Tipy Karsta S.S.S.R. (Types of Karst in the U.S.S.R.)* Nauka, Moscow.
PECHERKIN, I. A., 1962. Pritok karstovykh vod v gornyye vyrabotki Kizelovskogo kamennougol' nogo basseyna. (Flow of karst waters into mine workings in the Kizelovsk Coal Basin.) In: *Spetsial'nyye Voprosy Karstovedeniya. (Special Problems of the Study of Karst.)* Akad. Nauk S.S.S.R., Moscow.
PECHERKIN, I. A. and KARZENKOV, G. I., 1964. Podzemnyye i shakhtnyye vody Kizelovskogo karstovogo rayona. (Subsurface and mine waters of the Kizelovsk karst region.) In: *Karst i ego Narodnokhozyaystvennoye Znacheniye. (Karst and its Significance for the National Economy.)* Nauka, Moscow.
PITTER, T. M., 1962. Karst Tsentral'nogo Salaira. (The karst of central Salair.) In: *Novyye Dannyye po Geologii Altayskogo Kraya. (New Data on the Geology of the Altaysk Kray.)* Nauka, Novosibirsk.
POPOV, I. V., 1956. Obzor sostoyaniya izucheniya karsta v S.S.S.R. i za granitsey. (Review of the status of the study of karst in the U.S.S.R. and abroad.) *Tezisy Dokl. Soveshch. Izuch. Karsta*, 1.
POSLAVSKAYA, O. YU., 1954. Osobennosti karstoobrazovaniya v nekotorykh pustynnykh i polupustynnykh rayonakh Sredney Azii. (Characteristics of karst formation in some desert and semidesert regions of central Asia.) *Tr. Sredneaz. Gos. Univ., Geol. Nauki*, 1.
PROSKURYAKOV, P. S., 1889. Iyusskiye peshchery. (The Iyusskiye caverns.) *Izv. Vost. Sibirsk. Otd. Russk. Geograf. Obshchestva*, 20(2).
PROSKURYAKOV, P. S., 1892–1893. Torgashinskaya peshchera. (The Torgashinsk Cavern.) *Otchet Obshchestva Vrachey Yenisyskoy Gubernii 1892–1893 Gody.*
RAD'KO, N. I., 1967. Podzemnyye vody Zakarpatskogo vnutrennego progiba i vozmozhnostiikh ispol'zovaniya. (Groundwaters of the inner Carpathian Foredeep and the potential for their use.) In: *Osnovnyye Problemy Izucheniya Ispol'zovaniya Proizvoditel'nykh sil Ukrainskikh Karpat. (Basic Problems of the Study and Use of the Productive Capacity of the Ukrainian Carpathians.)* Kamenyar, L'vov.
RODIONOV, N. V., 1948. Nekotoryye zakonomernosti izmeneniya karbonatnykh porod v protsesse karsta. (Some mechanisms of the modification of carbonate rocks in the karst process.) *Tr. Lab. Gidrogeol. Probl., Akad. Nauk S.S.S.R.*, 3.
RODIONOV, N. V., 1949. Nekotoryye dannyye o skorosti rastvoreniya karsta v karbonatnykh porodakh. (Some data on the rate of solution of karst in carbonate rocks.) *Tr. Lab. Gidrogeol. Probl., Akad. Nauk S.S.S.R.*, 6.
RODIONOV, N. V., 1963. *Karst Yevropeyskoy Chasti S.S.S.R., Urala i Kavkaza. (Karst of the European Part of the U.S.S.R., the Urals, and the Caucasus.)* Gosgeoltekhizdat, Moscow.
ROMANYUK, A. F., 1967. Gidrogeologicheskiye usloviya Predkarpat'ya v svyazi s neftegazonosnost'yu. (Hydrogeologic conditions of the cis-Carpathian region in connection with the occurrence of oil and gas.) In: *Osnovnyye Problemy Izucheniya i Ispol'zovaniya Proizvoditel'nykh sil Ukrainskikh Karpat. (Basic Problems of the Study and Use of the Productive Capacity of the Ukrainian Carpathians.)* Kamenyar, L'vov.
ROZEN, M. F., 1954. Drevniye stoyanki cheloveka v peshcherakh Altaya. (Ancient camps of man in the caverns of the Altay.) *Priroda*, 2.
SADOV, V. P., 1937. Inzhenerno-geologicheskiye usloviya i karstovyye yavleniya rayonov Cheremkhovskogo promyshlennogo kombinata. (Engineering geology and karst phenomena of areas of the Cheremkhovsk industrial plant.) *Tr. Mosk. Geol. Razved. Inst.*, 6.
SALONENKO, V. P., 1960. *Ocherki po Inzhenernoy Geologii Vostochnoy Sibiri, (Outlines of the Engineering Geology of East Siberia.)* Vost. Sibirsk. Geol. Inst. Sibirsk. Otd. Akad. Nauk S.S.S.R. i Irkutsk. Gos. Univ., Irkutsk.
SAVARENSKIY, F. P., 1933. *Gidrogeologiya. (Hydrogeology.)* Gorgeonefteizdat, Moscow.
SEKYRA, Y., 1964. K problematice krasu Stredni Asie. (On the problem of the karst of central Asia.) *Česk. Kras*, 16.

SHCHERBAKOV, A. V., 1961. *Karst in Precambrian Rocks of the Iron Ore Basin of Krivorog–Kremenchug.* Akad. Nauk S.S.S.R., Moscow. (In Russian.)

SHCHERBAKOVA, YE. N., 1961. O karste Minusinskoy vpadiny. (On karst of the Minusinsk Depression.) In: *Regional'noye Karstovedeniye. (The Regional Study of Karst.)* Akad. Nauk S.S.S.R., Moscow.

SHCHUKIN, I. S., 1964. *Obshchaya Geomorfologiya, 2. (General Geomorphology. 2.)* Moskovskogo Universiteta, Moscow.

SKUODIS, V. P., 1959. About the ancient excavated valley in the region of the junction of the Lautse River, in the Daugava River. In: *Geology of the Valley of the River Daugava.* Nauka, Riga. (In Russian.)

SKVORTSOV, T. G., 1959. *Gidrogeologicheskiye i Inzhenerno-Geologicheskiye Usloviya Osvoeniya Bogatstv Vostochnogo Sayana. (Hydrogeology and Engineering Geology of Development of the Resources of East Sayan.)* Materialy Buryatskogo Regional'nogo Soveshchaniya po Razvitiyu Proizvoditel'nykh sil Vostochnoy Sibiri, Ulan-Ude.

SLAVIN, V. I., 1953. Karpatskaya mramoronosnaya provintsiya. (The Carpathian marble province.) *Tr. L'vovsk. Geol. Obshchestva*, 3.

SOKOLOV, D. S., 1951. Osnovnyye usloviya razvitiya karsta. (Fundamental conditions for the development of karst.) *Byul. Mosk. Obshchestva Ispytateley Prirody*, 2.

SOKOLOV, D. S., 1962. *Osnovnyye Usloviya Razvitiya Karsta. (Fundamental Conditions for the Development of Karst.)* Gosgeoltckhizdat, Moscow.

SOKOLOV, N. I., 1937. K voprosu o tektonike Samarskoy Luki. (The problem of the tectonics of the Samarskaya Luka.) *Byul. Mosk. Obshchestva Ispytateley Prirody, Otd. Geol.*, 11(3).

SOKOLOV, N. I., 1959. Nekotoryye novyye dannyye o Vorontsovskikh peshcherakh. (Some new data on the Vorontsovsk caverns.) In: *Speleologiya i Karstovedeniye. (Speleology and the Study of Karst.)* Moskovskoye Obshchestvo Ispytateley Prirody, Moscow.

SOKOLOV, N. I., 1960. Tipologicheskaya klassifikatsiya karsta. (Typological classification of karst.) *Materialy Komis. Izuch. Geoi. Geograf. Karsta, Inform. Sb.*, 1.

STUPISHIN, A. V., 1967. *Ravninny Karst i Zakonomernosti ego Razvitiya na Primere Srednego Povolzh'ya. (Plains Karst and the Mechanisms of its Formation, with the Example of the Central Povolzh'ye.)* Kazanskogo Universiteta, Kazan.

TATARINOV, K. A., 1965. Karst. In: PRIDNESTROV (Editor), *Karst Types of the U.S.S.R.* Nauka, Moscow. (In Russian.)

TIKHOMIROV, N. K., 1934. Znacheniye karsta v gidrogeologii. (Significance of karst in hydrogeology.) In: *Vodnyye Bogatstava Nedr Zemli na Sluzhbu Sotsialisticheskogo Stroitel'stva*, 7. *(Water Resources of the Earth's Interior in the Service of Socialist Construction*, 7.) Akad. Nauk S.S.S.R., Moscow.

VARSONOF'YEVA, V. A., 1915. Karstovyye yavleniya v severnoy chasti Ufimskogo ploskogor'ya. (Karst phenomena in the northern part of the Ufimsk Plateau.) *Zemlevedeniye*, 22(4).

YAKUSHOVA, A. F., 1949. Karst paleozoyskikh karbonatnykh porod na Russkoy ravnine. (Karst in Palaeozoic carbonate rocks on the Russian Plain.) *Uch. Zap. Mosk. Gos. Univ.*, 136(3).

ZAYTSEV, I. K., 1940. *Voprosy Izucheniya Karsta S.S.S.R. (Problems of the Study of the Karst of the U.S.S.R.)* Gosgeolizdat, Moscow-Leningrad.

Chapter 13

Karst of Great Britain

M. M. SWEETING
St. Hugh's College, Oxford (Great Britain)

INTRODUCTION

Limestones, including chalk, cover a good part of Great Britain, but not all of these are karstified. They occur in many of the stratigraphic units of Great Britain from the Cambrian to the Cretaceous and are indicated on Fig. 1. The texture, density and porosity of these limestones are extremely varied; the morphology of the limestone areas is also diverse.

Rocks giving rise to the British karst areas are of variable thickness, some only a few metres thick, others being over 2,000 ft. (600 m). The areal extent of each of the karsts is small; all are shallow karsts, and all intimately associated with the surrounding areas of non-limestone rocks. The British limestones become in general less porous and less permeable with age; karst landforms thus tend to appear more frequently in the older limestones.

The main areas of karst in Britain occur on the outcrops of the Carboniferous limestones, these being the only rocks which have the strength, compactness, and purity necessary for the development of karst landforms. Two older Palaeozoic limestones, the Durness (Cambrian) and the Devonian are also strong and compact enough to be karstified. However, the largest areal extent of limestones in Britain occurs on the Mesozoic rocks including the dolomites of the Permian, the oolites of the Jurassic and the chalk of the Cretaceous.

Karstification implies the development of underground drainage and the disorganisation of surface streams by dolines and swallow holes. The Mesozoic limestones are in the main chemically very pure, yet they are relatively unkarstified. These rocks are in fact more readily dissolved than the older limestones. For instance, the average calcium content of waters in the Jurassic limestones is over 350 p.p.m. and in the chalk is 260 p.p.m.; whereas, the average value for waters in northwest Yorkshire—an area regarded as one of the most karstified in Britain—is only 160–180 p.p.m. Thus a study of the British areas confirms the fact that karstification does not depend entirely upon the rate of dissolution of the rock, but is more dependent upon its type and structure and the selectiveness of the solution.

Although the Mesozoic limestones possess little surface drainage, they are characterised essentially by fluvial land forms, particularly dry valleys. Compared with true karst terrain a definite water-table is present. Dolines and swallow holes

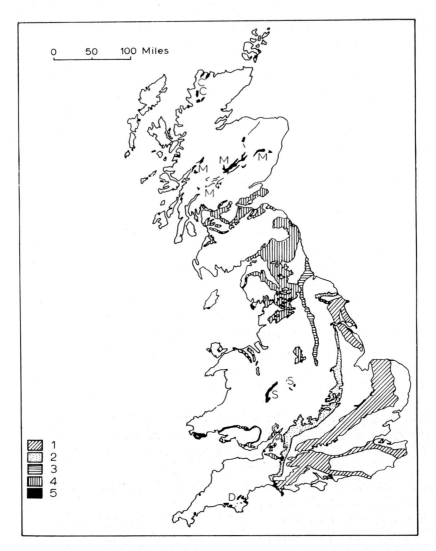

Fig.1. Map of Great Britain to show the outcrops of the rock systems which contain limestone. In the Carboniferous System, limestones are unimportant in the extreme north of England and in Scotland. *1* = Cretaceous; *2* = Jurassic; *3* = Permian; *4* = Carboniferous; *5* = Lower Palaeozoic: *D* = Devonian; *S* = Silurian; *M* = metamorphic limestones (Dalradian); *C* = Cambrian. (After CULLINGFORD, 1962.)

occur but are relatively infrequent. Furthermore, since the Mesozoic limestones are weak, caves are not formed. Owing to their high porosity and great development of pore space, solution in these limestones is diffused throughout the rock. As a result, locally accelerated corrosion does not normally take place and the characteristic karst landforms and hydrology are absent [GAMS, 1965]. Hence, in this chapter, a study of the landforms of the Mesozoic limestones will *not* be regarded as relevant to this review.

Recent work on the landforms of the Mesozoic limestones shows that they have been produced substantially by periglacial action.

THE KARST AREAS OF THE CARBONIFEROUS LIMESTONE

These form the most important karst areas in Britain and occur in northwest Yorkshire, the Morecambe Bay district, Derbyshire, the Mendip area of Somerset, and south and north Wales.

Stratigraphic, sedimentologic and tectonic aspects

The Carboniferous limestone forms the lowest division of the Carboniferous System in Britain. In the southern part its consist mainly of limestones but in the north, particularly in Scotland, it becomee less pure [GEORGE, 1958]. The facies are variable and are made up of reef, basin and massive limestones. The reef limestones form generally unbedded calcite mudstones and are often fine-grained; they are not important in the development of karst landforms though they sometimes contain extensive caves, as in Derbyshire [WARWICK, 1955]. Some of the basin facies are very pure and give rise to important karst features in Derbyshire and the Mendips; rhythmic sedimentation is unimportant in these areas. The massive limestones are well bedded and well jointed and are compact and coarsely crystalline. Some of the massive limestones are of shelf facies, in which rhythmic sequences occur; these show marked

Fig.2. Variations in the Carboniferous limestone, Colt Park, northwest Yorkshire.

vertical variation but remarkably little lateral variation. In any one vertical sequence micrites, biomicrites, sparry limestones and marly or shaly bands are represented (Fig. 2).

In northwest England the two main areas (northwest Yorkshire and Morecambe Bay) belong to different facies groups.

Northwest Yorkshire

The limestones are of shelf facies illustrating to an excellent degree the marked vertical variation and the relative lack of lateral variation. The Great Scar Limestone, as it is known in Yorkshire, is only from 500–700 ft. (180–200 m) thick [GARWOOD and GOODYEAR, 1924]. The lower part in the Visean is generally less sparry and less massive with a high frequency of jointing; the upper part in the Dinantian is more sparry and massive with a lower incidence of jointing [DOUGHTY, 1968]. The limestones rest upon a rigid floor of pre-Carboniferous rocks known as the Askrigg Block; they have a gentle dip of 1°–2° to the northeast, in which direction they are overlain by another rhythmic series, the Yoredales. The block is sharply defined on the west and south by two fault systems, the Dent faults and Craven faults respectively (Fig. 3). In the north and east the floor of the block falls away under an increasing thickness of newer rocks. Both fault systems are of Hercynian age, but have also been rejuvenated during the Late Tertiary and possibly even in the Quaternary [HUDSON, 1933]. The total amount of throw believed to have taken place along the south Craven fault is of the order of 6,000 ft. (2,000 m). The jointing in the Great Scar Limestone is related to the faulting and is probably of Hercynian age [WAGER, 1931].

Morecambe Bay district

In this area the limestones are more than 1,000 ft. (300 m) thick and are of basin facies. They illustrate great lateral variation but little vertical variation. They also rest upon a pre-Carboniferous floor but have been broken up into separate blocks by faulting and erosion, as is shown in Fig. 4. Some of the blocks consist almost entirely of pseudo-brecciated limestones—as at Hutton Roof Crag; some of micrites—as at Meathop; and others of sparry limestones—Arnside Knott (Fig. 5).

Derbyshire

The Carboniferous limestone outcrops in a roughly lozenge-shaped area approximately 18 miles (40 km) by 12 miles (25 km). It is of D_1 age and belongs essentially to a massive basin facies of the central province and is over 2,000 ft (800 m) thick. It occurs as an upraised block or limb of an anticline sloping gently to the southeast with steeper dips on the west and north (Fig. 6). A series of transverse folds and shallow anticlines run in an east–west direction across the main anticline and these folds have been invaded by volcanic material (lava flows, ash, tuff, etc.) which interfere with the karst development of the area. These volcanic deposits are common in the Matlock area in the south. The bedding of the limestone is variable; some of it is

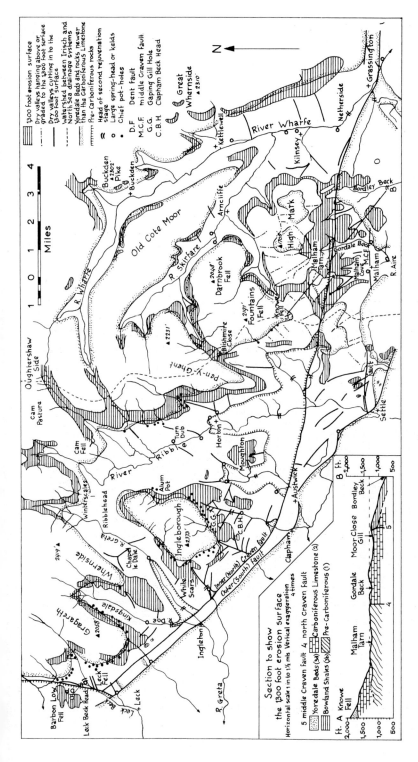

Fig.3. Map of the Ingleborough district of northwest Yorkshire. (After SWEETING, 1950, fig. 4.)

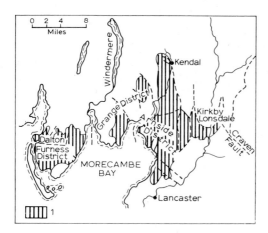

Fig.4. Carboniferous limestone areas of the Morecambe Bay district. 1 = Carboniferous limestone. (After GARWOOD, 1912.)

Fig.5. Pseudo-brecciated limestones near Arnside.

massive, but in other parts it is thinly bedded and cherty [FEARNSIDES, 1932]. Unbedded reef limestones occur along the west and north sides of the area, and can be seen at Castleton. The Derbyshire limestones are mineralized and veins (rakes) occur frequently.

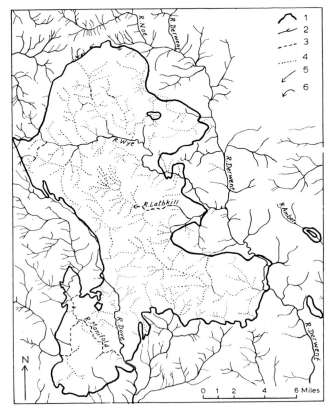

Fig. 6. Dry valley systems of the Carboniferous limestone in Derbyshire. *1* = Limestone boundary; *2* = permanent streams; *3* = temporary streams; *4* = dry valleys; *5* = active effluent cave; *6* = active influent cave. (After WARWICK, 1964.)

The Mendips

The Carboniferous limestone of the Mendips belongs to the basinal facies of the southwest province. It is from 1,700–3,000 ft. (550–1,000 m) thick, and is more thinly bedded and less crystalline in character than the rocks in Yorkshire or Derbyshire[1]. The basal beds in the k-zone are more shaley than those in the upper zones, but there are some thick limestone members. The Clifton Down Limestone is over 700 ft. (200 m) thick and the Black Rock Limestone over 1,000 ft. (300 m) — see GREEN and WELCH[1965].

The structure of the Carboniferous limestone of the Mendips is complex and is made up of four periclines folded during the Hercynian orogeny. The periclines are arranged en-echelon from east-southeast to west-northwest and are known as Beacon Hill, Pen Hill, North Hill, and Blackdown Hill. The cores of the periclines are formed

---

[1] The original zoning of the Carboniferous limestones was carried out in this area by VAUGHAN [1905].

of Old Red Sandstone, while on the flanks are Triassic sandstones, marls and conglomerates. The conglomerates consist of screes formed largely by the break-up of the limestone during Triassic times and now cemented; they are known as the dolomitic conglomerate and behave similarly to the limestone. The eastern end of the Mendips is partly blanketed by Mesozoic deposits. The dip of the limestone around the periclines ranges from 30°–85° (Fig. 7).

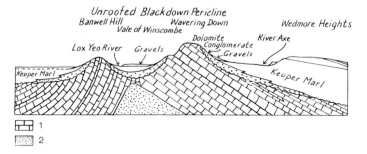

Fig.7. Idealized section from Mendip Hills to the Somerset levels. *1* = Carboniferous limestone; *2* = Old Red Sandstone.

South Wales

The Carboniferous limestones in south Wales are also associated with the basin facies of the southwest province. They outcrop in a great arc running from east to west along the northern edge of the south Wales syncline. The beds range in thickness from about 1,000 ft. (300 m) in the Neath Valley to less than 100 ft. (40 m) near Pontypool. The width of the outcrop is usually less than one mile (1.6 km). The facies show much lateral variation and are partly dolomitic. Structurally the rocks follow the east–west Hercynian fold lines. They are cut by several faults which run mainly north–south in the west and from northwest to southeast in the east. At the head of the Neath and Tawe valleys two large faults run northeast–southwest. In consequence, the outcrop of the limestone is complicated and the general dip is from 5°–15° to the south (Fig. 8).

North Wales

The Carboniferous limestones in north Wales outcrops in detached blocks in many ways similar to those of the Morecambe Bay district. The rocks accumulated in a number of partly connected troughs and in consequence are of varying facies and thicknesses; as much as 3,000 ft. (1,000 m) is found in the east. The basal beds tend to contain much non-calcareous detritus, but the upper beds are purer.

Phases of karstification

The two main phases of karstification affecting the Carboniferous limestone occurred

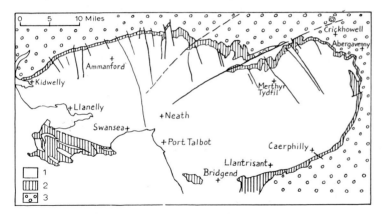

Fig. 8. Distribution of the Carboniferous limestone in south Wales. *1* = Rocks newer than Carboniferous limestone; *2* = Carboniferous limestone; *3* = pre-Carboniferous rocks.

during the Tertiary and the Quaternary, though it is possible that some areas were karstified in earlier periods.

In the Tertiary the climate of Great Britain is believed to have been warmer than it is today. Landforms resulting from the erosion during this period can be recognized in the western part of the Mendips, Derbyshire, and in the Morecambe Bay district.

In the Quaternary, the Carboniferous limestone areas were subjected to an oscillating climate which included periods of severe cold and glaciation. Much of the limestone in northern England became exposed during this time. All areas, except the Mendips, were at some time covered by ice. There were at least three main phases of glaciation with intervening periods of climate resembling that of today. Furthermore, the areas were not equally glaciated; for instance, it is believed that the ice of the Last Glacial Period did not cover parts of the Derbyshire limestones [PIGOTT, 1965].

In general, karstification was intense during phases of recession and melting of the ice sheet and was arrested by stages of severe cold. Since the end of the Quaternary, the climate has become drier and karstification has become less pronounced; there is relatively little erosion of the limestone though some solution is taking place [SWEETING, 1965].

Morphologic features

As a result of the Quaternary glacials much of the karst in Britain is of Quaternary age and only relatively small areas owe their morphology to the Tertiary or earlier periods. In areas outside the glaciations, the former Tertiary landforms have been considerably altered by periglacial activity.

Northwest Yorkshire

Here the main morphologic effects have been produced by the most recent rejuvenation along the Craven and Dent fault systems which has uplifted the Askrigg Block in relation to the surrounding country, so that the limestone forms relatively high ground and is perched above the surrounding lowland. It is therefore a free-draining karst. The average height of the limestones is from 800–1,300 ft. (250–400 m) above sea level, but they nowhere form the highest ground and are capped by residual hills. In the neighbourhood of Malham Tarn the highest beds of the limestone reach more than 1,600 ft. (500 m).

The most characteristic feature of the Carboniferous limestone in northwest Yorkshire is its rhythmic succession which is responsible for the development of the shelves or platforms surrounding the residual hills (Fig. 9). There is also a slight lateral variation from west to east across the area, the limestones in the west containing a higher percentage of sparry calcite than those in the east. Experiments have suggested that there is some differential solubility of these beds, those with the greater proportion of sparry calcite being the less soluble [SWEETING and SWEETING, 1968].

The area lies astride the main water-shed of northern England and has been greatly dissected by fluvial erosion and more recently by glaciation. The drainage is allogenic and has cut down to the level of the pre-Carboniferous floor in the south and west [HUDSON, 1933]. As a result of the rejuvenation of the Craven fault system the most youthful karst relief is also in the south and southwest. This rejuvenation was less marked in the eastern part of the area where the Carboniferous limestone was probably first exposed; the oldest relief is therefore in the eastern part.

The limestones of northwest Yorkshire first became exposed some time during

Fig.9. The limestone platforms surrounding Ingleborough Hill, Chapel-le-Dale, northwest Yorkshire.

the Tertiary period. However, the main stripping of the Yoredales and Millstone Grit overlying these rocks took place during the Quaternary. In the southeast between Arncliffe and Malham there exists an earlier Tertiary phase of karstification, and it is in this part of the area that the only true karstic hollows occur. Some of these are multiple and resemble uvalas and have been mapped by MOISLEY [1955]; they usually contain glacial deposits (Fig. 10). The area around Malham Tarn may be an example of a marginal (or rand) polje where solution planation has taken place [SWEETING, 1964]. Erosion along the mid-Craven fault line scarp has been accompanied by springhead recession and cave collapse, and two very fine *reculées*, one at Malham Cove and the other at Gordale Scar have been formed.

Weathering of the Carboniferous limestone during the Late Tertiary gave rise to a subdued alternation of cliffs and platforms [CLAYTON, 1966]. This step-and-stair relief was further emphasized by glaciation forming broad stripped limestone surfaces, such as are seen around Ingleborough Hill [SWEETING, 1950; Fig. 3,11]. The amount of stripping by the last glaciation may be estimated by the ring of abandoned swallow holes now considerably in advance of the existing Carboniferous limestone–Yoredale boundary. Such depressions may be as much as half a mile (about 800 m) from that boundary.

As a result of glacial scour upon relatively flat-lying beds, limestone pavements are an important feature of the karst relief of northwest Yorkshire; at the end of the Last Glacial Period the area must have been a good example of a *Schichttreppen karst* [BÖGLI, 1964]. These pavements have become modified during the Atlantic and sub-

Fig.10. Uvala, Malham Tarn area.

Fig.11. Glacially-stripped limestone surface, Sear Close, northwest Yorkshire.

Fig.12. Dissected limestone pavement near Alum Pot-hole.

Fig.13. A grike, Scar Close.

Atlantic phases of the Postglacial by the growth of peat and vegetation: ice-scraped surfaces have become grooved and rounded and much reduced in area (Fig. 12.) At some stages in the Postglacial, the pavements probably supported light forest and shrubby vegetation, but as a result of the cutting-down of the trees by man and the effects of grazing by sheep and cows much of the pavement has become re-exposed. The pavements are cut into by deeply-weathered fissures and joints, locally called *grikes* (Fig. 13), the intervening rock outcrops being known as *clints* [KENDALL and WROOT, 1924]. The finest pavements occur on massive sparry limestones. The pavements are relict landforms and when quarried away they become covered with turf.

Because of the perched nature of the beds and the available relief, about 1,000 ft. (300 m), the descent of water through the limestone is rapid; but, the lateral movement is irregular and confined, so that water may take up to fourteen days to travel about two miles (3 km) from swallow hole to resurgence [CARTER and DWERRYHOUSE, 1904]. Another main feature of the karst in northwest Yorkshire is the presence of deep shaft-like swallow holes, known locally as *pot holes*. These are particularly important in the upper part of the limestone and originate from several factors including the type of rock, jointing, available relief and the effects of faulting, and the effects of glacial melt water. Gaping Gill Hole is the most famous and is situated on the southern slopes of Ingleborough where Fell Beck disappears over a waterfall 360 ft. (120 m) high. Detailed mapping of the pot holes shows that many are related to the disposition of the boulder clay (glacial till); others have been formed in association with the ice-front during the Glacial Periods. Pot holes are more numerous in the south and west of the area.

Where superficial material covers the limestone, and enlargement of fissures in the rock has taken place, this material subsides to form small crater-like hollows; these resemble alluvial dolines and are known locally as *shake holes*.

The caves in the western part of the area are generally still active and are of relative recent origin; they have been formed in close association with the pot holes by glacial melt water during certain stages of the Quaternary and are predominantly of vadose origin. They consist of relatively horizontal passages separated by vertical shafts. It can be shown that various levels of cave formation exist and that these are not always related to variation in the lithology of the limestones [SWEETING, 1950]. Fig. 14 shows the succession of levels in the Ease Gill cave system which is believed to have been formed as a result of fluctuating water conditions during the Quaternary [EYRE and ASHMEAD, 1967]. There is no true water-table known in northwest Yorkshire. Underground-water follows particularly the major lines of jointing and resurgences tend to be related to the height of the pre-Carboniferous floor. In the eastern part of the area most of the caves are inactive and are partly relict landforms. A Watham has recently described systems of phreatic caves in this area.

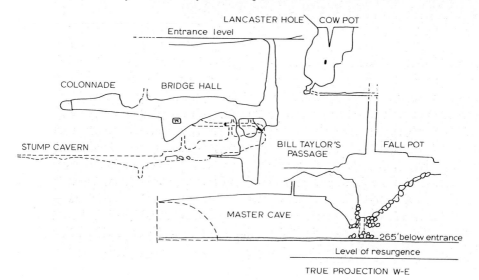

Fig.14. Outline plan of the Ease Gill cave system. (After EYRE and ASHMEAD, 1967.)

Morecambe Bay district

The Carboniferous limestone in the region surrounding Morecambe Bay is made up of faulted blocks resting upon a floor of pre-Carboniferous rocks. The blocks give rise to small hills which vary in altitude from about 300 ft. (100 m) to about 1,200 ft. (400 m). The relief of each block is somewhat different partly due to the differences in lithology and dip. The hills are conical or roof-like in form, as for example at Hutton Roof Crag, and show evidence of distinct phases of karstification. Arnside Knott for instance has smooth and weathered limestone-walls accompanied by a red clay; iron ore also occurs in pockets representing old swallow holes in the area known as Furness (DUNHAM and ROSE, 1949]. This evidende supports the idea brought forward by CORBEL [1957] that these hills are relics of a tropical Tertiary karst.

A later phase of erosion is represented by the immense deposits of layered scree which are piled up against the hills and indicate a long period of formation. The region was also affected by glacial scour and possesses some good pavements on steeply dipping beds. These pavements are dissected by rund and rinnen karren. The best examples of these in the British Isles are to be seen at Hutton Roof Crag in Westmorland, where they are up to 13 m long (Fig. 15). The summits of the hills have been so scoured by ice that there is little evidence of the pre-glacial development of enclosed hollows and there is nothing comparable to the large hollows seen in the Malham district of northwest Yorkshire.

Fig.15. Rinnen-Karren, Hutton Roof Crag, Westmorland.

Derbyshire

In many ways this is the most complex karst in Britain. The Carboniferous limestone is entirely surrounded by impermeable rocks and the region is an example of a *karst barré*. The presence of drainage from the impermeable rocks and from the intercalated volcanic tuffs has caused the limestone to be more water-logged than in any other area; this is often referred to in the history of mining in this area.

Fig. 6 shows the outcrop of the Carboniferous limestone. The stripping of the overlying rock probably took place during the uplift of the south Pennine anticline in the Tertiary Period. This anticline is aligned from north-northwest to south-southeast and it is along this axis that the oldest landforms have developed. Associated with this axis and also with the dolomitized limestone in the southeast are the so-called *pocket deposits* (Fig. 16). These occur in the hollows in the limestone surface and are regarded by YORKE [1954–1961] as old dolines or collapsed caverns, suggesting an early period of karstification. In the same area conical hills, about 1 km long and 50 m high, occur and these resemble the conical karst hills of the Mediterranean. In Derbyshire these hills are called *lows*, an example being End Low. Further work on the solution subsidence features has been done by T.D. Ford and R.J. King.

Towards the periphery of the limestone, nearer the impermeable rocks, the karst features are newer. Much of the stripping of the overlying rocks took place during the Quaternary; blind valleys, like Perryfoot, and caves associated with this stripping are

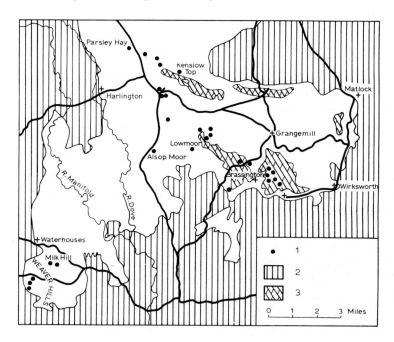

Fig.16. Distribution of sink hole deposits in the southern Pennines. *1* = Sink hole with Triassic beds; *2* = limits of Carboniferous; *3* = dolomitized limestone. (After KENT, 1957.)

therefore of relatively recent age. Some evidence for this fact has been demonstrated by A. F. Pitty (personal communication, 1967)[1].

It will be seen from Fig. 6, showing the drainage and dry valley network, that this is superimposed and let down from the overlying rocks. As a result of the generally easterly tilt of the Pennines, the main drainage is allogenic and towards the east and southeast. High level erosion surfaces accompanied the early development of this drainage. The upper parts of the valleys in particular tend to be wide and shallow and gently graded, often consisting "of a wide *bowle*, oval or circular in shape" [WARWICK, 1964]. Some of these bowles have been altered both by the presence of the interbedded volcanics and also by karstification; the presence of these rocks within the limestone gives rise to the conditions necessary for the impeding of the karst process and the development of poljes. An example of this may be the conspicuous high level flat occupied by the village of Chelmorton.

The lower parts of the valley are deeply cut, "with steep rocky walls and steeper gradient" [WARWICK, 1964]. These deeper valleys and the dry valleys were probably cut by water action in the Quaternary. Resurgences are also common, a particularly good example being that associated with the Lathkill River.

The effects of Quaternary periglacial action are also very evident. Many slopes consist of scree, the result of this action. The direct effect of glaciation upon the limestone is less important here than in northwest Yorkshire, and there are few if any limestone pavements. PIGOTT [1962] has deduced that most of the limestone surfaces of Derbyshire were covered with loess during the Last Glacial Period.

The caves of the Derbyshire area tend to be distributed around the edges of the limestone outcrop, and "with the exception of the Castleton Group are mainly found in the sides of valleys or at river level" [CULLINGFORD, 1962]. There are three main concentrations of caves—at Castleton, Matlock, and in the Manifold—Dove valleys. In the Castleton area streams draining off the shales pass underground into caves soon after reaching the limestone. Some of the caves are associated with mineral veins. Around Matlock the caves are small and are largely open fissures. In the Manifold–Dove area they are related to old valley floor levels; they occur mainly in the reef limestones. Some of the oldest caves of the area lie along the crest of the anticlinal axis already referred to; these caves lie high above the present water level and belong to the earliest stages of the karstification.

The Mendips

The four periclines forming the Mendip Hills stand out above the surrounding countryside as a well defined and flat-topped ridge varying in height from about 800–1,000 ft. (250–300 m) above sea level. Both the steeply dipping Old Red Sandstone and the Carboniferous limestone have been markedly planed across. The age and origin of this planation is disputed, but it may have originated by marine abrasion during the

---

[1] This sequence is not invariably true as in the region of Buxton and Castleton, Early Pliocene karst features occur as has been proved by cave fauna.

Mesozoic and later resurrected and trimmed by further erosion in the Tertiary. Owing to the fact that the limestone weathers more rapidly by solution, the Old Red Sandstone stands out as low hills. The Mendips slope abruptly down to the surrounding plain at an angle of 15°.

It is difficult to ascertain the extent to which the Triassic erosion and scree formation has influenced the present-day karst relief of the Mendips. A. Frey (personal communication, 1965) has shown that there is a relationship between the Triassic dolomitic conglomerate and certain elements in the present-day drainage network. Otherwise, the most important phase affecting the development of the landforms in the Mendips occurred in the Quaternary. The area was one of intense periglacial activity and it is probable that the dry valleys were formed by surface run-off when the sub-soil was frozen and impermeable. Much of the scree in the gorges of Cheddar and Ebbor is of periglacial origin. Moreover, since the area was not glaciated there are no limestone pavements (Fig. 17).

Streams originate on the higher ground of the Old Red Sandstone and sink into swallow holes *(swallets)* at or near the contact of the sandstone with the lower beds of the limestone. Water issues at the foot of the Mendips in a series of springs of which those at Cheddar and at Wookey Hole are the best known. The rates of solution of the Mendip waters have been studied for many years by the University of Bristol [SMITH and MEAD, 1962].

Dry valley-networks radiate from the centre of the periclines. The chief dry-valley system is that which gathers into Cheddar Gorge, one of the most impressive features in the British karsts; this gorge has often been regarded as a collapse cavern,

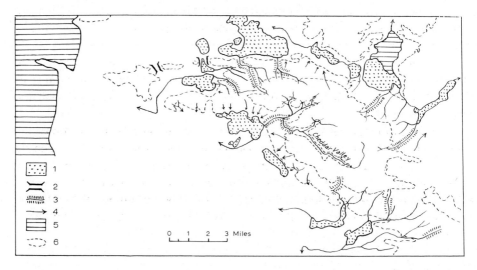

Fig.17. Periglacial gravels around Mendip. *1* = Periglacial gravels; *2* = Trias floored gaps; *3* = gorges and deep combes; *4* = principal valleys; *5* = seas and reservoirs; *6* = 250-ft and 800-ft. contours. (After A. Frey, personal communication, 1965.)

but this is highly unlikely [FORD, 1965]. Blind valleys also occur, as at Eastwater Swallet, where the blind valley end is about 40 ft. (12 m) high.

Dolines, of varying form and dimensions, are a characteristic feature of the Mendips. They frequently occur in the floors of the dry valleys and FORD [1965] has shown how their distribution is related to the speed of run-off of the drainage; they occur in those locations where runoff is impeded. There is evidence which shows that dolines are forming at the present time. The Mendips also have some larger closed depressions which were probably initiated in the Tertiary.

There are four main groups of caves:

(1) Swallet caves; these are formed at the contact of the sandstone (or the basal limestone shales) with the limestone, and include some of the most interesting caves of the area, as for example Swildon's Hole. Many of them are over 400 ft. (150 m) deep and have been explored for many thousands of feet.

(2) Doline caves; there are only a few of these and they vary in depth from 80–200 ft. (20–70 m).

(3) Interrupted caves; these were formed before the cutting of the dry valleys in the Quaternary; an example is Goatchurch Cave which has been cut across by the development of the dry valley of Burrington Coombe.

(4) Resurgent caves; such caves are associated with the uprising of water around the flanks of the Mendips, the best known examples being Wookey Hole and the Cheddar caves. The average daily flow of the rising at Wookey Hole is about 23 million gallons.

A model for the origin of the main caves of central Mendip was recently constructed [FORD, 1965]. This, in many ways, resembles that devised by BÖGLI [1966] for the formation of the Hölloch in Switzerland. Distinct phases of erosion and deposition can be recognized and these may relate to the various stages of the Quaternary.

Until recently the connections between the swallow holes and risings of the Mendips are unknown. However, they have now been proved by work undertaken by the University of Bristol in co-operation with the City of Bristol Waterworks [ATKINSON et al., 1967].

There is a contrast between the eastern and western parts of the Mendips. This is mainly due to the fact that the limestone was exposed earlier in the west and the karst is therefore older. In addition, the western part was affected by the warmer conditions of the Tertiary Period, and it is possible that relic tropical landforms can be recognized. In the east the limestone only became exposed during the Quaternary and has been solely affected by the cooler conditions prevailing during that period [CORBEL, 1957].

South Wales

Despite the relatively narrow outcrop of the Carboniferous limestone in south Wales, the area shows some of the most impressive karst features of Britain. The main rivers rise on the Old Red Sandstone to the north and cut across the limestone in a series of gorges. The river valleys are normally dry in summer but are flooded in winter; a study of the surface and underground flow of the river Mellte has been made by GROOM and WILLIAMS [1965]. The most important caves are associated with these rivers, those in the western part of the area being in general longer than those in the east. They show many features typical of their phreatic origin but they have been much modified by the corrosive and erosive action of the streams. A number of new explorations have taken place in this area, notably at Dan Yr Ogof; this cave has now been followed for over 6 miles (10 km) and contains some fine straw stalactites.

South Wales was glaciated and poorly preserved limestone pavements occur at intervals. Compared with Yorkshire the limestone is much more fractured and thinly bedded so that the conditions for the development of pavements are not so favourable. Much of the area is covered by glacial deposits and peat; acid waters which drain from these have accomplished much solution of the limestone.

A feature of the karst of south Wales is the occurrence of shafts formed by solution beneath a cap rock. THOMAS [1954] drew attention to the large number of doline-like hollows which are found not only on the limestone but also on the overlying Millstone Grit. A recent study by BURKE [1967] has shown how solution takes place underneath the cap rock of Millstone Grit producing shafts up to 100 ft. (30 m) deep in the limestone. Similar features have also been described in the karst of Kentucky. Widespread solution subsidences of the Millstone Grit on to the Carboniferous limestone have also taken place and these indicate that large amounts of water must travel down the dip of the rocks [THOMAS, 1963; Fig. 18].

North Wales

The karst here resembles that of the Morecambe Bay district. There is evidence of a palaeo-karst developed in the Late Tertiary, as for instance in the Elwy Valley near the Cefn caves. However, the entire karst area of north Wales was glaciated so that glacial and periglacial features are superimposed on to the relic forms. Limestone pavements occur, though they are not well developed and have been much dissected by the solutional action of vegetation.

There is relatively little doline development. The most important part of the north Wales karst occurs east of the Clwydian hills, in the region of Maeshafn in Flintshire where there is some important cavern formation.

THE KARST AREAS OF OLDER PALAEOZOIC LIMESTONES

These areas form minor karstic regions and are developed on the Durness Limestone in Scotland and on the Devonian limestone in southern England. They do not possess in any degree such a wide range of karst landforms as has been described for the younger Carboniferous limestones.

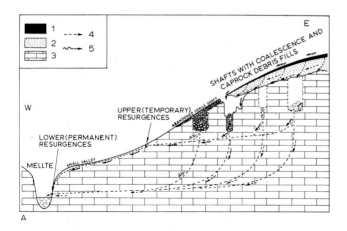

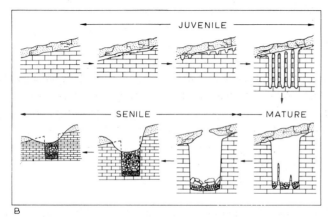

Fig.18. The formation of shafts below a caprock in the Ystradfellte district of south Wales. *1* = Peat; *2* = caprock; *3* = limestone; *4* = normal flow; *5* = storm flow; *Ab* = abandoned. (After Burke, 1967.)

A. Surface and ground water feeds to shafts under different conditions of precipitation, and their relation to resurgences.

B. Diagrammatic west–east sections of a typical simple shaft illustrating successive morphological stages in the sequence of development.

The Durness Limestone area

The Durness Limestone outcrops along the line of the great Moine thrust in northwest Scotland from Loch Kishorn to the coast of Durness. It is also found on Lismore Island in Loch Linnhe. The limestone structurally forms part of the imbricate zone associated with the Moine thrust and is much faulted.

The Durness Limestone which is of Cambro-Ordovician age is formed of six main sub-groups and is over 1,500 ft. (500 m) thick. It is predominantly fine-grained and dense but is far from homogeneous and contains many chert and dolomitic layers.

Little is known of the phases of karstification of the limestone which must have been exposed to erosion for a considerable period, but the greatest effect upon the land forms has been caused by glacial conditions in the Quaternary.

Glacially scoured surfaces occur in the area, but they are not so well developed as in Yorkshire; this is partly due to the less massive nature of the limestone and also due to greater frost action. Indeed, this area illustrates particularly well the erosive action of frost and snow upon limestone. Numerous small hollows have resulted from solution under snow; these resemble the *kotliči* of Slovenia.

Underground water courses and caves are relatively small and undeveloped and indicate that they belong to a recent phase of karstification. The largest cave is in Traligill near Inchnadamph and is about 700 ft. (230 m) long. On the north coast near Durness both karstic and marine processes have contributed to form the landscape and an example is seen at Great Smoo Cave [FORD, 1959].

The Devonian limestone area

A second minor area of karst in Great Britain occurs in the Devonian limestone of south Devon; it is about 250 ft. (80 m) thick and is generally massive and unstratified and built up of corals and stromatoporoids. The dip is variable. The most important development in this area is the caves; these are numerous and have an interesting series of stalagmitic and other deposits[1]. The caves are of phreatic origin. The limestone is relatively near sea level and has been affected by recent changes in base level during Glacial and Postglacial times; many of the caves are associated with river terraces, though strong lithologic control also exists.

HYDROGEOLOGIC CONDITIONS WITH SPECIAL REFERENCE TO THE CARBONIFEROUS LIMESTONE AREAS

There are great variations in the hydrology of the different areas, which are dependent upon structural, lithological and morpho-climatic factors. The most common feature is the relative youth of the hydrologic pattern which is due to the generally recent evolution of the landforms during the Quaternary. This has two main hydrogeologic consequences:

(*1*) Underground water flow is in distinct and discrete channels of varying grade. The existence of several underground basins each feeding different spring heads, as in Yorkshire and in the Mendips, suggest that there is little mixing of the different water bodies underground. In the Settle area of northwest Yorkshire three bore holes for water were recently drilled within 100 m of each other; one of these yielded water at 88 ft. (28 m) whereas the others gave little or no water at an elevation of over 300 ft. (100 m). This is but one example of the unpredictability of the occurrence of water in the Carboniferous limestone.

---

[1] The cave deposits of this area are the most important of their kind in Great Britain.

The only area where the hydrologic conditions suggest the presence of a water-table is in parts of Derbyshire; this is due to a certain extent to the structure and to the longer period of time that the Derbyshire limestones has been exposed.

(2) The rate of water movement is generally rather slow, particularly when compared with such regions as the Canadian Rockies and Jamaica. In 1904, The Yorkshire Geological Society found that water took five to eleven days to travel two miles from Gaping Gill to its resurgence [CARTER and DWERRYHOUSE, 1904]. PITTY [1966] has found more recently that water may take up to 84 days to travel about 100 m. In the Mendips where tectonic conditions are more favourable, water movement is more rapid being about 800 ft. (260 m) per hour [DREW, 1967].

These facts show that the water conditions in the Carboniferous limestone are tight or obstructed; this is borne out by the sudden flooding of underground cave passages or pot holes after heavy rain.

PRACTICAL PROBLEMS AND THEIR APPROACH

The practical problems associated with the karst areas resolve largely into two main groups, those affecting water supply and those affecting land use.

Problems of water supply

The karst areas are all regions of relatively high rainfall (over 40 inches—100 cm—per year), and until recently it was thought that there was no real problem of water supply. However, with greater demand both for domestic and industrial uses, more efficient employment needs to be made of British water resources. For this, a detailed knowledge of the hydrology of the karst areas is required. Though the Yorkshire Geological Society pioneered one of the first experiments in water tracing, there has been little subsequent interest in karst hydrology by geologists in Great Britain. Supplies for many rural areas are based on very local catchments as exemplified by one small village in the Settle district of Yorkshire where there are no less than six different sources. Thus, concerted efforts should be made to find out more precisely the paths of underground water flow, the nature of that flow, and the length of time taken by the water to travel underground; for this work radioactive isotopes could be used with profit. The recent collaboration between the University of Bristol Speleological Society and the City of Bristol Waterworks illustrates the type of experiments that can be done [DREW, 1967; Fig. 19]. Until more accurate information about the water budget of each particular input–output system of swallow hole to resurgence is known, little development and conservation of water resources can be undertaken. The installation of autographic rain gauges and flood gauges has taken place only in the last few years. From the point of view of the use of underground water in karst areas Great Britain is almost completely undeveloped.

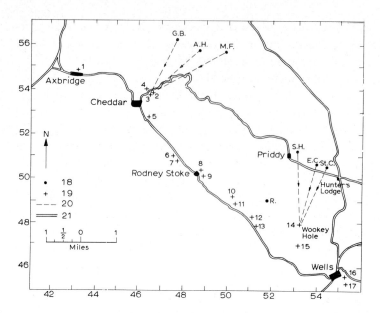

Fig.19. Established flow lines in the central Mendip. *1* = Axbridge; *2* = Cheddar First Feeder; *3* = Cheddar Second Feeder; *4* = Cheddar Lake Springs; *5* = Laubram Batch; *6* = Barnet's Well; *7* = Honeyhurst Borehole; *8* = Rodney Stoke; *9* = Springhead; *10* = Westbury Main; *11* = Westbury (Railway Inn); *12* = Hollybrook; *13* = Easton; *14* = Wookey Hole; *15* = Glencot; *16* = St. Andrews Well, Scotland Rising; *17* = St. Andrews Well, Main Rising; *18* = sink; *19* = rising; *20* = trace line; *21* = road; *G.B.* = G.B. Cave; *A.H.* = Longwood Swallet; *M.F.* = Manor Farm Swallet; *S.H.* = Swildons Hole; *E.C.* = Eastwater Swallet; *St.C.* = St. Cuthberts Swallet; *R.* = Ramspit. (After ATKINSON et al., 1967.)

Problems of land use

The karstlands of Great Britain also illustrate many of the problems associated with the competition for land use; in a small country with such a high density of population further land planning is essential. Road and rail communications are good for these upland areas. There are three main aspects of land use which need to be considered, those connected with agriculture, with quarrying and mineral extraction, and with recreation. These uses are frequently conflicting.

At present, the main type of farming is live-stock production and these parts of highland Great Britain form great rearing grounds for young animals, particularly sheep and cows. However, the pastures could be much improved by grassland improvement schemes and by the removal of surface layers of limestone pavements. As a result of deforestation in these areas, they are now open moorlands. It can be shown that if grazing is properly controlled, trees will regenerate; however, afforestation schemes in Britain are often bitterly opposed.

As in other upland areas of western Europe, the karstlands have experienced a certain amount of rural depopulation which has allowed much land to degenerate. Capital investment in sensible land schemes is important for their future development.

The karst areas contain some of the most important reserves of limestone used in the production of iron and steel, fertilisers and road metal. Some of the largest limestone quarries in the world are at Buxton in Derbyshire. In all areas limestone quarrying is one of the major occupations and as a consequence large tracts are taken up by this pursuit. In addition, mining for fluorspar and barytes is of much significance in Derbyshire and Yorkshire.

The quarrying and also to a certain extent the agricultural interests of the karst areas conflict with their use as recreational and scenic countryside. This is particularly so in northwest Yorkshire and in Derbyshire, both of which have been designated National Parks. The preservation of even such relatively small areas from industrial and residential development raises many difficult problems.

The limestone pavements in northwest Yorkshire are conspicuous scientific and scenic features, but from the point of view of the farmer they prevent the maximum use of the land. In the latter part of the nineteenth century many acres of pavements were removed, thereby improving the quality of the land (Fig. 20). One of the objectives of the National Park is to preserve these highly interesting landforms and this conflicts with efforts made to farm the land to its maximum intensity. The demand for limestone in many sectors of industry has increased enormously in recent years. In order to preserve as much of the fine scenery as is possible, careful selection of sites for quarrying is essential and these may not always be the most suitable economically. Moreover, the need for housing and similar developments are frequently opposed to the ideals of a completely "natural" National Park.

Fig.20. Present-day removal of limestone pavements.

Thus the problems of the karstlands of Britain are in many ways very different from those of the classical karstlands of Yugoslavia, yet at the same time present difficulties peculiar to all limestone areas.

REFERENCES

ATKINSON, T., DREW, D. P. and HIGH, C., 1967. Mendip karst hydrology (research project). *Wessex Cave Club*, 2(1):1–38.
BÖGLI, A., 1964. Le Schichttreppenkarst. *Rev. Belge Géograph.*, 88:63–82.
BÖGLI, A., 1966. Karstwasserfläche und unterirdische Karstniveaus. *Erdkunde*, 20(1):11–19.
BURKE, A. R., 1967. Geomorphology and spelaeogenesis of vertical shafts in Carboniferous limestone at Ystradfellte, Breconshire. *Proc. Ann.Conf. Brit. Speleol. Assoc.*, 5:17–46.
CARTER, W. L. and DWERRYHOUSE, A. R., 1904. The underground waters of northwest Yorkshire, 11. The underground waters of Ingleborough. *Proc. Yorkshire Geol. Polytech. Soc.*, 15(2): 248–292.
CLAYTON, K. M., 1966. The origin of the landforms of the Malham area. *Field Studies*, 2(3):359–384.
CORBEL, J., 1957. Les karsts du nord-ouest de L'Europe. *Rev. Géograph. Lyon, Mém. Doc.*, 12:514 pp.
CULLINGFORD, C. H. D. (Editor), 1962. *British Caving*. Routlegde and Kegan Paul, London, 592 pp.
DOUGHTY, P. S., 1968. Joint densities and their relation to lithology in the Great Scar Limestone. *Proc. Yorkshire Geol. Soc.*, 36:479–512.
DREW, D. P., 1967. *Aspects of the Hydrology of the Eastern Mendips*. Thesis, Bristol University, 119 pp., unpublished.
DUNHAM, K. C. and ROSE, C. C., 1949. Permo-Triassic geology of South Cumberland and Furness. *Proc. Geologists Assoc. (Engl.)*, 60:11–40.
EYRE, J. and ASHMEAD, P., 1967. Lancaster Hole and the Ease Gill Caverns, Casterton Fell, Westmorland. *Trans. Cave Res. Group, Gt. Brit.*, 9(2):65–123.
FEARNSIDES, W. G., 1932. Geology of the eastern part of the Peak district. *Proc. Geologists Assoc. (Engl.)*, 30:152–191.
FORD, D. C., 1965. The origin of limestone caverns: a model from the Central Hills, England. *Bull. Natl. Speleol. Soc. Am.*, 27(4):109–132.
FORD, T. D., 1959. The Sutherland Caves. *Trans. Cave Res. Group, Gt. Brit.*, 5(2):139–190.
GAMS, I., 1965. Types of accelerated corrosion. *Proc. Intern. Speleol. Conf., Brno, 1964*, pp. 133–139.
GARWOOD, E. J., 1912. The Lower Carboniferous in the northwest of England. *Quart. J. Geol. Soc. London*, 68:451.
GARWOOD, E. J. and GOODYEAR, E., 1924. The Lower Carboniferous succession in the Settle district and along the line of the Craven faults. *Quart. J. Geol. Soc. London*, 80:184:273.
GEORGE, T. N., 1958. Lower Carboniferous palaeography of the British Isles. *Proc. Yorkshire Geol. Soc.*, 31(3):227–318.
GREEN, G. W. and WELCH, F. B. A., 1965. Geology of the country around Wells and Cheddar. *Geol. Surv. Gt. Brit., Mem. Geol. Surv. Gt. Brit., Engl. Wales*, 280:1–225.
GROOM, G. E. and WILLIAMS, V. H., 1965. The solution of limestone in south Wales. In: *Symposium on Denudation in Limestone Regions—Geograph. J.*, 131(1):37–41.
HUDSON, R. G. S., 1933. The scenery and geology of northwest Yorkshire. *Proc. Geologists Assoc. (Engl.)*, 44:228–255.
KENDALL, P. F. and WROOT, H. E., 1924. *Geology of Yorkshire*. Hollinek, Vienna, 995 pp.
KENT, P. E., 1957. Triassic relics and the 1,000-ft. surface in the southern Pennines. *East Midland Geographer*, 8:3–10.
MOISLEY, H. A., 1955. Some features in the Malham Tarn district. *Field Studies*, 1953–1954: 33–42.
PIGOTT, C. D., 1962. Soil formation and development on the Carboniferous limestone of Derbyshire, 1. Parent materials. *J. Ecol.*, 50:145–156.
PIGOTT, C. D., 1965. The structure of limestone surfaces in Derbyshire. In: *Symposium on Denudation in Limestone Regions—Geograph. J.*, 131(1):41–44.
PITTY, A. F., 1966. An approach to the study of karst water. *Univ. Hull, Occasional Papers Geograph.*, 5:1–70.

SMITH, D. I. and MEAD, D. G., 1962. The solution of limestone with special reference to Mendip. *Proc. Univ. Bristol Speleol. Soc.*, 9(3):188–211.

SWEETING, G. S. and SWEETING M. M., 1969. Some aspects of the Carboniferous limestone in relation to its landforms. *Mediterranée*, 7:201–209.

SWEETING, M. M., 1950. Erosion cycles and limestone caverns in the Ingleborough district. *Geograph J.*, 115:63–73.

SWEETING, M. M., 1964. Karst features of the Ingleborough Settle area. Paper given to the Karst Symposium Intern. Geograph. Congr., Settle, 000 pp., unpublished.

SWEETING, M. M., 1965. The weathering of limestones. In: G. H. DURY (Editor), *Essays in Geomorphology*. Heineman, London, pp.1–210.

THOMAS, T. M., 1954. Swallow holes on the south Wales coalfield. *Geograph. J.*, 120:468–475.

THOMAS, T. M., 1963. Solution subsidence in southeast Carmarthenshire and southwest Breconshire. *Trans. Inst. Brit. Geograph.*, 33:45–60.

VAUGHAN, A., 1905. The palaeontological sequence in the Carboniferous limestone of the Bristol area. *Quart. J. Geol. Soc. London*, 41:181–305.

WAGER, L.R., 1931. Jointing in the Great Scar Limestone of Craven and its relation to the tectonics of the area. *Quart. J. Geol. Soc. London*, 88:392–424.

WARWICK, G. T., 1955. Polycyclic swallow holes in the Manifold Valley, Staffordhire. *Proc. Intern. Speleol. Congr., 1st, Paris, 1953*, 11:59–68.

WARWICK, G. T., 1964. Dry valleys of the southern Pennines, England. *Erdkunde*, 18(2):116–123.

YORKE, C., 1954–1961. *The Pocket Deposits of Derbyshire*. Privately printed, Birkenhead, 86 pp.

Chapter 14

Karst of Jamaica

H. R. VERSEY[1]
Geological Survey Department, Kingston (Jamaica)

GENERAL GEOLOGY

Jamaica is one of the two most westerly islands of the Antillean Island Arc. The other, Cuba, lies 100 miles to the north and is separated from Jamaica by the Cayman Trench reaching a depth of over 2,500 fathoms in the Oriente Deep. The southern peninsula of Haiti lies 120 miles to the east.

The island has two broad physiographic and structural units which ZANS [1958, p. 34] named the "Eastern Mountain Mass" and the "Main Block". The former consists of Cretaceous and Lower Eocene clastic rocks with acid and basic intrusions: limestones occur only peripherally and karst is but poorly developed within them. It is in the Main Block, comprising two-thirds of the island, that limestones have their maximum development and the karst its maximum expression (Fig. 1). The block consists of dissected limestone plateaus with inliers of the non-karstified Cretaceous rocks.

The Cretaceous core of the island is made up predominantly of pyroclastic rocks with subordinate conglomerates, shales and limestones. At the end of the Cretaceous, these rocks were folded and uplifted and intruded by several granodiorite bodies. As a result, there is a strong unconformity between the Cretaceous and Tertiary strata. The only part of the island to remain beneath the sea at this time was the northwest–southeast Wagwater Belt into which poured clastic material from the uplifted Main Block and from the advancing Blue Mountain geanticline. Further tectonism transformed this foredeep into a highly folded mountain range, the Port Royal Mountains, which now constitutes the western element of Zans' Eastern Mountain Mass. As already explained, this mass is of little interest as far as karst is concerned and we shall confine our study to the Main Block, concentrating on that part which also excludes the western end of the island and has been called the Clarendon Block.

During the Middle Eocene, the sea transgressed from the north slowly and irregularly over the Main Block resulting in the deposition of the Yellow Limestone Formation. The first marine incursion reached half way across the block and a lower limestone (frequently with a basal grit) was laid down. Following a minor regression, a series of estuarine sands and clays with frequent carbonaceous beds were laid down on this lower limestone. The next transgression was much more complete

---
[1] Present address: 1 Gledhow Court, Leeds L.S.7, 4N.L., Great Britain.

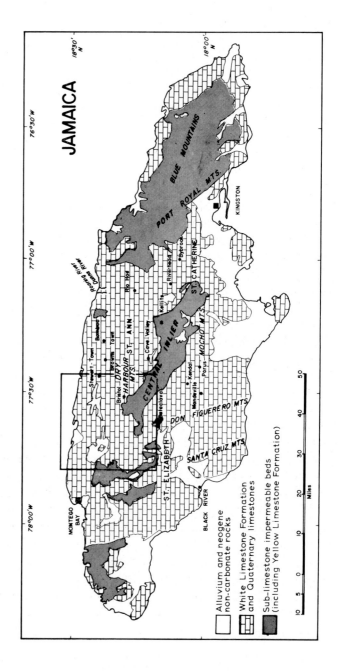

Fig.1. The limestone outcrop and localities referred to. Inset area is shown in more detail in Fig.2.

covering virtually the whole of the block and resulting in the deposition of the upper limestone of the formation. Littoral clays and sands were still being laid down at this time in the southeast part of the area. Along the north coast the formation is represented by an offshore facies of pelagic marls.

With the transgression complete and the influx of detrital materials stopped, the White Limestone Formation was deposited conformably above the Yellow Limestone. Its age ranges from Middle Eocene to Lower Miocene. While not conforming to international stratigraphic practice, the hallowed name of the White Limestone offers a fairly sound description of the rocks involved. The non-carbonate fraction seldom exceeds 2% and is generally less than 0.5%; the only departure from very pure whiteness of the rocks is in the dolomites and recrystallized limestones at the base. The formation is over 5,000 ft. thick where fully developed but discontinuities of deposition and the absence of higher beds leaves a maximum thickness over most of the block of about 1,500 ft. The White Limestone carries the most extensive and intensive karst development; where in subsequent pages a limestone is left unnamed, this formation is being referred to.

A division of the formation into members and faunal zones was made by HOSE and VERSEY [1956] and revised by VERSEY [1962]. Subsequent work notably by E. Robinson and by R. M. Wright indicates that further revision is required. For purposes of discussing karst hydrology, only three lithological divisions are necessary, each having an individual hydrologic regime. They will be described in turn. It may be borne in mind that they are not intended as stratigraphic terms.

The Montpelier Chalks

Along the north coast and within a north-northwest–south-southeast belt stretching on the west of the Clarendon Block, from Montego Bay to Black River the whole of the White Limestone Formation is represented by well-bedded chalks, with nodules and in places lenticular beds of chert. This facies represents sedimentation in an outer neritic environment which only the finest reef-derived material could reach; the bulk of the sediment is made up of planktonic microfauna. These rocks are here called the Montpelier Chalks and include the Montpelier Member of VERSEY [1962] and Bonny Gate Formation of ROBINSON [1967]. The chalks have reacted to stress by folding rather than faulting; large-scale fracturing is not characteristic of them as it is of the other divisions. They are characterized by an initial (rather than a diagenetic) porosity. These two features, the porosity and the general absence of fractures result in primary permeability exceeding any secondary permeability. The groundwater moves through the body of the rock rather than through specific channels.

It is only the largest structures which have the effect of acting as definite lines of underground water flow. In their absence, transmissibility values are low and along the north coast this low permeability plays its part in effecting a barrier to the northerly groundwater flow.

The recrystallized limestones

Over a large part of central Jamaica the lowest 100–150 m of the White Limestone Formation is made up of recrystallized limestones, representing the Troy and part of the Claremont Formation [VERSEY, 1962]. The former is generally made up entirely of diagenetic dolomites while the latter is simply recrystallized. In each, there is no primary porosity or permeability, groundwater movement being entirely through fissures. As the developments of jointing in the Troy at least was probably coincidental with the penecontemporaneous dolomitization, subsequent movement may well have occurred by adjustment along these already existing joints, further enlarging them. These two limestone types are thought to represent low-energy, shoal-water deposits laid down over the Clarendon Block.

The rubbly limestones

The rest of the White Limestone, including seven of the members erected on biomicrofacies, consists of more or less rubbly limestone with occasional hard bands. A typical road section will show 3–8 ft. beds of limestone rubble interbedded with 6–12 inch beds of hard fine-grained limestone. It appears therefore that the widespread nodular texture is an original sedimentary feature and not simply the result of tectonism. This texture is, however, accentuated in broad zones along major faults. The limestones carry a cuirass of induration giving them an appearance of solidity but when this is stripped off, the rubbly nature becomes apparent and the limestone can be quarried with pick and shovel. Despite the loose character, vertical quarry faces are quite stable. This varied assemblage of limestones represents local reef deposits, reef debris, back-reef or shoal-water lime muds, all having in common a high content of benthonic Foraminifera, in particular of orbitoids, nummulitids, peneroplids and dictyoconids.

The limestones may be divided tentatively into two parts, a lower one of Eocene age, consisting of the Swanswick and part of the Claremont members of VERSEY [1962] as redefined by WRIGHT [1967] and separated from it by an unconformity, and an upper one of Late Eocene–Miocene age, including the Gibraltar, Somerset, Walderston, Brown's Town and Newport of VERSEY [1962]. The lower division is more consolidated while the upper tends to be more soft and powdery.

Their distinctness, the one from the other, is far from clear and it is likely that there is a gradually changing sequence of lithologies from a fine, powdery rock to a limestone conglomerate, the former nearly impermeable and the latter highly permeable.

Limestones younger than the White Limestone occur around the coast but are seldom hydrologically distinct from the main karst.

Tectonic setting

The main structural elements of the island are the Main Block making up the western and central part of the island and the fold mountains making up the eastern one-third. This eastern zone can itself be divided into two: the Blue Mountain zone, forming the highest mountains, advanced as a geanticline from the northeast, tightly folding the Eocene sediments deposited in its foredeep. The latter are now represented by a subsidiary range lying to the southwest of the Blue Mountains: they have been termed the Port Royal Mountains.

The main tectonic elements affecting the outcrop of the White Limestone of the Main Block are two series of faults each having a strong topographic expression. An east–west series is probably isostatically controlled and is one with the steps by which the ocean floor descends to the depth of the Oriente Deep. Among the faults of this series are those of the Duanvale System which extends from Bamboo, St. Ann to Montego Bay, a distance of 45 miles. The system parallels the coast at about 6 miles inland. At about 6 miles offshore a similar system is indicated by a number of earthquake epicentres. The south coast is also bounded by faults having the same trend. That the Duanvale System clearly affects the Lower Tertiary facies distribution indicates that it is at least as old as the Early Eocene movements. In that the faults are subparallel to the trend of the whole island and of the island arc, it may be thought that they originated as major crustal dislocations even earlier.

The other set of faults is nearly perpendicular to the first set and is subparallel to the tectonic trend in the fold mountains of the eastern end of the island. It seems, however, improbable that the faults cutting the Tertiary rocks are rejuvenated faults of a Laramide or pre-Laramide System; they are more probably the product of the warping of the Cretaceous basement, taking only their pattern from an underlying structure. It may be noted, however, that two of these faults are nearly coincident with old features which have influenced the facies distribution of the White Limestone itself. The faults trend between north-northeast–south-southwest and north-northwest–south-southeast.

In the south central part of the island the faults of this system play an especially marked role. The Don Figuerero and Santa Cruz Mountains each constitute an easterly dipping fault block bounded on the west by a fault complex, with a throw of over 3,000 ft. There is some evidence that these structures are very young. These two great fault blocks each with a surface area of over 100 square miles are separated from the rest of the Clarendon Block by a hinge structure extending from Porus to Balaclava. Beyond having the general form of a syncline with oblique cross-faulting the character of this hinge is not clear. Within the fault blocks the north–south faults are perpendicular to the general direction of groundwater flow and do not appear to affect it greatly. Elsewhere, however, where the regional flow is towards the north and south from the island's main east–west axis and the faults of the north–south system are parallel to the direction of flow, they influence the hydrological pattern considerably by acting as lines of preferential flow.

The faulting has the general effect of breaking up any hard limestone beds and increasing the permeability. Once this is achieved, the greater volume of water passing perpetuates and increases this higher permeability. Instances are known, however, where the limestones are initially so incoherent, that the faulting actually decreases the permeability. Evidence of this is provided by extremely poor water wells and by fault line ridges which have resisted solution.

Faulting gives way to folding in the outcrop areas of the Montpelier Chalks. Even major faults with clear topographic expression over the outcrops of other facies lose this sharp definition where they trend into the Montpelier outcrop. They can, in some cases, be seen to pass into steeply dipping monoclines. The other facies rarely exhibit clear dips and where they do, as in parts of the Claremont Limestone, the beds are thick and the angle of dip is generally small ($15°$): steeper dips invariably indicate faulting nearby. The Montpelier Chalks are consistently well-bedded and though the dips are also usually gentle, quite tight folding without rupture occurs in some places; that the chalks react by folding may be attributed to adjustment gliding along bedding planes. The absence of well developed bedding in other facies has not permitted this reaction.

PHASES OF KARSTIFICATION

It is fairly clear that throughout the deposition of the White Limestone the central part of Jamaica constituted a fault-bounded shoal on which deposition of biogenic limestones was taking place. Now this shoal has been uplifted, its sediments lithified: it has been extensively block-faulted and its surface deeply dissected by karst. However, the course of events which filled the gap between these two conditions is poorly known.

Terraces are locally well developed but remain undated; there is a fairly conspicuous concordance of summit level in some areas, the age of the planation being unknown. Over large areas the karst valleys are filled with bauxite which is believed (although this is, of course, also disputed) to represent an alluvium laid down by karst springs and, with subsequent uplift, lateritized. How long this process took and when it started is not known. What follows in this section is therefore more speculative than the writer (and the reader) would wish.

It is certain that the White Limestone did not have imposed on it a drainage pattern inherited from overlying strata. It was uplifted above the sea in the same manner as a raised calcareous shoal (although accompanying tectonism must have occurred) and the form of its karst features must find their origin solely in the character and attitude of the limestones and the manner of the uplift and of the tectonism. We have already noted briefly the character of the limestones.

The conformity of summit levels is marked but inadequately studied. This conformity is not necessarily to horizontality but frequently to the plane of tilt of individual fault blocks. The inference is strong that the uplift had proceeded some way allowing a degree of dissection before the breaking up of the block into the present smaller blocks took place. At least a pattern of karst had been established, directed by an existing joint system. This initial uplift was the response to the release of that pressure which, since the Middle Miocene, had caused progressive downwarping and had allowed the deposition of several thousand feet of shallow-water limestones. It may not, therefore, have been accompanied by tectonism, other than movement of the isostatic control faults along the north and south of the island. If this was the case, then the rate of dissection of the karst would depend on the difference between the rate of erosion and the rate of uplift. All the present major features are related, however, to younger structural features and the great block-faulting episode in the Late Tertiary, while not initiating the karst process has been mainly responsible for the variety of its present forms.

To speak of karst phases is therefore as unreal as to speak of the karst cycle. If our history of the tectonism is approximately correct we may presume that in the early phase of the uplift, drainage was entirely subterranean, as it is for instance in the relatively recently uplifted 300–400 ft. terrace (including Hellshire Hills, Portland Ridge, Braziletto Savannah) along the south coast. The complexities introduced by subsequent faulting, throwing up drainage barriers and creating distinct catchments within the karst, together with the erosion to base level of some streams has resulted in major modification of the earlier entirely subterranean flow.

A major event in this little known history, was the erosional breaking of the Tertiary limestone cover and exposure of the underlying Cretaceous pyroclastic rocks in several inliers. Drainage from these inliers was onto and almost immediately beneath the karst plateaus surrounding them. Whereas previously the erosional force had been pure water, a new factor of underground abrasion by sediment was now brought into play. It may be presumed that the enlarging of the inliers increased the part which abrasion played in erosion of the plateaus. This will be referred to again in the discussion of morphology of the karst. The sediment is still being transported by the karst streams flowing beneath the plateaus and while consisting mostly of clay, includes a high proportion of quartz grains in some areas and of magnetite and ilmenite sand in others. The sediment after passing beneath the karst plateaus either continues to the sea or is trapped in interior valleys.

HYDROLOGY

From the description of the geology of the White Limestone we have seen that the variability of the limestone lithology carries a comparable range of permeabilities. The finer grained limestones have a low permeability and in certain places provide confining condition on deeper and more permeable horizons. The massive limestones towards the base of the formation have no primary permeability but a high secondary permeability combined with a low storage factor. A large part of the formation, consisting of loose, rubbly or broken, limestones, has a high to very high permeability with a moderate storage factor.

The generalizations in the preceding paragraph do not take account of cavern development although as we have seen this constitutes an important part of the underground drainage, the conduits acting as trunk canals draining the limestones enclosing them.

The development of the drainage pattern was controlled by the permeability variations in the limestones. The regional groundwater gradient would have had superimposed on it a pattern dictated by the permeability of the limestone. That the pattern is clearly rectilinear (if the surface indications are a valid parallel) shows that a tectonic cause of permeability variations is more important than the original variation due to differing character of the limestones. Thus, beneath the Cockpit plateaus the flow is essentially through joints and crush belts which, perhaps originally merely more permeable zones, have been made more open and in fact perpetuated by the continuing passage of groundwater. The largest of these lines of flow will of course be the down-gradient trunk conduits which receive the flow from the lesser conduits within their catchment. Examples of springs which discharge from a system such as that described are Gineppa Spring and Clear Spring (see Fig. 2) along the northern edge of the Cockpit Country, and Seven Rivers at its southern edge. These springs have in common, gradual fluctuation in flow and an absence of turbidity. Their catchment is entirely within the limestone plateau.

In marked contrast are the springs fed by systems in which the main lines of flow also carry drainage from the inliers of sub-limestone clastic rocks. The run-off from these catchments sinks at the base of the limestones to join the karst drainage, greatly modifying it in the process. The precipitation over the inliers, which are at nearly 3,000 ft. above sea level, is often in the form of brief high-intensity rainfall which loses a relatively small percentage to infiltration and consumptive use. The resulting flashy floods of heavily turbid surface water are forced into the karst system and have a very powerful erosive effect usually along a preferred channel of flow. For a few hours this channel must take far more water, abrasive and aggressive water at that, with greater velocity and under greater pressure than the system, of which it is a part, was designed for. The result is very rapid erosion by the corrasive, corrosive and hydraulic strength of the floods along a particular conduit or groups of conduits. The flood waters surging far above the normal water circulation level weaken the conduit roofs. At the same time the waters emerge at the surface attacking the conduit roofs from above. ZANS [1951] notes instances in which the depressions are flooded to a depth of 30 ft. following several days of heavy rain. Examples of springs which are fed by modified systems of this type are Deeside Rise, Windsor Spring, Fontabelle Lagoon and Dornock Riverhead at the northern edge of the Cockpit Country and Oxford Rise and Mexico Riverhead at the southern edge. They are all characterized by rapid fluctuations in flow and high turbidity during high stage (quartz and magnetite sand are deposited by the flood waters). Many of the largest caves, Windsor, St. Clair, Ipswich, Mexico, Mouth River are located close to (or actually carry) the water of this type of spring. The accessible levels clearly represent past drainage channels of the

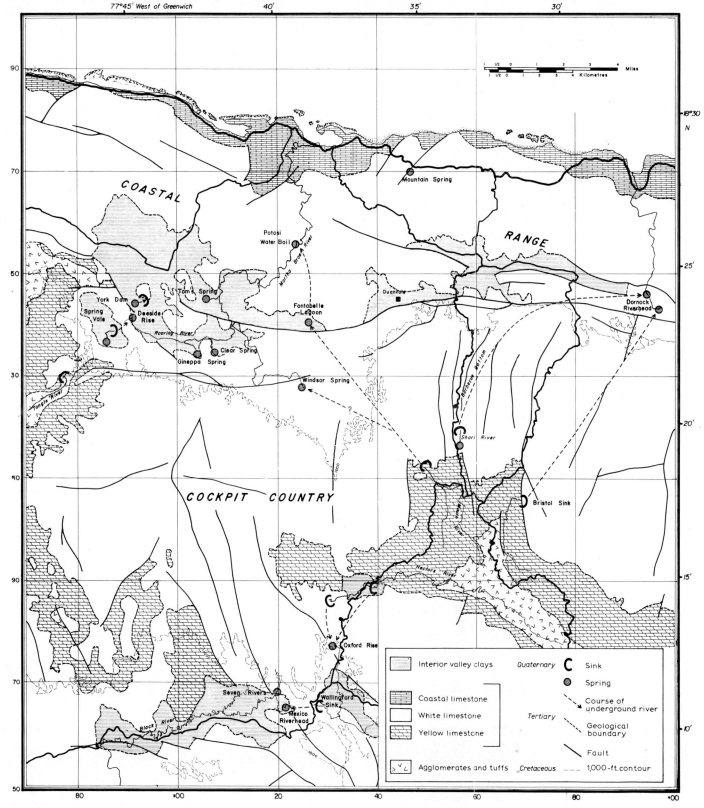

Fig.2. Geology and drainage of the Cockpit Country.

Fig.3.A. A small part of the Cockpit Country, Trelawny. B. Barbecue Bottom, Trelawny.

same type. Some of the larger karst valleys such as Barbecue Bottom (Fig. 3) and the valley south of Stewart Town appear to be due largely to the erosive effects (possibly initiating collapse) of the underground floods, although faulting is contributary and the extent of its influence is not clear.

During the flood rains over the Cockpit Country and its contributary inliers, the storage within the limestones will rise and the flow of springs draining it will increase and for a long time maintain that increase. The floods from the inliers pass through rapidly (although adding some recharge to the limestones which would be rapidly discharged after the flood has passed) and the springs which they feed will show a very rapid increase in flow followed by nearly as rapid a decrease. This is followed by a slower decline as the local recharge from the flood is voided (bank storage both in the inlier and within the limestones), this decline passing into the recession curve of the limestone storage. The low flow of springs of this type is entirely from the storage beneath the karst plateaus. A probable exception is the Fontabelle Lagoon which is always turbid. This spring has been traced back during low flow to the Mouth River in the northwest part of the Central Inlier, and it may well be that the water flows beneath the karst under confined conditions, possibly within the Troy Dolomite into which it sinks, and has little connection with the limestone storage beneath or through which it flows.

These two types of springs representing two types of karst groundwater flow are represented in each of the major karst areas. Thus, the ever-clear, non-fluctuating Dunn's River and Roaring River draining the limestones of eastern St. Ann can be contrasted with the greatly fluctuating, frequently turbid Rio Hoe draining the inlier of Middlesex. In St. Thomas in the Vale, the Bybrook Spring has similarly contrasting relation with the Riverhead Spring which carries the floodwaters from the Kellits area of the Central Inlier.

The volume of storage beneath the limestone plateaus is very great. The very slow fluctuation in flow of the springs draining it demonstrates this. Springs such as Dunn's River, St. Ann, Bybrook Spring, St. Catherine and Seven Rivers, St. Elizabeth each have a yield of over 50 cubic ft./sec with very little variation (none actually measured). Water level fluctuations in a test well at Kendal were measured by Alcan (Jamaica) Ltd., over a period of three years. The results are plotted in Fig. 4.

GEOMORPHOLOGY

The main element in the physiography of the limestone outcrop is the block faulting, young and more pronounced in the south of the island, older but still marked in the north. The karst itself is superimposed on this pattern and its form is guided by it; in some cases the karst processes have accentuated structures, as in the case of the Barbecue Bottom and Bristol–Stewart Town faults; in others, such as the alluviated interior valleys (Fig. 3), much of the structure has been masked.

Through his studies in various parts of the Jamaican karst lands over twenty

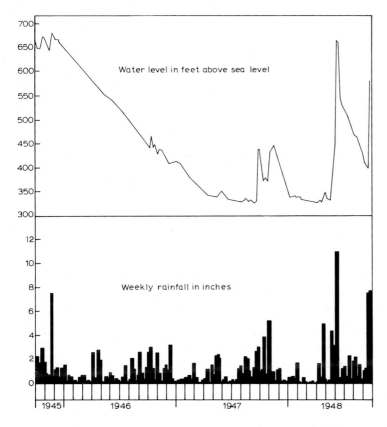

Fig.4. Record of water level in Kendal borehole and of rainfall on watershed.

years, the writer has been led to the firm belief that while the karst represents a continuum resulting from the interaction of multiple processes, the most meaningful subdivision of the karst into "types" should be based on the degree to which groundwater circulation has influenced the physiographic development. It may have had no effect being too deep or it may have pronounced effect where the water table is near to or intersects the land surface. This will be explained in more detail in what follows. The writer wishes it to be understood that this view is not fully accepted by other authors; that, while it is based on a lengthy experience of, and intimate acquaintance with the problem, it remains to be properly substantiated by detailed study.

The karst uplands

The landform which has been most discussed in the literature on Jamaican karst is the cockpit. This is a round or more commonly lobate depression with steep to vertical sides and with a depth of hundreds of feet (Fig. 3B). The impression is given of a very wide sinkhole (doline) depression which has suffered an over-deepening. The cause of this over-deepening has been ascribed to subterranean erosion by groundwater, to concentration of the rainfall in the centre of the cockpit by the overhanging forest canopy, or simply to the volume of precipitation. Cockpit karst is found generally in areas of high rainfall where the limestones are fairly hard, where water circulation cannot be deep and where there is an absence of soil cover. The relative importance of these factors is debated. The writer holds the view that the great rise in water level during and after heavy tropical rains brings the groundwater circulation close beneath the bottom of the cockpits, eroding them. Subsequent collapse will then deepen the depression and remove the soil cover. Many cockpits are floored with boulders; others, with solid floors, frequently have vertical sinkholes in the bottom which connect with deeper horizontal caves. For this mechanism to be continually effective (and in the Cockpit Country it appears that the cockpit deepening has been active for a long time) the rate of erosion has to be matched by the rate of uplift. If the uplift is too slow a base level for erosion will be established, if too rapid, the water circulation will become too deep to be effective. For the latter case, solution will then take over as the main erosive force and soil will encroach over the floor and sides of the valleys. This will produce topography such as that of the Dry Harbour Mountains which SWEETING [1958] called "degraded cockpit karst" (Fig. 5A).

C. F. Aub (personal communication) from his unpublished study of a cockpit karst area found little evidence of groundwater circulation below the level of the cockpits and ascribed the overdeepening to the effects of solution by rain water, concentrated by wind action, by the forest canopy and by the few fissures available for draining off this water. He considers that an important difference between cockpit karst and karst elsewhere in the island with less relief is that in the former there has never been a bauxite cover so that solution has a different and more powerful effect than in the latter where the bauxite valley-filling impedes or prevents the deepening. This explanation of the origin of cockpits precludes the idea that the bauxite is the soil derived from the limestone.

Although the sides of the cockpits are steep with overhangs in places, the top of the intervening hills are rounded though with rocky pinnacles. This has led to the Cockpit Country being described as a typical *Kegelkarst*. As we have seen the modifications imposed on the kegel karst by over-deepening of the depression makes the use of this term inappropriate.

It is more aptly applied to the terrain in which over-deepening has not taken place, the gentle (up to 30°) slopes extending from hilltop to valley floor. Here the depressions are filled with bauxite up to tens of feet thick. The floor of the valley usually slopes towards one side where storm drainage sinks at the foot of a limestone hill; discrete sinkholes are, however, uncommon. This type of karst results solely from solution by percolating rainfall and is typical of the uplands such as the Santa Cruz, Don Figuerero and Mocho Mountains which have been rapidly uplifted and where groundwater circulation is at depth. It is also found over the Montpelier Chalks.

These then are the two main terrain types of the karst uplands, the cockpit karst

Fig.5. A. Degraded cockpit karst, Gibraltar St. Ann. B. Eastern end of the interior valley of St. Thomas in the Vale, St. Catherine.

and the kegel karst. The "degraded cockpit karst" of SWEETING [1958] represents a cockpit karst which has suffered either uplift or freeing of its underground drainage. Modifications due to age, structure, variations in limestone lithology, etc., appear to be minor.

Although the interior valleys are described below, it may be noted that these also develop within the karst uplands where the water table intersects (at least at times) the ground surface. Cave Valley (altitude 2,000 ft.) is of this type. The Cave River flowing off the Central Inlier enters the valley from the west. During flood the karst drainage is inadequate to take all the water so that flooding and alluviation result.

Interior valleys (poljes)

The interior valleys have been depressed structurally; there is no evidence of their being produced by erosion. They are generally down-faulted on at least one side, the character of the fault being such that underground drainage is impeded. Groundwater is thereby forced to the surface during high water stages, causing solution at the foot of the surrounding hills and frequently causing widespread alluviation by the turbid conduit water, the alluviation producing the characteristically flat floors. A common feature is the isolated hill piercing the alluvium.

The impedance of the drainage may be caused by a reversal of dip, by the upfaulting of the karst base or by a change in limestone facies from a permeable to an impermeable phase. In the case of some valleys all these factors play a part.

These valleys are generally linear (although with notable exceptions), the long axis being perpendicular to the overall direction of underground drainage. Some of the valleys are being actively alluviated, some only in specific low-lying areas, while others, older ones already uplifted to the point that groundwater no longer reaches the surface are suffering resection. In some instances, the up-gradient side of a valley may be periodically flooded while the down-gradient side, carrying the flood water underground again through sinkholes is being resected.

The Nassau Valley

About seven miles long and two miles across at its widest point, this valley is part of the drainage area of the Black River.

This river with its headwaters in the western end of the Central Inlier of the island, sinks near to Troy, flows underground to emerge in the Oxford interior valley whence, after crossing that valley, it sinks again into a cave at Wallingford to rise at the eastern end of the Nassau Valley in Mexico Riverhead. This spring may be regarded as the flood water head of the valley's drainage system, although in dry weather its flow is much less than that of Seven Rivers Spring about 0.5 mile to the north of it. The latter emerges in several main heads from multiple holes in the limestone at the base of a steep hill. It is of constant flow and low turbidity (in marked contrast to Mexico Spring with its great fluctuations in flow and frequently high turbidity) and is presumed to have its catchment entirely within the karst of the southern part of the Cockpit Country.

There are other smaller springs within the valley but these appear to have their catchments in nearby outcrops of Yellow Limestone and have little relevance to the karst. There are at least two distributaries within the valley which are fed by flood flow, the water overflowing into deeper areas of the valley (the river being perched on its own alluvium) and disappearing down sinkholes at the foot of the Nassau Mountains to reappear on their southern side in Bogue and Elim springs.

While any evidence is concealed beneath the valley's alluvial cover, it is certain that faulting plays an important part in bringing up the basal part of the White Limestone on the south side of the valley, forming a barrier to any southerly drainage. Halfway along the length of the valley is a suballuvial anticline (again faulted) of Yellow Limestone. This second barrier together with that just described has affected a near complete impedance of underground flow and forced the groundwater to the surface. Of this groundwater, that draining the Cretaceous inlier at its headwaters and issuing into the valley through Mexico Cave is, during flood, highly charged with sediment and is responsible for the thick alluvial cover of the valley.

Whether the area limited by the Nassau Mountains and the Yellow Limestone anticline ever experienced lacustrine conditions is unknown but the obvious impedence of drainage makes it likely that intermittent flooding and even complete inundation were an active element in the deposition of the alluvial cover. When this cover became so thick that the river was able to spill over the Yellow Limestone barrier into the western end of the present valley, this latter suffered a comparable history until, with the breaching of the gap through the limestone it was possible for the river to flow freely at surface from Maggoty to Newton and the interior valley was opened. There has been little resection of the topography since this freeing of the drainage and local flooding is still common particularly in the area of Appleton.

The interior valleys of northern Trelawny

This series of valleys bounded on the south by the Cockpit Country and on the north by this coastal range was well described by ZANS [1951]. In the west, two or three west-northwest–east-southeast trending valleys have coalesced to form the large Queen of Spains Valley, some five miles across. Other valleys to the east have remained clearly distinct. They include Duanvale, Hamshire Valley and the Cambridge Valley. This second group while heavily alluviated now carry no surface drainage (except their own run-off into nearby sinkholes) and have clearly been uplifted relative to the groundwater circulation. In the Queen of Spains Valley (with its offshoot valleys) on the other hand are numerous springs feeding the Roaring and Windsor rivers which unite to form the Martha Brae River which drains the entire valley. The head of each of these two rivers is in a large spring and the catchment of each is partly in the inliers of Cretaceous rocks. They thus have the dual character already described taking their low flow from storage within the limestone and their floods which can be great, from the rapid run-off from the inliers.

Alluviation of the Queen of Spains Valley is thus still active. A third spring at Fontabelle Lagoon also derives its flood flow from an inlier and has a similar character to the two already described except in that it rises through a thick blanket of clay and sinks further west along the Fontabelle Valley. Other springs rising at the southern side of the valley, Gineppa Spring and Clear Spring, are never turbid to any degree and may be presumed to have their catchments entirely in the limestone outcrop of the Cockpit Country. Their fluctuation in flow is also much less rapid.

There are two springs within the valley itself, York Dam Spring south of Gales Valley and Tom's Spring, north of Friendship. York Dam has been impounded to serve a nearby sugar factory and does not normally flow but in flood it overflows to a sinkhole to the east. Tom's Spring rises about 0.5 mile west of the Martha Brae River into which it flows. Neither spring is ever turbid and their fluctuation in flow is gradual. A contour map of water levels in wells in the valley indicates groundwater flow from both northwest and southwest towards Tom's Spring.

From the point where the Martha Brae leaves the valley and enters a gap through the coastal range, it loses a large part of its dry weather flow to groundwater. Three miles from the same point is a large spring, the Potosi Waterboil which rises in the river. This has been shown to be in part at least, the resurgence of the water sinking in Fontabelle Lagoon.

The barrier which impedes the northerly flow of groundwater beneath the valleys of northern Trelawny is probably a combination of an upfaulted block of Yellow Limestone beneath the valley and the change of White Limestone facies from crystalline or rubbly phase limestone to Montpelier Chalks at the southern margin of the valley.

PRACTICAL ASPECTS

The rural population of Jamaica does not live in villages: each farmer builds a home on the land which he cultivates. Such villages as there are remain as mere service centres with a shop, post office, police station, clinic and little else The cost of pipelines for water supply installations are correspondingly heavy. The small farmers are in general restricted to high ground, the more fertile, more arable, valley lands having remained in the hands of large estates. There being no water on this high ground, pumping costs for water supply are also heavy. As a result of the combination of these factors piped water supplies over the limestone plateaus, even over those such as the Dry Harbour Mountains which are heavily populated, are virtually non-existent. Recourse is had to concrete rainwater catchments and storage tanks. During even a moderate drought this system fails and water has to be trucked to the farmers.

A few wells have been drilled in these high-level plateaus but none has been successful. The water level is deep but more important is that the permeability encountered is poor. A well at Huntley, 1 mile north of Brown's Town at an elevation of about 1,050 ft. was discontinued at a depth of 595 ft., no water having been found. A well at Marshalls Pen near Mandeville at an elevation of over 2,000 ft. reached water at a depth of 350 ft. but the yield was negligible. It seems likely that the steep hydraulic gradient of the relatively free underground drainage beneath the plateaus tends towards the creation of restricted zones of higher permeability, leaving the bulk of the limestone as it were "undeveloped". In the low-lying areas where gradients are much lower, regional high permeabilities are found; this is not the case beneath the plateaus.

There has been no attempt to create artificial storage on or in the limestones. A possible dam site in a limestone gorge east of Bog Walk was test-drilled but large cavities were encountered and the site was rejected. A far greater understanding of the conditions of occurrence and movement of groundwater and of the geological structure would be needed before any attempt could be made to create artificial underground storage of the excess flows which now go to waste. However, this is a possibility which, with increased development, will require study.

With the growth of a new well-field the natural pattern of flow of groundwater is markedly altered. Since it is the passage of groundwater which improves the permeability, any such change in the pattern may have a permanent effect on the distribution of porous or permeable zones within the limestone aquifer. This could become especially important in areas of salt water encroachment where during low water levels the inflow from the saline source is increased. This channel of flow (usually along a fault) may be progressively freed and the salinity problem become progressively worse irrespective of the water balance.

The seasonal character and the incidence of high-intensity precipitation makes water balance studies difficult. The delineation of groundwater catchment divides becomes meaningless when underground flood waters are directed via old superphreatic drainage trends across the divide into adjacent catchments. The crossing of underground drainage lines recently shown in the catchment of the Y.S. River using *Lycopodium* spores makes it nearly impossible to delineate the area which is drained to any given stream discharge measurement station within that catchment. This characteristic (of the catchment divide ranging drastically with underground stage) is suspected in other less well studied parts of the island (such as Lluidas Vale) but is yet to be proven. It is known from the unpublished work of C. B. Brown (personal communication), of McMaster University that *Lycopodium* spores fed into Mouth River in southern Trelawny at low stage reappeared at Fontabelle Lagoon and were not recovered at Windsor Spring. Yet it is not possible that the turbid waters issuing during flood from the latter can have their source anywhere but in the same Mouth River. Thus, while the low flow takes one course, the flood overflow takes a different course to reappear nearby 3 miles away from the reappearance of the low flow. All these factors have to be taken into account in relating discharge measurements to the area which they are presumed to drain.

REFERENCES

Hose, H.R. and Versey, H.R., 1956. Palaeontological and lithological divisions of the Lower Tertiary limestones of Jamaica. *Colloq. Geol. Min. Res.*, 6:19–39.
Robinson, E., 1967. Submarine slides in White Limestone Group, Jamaica. *Bull. Am. Assoc. Petrol. Geologists*, 51(4):569–578.
Sweeting, M.M., 1958. The karstlands of Jamaica. *Geograph. J.*, 124:184–199.
Versey, H. R., 1962. Older Tertiary limestones. In: V. A. Zans (Editor), *Synopsis of the Geology of Jamaica*. Geol. Surv. Dept., Kingston.
Wright, R. M., 1967. *Biostratigraphical Studies on the Tertiary White Limestone in Parts of Trelawny and St. Ann, Jamaica*. Thesis, University College, London.
Zans, V. A., 1958. On karst hydrology in Jamaica. *Soc. Intern. Hydrol. Sci.*, 2:267–279.
Zans, V. A., 1958. Major structural features of Jamaica. In: *Caribbean Geol. Conf., Rept. 1st Meeting, Antigua, B.W.I.—Geol. Surv. Georgetown*.

Chapter 15

Karst of the United States

W. E. DAVIES AND H. E. LEGRAND

U.S. Geological Survey, Washington, D.C. (U.S.A.)
U.S. Geological Survey, Raleigh, N.C. (U.S.A.)

FOREWORD[1]

In response to an invitation from Prof. Dr. Milan Herak, Zagreb, Yugoslavia, to prepare a chapter on karst of the United States, I requested two of my colleagues, H. E. LeGrand and W. E. Davies in the U.S. Geological Survey to prepare the following report. We are pleased to have this opportunity to contribute to this book and are indebted to all who assisted in the preparation of this chapter especially Miss Beulah Hilliard, and Mr. John Teel, U.S.G.S. for their work on the final draft of the manuscript and illustrations.

The Atlantic Coastal Plain is the easternmost province of the four major physiographic units of the United States (A in Fig. 1). The karst in this province is in carbonate rocks of Tertiary age except in southeastern Florida where Late Tertiary and Pleistocene rocks are karstified. The carbonate rocks include two of the most productive aquifer systems known. One extensive aquifer system is artesian in much of the region. It is the source of thousands of artesian wells and the largest limestone springs in the United States. Silver Springs, near Ocala, Florida with an average discharge of 808 cubic ft./sec (22.9 m^3), apparently is the third largest limestone spring in the world.

The Appalachians are the eastern part of another large groundwater province (B in Fig. 1) which is adjacent to the Atlantic and Gulf Coastal Plain in part of the region. The Appalachians include karst areas in steeply folded carbonates of Palaeozoic age.

The karst region of central Kentucky and southern Indiana are in the Interior Low Plateau region and are underlain by Palaeozoic rocks, which are essentially horizontal in contrast to the Appalachians. The area in Kentucky includes Mammoth Cave and many other well known caves. Also, the karst area of southern Indiana includes many caves. The karst of the Pecos Valley area in southern New Mexico in the Great Plains region is in Palaeozoic carbonates which include anhydrite, gypsum and salt. The formations dip gently to the east. Two well-known features of the region are the Carlsbad Caverns and the very productive artesian aquifer, in the Roswell

---

[1] Written by V.T. Stringfield, Department of the Interior, U.S. Geological Survey, Washington, D.C. (U.S.A.).

area, known as the Roswell artesian basin which refers to the topography rather than the geologic structure.

Other karst regions include:

(*1*) The Ozark plateaus in Missouri and Arkansas underlain by Palaeozoic carbonates dipping gently on the flanks of an asymmetrical dome.

(*2*) The Nashville and Lexington plains in central Tennessee and central Kentucky, which are broad lowlands formed on gently arched lower Palaeozoic carbonates. The solution and erosion removed the cover on the Nashville Dome forming a topographic basin which is known as the Nashville Basin.

(*3*) The Edwards Plateau in Texas is underlain by Cretaceous carbonates with low dip toward the Balcones fault zone which separates the plateau from the (A) Gulf Coastal Plain. The Edwards Limestone and associated formations form one of the most productive aquifer systems in the United States. Water enters the limestone through sinkholes and other karst features where it is on or near the surface and where streams cross the outcrop. The water moves down gradient approximately at right angles to the fault zone and then moves along the fault zone to large springs.

In addition to these very productive aquifers, carbonate rocks are a source of water in all of the major groundwater provinces. In some regions they are the principal source of groundwater.

As mentioned above, four carbonate aquifer systems: (*1*) the shallow aquifer (Biscayne aquifer, of Tertiary and Quaternary age) in southeastern Florida; (*2*) the principal artesian aquifer of Tertiary age (known as Floridan aquifer in Florida) in southeastern Georgia and Florida; (*3*) the Edwards Limestone aquifer of Cretaceous age in the Edwards Plateau, Texas; and (*4*) the artesian aquifer of Permian age in the Roswell artesian basin in southeastern New Mexico are among the most productive aquifer systems in the United States. Although large parts of three of the systems are artesian, the available information indicates that the permeability is chiefly secondary and that it developed chiefly in the upper part of the zone of saturation under watertable conditions as the water rose or fell with changes in base level. A large part of the secondary permeability is due to buried karst in which the permeability was not destroyed in the burial by overlying formations. The present cycle of karstification renewed circulation of water, and developed additional secondary permeability by solution and by removal of unconsolidated deposits in the buried karst. In some coastal areas the circulation has not yet flushed out all of the salty water in the aquifers.

Many of the solution cavities in the two carbonate aquifer systems in Florida are filled with relatively permeable quartz sands, which reduce the chances of pollution commonly present in carbonate aquifers under water-table conditions. In addition to serving as a filter, the sands in cavities in the shallow aquifer in southeastern Florida have reduced the permeability sufficiently to provide a fresh water head which prevents encroachment of sea water. If the cavities were not filled with sand the freshwater head would be so small that sea water would move freely into the aquifer, as in Bermuda.

The relation of salt water to fresh water in carbonate aquifers in coastal areas is similar to that in other aquifers where the Ghyben–Herzberg principle can be used to indicate approximately the position of the interface between the fresh water and salt water. However, sinkholes or vertical solution shafts or other openings which extend below sea level through relatively impervious parts of the aquifer into a cavernous zone and which are open to sea water at the surface may serve as one leg of a natural "U"-tube in which sea water is in dynamic equilibrium with fresh water. For example, at Tarpon Springs, in Pinellas county, Florida, one leg of the natural "U"-tube is open to sea water through a shaft in Spring Bayou, and another leg is a sinkhole or shaft open to Lake Tarpon which is a fresh-water lake in balance with sea water. When the lake level is relatively low, during the effect of a high tide in Spring Bayou the head of the salt-water column exceeds the fresh-water level of the lake, and sea water flows through the U-tube into the lake until the fresh water head in the lake equals to or exceeds that of the salt water. Only recently, after studying sinkholes and solution shafts in karst areas, it was realized that vertical shafts or sinkholes which pass through relatively impervious parts of the aquifer are an essential part of such circulation of salt water and fresh water as found at Tarpon Springs. Such drowned karst features may be expected in coastal areas of carbonate rocks. They were formed during a low stand of the Pleistocene sea when the top of the zone of saturation in the limestone was at a lower level. With the rise of the sea to its present level, some of the shafts near the coast as at Tarpon Springs are now a few feet below sea level.

The first section of this chapter gives the general distribution of karst areas in the United States except Alaska where karst is not well known. The State of Hawaii outside of the North American Continent is underlain by volcanic rocks. The northern coast of the Commonwealth of Puerto Rico is underlain by Tertiary limestone dipping gently to the north. The karst features in general resemble the classic Cockpit Country of Jamaica, W.I., described by H. R. Versey in the chapter on that country. However, faulting has not affected the karstification as it has in Jamaica. This accounts in part for the absence of poljes in Puerto Rico. The hydrology of the carbonate rocks is not well known. Some of the water is under relatively high artesian head. The general recharge and discharge relationship appears to be similar to that in Jamaica, W.I. Due to the limited space available for this chapter, it is not possible to include Puerto Rico in the following discussions.

GENERAL AREAS[1]

About 15% of the continental United States, exclusive of Alaska, has limestone, gypsum, or other soluble rocks at or near the surface. Fig. 1 shows the distribution of karst and carbonate areas of the United States. In the eastern part of the country, belts of limestones occur throughout the Appalachian Mountains, the Allegheny Plateau, and the

---

[1] Written by W.E. Davies.

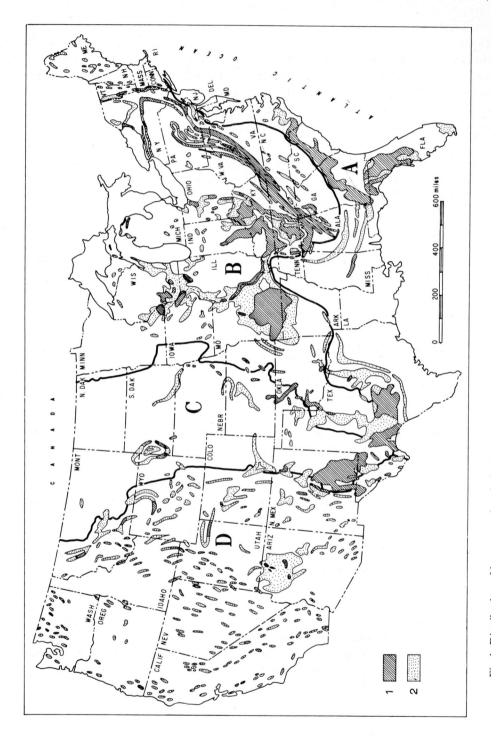

Fig.1. Distribution of karst areas in relation to carbonate and sulphate rocks in the United States. A = Atlantic and Gulf Coastal Plain region; B = east-central region of Palaeozoic and other old rocks; C = Great Plains region; D = western mountain region; I = karst areas; 2 =

Interior Lowlands (*B* in Fig. 1). The major limestone formations in these areas are Palaeozoic in age. Cambrian and Ordovician carbonate rocks are dominant along the east side of the Appalachians and Silurian and Devonian limestones, along with some Cambrian and Ordovician carbonate rocks, in the mountainous part of the Appalachians, and Mississippian limestones in the plateau to the west. The Interior Lowlands have extensive areas of Ordovician, Silurian, Devonian, and Mississippian limestones in which karst features are well developed.

In the Atlantic Coastal Plain, Tertiary limestones are at the surface in a broad belt from North Carolina to Florida, but in the coastal plain along the Gulf of Mexico, Tertiary and Cretaceous limestone and chalk lie along a narrow band at the inner edge of the plain (*A* in Fig. 1).

In Texas a large expanse of Cretaceous limestone forms the plateau in the central western part of the state. Carboniferous limestones form extensive plateaus in central and southern New Mexico and northwestern Arizona. In the Rocky Mountains, limestone beds are not extensive except for Carboniferous limestones in Wyoming and Montana. Throughout the area west of the Rocky Mountains there are numerous short, discontinuous, narrow bands of limestone, primarily Late Palaeozoic and Triassic in age.

Areas of Permian gypsum beds of significance in karst development occur in central and northern Texas, southeastern New Mexico, western Oklahoma, and central Kansas.

The degree of development of karst features in the carbonate and sulfate rocks varies greatly, and in many areas these rocks show no large-scale karst features. The major karst areas of the United States are in the Appalachian Mountains from Pennsylvania to Alabama, the Interior Lowlands and plateaus of Kentucky, Indiana, and Tennessee, the coastal plain of Florida and Georgia, the Ozark Highlands of Missouri and Arkansas, and the plateaus and lowlands of southern and western Texas and eastern New Mexico.

The karst provinces in the United States are so numerous and diverse that it is not possible to describe them adequately in a short summary. Accordingly, two classic areas, the central Kentucky karst and the Appalachian area along with the Atlantic and Gulf Coastal Plain and the Pecos River Valley of New Mexico and Texas are covered in some detail. For the remainder of the United States the more important karst areas are summarized in Table I.

There are approximately 7,500 caves in the United States of which at least 35 have 8 km or more of mapped passages, 12 have 16 km or more, and 3 have more than 32 km. Although this discussion is limited solely to karst areas, pseudokarst forms are extensive in the United States and include suffosional karst (mostly in clays), lava pseudokarst, glacier pseudokarst, littoral pseudokarst, and deflation pseudokarst.

TABLE I

MAJOR KARST AREAS OF THE UNITED STATES

Karst area	Location	Characteristics
Southeastern coastal plain	South Carolina, Georgia	rolling, dissected plain, shallow dolines, few caves; Tertiary limestone generally covered by thin deposits of sand and silt
Florida	Florida, southern Georgia	level to rolling plain; Tertiary, flat-lying limestone; numerous dolines, commonly with ponds; large springs; moderate sized caves, many water filled
Appalachian	New York, Vermont, south to northern Alabama	valleys, ridges, and plateau fronts formed of Palaeozoic limestones, strongly folded in eastern part; numerous large caves, dolines, karst valleys, and deep shafts; extensive areas of karren
Highland Rim	central Kentucky, Tennessee, northern Georgia	highly dissected plateau with Carboniferous, flat-lying limestone; numerous large caves, karren, large dolines and uvala
Lexington–Nashville plains	north-central Kentucky, central Tennessee, south-eastern Indiana	rolling plain, gently arched Lower Palaeozoic limestone; a few caves, numerous rounded shallow dolines
Mammoth Cave–Pennyroyal Plain	west-central, south-western Kentucky, southern Indiana	rolling plain and low plateau; flat-lying Carboniferous rocks; numerous dolines, uvala and collapse sinks; very large caves, karren developed locally, complex subterranean drainage, numerous large "disappearing" streams
Ozarks	southern Missouri, northern Arkansas	dissected low plateau and plain; broadly arched Lower Palaeozoic limestones and dolomites; numerous moderate-sized caves, dolines, very large springs; similar but less extensive karst in Wisconsin, Iowa, and northern Illinois
Canadian River	western Oklahoma, northern Texas	dissected plain, small caves and dolines in Carboniferous gypsum
Pecos Valley	western Texas, southeastern New Mexico	moderately dissected low plateau and plains; flat-lying to tilted Upper Palaeozoic limestones with large caves, dolines, and fissures; sparse vegetation; some gypsum karst with dolines

TABLE I *(continued)*

Karst area	Location	Characteristics
Edwards Plateau	southwestern Texas	high plateau, flat-lying Cretaceous limestone; deep shafts, moderate-sized caves, dolines; sparse vegetation
Black Hills	western South Dakota	highly dissected ridges; folded (domed) Palaeozoic limestone; moderate-sized caves, some karren and dolines
Kaibab	northern Arizona	partially dissected plateau, flat-lying Carboniferous limestones; shallow dolines, some with ponds; few moderate-sized caves
Western mountains	Wyoming, northern and western Utah, Nevada, western Montana, Idaho, Washington, Oregon, California	isolated small areas, primarily on tops and flanks of ridges, some area in valleys; primarily in folded and tilted Palaeozoic and Mesozoic limestone; large caves, some with great vertical extent, in Wyoming, Utah, Montana, and Nevada; small to moderate-sized caves elsewhere; dolines and shafts present; karren developed locally

ATLANTIC AND GULF COASTAL PLAIN[1]

Abstract

Nearly flat-lying limestone formations extend from North Carolina to Alabama and tend in a coastal direction to: (*1*) dip gently; (*2*) thicken slightly; and (*3*) be covered by younger sands and clays. A general monoclinal structure is broken by a few scattered minor faults and by a subdued arch in central Florida. Limestones are chiefly of Eocene to Miocene age, but are Pleistocene in southern Florida. On nearly flat interstream areas limestone is commonly concealed by sands and clays of younger formations, but sinks (some of which contain lakes) are common over large areas where the cover of overlying material is less than 25 m. Surface streams are scarce only in north and central Florida; positive elements, such as mogotes, are absent or are concealed below overlying materials, and the few escarpments have moderate topographic relief. Nearly even seasonal distribution of more than 100 cm of rain each year keeps the water table and other groundwater levels near the land surface in most of the region. Permeability ranges from poor in marls and poor in parts of some other formations to extremely good in other places. Despite the range in permeability, water moves almost en masse through the limestone. There is gross coastward movement of water through the homoclinal artesian systems, and some artesian discharge is upward

---

[1] Written by H.E. Le Grand.

through confining beds in coastal regions. Great discharge occurs as diffuse seepage and as large springs in streams entrenched in limestone. Subdued karst with a cover of sand and clay tends to reduce the economic problems that are common in most karst areas. Topography is acceptable for most uses—subsidence of ground in vicinity of sinks is locally troublesome but is not yet a major problem. Soil lies almost everywhere on karst, and where sandy the soil aids recharge of water. The soil cover and some sand-filled sinks: (1) promote more regular movement of water than occurs in many karst regions; and (2) reduce likelihood of contamination of water in the aquifers. Fresh water is widespread, but salty water is prevalent in coastal areas under the freshwater zone. Limestones of the region represent some of the best aquifer systems of the world.

Geologic setting

Limestone formations of Tertiary age underlie all of Florida and parts of adjacent areas in Alabama and Georgia (see Fig. 2). The formations extend northward through Georgia and South Carolina and into North Carolina. They are at or near land surface through much of the coastal plain province in these states. Structurally the formations are nearly flat, dipping gently and monoclinally toward the coast. The limestone formations, as well as some underlying and overlying sand and clay formations, tend to be wedge-shaped and are thicker toward the coast; the limestone is more than 300 m thick in the karst land of central Florida and is much thicker to the south, east and west. The Tertiary limestone formations are commonly covered by sands and clays that range in thickness from a few metres near the inner margin of the coastal plain to several tens of metres along the coast.

No sharp tectonic action has been imposed on the limestone. Some small normal faults occur, but in general they are not significant in the development of karst. A broad gentle arch in the overall monoclinal structure occurs in central Florida, at which place the limestone is at the land surface and around which it has a very gentle radial dip.

The Tertiary limestone system includes as many as seven geologic formations in Florida [STRINGFIELD, 1964, p. 164] ranging in age from Middle Eocene to Middle Miocene, but farther northward and inland as the entire system becomes thinner only one or two formations are present. Some of the limestone is rather hard and consolidated whereas other parts are semi-consolidated and marly. Unconformities separate several of the limestone formations. Reconstruction of the geologic and hydrologic history indicates that at the times when the unconformities were developed the exposed limestone was subjected to vigorous circulation by near-surface meteoric water and at one time or another attained various stages of karst development. Each emergence of the land and the related development of karst was followed by marine submergence and the deposition of later limestone formations. As a result of these changing conditions during Eocene and Miocene time, differences in lithology and texture were inevitable, and some clay and sand beds were deposited; yet in aggregate the entire

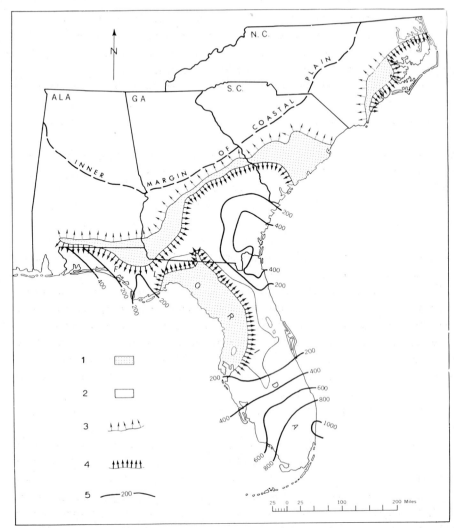

Fig.2. Map showing outcrop areas of Tertiary limestone. Contours show top of Tertiary limestone in the subsurface. *1* = Tertiary limestone at or near land surface; *2* = principal area in which sinks breach the Hawthorn Formation; *3* = line north and west of which some thin patches of Tertiary limestone may occur near land surface; *4* = line beyond which limestone thickens and is more deeply buried; *5* = top of Tertiary limestone, in feet below sea level. (After STRINGFIELD and LEGRAND, 1966, p.4.)

unit may be considered as limestone, which now acts as a single hydrologic system. During Middle and Late Miocene time some beds of clay were deposited over the preexisting limestone, and today these beds of clay conceal much of the limestone in the region. Fluctuation of sea level occurred in Pleistocene time, which ranged from as much as 80 m above present level to as much as 150 m below sea level; this change in sea level has had a significant effect on present aspects of the karst.

Limestone of Pleistocene age, which is especially prominent in southern Florida, is separated from the Tertiary limestone system by Miocene clay and sand in some coastal areas. The Pleistocene limestone is almost horizontal and is largely covered by Pleistocene and Holocene sand and clay.

The climate is temperate and humid. Annual rainfall ranges from about 95 cm at the southern tip of Florida to about 120 cm in most places; the rainfall is fairly evenly distributed during each year.

Karst features

A variety of karst features occur in the coastal plain province, but many of these features are masked or subdued by a cover of overlying materials—Miocene sand and clay or a thin veneer of Pleistocene sand and clay. Fig. 2 shows general areas where the limestone is sufficiently close to the land surface to cause sinks and related karst topography. Outcrops of bare limestone are rare except along some streams that are entrenched into limestone.

The most common type of karst feature is gently rolling topography lacking open drainage to streams and consisting of gently sloping depressions, or sinks. Some of the sinks contain water, as in the lake country of central Florida (Fig. 3). Sharp local topographic relief does not exist. Positive elements such as mogotes are absent as surface features; beneath the cover of overlying materials, however, relief on the limestone may be sharper and mogotes may be present. Conspicuous interior valleys, or poljes, are absent, although there may be buried valleys. The Flint River Valley in southwest Georgia resembles a large polje in some respects; this valley-flat environment has some sinks and is bordered by a subdued escarpment (Fig. 4). Remnants of this es-

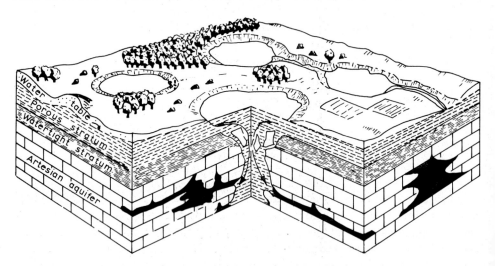

Fig.3. Sketch showing recharge of water through sinkholes, a condition common in central Florida. (After COOPER and KENNER, 1953.)

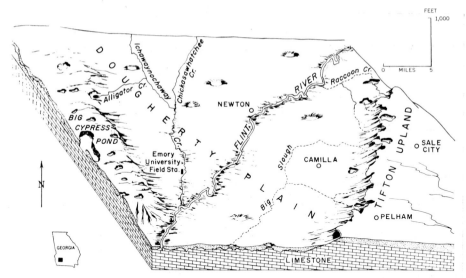

Fig.4. Sketch of Flint River Valley in southwestern Georgia. (After HENDRICKS and GOODWIN, 1952; HERRICK and LEGRAND, 1964.)

carpment extend in a complicated fashion into parts of northern Florida, approxfimating the edge of the Miocene Hawthorn Formation, which is composed chiefly o-interbedded clay, sand, marl and limestone.

Lake basins formed by solution of limestone occur chiefly in areas where the limestone is at or near the surface. Lakes are especially common in central and west-central Florida, according to STRINGFIELD [1966, p.86]; they represent sinkholes that were formed during a low stage of Pleistocene sea and have become lakes as sea level rose to present positions. The fairly even distribution of rainfall during the seasons causes fairly high groundwater levels, and consequently many of the sinks are perennial lakes. Most of the lakes are relatively shallow except those in natural wells and sinkholes which extend into the limestone below. The lakes range in size from mere ponds to bodies of water such as Lake Apopka, which is 14 km wide [COOKE, 1939, p.102].

Hydrogeology

The hydrogeology of this region has been summarized by STRINGFIELD [1966], and by STRINGFIELD and LEGRAND [1966]. The Tertiary limestone is an extensive aquifer system, and is under water-table conditions in some places where the limestone is near the ground surface and is under artesian conditions elsewhere. A generalized map of the piezometric surface of water in the limestone aquifer in Florida and Georgia (Fig.5) indicates the general direction of movement of water as well as areas of recharge and discharge. In general, water enters the aquifer where the piezometric surface is relatively

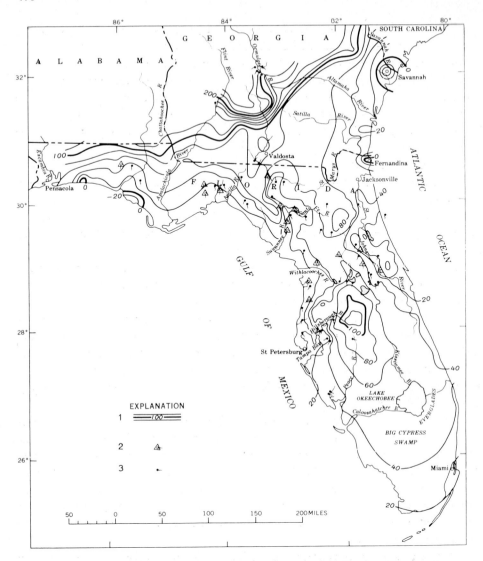

Fig. 5. Piezometric map of water in the principal artesian aquifer. *1* = Piezometric contour (contour interval 20 ft.; datum is mean sea level); *2* = major spring with flow of 100 cubic ft./sec or more; *3* = large spring with average flow less than 100 cubic ft./sec. (After STRINGFIELD, 1964.)

high and is discharged where the surface is low. The highest part is as much as 70 m above sea level in areas of outcrop of the aquifer extending across Georgia. Water moves from that general recharge area toward the south and the Atlantic coast. Much natural discharge occurs offshore. In Florida west of the Suwannee River the general movement is south toward the Gulf of Mexico. Recharge occurs in areas where the aquifer is near the surface (except in the stream valleys) and through sinkholes which penetrate the overlying Miocene clay. Discharge from the aquifer into the Suwannee and Santa Fe rivers causes a large valley in the piezometric surface where these surface streams flow in channels that are cut into the limestone (Fig. 5). In the lake region of south-central Florida, where the piezometric surface is as much as 30 m above sea level, the water moves laterally in all directions from this recharge area. Part of the water on the north, northeast and west is discharged through large springs. To the south, southeast and southwest the water moves many miles to the coast and beyond. The water-level gradient to the coast suggests that discharge occurs by upward leakage through relatively permeable beds in coastal areas. Movement of water in the limestone in South Carolina and North Carolina has not been clearly defined, but much of the water moves from inland recharge areas to streams entrenched in the limestone in coastal regions.

The rate of movement of water ranges greatly from place to place in the Tertiary limestone system. HANSHAW et al. [1968, p. 601] show, through radiocarbon dating, that velocities range from 2 to 12 m/year in the fresh-water part of the Tertiary artesian limestone system in south-central Florida. In parts of the uppermost beds, especially where discharge occurs, velocities may be many times greater, and in some basal salt-water beds velocities are many times slower.

The discharge of springs from the Tertiary limestone aquifer ranges from a few litres per minute to more than 28 m^3/sec. STRINGFIELD [1966] has described the largest springs, and their locations are shown in Fig. 5. Almost all of the springs yield clear water continuously, a fact that may be attributed to the prevalent sandy cover that allows filtering of water into the limestone. Also, many of the solution cavities are filled with permeable sand.

The permeability of the aquifer system differs greatly. The Cooper Marl near Charleston, South Carolina, has extremely low permeabilities. Overall, however, the limestone is moderately permeable to very permeable and tends to transmit water through a close network of solution cavities. The Pleistocene limestone in southern Florida is extremely permeable, and it is one of the most prolific aquifers in the world. Although large caverns may exist in the Tertiary limestone, the overall permeability is represented by a network of small solution openings, some of which are filled with permeable sands.

The factors of: (*1*) considerable annual precipitation; (*2*) the existence of a cover of permeable sand; and (*3*) subdued topographic relief result in considerable recharge, and cause the water level related to the limestone to be near the ground surface in many areas. Where the limestone is covered by Miocene clay, the aquifer is completely filled with water and the piezometric surface is near the land surface; the water table in the surface materials may also be near the surface. The piezometric surface is above land surface in many coastal areas, and here flowing artesian wells are common.

In some areas, especially inland, circulation of the water has been moderately good and the water is potable. This potable water is moderately low in mineral matter with the exception of a calcium bicarbonate hardness. Deep-lying parts of the aquifer, and to some extent almost all of the aquifer in some coastal regions, contain salty water that has not been flushed from the beds. Some of the salty water may have existed from Miocene time, and some from the times of Pleistocene submergence.

Development of permeability in the Tertiary limestone system is in part modern and in part historical. Where water-table conditions occur or have occurred in the past, circulation of water and solution activity tend to be greatest in the upper part of the zone of saturation. The ease with which water discharges from the limestone controls circulation and solution under both water-table and artesian conditions. The current discharge areas in general are easy to determine, but former discharge areas and areas of greatest circulation and solution must be inferred from a study of the geologic and hydrologic history.

Practical considerations

Limestone terranes and related karst topography tend to have some aspects which are liabilities and others which are assets insofar as practical use by man is concerned. The following practical considerations relate to the coastal plain region:

(1) The limestones of the coastal plain of the southeastern United States in general have no undesirable topography because topographic relief is low and generally the sinks are subdued and have gentle slopes.

(2) Although not yet a serious problem in much of the area, subsidence of the ground to form sinks does occur in some places where the limestone is near the land surface; excessive pumping of water from wells tends to aggravate this condition in a few areas.

(3) In typical karst regions where solution has increased the permeability in the limestone some of the surface streams go underground. This condition exists in parts of central Florida. The scarcity of surface streams causes some problems insofar as waste disposal is concerned.

(4) Unlike typical karst regions the limestone formations of the coastal plain are covered by soils almost everywhere. Some of the soils are related to underlying younger marine formations, and others are residual soils formed on the limestone. The soils of Florida are commonly yellow and gray, although the deep red *terra rossa* of tropical karst occurs in the outcrop areas of the limestone in central Georgia.

(5) Although highly mineralized water occurs in the deeper parts of the Tertiary limestone in the coastal plain, this mineralized water is overlain by large quantities of acceptable fresh water. The problem of disturbing the fresh water–salt water boundary and thus causing salt-water contamination of the aquifer is a problem in parts of the region.

(6) A characteristic of limestones is their great range in permeability, and this is true also with the Tertiary and Pleistocene limestones of the coastal plain. Extremely poor permeability reduces the water-bearing capacity of the limestone, and extremely good permeability may cause the water table to be so low that underground storage of fresh water is slight. These extreme conditions in permeability in the coastal plain are

less troublesome than in many other karst regions. In fact, the extremely low permeability of the limestone in a part of South Carolina permits the use of this limestone as a relatively tight underground conduit to bring water many kilometres through a man-made unlined tunnel to Charleston, South Carolina. The extremely good permeability of some of the Tertiary limestone and of the Pleistocene limestone in the Miami area is lessened somewhat by the filling of the solution openings by sand—a fortunate situation insofar as regulating the movement of underground water and preventing contamination by waste water which enters the fresh-water aquifer from the surface. The two limestone aquifer systems are among the most productive sources of groundwater in the world.

Methods of studying the karst geology and hydrology of the coastal plain are chiefly conventional. Field observations have been useful but are somewhat limited because of a cover over many of the karst features. As a result, much reliance has been placed on a study of subsurface geology utilizing data, such as well logs. Surface geophysical methods, especially electric resistivity techniques, have been used sparingly to estimate the fresh and salt water boundary in a few places; thus far, surface geophysical methods have been of only minor importance. The conventional hydraulic techniques of testing aquifers have been successfully used, because permeable limestone is widespread.

Perhaps in no other karst area has as much use been made in studies of such an expansive water-level surface as that characterizing the Tertiary limestone aquifer of Florida and Georgia (see Fig. 5).

APPALACHIANS[1]

Introduction

The Appalachians consist of an extensive region of hills, mountains, and plateaus extending from the New England states south to Alabama and Georgia. The mountainous area in New England lies along and to the east of the Hudson River–Lake Champlain Lowland. To the south it is 96–400 km inland from and roughly parallel to the Atlantic coast. The plateau lying to the west of the mountains covers an area 48–80 km wide in Tennessee and Alabama increasing northward to a width of more than 320 km in Pennsylvania (Fig. 6).

South of New England the mountainous part of the Appalachians embraces the Blue Ridge and Great Smokies in the east and the Valley and Ridge section to the west. The Blue Ridge and the Great Smokies are complex maturely dissected mountains formed of metamorphic and intrusive igneous rocks. The ridges are 300–600 m high in Pennsylvania. Altitudes increase southward and culminate in Mount Mitchell (2,013.3 m) in northwestern North Carolina.

---

[1] Written by W.E. Davies.

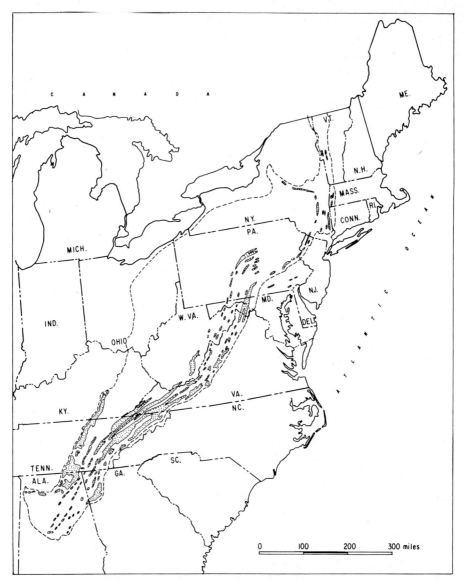

Fig.6. Appalachian karstlands. Stippled = karst areas.

The mountains in the Valley and Ridge section are long even-crested parallel ridges separated by narrow valleys. The summits of major ridges are 600 to nearly 1,500 m high with numerous subordinate ridges at lower altitudes. Along the eastern side of the Valley and Ridge section is a broad lowland, the Great Valley. The lowland is rolling to hilly; in the headwater areas of major rivers low discontinuous ridges disrupt it.

The Appalachian plateaus are maturely dissected with narrow, sinuous, steep-walled valleys, in many cases more than 200 m deep, cut into the upland. Along their east front the plateaus range from 600 to nearly 1,500 m. To the west the upland altitudes are less, and the plateaus grade into the central lowland in Ohio, Kentucky, and Tennessee.

In New England and northeastern New York, the Taconic and Green Mountains consist of irregular ridges and rounded peaks with altitudes of 360 to more than 1,200 m separated by short interconnected valleys. West of the mountains are the Adirondacks, an area of complex hills and ranges with an altitude of 600 to more than 1,500 m. On the north flank of the Adirondacks is the St. Lawrence Valley, a rolling lowland 60–150 m above sea level.

Pleistocene glaciation extended over all of New England, New York, and the northeastern and northwestern parts of Pennsylvania. In these areas the topography has been rounded and subdued along with extensive disruption and rearrangement of drainage.

The major physiographic provinces of the Appalachians reflect differences in rock type and geologic structure. The Blue Ridge, Great Smokies, and the Green Mountains in Vermont are primarily non-carbonate metamorphic and igneous rocks which are of no significance in karst development. In the western part of the Great Smokies in Tennessee, there are small enclaves of Lower Palaeozoic limestone similar to those in the Great Valley to the west that are exposed in "windows" eroded through the overthrust non-carbonate metamorphic rocks.

The Valley and Ridge sections are made up of a sequence of Palaeozoic rocks as much as 18,000 m thick ranging from Cambrian to Pennsylvanian (Carboniferous) in age. These rocks are exposed in a series of successive open folds cut by numerous thrust faults. In the Great Valley more than 70% of the lowland is limestone and dolomite of Cambrian and Ordovician age. These rocks along with complementary shale and sandstone formations at the top of the Ordovician and base of the Cambrian are intensely folded, and in the eastern half of the Great Valley recumbent folds are common.

West of the Great Valley clastic rocks dominate in the Valley and Ridge section. Limestone formations with thickness as much as 300 m occur primarily in the Lower Silurian and in a zone straddling the Silurian and Devonian boundary. Cambrian and Ordovician limestones and dolomites are at the surface in a few large anticlinal lowlands in central Pennsylvania and eastern West Virginia.

In the Appalachian plateaus the rocks are mainly Mississippian to Permian in age. Their gentle westward dips are locally disrupted by long, broad anticlines trending northeast–southwest. In the Mississippian System sandstone and shale are the main lithologic types, but along the eastern part of the plateaus, south from southern Pennsylvania, limestone is present in the Lower Mississippian.

The Pennsylvanian and Permian systems consist of a series of cyclothems formed of repetitive sequences of sandstone, shale, limestone, clay, coal and siltstone. The limestones are generally thin and of little speleological significance.

In New England and adjacent parts of northeastern New York the rocks are strongly folded and metamorphosed sediments of pre-Silurian age. Limestone occurs in the Taconic Mountains and along the Hudson River–Lake Champlain Lowland. In the St. Lawrence Lowland most of the rocks are folded limestone of Ordovician age.

The great latitudinal extent of the Appalachians involves climate varying from warm and humid in the south to cold and humid in the northern part of New York and New England. Average annual temperature is 60°F (16°C) in the southern Appalachians dropping to 40° and 45°F (4° and 7°C) in northern New York, New England, and on the high mountainous area as far south as southern Virginia. Average annual precipitation is 875–1,400 mm/year.

The mountains and plateau uplands from Pennsylvania south are now covered by temperate deciduous forests. To the north this forest is replaced by Boreal coniferous forest. Most of the lowlands are cleared and are used for cultivation or pastures.

Great Valley

The Great Valley is essentially a limestone–dolomite lowland throughout its extent from New Jersey to Alabama. Solution erosion of these rocks has been great, and the Great Valley contains the most extensive karst and cavern area in the eastern United States.

The carbonate rock sequence is almost unbroken from Upper Cambrian to Upper Ordovician. In the north the Cambrian carbonate rocks are highly magnesian; the Ordovician rocks are high calcium limestones. From Viriginia south to Alabama, however, magnesian content increases in the Ordovician, and non-magnesian limestones increase in the Cambrian. Impurities in the form of chert and other siliceous material are present. However, in the Lower and Middle Ordovician there are several extensive formations of high calcium limestones that are almost free of impurities. The carbonate sequence in eastern Pennsylvania is about 1,200 m thick; southward the thickness of carbonate rocks increases to more than 3,300 m in south-central Pennsylvania, Maryland, and Virginia, decreasing to 2,300 m in Tennessee and 3,050 m in Alabama. The rocks are strongly folded and are cut by numerous low-angle thrust plane faults interrupted by normal offset faults. Along the eastern side of the Great Valley the carbonate rocks have large recumbent folds overturned to the west and locally, primarily in Tennessee, the rocks of the Blue Ridge have been thrust across large areas of limestone.

The surface features of karst in the Great Valley, while extensive, are limited to a few forms. Dolines and uvalas are present in areas of hills, lowlands, and valleys. Where the lowlands are extensive in Pennsylvania, Maryland, and northern Virginia the shallow dolines form a billowy to rolling surface. Most of the dolines are from 6 to 600 m in diameter and 3 to more than 36 m in depth. Slopes are 10°–30° and the floors of most dolines are flat with alluvial fill. The dolines are in bands and most are elongated slightly in the direction of the strike of the rock. The alluvial floor commonly covers open fissures and small ponors in the bedrock. The fill is subject to erosion by both descending surface water and fluctuating groundwater, and collapse of the alluvial floor occurs occasionally. Because the Great Valley is intensely developed for agriculture and industry, the collapses cause considerable economic loss as well as loss of life.

Karren ridges are present throughout most of the Great Valley (Fig. 7). The ridges are 0.6–1.8 m high and are in parallel bands trending along the strike. The karren ridges are separated by areas of soil as much as 9 m deep and as much as 12 m wide. The karren contains rills and shallow solution fluting; chert blocks in the limestone are etched out and stand in low relief on the sides and much of the top of the karren.

Soils in the karst area of the Great Valley vary considerably. North of Tennessee they are red brown to brownish yellow, well drained and oxidized, granular silty clays and clays. They are from 5 cm to more than 9 m deep. Because of the heavy rainfall most of the carbonate material has been leached from the soil, and the soils are acid to within a few inches of bedrock. Angular chert and fragments of siliceous limestone occur throughout the soil.

From Tennessee south the soils are deep red brown to reddish purple with subordinate areas of yellowish-brown color. Stiff red clays occur within the dominant silty clay and stony silt soils. Depth of soil is similar to that in the area to the north.

In the Great Valley there are more than 2,100 caves of which 1,100 are in Virginia. Most of the caves are simple in pattern consisting of one or several parallel passages trending along the strike. Passages at multiple levels are common, and generally, successive levels are offset downdip. Domepits as much as 30 m deep are common, some extending to the surface and serving as entrances to caves. This type of cave has been designated as *Appalachian* type [BARR, 1961; Fig. 8A]. About 70% of the caves in the

Fig.7. Karren in the Great Valley, Whitings Neck, West Virginia.

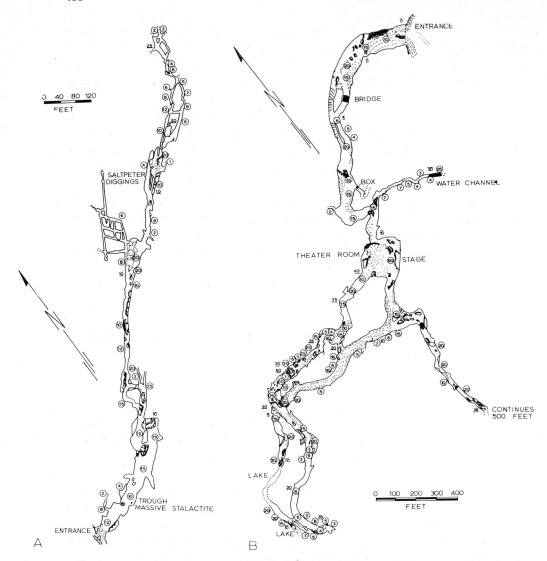

Fig.8. Plan of typical caves. (Numbers indicate ceiling height.) A. *Appalachian* type, Trout Cave, West Virginia. B. *Allegheny* type, Laurel Creek Cave, West Virginia. (After DAVIES, 1965.)

Great Valley have less than 150 m of passage, 20% have 150–600 m, and only 10% have more than 600 m.

Speleothems are common in most caves and are often large and prolific enough to be an attraction to tourists. Twenty caves in the Great Valley are now operated as commercial ventures catering to tourists. Gypsum deposits are absent from most caves in the Great Valley. Most caves contain fills formed of clastic material and clay. The colors of the fills are yellow brown to gray north of central Virginia. Red-colored fills, however, are common to the south.

Hydrologic conditions in the folded limestones of the Great Valley are complex and have received very little study. The average depth to water-bearing cavities is about 45 m below the ground surface. However, the depth of water-bearing cavities ranges from 6 to more than 500 m. Yields are as much as 680 gallons (2,574 l)/min, but more than half the wells are less than 10 gallons (37,8l)/min. Large springs are common with flows as much as 9,600 gallons (36,400 l)/min. [CADY, 1936; DeBuchananne and Richardson, 1959]. A few ebbing and flowing springs occur in which short periodical flows of a much as 1,000 gallons (3,785l)/min. are followed by dormant periods ranging from one hour to several days. Hydrographs of the flows indicate that the discharge is controlled by siphons in the cavernous rock.

Recharge of the groundwater in the Great Valley is primarily from direct infiltration of precipitation. In addition seepage from some headwater streams and recharge from water in thick gravel and talus deposits along the mountain ridges bounding the valley add significantly to the water in the limestone. Some long-range migration also may occur as artesian flow along thrust fault planes.

Exploration work for Tennessee Valley Authority dams in the Great Valley of Tennessee showed that solution cavities occur as much as 36 m below the river bed and as much as 76 m below the surface of the ground. At shallow depths the cavities are filled with sand and clay, but at depths below 9 m the cavities are open and commonly water filled [Anonymous, 1949].

Valley and Ridge section

Carbonate rocks form only a small part of the Valley and Ridge section. Limestones in the Upper Silurian and Lower Devonian are about 250 m thick in central Pennsylvania and Maryland, but all of the carbonate formations thin rapidly to the south and are replaced by clastic sediments south of central Virginia. The limestones are impure, generally containing small to moderate quantities of shale and chert. The limestones, along with a great thickness of clastic rocks stratigraphically above and below, outcrop in long ridges reflecting the nearly symmetrical broad folds of the region. Dips are extensive. Thrust faults are present in the folded belt, but are less extensive than in the Great Valley to the east. Commonly the Silurian and Devonian limestones are ridge-forming rocks and outcrop on the summit and dip slopes of subordinate ridges.

Cambrian and Ordovician limestones and dolomites, similar to those in the Great Valley, outcrop extensively in the Valley and Ridge in southern Virginia, Tennessee, Alabama, and Georgia, primarily in a narrow belt to the west of the Great Valley. In Pennsylvania and West Virginia the Cambrian and Ordovician rocks are brought to the surface in large anticlines that have been eroded into broad lowlands.

In the Valley and Ridge section, karst features are not as extensive nor as varied as in other parts of the Appalachians. Where the limestones form the tops and flanks of ridges, dolines are small, seldom exceeding 30 m in diameter or more than 9 m in depth. The dolines are widely separated along the limestone ridges and some contain thick, impervious alluvial fills on which small ponds exist (Fig. 9). Ponors, as much as 82 m deep are common, especially on the flanks of ridges. On some limestone ridges, primarily in northern West Virginia, the dolines formed on limestone summits have been enlarged so that they cut saddles through the ridge forming a series of low, rounded knobs separated by steep-sided, narrow gaps.

Soil is thin with bedrock generally less than 0.6 m below the surface. Karren, however, is rare and where present the karren ridges are generally less than 0.5 m high. At the base of steep slopes, limestone talus is extensive, and the rock fragments forming the talus are etched from solution. The soil is light brown to brown in color and ranges from silty clay to friable, fine sandy silt. Plastic clay is present at the soil–bedrock contact. Most of the soil is stony and contains a large quantity of silicified fossils, chert, and sandstone from adjacent clastic formations.

About 750 caves are known to exist in the mountainous part of the Valley and Ridge section (Fig. 10). The patterns are simply consisting of one or two parallel passages developed along the strike of the limestone (Fig. 8A). Multiple level caves are common with most levels developed at correlative altitudes. Pits, some as much as 70 m deep, are common especially in caves developed in large anticlinal lowlands containing Cambrian and Ordovician limestones and dolomites. About 72% of the caves are small with total length of all passages less than 150 m; 22% are as much as 600 m and 6% are more than 600 m. In Virginia several caves are 6,000–15,000 m in total length of all passages.

Most of the caves in the mountainous region contain relatively few speleothems.

The hydrologic regime in areas where limestone is on ridges is relatively simple. Most of the water entering the limestone is from direct infiltration of precipitation.

Fig.9. Upland doline and pond, Valley and Ridge section, Knobly Mountain, West Virginia.

Fig.10. Entrance to Trout Cave, a large *Appalachian* type cave in the mountainous part of the Valley and Ridge section of West Virginia.

Where sandstone beds overlie the limestone, recharge is from seepage from the water in the sandstone. Dolines on the summit and flanks of ridges serve as small catchment basins. There are few cavern streams and all are small, being fed by seepage and subterranean discharge from upland sinkholes. Small surface valleys commonly end downslope in ponors, but they generally are dry except during periods of heavy rain. Yield from wells is small, generally less than 5 gallons (18.9 l)/min.; only a few springs yield more than 1,000 gallons (3,785 l)/min. and the maximum is 7,700 gallons (29,150 l)/min [Cady, 1936; Price et al., 1936; Lohman, 1938]. One ebbing and flowing spring is in the mountainous part of the Valley and Ridge section in central Pennsylvania.

Where the limestones are in anticlinal valleys the caves, fissures, and solution-enlarged joints carry most of the drainage, and large areas lack permanent surface streams. Most of the recharge is from surface and groundwater moving from the ridges and talus deposits at their bases into limestone. The surface streams, many of which are intermittent, drain into dolines and ponors. The groundwater level is generally within 30–60 m of the surface, although principal water-bearing channels occur as much as 91 m below the water table.

Springs are numerous and at least 100 springs in the anticlinal limestone lowlands of the Appalachians discharge more than 500 gallons (1,893 l)/min. [Reeves, 1932]. Maximum yield is from a spring at Bellefonte, Pennsylvania, which discharges 14,000 gallons (53,000 l)/min.

In west-central Virginia and adjacent parts of West Virginia there are about 40 thermal springs issuing from the limestones in large anticlinal valleys [Waring, 1965]. Temperatures range from 61° to 106°F (16°–41°C). The thermal conditions have been ascribed to heat acquired from deep circulation of water in the limestone. The temperature has also been related to possible residual heat surrounding small Mesozoic diabase dikes that intrude the rocks of the region.

Appalachian plateaus

The rocks of the Appalachian plateaus are primarily cyclothems of clastic rocks and coal of Carboniferous age. The only limestones that are significant from a karst viewpoint are those of Mississippian age in the central and southern Appalachians and small areas of Devonian and Silurian limestones along the edge of the plateau in New York.

The Mississippian limestones outcrop along the eastern part of the plateau from southwestern Pennsylvania to Alabama and along the western part in central Kentucky, Tennessee, and northern Alabama. The limestones constitute six or seven formations or members of a succession of carbonate and clastic rocks. In Alabama, the succession is about 480 m of which limestones form about 420 m. Northward the limestones are of less thickness and are reduced to 360 m in southern West Virginia and are less than 30 m in southwestern Pennsylvania. Most of the limestones are shaly or cherty. The dips of formations are gentle to the west, but along the eastern part of the plateaus there are several broad anticlines that disrupt the structure, bringing the limestone to the surface.

In Pennsylvania, northern West Virginia, and Kentucky the limestone outcrops are primarily along the flanks of low ridges and on steep valley walls. In central and and southeastern West Virginia there are several large areas of low plateau where the limestone forms the surface. In Tennessee and northern Alabama the plateau upland is formed by limestone. Where the limestones are exposed on the upland surfaces of the high and low plateaus karst features are extensive. Large dolines and uvalas are so numerous that in many areas they coalesce forming a billowy surface with few streams or valleys (Fig. 11). The dolines are from 3 to more than 150 m in diameter and as much as 60 m deep. Side slopes of 30° are common and most dolines have alluvial floors.

Uvalas are commonly elongated in the direction of the strike of the bedrock and form karst valleys as much as 4 km long and 1 km wide. Most of the karst valleys are steep walled with slopes from 45° to vertical, and they are from 15 to more than 213 m deep.

In southern West Virginia subordinate karst valleys also trend across the strike and carry surface drainage from higher parts of the plateau to subterranean channels. Some of these valleys are as much as 3 km long and from 60 to 600 m wide. They are generally less than 60 m deep. Ponors are numerous especially where the limestone is along valley walls. Many of the ponors are more than 60 m deep, and one in Alabama reaches 120 m in depth.

Along the periphery of the limestone upland in parts of West Virginia, Tennessee, and Alabama there are many short, steeply sloping karst valleys that are cut into the

Fig.11. Coalesced dolines on the low plateau, Greenbrier county, West Virginia.

Fig.12. Terraces along a karst cove, Herbert Hills Creek, West Virginia.

flanks of higher parts of the plateau. Locally they are designated coves. The coves are from 30 to more than 210 m deep and from 300 to 1,500 m wide. Most of them are less than 3,200 m long. The bottom of the coves are formed of small irregular dolines and ponors covered in part with alluvial and talus deposits. A few of the coves contain well-developed alluvial terraces reflecting impoundments when subterranean drainage was blocked (Fig. 12).

In east-central Tennessee there are several large uvalas that approach poljes in size. Grassy Cove, the largest, is 8 km long and 3 km wide. Its alluvial floor is 180–360 m below the surrounding upland (Fig. 13).

Soils on the upland surfaces of the plateaus are deep, well drained, and moderately permeable. Silty clay and clay are dominant. Color is mostly yellow brown to brown north of Tennessee. In Tennessee and Alabama the color is darker with many areas of reddish brown to deep red purple colored soils. The upland soils are 1.5–3 m deep, but ledges of limestone forming small areas of karren occur throughout the plateau. In areas of steep slopes the soil is generally only 0.3–0.6 m thick, and karren formed by low ledges of bedrock are common (Fig. 14).

Fig.13. Grassy Cove, Tennessee. Flooding is from impoundment because of insufficient capacity of cavern streams to carry spring run-off.

Fig.14. Karren on Mississippian limestone in the high plateau of central West Virginia.

There are about 900 caves known in the plateau area. About 62% have less than 150 m of passage, 15% are of moderate size with as much as 600 m of passages, and 23% have more than 600 m of passages. Cavern plans vary greatly. Most caves consist of random successive branching passages, a form designated as *Allegheny* type [BARR, 1961; Fig. 8B], but maze caves formed of numerous intersecting cross passages and sinuous single passage caves are common. Multiple level caves are prevalent in thick limestones. Many caves contain deep pits with depths of 30 to more than 121 m. The deepest pit is the 130 m Surprise Pit in Fern Cave, northern Alabama [TARKINGTON et al., 1965].

Speleothems are not extensive in most caves, but gypsum is common in the cave fills and locally as speleothems. The source of the gypsum is probably in the coal beds and Carboniferous shales that are stratigraphically above the limestone with seepage or stream invasion carrying the gypsum into the cave. *Allegheny* type caves also contain large areas of breakdown (ceiling collapse).

Hydrologic conditions in the plateau limestones vary greatly from one area to another. Deep drilling in the plateau has revealed no significant cavities in the limestone where the overburden of younger clastic rocks are more than 90 m thick. Cavern development in the limestones also reflects this in that caves in areas of deep clastic cover seldom extend more than 60 m normal to valley walls. However, many caves have extensive passages parallel to valley walls and through spurs of the plateau.

In northern West Virginia and Pennsylvania, the limestones contain very little water, and recharge is by seepage from overlying clastic beds.

In areas where the limestones are at or near the surface on the upland, recharge is directly from precipitation and from seepage and streamflow from higher parts of the plateau adjacent to the limestone areas. The streams enter the limestone areas in karst valleys and disappear underground in shallow dolines or at steepwalls that terminate the valleys. Large tubular springs and caves with open channel flow are common with

Fig.15. Piercys Mill Cave Spring, West Virginia. Discharge is about 6,000 l/min.

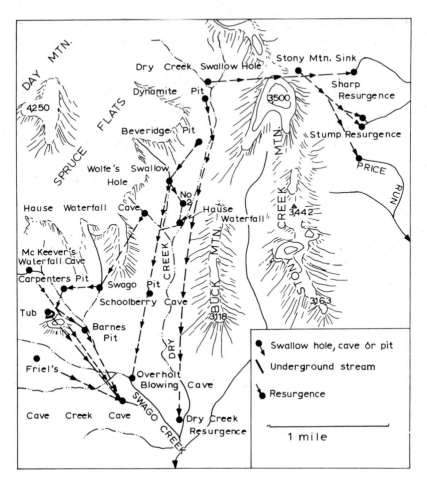

Fig.16. Traces of subterranean streams, Swago Creek, West Virginia. (After ZOTTER, 1965.)

discharges ranging up to 1,000 gallons (3,785 l)/min (Fig.15). Maximum flow is 35,000 gallons (133,000 l)/min [JOHNSTON, 1933].

Very little is known about the subterranean drainage nets, but the data available indicate they are extremely complex. The groundwater conduits generally do not follow simple, direct courses but commonly are along sinuous, indirect paths, at times passing above or below other conduits and in some sections doubling back along the general trend of subterranean drainage. Conditions at Swago Creek, in central West Virginia (Fig.16) are illustrative of the complexity [ZOTTER, 1965].

Other karst areas

In the Great Smoky Mountains in eastern Tennessee there are several enclaves of Ordovician limestone exposed where erosion has cut through the overthrust crystalline rocks of the Blue Ridge. Karst features similar to those in the Great Valley are present. Caves are numerous and large. Many have great vertical extent and consist of a series of pits as much as 45 m deep, offset by short sloping passages. Bull Cave in Cades Cove has a vertical extent of 200 m and is the deepest in the eastern United States. Water from rocks and streams of the adjacent mountains invades the limestone and discharges in large tubular springs with flows as much as 9,600 gallons (36,300 l)/min.

In the Hudson–Champlain Lowland and the St. Lawrence Lowland there are extensive areas of Ordovician limestone and marble. Karst development on these rocks is poor and consists of small isolated dolines and ponors. Karren, however, is extensive in the form of low parallel ridges. Caves are few and are less than 90 m long consisting generally of narrow fissure passages connecting with small chambers [SCOTT, 1959]. Similar karst conditions are in the limestone areas of the Piedmont in southeastern Pennsylvania.

The Appalachian karst presents few serious economic problems. Erosion is present but the only areas where it is extensive is on open hillsides where sheet wash removes much of the soil. Road and other construction work is complicated by the extreme relief at the rock–soil interface. Most cuts and excavations encounter alternate bands of shallow bedrock and deep soil.

Surface and groundwater are copious in the Appalachian limestone areas, but pumping of large quantities of water from deep wells in limestone lowers the water table over large areas causing widespread failure of shallow walls.

Locally, as at Staunton, Virginia, and Lewisburg, West Virginia, subsidence into sinkholes has caused extensive damage to buildings, streets and utility lines.

CENTRAL KENTUCKY AND SOUTHERN INDIANA[1]

The Interior Lowlands of central Kentucky and southern Indiana contain extensive areas of Mississippian (Carboniferous) limestone. These limestones are at the surface from southwest Kentucky, eastward to the center of the state where they swing to the north cutting through the center of the state and extending into the south-central part of Indiana (Fig. 17). To the north they are buried by Pleistocene glacial deposits, and at the southwest Cretaceous and younger deposits of the Mississippi embayment cover the limestones.

The karstlands in Kentucky developed on these limestones form two distinct landform units. Along the southern and eastern part of the limestone area is a sinkhole plain of low relief, the Pennyroyal, bounded on the north and west by a somewhat higher plain or low plateau, the Mammoth Cave Plateau.

In Indiana the low plain is more extensive and the higher karst plateau is not developed to the point that it can be recognized as an extensive karst unit.

The Mississippian limestones in the Kentucky–Indiana area are about 360 m

---

[1] Written by W. E. Davies.

Fig.17. Karst areas of central Kentucky and Indiana.

thick and contain sandstone and shale formations along with the limestone. The Pennyroyal Plain and the karst plain in Indiana are formed on the Warsaw, St. Louis, and Ste. Genevieve limestones which have a total thickness of 100–200 m. The plateau area includes the Girkin Limestone (30–50 m) overlain by the Big Clifty Sandstone (20–25 m), the latter forming the cap rock over much of the area. Above the Big Clifty are a series of limestones, sandstones, and shales with a maximum aggregate thickness of about 60 m. In the plateau these are present in only a few small areas in the central part of the karst area and along the northern edge.

The karst is in an area of relatively flat-lying rock with a dip of 6 m/km to the northwest. A low anticlinal dome trends east-northeast across the southern part of the plateau; along the eastern edge of the karst area the anticline trends north-northeast. A few minor, widely spaced, normal faults cut across the anticline.

The boundary between the Pennyroyal and Mammoth Cave Plateau is formed by the Dripping Springs escarpment. On topographic maps it is a distinctive feature with relief of 65–130 m. From the ground, however, it is less distinct because of the high degree of dissection that has altered much of the front into a series of low rounded spurs separated by short irregular karst valleys,

The Pennyroyal Plain contains numerous karst features of small to moderate size. The entire surface of the plain is formed of small dolines that are closely spaced. They are generally circular in plan and from 3 to over 1,600 m in diameter. Most of the dolines are less than 10 m deep and slopes of the sides are smooth and gentle; the floors are alluviated, forming a flat center. A few of the dolines contain ponds but, in general, there is an absence of surface drainage. In the northern half of the Pennyroyal all streams are subterranean. A few small streams are at the surface in the southern part of the area. These streams are up to 16 km long, draining areas of 76 km^2. However, all of them terminate in swallow holes near the end of shallow, blind, karst valleys. Their subterranean courses are mysteries, but the little data available indicate that some of the subterranean drainage transects the Mammoth Cave Plateau and reaches Green River. Typical of these streams is Little Sinking Creek lying 6 km south of the Dripping Springs escarpment which has a surface drainage basin of 20 km^2. After flowing on the surface for about 10 km the stream plunges down a small swallow hole at the end of a large karst valley. At the swallow hole the floor of the valley is covered by alluvium over 6 m deep. The apparent subterranean course is northwards under the Dripping Springs escarpment for 5 km to Mill Hole, a 240 m long by 150 m wide gulf (uvala), where the stream resurges for a distance of 80 m (Fig. 18, 19). At the center of the gulf the stream flows under a ledge of limestone and follows a subterranean course for 5 km to the northwest where it resurges in another gulf, Cedar Sink. After flowing on the floor of the gulf for 80 m it again enters a subterranean channel flowing north for 2.5 km to Turnhole Bend, a large flooded doline connected to Green River.

The gulfs, which are primarily in the western part of the Mammoth Cave Plateau are up to 45 m deep and some reach a length of 1.6 km. The walls surrounding the gulf are vertical or very steep. The floors of the gulfs contain an alluvial fill 10 m or more thick. During floods the water backs up in the gulfs and inundates most of the floor, and in extremely high floods some are filled to their brim from combined subterranean flow and surface run-off.

In the eastern part of the Mammoth Cave Plateau subterranean drainage is less conspicuous. In Mammoth Cave some of the karst water is from inflow and back-flooding from Green River and some from infiltration from the plateau [WATSON, 1966]. Apparently most water moving through the caves in the eastern part of the plateau is from local surface run-off. An exception to this is Horse Cave 16 km east of Mammoth Cave where a subterranean stream flows westward from the Pennyroyal towards the plateau. The course of this stream and its discharge point are not known.

The karst features on the Mammoth Cave Plateau contrast sharply with those of the Pennyroyal. Large, steep-sided dolines are common features on the plateau. They are commonly more than 30 m in diameter and some are over 1,200 m on a side. Most of the large dolines extend to a depth of 30–95 m below the level of the plateau upland.

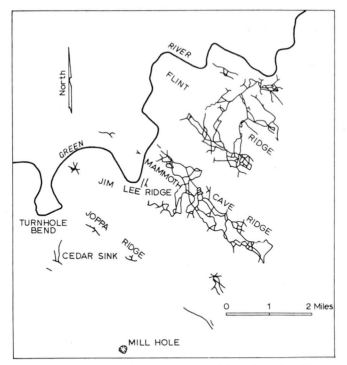

Fig. 18. Major cave systems Mammoth Cave, Kentucky. (CAVE RESEARCH FOUNDATION.)

Fig. 19. Resurgent point in Mill Hole, Mammoth Cave area, Kentucky.

All of the dolines have alluvial floors with as much as 15 m of silt and sand.

In the eastern part of the plateau there are several large karst valleys which are up to 3.2 km in length. These valleys are up to 800 m wide and are 45–60 m deep. Valley walls are steep to vertical and most of the valley floors lack permanent streams.

Between the valleys are ridge-like remnants of the plateau surface. Small dolines, up to 100 m in diameter and less than 10 m deep, occur on this upland. Most of the surface, however, is rolling with little indication of the vast net of cavern passages that lie beneath it.

There are more than 80 caves in the Mammoth Cave Plateau. Many of these are less than a hundred metres long but two of the cave systems, Mammoth Cave and Flint Ridge Cave, each have more than 50 km of surveyed passages, ranking them among the longest in the world [SMITH, 1964; Fig. 18].

Mammoth Cave contains 48 km of major passages that have been mapped and about an equal length of small secondary passages. The main passages are long galleries, 6 m to more than 15 m wide and high extending to over 1 km in length. These passages connect with subordinate passages grading from small crawlings and narrow fissures to walkways 3–6 m wide and high.

The cave is developed along joints with the pattern modified by arcuate fractures resulting in a complex of interconnected, curving passages. The pattern is more complex than appears in a plan view because the passages are at different elevations with some grading from one level to another and passing beneath or above other passages without connecting. Most of the passages lie in one of three vertical zones The floor level of the highest zone is 175–190 m, the floor level of the middle zone is 150–170 m, the floor level of the lowest zone is 130–135 m above sea level. The highest zone is in the Girkin Formation and the uppermost part of the Ste. Genevieve Limestone.

The majority of the cave lies in the 150–170 m zone which is in the upper part of the Ste. Genevieve Limestone. Most of these passages are low galleries up to 30 m wide and 6 m high and fissures up to 6 m wide and 25 m high along with numerous walkways of smaller dimensions. In the westernmost (New Discovery) section the passages in this zone contain about 1,220 m of large galleries similar to those in the the highest zone elsewhere in the cave.

The lowest level passages contain streams and pools and are subject to periodic flooding. The passages in this level are 3–6 m high and 3–15 m wide, but they are somewhat larger where they grade with passages in other zones. The lowest passages are probably in the lower part of the Ste. Genevieve Limestone.

The passages have been greatly modified by rockfalls containing blocks of limestone up to 9 m on a side mixed with masses of smaller rubble. The falls have greatly reduced the size of passages and increased the elevation of the floor.

Domepits, vertical shafts up to 60 m deep, are well developed in the Mammoth Cave and Flint Ridge Cave. The domepits are up to 9 m in diameter and their walls are commonly grooved from the solution action of descending water.

Earth fills, consisting of gravel, sand, and silt topped by clay, packed to the ceiling are common at all levels of the cave. The gravel contains quartz pebbles and sand from the Big Clifty Sandstone along with chert and limestone pebbles. Some of the pebbles are frosted while others bear a high polish. The deposits are intricately cross-bedded and contain other evidence of rapid deposition from streams invading the cavern.

Mammoth Cave contains only a few areas in which there are stalactites, flowstone, columns, and other speleothems. Minerals other than calcite and gypsum are abundant in the cave but are seldom seen since they are disseminated in the cave fills. Barite is present throughout the cave; celestite is in the fills in the intermediate and highest zones. Several mineral forms of manganese dioxide are also present. A hydrous sodium sulfate is in the cave fills but the inherent instability of the material makes it difficult to identify the specific mineral present. Small percentages of detrital garnet, chromite, goethite, leucoxene, collophane, brookite, anatase, rutile, tourmaline, and zircon are in the fills.

The cavern system in Flint Ridge is similar in form to that of Mammoth Cave. It consists of alternating large galleries and smaller passages developed in four zones. This system consists of four caves formerly believed to be separate but now known to form one vast system with more than 64 km of passages that have been mapped [BRUCKER and BURNS, 1966]. In addition, one other cave, Great Onyx, is apparently a part of the system but as yet no actual connection has been explored.

In Joppa Ridge, conditions are quite different from Mammoth and Flint ridges. Here the cavern system is primarily a subterranean trunk stream that lies at or below the groundwater level. Only short high-level segments of this cave system, primarily in Cedar Sink and the outlet cave at Turnhole Bend, have been explored.

The cave patterns of the Mammoth Cave Plateau were probably formed in pre-Pleistocene time and developed into mature cavern systems in Early Pleistocene. The initial pattern developed along arcuate fractures and joints when the limestone was deep beneath the water table. Integration and maturity of the system took place in a restricted zone at the top of the water table.

During the concluding phase of solution enlargement and integration, large quantities of clastic sediments were brought into the cave by invading surface drainage under alternate phreatic and vadose conditions. As the passages became dry with the lowering of the water table by downcutting of major surface drainage-ways, they were greatly modified by rockfall, erosion by free-flowing streams and vertical solution forming pits and domes. The youngest deposit in the caves is guano, developed from bat droppings, which is 0.3 m or more deep in some galleries. In Mammoth Cave these deposits have been dated by radiocarbon methods as over 38,000 years old.

Karren are not present in most of the Kentucky karst except for a few small areas where ledges of limestone occur along the sides of dolines. The soil is sandy clay, brown, and light red to deep red in color. The residual soil on the Pennyroyal is up to 6 m deep. The upper 15 cm is leached and is light gray in color. The Pennyroyal is farmed extensively while the plateau is heavily wooded. Much of the plateau in the

Mammoth Cave area is included in the Mammoth Cave National Park.

Dependable water supplies are a problem. Wells on both the Pennyroyal and the plateau are difficult to locate and generally yield only small supplies [CUSHMAN et al., 1965; BROWN, 1966]. Pollution is a serious problem adjacent to small towns where numerous septic tank fields furnish effluent that often reaches subterranean water courses. Soil erosion is serious in the Pennyroyal area where extensive farming is carried on. The flanks of dolines and karst valleys erode rapidly with consequent silting and plugging of the areas near swallow holes [DICKEN and BROWN, 1938].

Tourists to the Mammoth Cave are a major factor in the economics of the region. Formerly eight caves in the area were open for public visit. In recent years this number has decreased to three, because of inclusion of some of the caves in the National Park system and the financial failure of others. Conservation at Mammoth Cave meets with difficult problems at present. The large influx of visitors to the cave and the extensive alteration of the surface to provide roads and accommodations threaten the natural environment of the cave, and the problem is now receiving attention from both government and private sources.

In Indiana the karstlands reflect conditions in Kentucky. The Mitchell Plain forms the east part of the area and resembles the Pennyroyal. The Crawford upland is the equivalent of the Mammoth Cave Plateau but it is more highly dissected consisting of low rounded hills. Karst features on the upland are confined mainly to the eastern edge.

The Mitchell Plain is noted for its underground drainage features. Small dolines form much of the surface and several large subterranean stream channels carry the drainage across the karst belt. The largest is Lost River which is about 11 km long. There are about 400 caves in the Indiana karst [POWELL, 1961] of which over 300 are in the Mitchell Plain. Most of the caves are on a single level and patterns are generally simple consisting of a single passage with numerous changes in direction. Larger caves have curving, branching patterns with many of the passages interconnected. At least 40 caves are over 300 m long and several caves contain more than 1.6 km of passageways. Gypsum and other sulphate minerals are common in many of the caves.

THE PECOS VALLEY OF NEW MEXICO[1]

The Pecos River drainage basin, extending southward through east-central New Mexico and into Texas, does not appear to be extensively karstified if viewed from above the ground. Yet, the volume of soluble material that has been removed by solution has been considerable. The Pecos River cuts into Triassic and Permian rocks that have a gentle eastward dip. Rocks of several formations change laterally according to their depositional environment, limestone, dolomite, and anhydrite being prevalent. Westward from the Pecos River the land rises somewhat gradually until peaks of the

---

[1] Written by H.E. LeGrand.

Guadalupe Mountains are reached, where reef limestone is boldly exposed at elevations greater than 2,500 m. The climate is semi-arid, and rocks denuded of soil abound everywhere, although a broad belt of alluvium borders the river and covers low re-entrants into the canyoned upland areas.

Much of the solution in the evaporite beds has been along the bedding planes, especially west of the river where the land surface crudely parallels both the dip slope and the gross movement of groundwater. Some of the evaporite beds west of the river have been removed by sheet-like solution, but the extent of lowering of land surface by this solution is not easy to distinguish from lowering caused by physical erosion. Vertical permeability does not appear to be well developed in the evaporites, although along the escarpment on the east side of the Pecos River, near Roswell, deep sinks and sinkhole lakes 50–100 m wide and 30–65 m deep occur.

Sinks are not common in the Guadalupe Mountains, and one must look underground, as in Carlsbad Caverns, to see the effects of solution in the limestone. This large, spectacular cavern system lies under the mountains adjacent to the reef escarpment and adjacent to a lower plain. Carlsbad Caverns extend downward about 300 m below the present ground surface. The cavern system is extremely large, and piles of rock debris that have fallen from the roof are common on the floor. Solution at some places and deposition at others in the caverns lead to a magnificent display of stalactites, stalagmites, and other cave formations.

Groundwater tends to flow eastward from beneath the Guadalupe Mountains toward the Pecos River, and the river is a discharge area for much of the moving water. MOTTS [1968, p. 283] points out that the reef-zone facies is much more permeable than the other carbonates and evaporites. A combination of factors, including good dip-slope permeability and relatively easy means for water to discharge into the Pecos River has resulted in good circulation of potable water in the upper part of the zone of saturation west of the river. Some flowing wells for irrigation in the Roswell Basin were drilled before 1900. Wells tapping the artesian aquifer of Permian limestones originally had considerable head, and one had a natural flow in 1926 of 5,710 gallons/min [FIEDLER and NYE, 1933, p. 193]. The water levels have since declined, chiefly because of withdrawal of water for irrigation. In spite of legal restrictions on the use of water in this basin, the total discharge still exceeds the perennial yield. In contrast to conditions on the west side are conditions east of the river, where water is highly mineralized in rocks of the same formations and where facilities for water to move and discharge are not good except in the close vicinity of the river.

OTHER KARST REGIONS[1]

In addition to the karst regions in the United States that have been described in some detail, there are three other large regions of carbonate rocks that contain significant areas of karst.

---

[1] Written by W. E. Davies.

The Ozark plateaus in southern Missouri, northern Arkansas, and northeastern Oklahoma are formed mainly by limestones and dolomites of Palaeozoic age. These rocks have been uplifted into an asymmetrical dome. Minor folds and faults locally disrupt the rocks. The uplands of the Ozark plateaus are 425–520 m above sea level, and the major valleys are 90–150 m deep.

Dolines are widespread on the upland and are mainly smooth, shallow, flat-floored basins. Others are steep-sided funnel-shaped depressions, and a few are vertical-sided collapse sinks. The dolines are as much as 750 m in diameter and as much as 60 m deep. The abundance of residual chert derived from the carbonate rocks affects the development of topography by solution. The chert mantle is permeable and poorly compacted, and infiltration of rainfall on the upland as well as in the chert wash in valleys is very large. Erosion by corrasion is retarded by the chert cover whereas removal of material through solution is enhanced. Karst forms other than dolines are not common at the surface in the Ozark plateaus. Caves are numerous and generally extensive. About 1,000 caves are in the Ozarks. Most cave patterns are simple, consisting of one or two major passages irregularly alined along joints and fractures. Maze caves with as much as 6,000 m of passage are also present. Most of the caves are in one level only and are as much as 105 m in depth.

Many of the caves act as phreatic conduits for water, and large springs are characteristic of Ozark karst. The largest, Big Spring, discharges an average of 250,000,000 gallons (945,000 m^3)/day, or about 12.3 m^3/sec.

The Edwards Plateau in southwest Texas is formed by a thick sequence of non-folded limestone of Early Cretaceous age. The plateau is about 600 km long east–west and is about 320 km north–south at its widest part. The northwestern part of the plateau stands 900 m above sea level, and the upland reflects the gentle dip of the beds to the southeast with the altitude of the upland at 300 m along the eastern and southeastern edges. The inner part of the plateau is in a youthful stage with the marginal areas in a stage of mature dissection. The northern edge of the plateau grades into the High Plains as the limestone is progressively covered by Tertiary clastic sediments. Along the southern margin is an irregular escarpment 90–300 m high.

The east half of the Edwards Plateau receives moderate amounts of precipitation, but the west part is an arid grassland. On the upland, karst features include widely separated shallow dolines, collapse sinks, and areas of barren ledges. In the valleys cut into the margin of the plateau, broad solution depressions are common at the junctions of tributaries. In the 56,000 km^2 of the plateau the caves have irregular patterns, and collapse areas are extensive. In the western part there are many large branching and multi-level maze caves.

Much of the eastern part of the plateau is cut by an extensive fault system that continues in the limestone to the east and south beneath the coastal plain. Large springs occur in the latter where the water is under artesian head. The daily yield of the springs is as much as 212 million gallons (805,360 m^3)/day.

Limestones adjacent to the southern and eastern borders of the Edwards Plateau cover about 40,350 km² and more than 600 caves occur, some of which have more than 6,000 m of passages. Dolines and collapse sinks are present locally.

The Nashville and Lexington plains in central Tennessee and central Kentucky are broad lowlands formed by gently arched Lower Palaeozoic limestones. The altitude of the lowlands is about 180 m with major streams entrenched to a depth of 30–45 m.

Shallow, smooth-sided dolines 30 to over 300 m form much of the rolling surface of the plains. Karren are on the Lexington Plain, and the surface is formed of rich but stony soil. In the area of the Nashville Plain, however, there are larger areas in which the limestone forms a bare surface pavement known as glades. Red cedar trees and lesser areas of hardwoods grow along the joints in the bedrock where meager accumulation of soil occurs.

There are about 80 caves in the limestones of the Nashville and Lexington plains. Most of these have simple patterns and consist of one or two relatively straight passages with total lengths of less than 300 m. In the southern part of the Nashville Plain, however, there are several large subterranean stream passages as much as 7,500 m long.

To the east the plains grade into the Highland Rim which is developed on Mississippian rocks and is a dissected plateau with uplands at 210–390 m altitude. Limestone is at or near the surface on much of the plateau, and karst features, primarily large dolines, uvalas, and short, steeply sloping karst valleys (coves) occur along the flanks of the upland.

In the Tennessee section of the Highland Rim there are 250 caves, many of them being more than 300 m long consisting of irregular patterns and some with multiple levels. In Kentucky the Highland Rim grades into the Pennyroyal Plain, which is primarily a doline plain extending east–west across the southern part of the state and adjacent parts of northern Tennessee.

REFERENCES

ANONYMOUS, 1949. Geology and foundation treatment. *Tennessee Valley Authority, Tech. Rept.*, 22:548 pp.
BARR JR., T. C., 1961. Caves of Tennessee. *Tenn., Dept. Conserv., Div. Geol., Bull.*, 64: 567 pp.
BROWN, R.F., 1966. Hydrology of the cavernous limestones of Mammoth Cave area, Kentucky. *U.S., Geol. Surv., Water Supply Papers*, 1837: 62 pp.
BRUCKER, R. W. and BURNS, D. P., 1966. *Flint Ridge Cave System*. Cave Res. Found., Washington, D.C., 3 pp., 31 maps.
CADY, R.C., 1936. Groundwater resources of the Shenandoah Valley, Virginia. *Virginia Geol. Surv., Bull.*, 45:137 pp.
COOKE, C. W., 1939. Scenery of Florida interpreted by a geologist. *Florida Geol. Surv., Geol. Bull.*, 17:118 pp.
COOPER JR., H.H. and KENNER, W.E., 1953. Central and northern Florida. In: *Subsurface Facilities of Water Management and Pattern of Supply-Type Area Studies*. U.S. Congr., House Comm. Interior Insular Affairs, Washington, D.C., pp.147–161.
CUSHMAN, R. V., KRIEGER, R. A. and MCCABE, J. A., 1965. Present and future water supply for Mammoth Cave National Park, Kentucky. *U.S., Geol. Surv., Water Supply Papers*, 1475-Q: 601–647.

Davies, W. E., 1965. Caverns of West Virginia (with supplement). *West Va. Geol. Econ. Surv., Bull.*, 19–A:330+72 pp.

DeBuchananne, G. D. and Richardson, R. M., 1956. Groundwater resources of eastern Tennessee. *Tenn., Dept. Conserv., Div. Geol., Bull.*, 58(1):393 pp.

Dicken, S. N. and Brown Jr., H. B., 1938. Soil erosion in the karst lands of Kentucky. *U.S. Dept. Agr., Circ.*, 490:61 pp.

Fiedler, A. G. and Nye, S. S., 1933. Geology and groundwater resources of the Roswell artesian basin, New Mexico. *U.S., Geol. Surv., Water Supply Papers*, 639:372 pp.

Hanshaw, B. B., Back, W. and Rubin, M., 1968. Carbonate equilibria and radiocarbon distribution related to groundwater flow in the Floridan limestone aquifer, United States of America. In: *Hydrology of Fractured Rocks; Proceedings of the Dubrovnik Symposium, October, 1965—Intern. Assoc. Sci. Hydrol., Bull.*, 2:601–614.

Hendricks, E. L. and Goodwin Jr., M. H., 1952. Water-level fluctuations in limestone sinks in southwestern Georgia. *U.S., Geol. Surv., Water Supply Papers*, 1110E:157–246.

Herrick, S. M. and LeGrand, H. E., 1964. Solution subsidence of a limestone terrane in southwest Georgia. *Intern. Assoc. Sci. Hydrol., Bull.*, 9(2):25–36.

Johnston Jr., W. D., 1933. Groundwater in the Palaeozoic rocks of northern Alabama. *Geol. Surv. Alabama, Spec. Rept.*, 16:414 pp.

Lohman, S. W., 1938. Groundwater in south-central Pennsylvania. *Penn. Geol. Surv., Bull. W*, 5:315 pp.

Motts, W. S., 1968. The control of groundwater occurrence by lithofacies in the Guadalupian Reef Complex near Carlsbad, New Mexico. *Geol. Soc. Am., Bull.*, 79(3):283–298

Powell, R. L., 1961. Caves of Indiana. *Indiana, Dept. Conserv., Geol. Surv., Circ.*, 8:127 pp.

Price, P. H., McCue, J. B. and Hoskins, H. A., 1936. Springs of West Virginia. *West Va. Geol. Econ. Surv., Bull.*, 6:146 pp.

Reeves, F., 1932. Thermal springs of Virginia. *Virginia Geol. Surv., Bull.*, 36:56 pp.

Scott, J., 1959. *Caves in Vermont*. Killooleet, Hancock, Vt., 45 pp.

Smith, P. M., 1964. The Flint Ridge Cave system: 1957–1962. *Natl. Speleol. Soc. Bull.*, 26(1):17–27.

Stringfield, V. T., 1964. Relation of surface-water hydrology to the principal artesian aquifer in Florida and southeastern Georgia. *U.S., Geol. Surv., Profess. Papers*, 501–C:C164–C169.

Stringfield, V. T., 1966. Artesian water in Tertiary limestone in the southeastern states. *U.S., Geol. Surv., Profess. Papers*, 517:226 pp.

Stringfield, V. T. and LeGrand, H. E., 1966. Hydrology of limestone terranes in the coastal plain of the southeastern United States. *Geol. Soc. Am., Spec. Papers*, 93:46 pp.

Tarkington, T. W., Varnedoe Jr., W. W. and Veatch, J. D., 1965. *Alabama Caves: Huntsville (Alabama) Grotto*. Natl. Speleol. Soc., Huntsville, Ala., 79 pp.

Watson, R. A., 1966. Central Kentucky karst hydrology. *Natl. Speleol. Soc. Bull.*, 28(3):159–166.

Waring, G. A., 1965. Thermal springs of the United States and other countries of the world; a summary. *U.S., Geol. Surv., Profess. Papers*, 492:383 pp.

Zotter, H., 1965. Stream tracing techniques and results, Pocahontas and Greenbrier counties, West Virginia, 2. *Natl. Speleol. Soc., News*, 21(12):169–177.

Chapter 16

Conclusions

M. HERAK AND V. T. STRINGFIELD

Geological and Palaeontological Institute, Faculty of Natural Sciences, Zagreb (Yugoslavia)
Department of the Interior, U.S. Geological Survey, Washington, D.C. (U.S.A.)

TERMINOLOGY, GENESIS AND CLASSIFICATION[1]

The preceding chapters show various approaches to and achievements in karst investigations in the countries selected up to the end of 1967. We hope that other countries with karst areas would fit into one of the types presented.

Though all the articles were planned along the same lines, the various types of karst, different practical needs and personal inclinations of the authors caused a variety of the data selected for presentation, as well as the differences of styles, methods and concepts. Therefore, it may be useful to select the main elements connecting the articles in order to integrate the partially heterogeneous content of the book.

At first, the terms *kras* or *krš* (Slavian), *carso* (Italian), and afterwards *Karst* (German) had a toponymic meaning defining geographic areas with specific morphologic and hydrologic features. However, over the years the terms were also used in an appelative sense, defining the sum of the phenomena characterizing the areas mentioned. Since the first theoretical concepts were established in the classic Dinaric karst, Yugoslavian expressions concerning individual characteristics were accepted as international terms (*doline, jama, uvala, polje, ponor*, etc.). However, there is also a tendency to replace Yugoslavian terms with expressions adapted to one's own language. Such a procedure can not avoid creating synonyms, homonyms and sometimes inadequate expressions. All this can increase misunderstanding not only among the experts speaking different languages, but also among those who, though speaking the same language, are inclined to " improve" the terms, to change the meaning partially or completely, etc. Therefore, the common scientific interest speaks in favour of priority in the karst terminology. Consequently, well defined Yugoslavian terms such as polje, uvala, doline, ponor and some others should be maintained as international expressions.

"Karstification" is a term whose meaning varies with the concepts held concerning karst development. It ranges from being almost a synonym of the solution process (corrosion) to denoting all processes responsible for the development of karst as a whole (besides the solution process, especially jointing, faulting, epeirogenic movements, mechanical erosion etc.). As the factors mentioned are not equally

---

[1] Written by M. Herak.

important in the formation of every karst area, the term itself can not be unanimously defined. Therefore, it is applicable only in a general sense in the meaning of the karst development itself, regardless of its stage and completeness. In order to be precise enough well defined elementary terms (solution or corrosion, mechanical erosion, jointing, faulting, etc.) are recommended even if the explanation requires the use of several of these.

The genesis of karst features depends, in general, upon the primary properties of rocks, their position in the stratigraphic sequence, tectonic disturbances and climate.

The primary properties of rocks were regularly taken into account in all papers, even in the oldest ones. The pure carbonate rocks (limestones and dolomites and their varieties) have been considered most suitable for karst development. Other rocks (such as gypsum and salt, etc.) lag far behind in the amount and distribution of karst phenomena.

Concerning the stratigraphic position, it is most important to establish whether or not intercalations of impervious beds are within the carbonate complexes. If present, they may be involved in very entangled hydrogeologic relations characterized by regular or irregular alternation of aquifers and aquicludes defining subsurface and surface water flows, and consequently influencing the genesis and distribution of solutional and erosional morphologic features. The presence of pervious and impervious horizons within a stratigraphic sequence determines karst zones between impermeable deposits (karst barré), facilitating the longitudinal flows or causing overflow of groundwaters, with development of corresponding morphologic phenomena. Large karst features in such areas are usually absent.

Tectonic disturbances influence the development of karst both generally and specifically; in general by changing the "normal" relations concerning the succession of lithostratigraphic units, their contacts, elevations, etc., and specifically by creating various systems of joints, fissures and fractures which transform homogeneous carbonate rocks into secondarily pervious complexes. The folded systems influence the morphologic features, accumulations of groundwater, and formation of barriers due to which the water is trapped, periodically rising or even overflowing. Tangential structures can intensify the disturbances with various already emphasized effects. Joints, fissures, fractures, unconformities and disconformities increase the heterogenety of the area. Diatrophic movements continuously changed the general conditions of sedimentation in the creation of geosynclines, epicontinental seas, mountain ranges with local sedimentary basins, etc. They governed conditions for intensive or minute sedimentation of carbonate rocks, for their possible alternation with impervious deposits (owing to oscillations), or for depositing of continuous carbonate series (owing to steady slow sinking of shallow marine environments). They also defined, in general, the phases of karstification which correspond more or less to those phases of different orogenies which followed the periods with intensive deposition of carbonate rock complexes.

The importance of climate has been pointed out in many papers concerning karst development. However, it is necessary to be aware that the climate exhibits more a quantitative than a qualitative influence on the solution process. Also, young solution processes have been recognized in the far north as well as in many tropical areas. Further, the distribution of the different climatic belts in the history of the earth was mobile due to the changes of the earth's pole position and probably because of the drift of continents. Consequently, the karst forms of a single area might have developed successively under different climatic conditions.

Some single karst forms are still the subject of discussion, while others are more or less firmly defined.

Karren (lapies) are commonly considered as a karst feature formed by differential solution processes on steep rock surfaces; they are influenced partly by joints.

Dolines are formed in jointed and fractured areas chiefly by solution processes, but partly by collapse and mechanical erosion.

Pits (jamas) are mostly connected with areas having local collapses. Corrosional and erosional forces are less important.

Ponors are closely linked to tectonically disturbed zones. The formation includes collapses, mechanical erosion and solution processes.

Dry valleys are signs of a more or less general sinking of the subsurface water level, or of a differential rising of land masses.

Uvalas imply fracturing and differential sinking or rising of smaller parts of land masses, then modelling due to solution of carbonate rocks. But, the forms within them can also be more or less inherited from the time before the formation of the uvala itself. This means that the previous erosional and corrosional processes can be at least partly responsible for the modelling of their margins and bottom.

Poljes are still a subject of controversy. Although most authors agree on the role of the tectonics in their formation, there are different opinions concerning the age of the tectonics responsible for the genesis of the polje, the morphologic stage of the land at that time, and consequently the importance of solution processes (corrosion) in modelling the depression itself and the *hums* (stony hills) within them. Even the concept itself is under discussion. The term polje is most often applied to the depression as a whole, including the margins and bottom. Sometimes it is restricted to the bottom only. Hence, the statement that the polje represents a strange element in a karst landscape (J. Roglić). Taking into account that the hydrogeologic regime of a polje, being an essential property, depends more on the subsurface relations than on the cover of the polje bottom, it seems more convenient, at least for hydrogeologists, to maintain the wider, complex meaning of the term polje.

The best known poljes (s.l.) are relatively young, being formed in Neogene–Pleistocene time. The depressions were primarily formed owing to large fault zones and differential block movements in the areas which were generally rising or sinking. Before that event the area was already morphologically more or less formed with smaller or larger peaks *(mosors)* and depressions (dolines or uvalas) which might be partly or completely covered by Neogene clastic deposits. The corrosional processes which followed had only to model and not to create the form of the polje as a whole and of its single features. Young sediments levelled the new polje bottom, filling out the depressions and isolating the already existing hills which became hums. Changes of climate during the Pleistocene caused the phases of sedimentation and evacuation of loose deposits and dissolved materials through the ponors, the consequences being preserved in recent relations. It is not my intention to neglect the role of the solution processes in forming the morphologic features of the poljes. The purpose of this discussion is only to point out that they are not the only processes responsible for it, and that processes before the creation of the polje depression should be treated separately, i.e., not together with those which were active after that event. The recent morphologic features of the poljes are consequences of both of them.

The role of solution process (corrosion) in the final formation of large *karst plains* is generally recognized. But the great differences among the plains in the area of geologic relations show that one should also take into account the previous geologic history characterized by diastrophic movements, erosional and possibly abrasional activities which preceded the final corrosional modelling of the plains.

The formation of *caves* and *caverns* is still a subject of uncertainty concerning the relation of the groundwater level to their formation. The role of fractures and joints in the initiation of water movements, solution and mechanical erosion is generally recognized.

For a long time efforts were made to establish the succession of cycles in the genesis of different karst features. But new experience shows clearly that the features were primarily determined by the thickness of carbonate deposits, diastrophic disturbances, possibilities for water flow, climate, and other natural conditions. The length of the process could intensify or shorten the formation of a given form, but rarely transforms one feature into another.

Karst areas with similar features can be formed in different ways, i.e., in different surroundings, which in general can be distinguished as (*a*) geosynclinal and (*b*) epicontinental.

(*a*) "Geosynclinal" karst areas are included in the rock sequences deposited during the evolution of a geosyncline up to the final diastrophic movements causing the rising of new mountain ranges. There are many lateral and vertical lithologic variations. The deposits were involved in several orogenic movements. A series of interbedded (impervious) and carbonate (secondarily pervious) complexes is more frequent than a deposition of a more or less continuous calcareous series. The deposition of carbonate rocks increased from Precambrian, with the first orogenies, to the Palaeogene when the Alpine orogeny was most intensive. Therefore, the best possibilities for development of karst forms of the geosynclinal type are within the Alpine (Mesozoic and Palaeogene) structures characterized by various and complicated disturbances (complex faults, folds, overthrusts, etc.). In this type of karst two varieties can be distinguished.

The first one is to be found within the disturbed geosynclinal belts with a series of aquicludes and aquifers. Karst barré occurs often. Larger depressions, especially true poljes, are usually absent. This type of karst, characterized by limited areas (horizontally and vertically), is to be found in the Alps, Carpathians, the Balkan, Atlas, Riff-Tell, the Ural, etc.

Extreme examples of geosynclinal sequences with many carbonate rocks disturbed by successive orogenic phases are in the Dinaric, Hellenic and Toros mountains. In spite of a similar common basis for karst development (a more or less continuous carbonate series, intensive disturbances, etc.), and although the orogenic belt is the same, the differences are apparent.

Especially interesting is the comparison between the Dinaric Mountains and the middle part of the Toros Mountains as peripheral parts of the belt under consideration. Both of them are governed by high mountains and large depressions (poljes) with linear distribution following the main tectonic trends. Both have a very complex hydrogeologic pattern characterized by differences between surface and subsurface watersheds, by estavelas, etc. The morphologic features are almost identical. The differences mainly concern hydrogeology, especially the existence of surface streams and submarine springs. As already stated, in the Dinaric Mountains the surface rivers are relatively rare, while submarine springs along the coast of the Adriatic Sea are numerous. The reason lies in the fact that Palaeogene flysch, being the barrier against sea water, is tangentially disturbed, cut off, overturned and overthrust within a complex tangential structure, allowing connections of fresh water and sea water below the flysch. The middle part of the Toros Mountains, on the contrary, is highly influenced by the Palaeozoic massif on the southern margin of the Toros karst area and by Neogene clastic deposits which dip against the Mediterranean Sea (and below it). The clastic and metamorphic rocks do not allow any connection of subsurface water courses with the sea water. Thus, the differences are clearly defined by geologic relations.

(b) "Epicontinental" karst areas are bound on sediment deposits mainly in epicontinental shallow seas and covering Precambrian, Caledonian, Hercynian and Alpine orogenic basements (possibly with palaeokarst). The presence of carbonate rocks in an epicontinental sequence, their possible alternation with clastic aquicludes, and relatively simple tectonic relations define the types of karst in general and development of individual forms. For the most part shallow karst is developed, or fluviokarst, or merokarst, or even karst barré. Examples are the karst areas in Carboniferous limestones in the depression of Moscow, in Great Britain, or in the U.S.A., in the Mesozoic limestones in Germany, France, Lebanon and Jordan, and in most of Neogene limestones. Since the first-mentioned are presented in the foregoing chapters, a few words on the karst of Lebanon may be added. The karst features are bound on Jurassic, Cretaceous and Tertiary carbonate (mostly calcareous) rocks. The carbonate sequence is not continuous because Lower Cretaceous and Senonian clastics are present. The Neogene is transgressive and as a whole impervious. Senonian and Palaeogene limestones make a continuous series. The whole sedimentary complex lies on a relatively disturbed Precambrian platform dominated by faults accompanied by gentle folds. In such relations the karst features in general are developed mostly in Palaeogene limestones and less in Turonian, Cenomanian and Jurassic, which are more or less confined by impervious horizons. Only deep faults connecting different karst horizons, and larger outcrops of older rocks may intensify karst development within them. Younger aquicludes impound the whole region against the sea. The dipping of beds towards the sea (and below it) furnishes the conditions for sporadic accumulation of artesian waters in karst cavities.

Exceptionally, also, in epicontinental carbonate series karst forms analogous to those of geosynclinal belts with thick carbonate series can be developed. A very ininstructive example is the karst of Jamaica, West Indies, which includes poljes, uvalas, cockpits, caves, caverns, estavelas, submarine springs, etc., developed in a Tertiary carbonate series. The depth of the carbonate sediments and tectonic disturbances enabled development of karst similar to the geosynclinal type.

Thus, a genetic classification of karst types lacks the firm and consistent criteria which can be regularly recognized over large areas and might make possible predictions concerning the type of a karst area. Therefore, we are still obliged to use descriptive units, more or less well defined, such as shallow karst, deep karst, merokarst, holokarst, fluviokarst, covered karst, denuded karst, etc., which are more concerned with the final effects than with the circumstances under which the karst was formed.

HYDROLOGY AND HYDROGEOLOGY[1]

The reports of many countries show that the importance of hydrology and hydrogeology of carbonate rocks is being recognized in many parts of the world. Also, information obtained by investigators of karst, including caves and other underground features of solution, is now being applied to good advantage in studies of hydrology and hydrogeology of carbonate rocks.

It is generally recognized that the principal hydraulic permeability in carbonate terranes is secondary and is caused chiefly by solution of the carbonate rocks, but there are differences of opinion concerning the conditions under which this permeability is developed and of the interpretation of the occurrence and movement of water in carbonate rocks.

The discussions of the physical and chemical hydrology and other factors such as differences in latitude, climate, temperature, altitude, forest cover, and the hydrogeology (including complex structure and stratigraphy which control or affect karstification and the development of secondary permeability) show the progress which is being made in research in all of these fields. However, some investigators overlook the significance of a circulation system and the principles which govern the circulation

---

[1] Written by V. T. Stringfield.

of water in carbonate rocks. Without circulation of water, secondary permeability will not develop even though all other factors are favourable. Other factors being equal, the maximum permeability is generally in recharge and discharge areas of karstified carbonates where circulation has been the most vigorous. The permeability is also proportional to the length of geologic time during which the water moved under unconfined conditions in the phreatic zone.

In addition to the factors which have been described in the above reports as being essential in the initiation of karstification and secondary permeability, some investigators have found evidence that a bare carbonate rocks surface with no overlying water-bearing deposits is less favourable for karstification than one with such deposits. Even though the carbonate rocks may have joints and fractures, rocks exposed to weathering without a cover of unconsolidated water-bearing deposits tends to become case hardened and resistant to solution and erosion. Recharge which may enter the rocks without a soil cover does not receive carbon dioxide from a soil zone and, therefore, is not very effective in dissolving the carbonate rocks. In a few places where the carbonate sequence includes anhydrite, gypsum, and salt it appears that even water containing little carbon dioxide may be quite effective in dissolving these rocks. Studies of palaeohydrology and hydrogeology show that some of the most extensive karst regions are those in which the rocks are essentially horizontal or dip gently and were covered with less soluble water-bearing deposits which play an important part in initiation of karstification and secondary permeability. These overlying deposits prevented the underlying carbonate rocks from becoming case hardened and resistant to solution and erosion before karstification began. Water from the surface and soil zone percolated downward to joints and other openings in the underlying soluble rocks forming the phreatic zone. As surface streams cut through the cover into the underlying rocks, water in joints and other openings in the carbonate rocks discharge into the streams, thus initiating a circulation system in the phreatic zone. In regions where the less soluble overlying deposits contain beds which are resistant to erosion and solution, solution escarpments were formed. Karstification with sinkholes, vertical shafts, and other solution features were initiated along these escarpments. As the escarpments retreated, removing the cover from the underlying rocks, a sinkhole plain remained. Although at the beginning the lateral circulation system is chiefly along joints and cracks, it forms a phreatic zone or zone of saturation, the top of which is the water table. The upper part of this zone is considered the most favourable for maximum solution by lateral movement of water. The most favourable zone of solution by downward percolation of water is in the zone of aeration above the water table, where solution features such as vertical shafts and sinkholes are formed. As the base level is lowered by down cutting of streams or other causes, the top of the zone of saturation is lowered. Other factors being equal, the maximum solution and permeability is developed in the upper parts of the phreatic zones in which the circulation continued for the longest period of time. As the top of the phreatic zone is lowered in response to changes in base level, caverns and cavities which formed in the phreatic zone are left in the vadose zone where they are modified by streams and circulation of water above the water table. These changes and modifications have caused some investigators to conclude that caves consisting of lateral passages are formed above the water table. There is general agreement that solution features as vertical shafts and sinkholes are formed above the water table. In coastal areas where base level is affected by changes in sea level, the top of the zone of saturation rose and fell with changes in sea level during Pleistocene time. These changes in sea level affected the circulation system and the permeability of carbonate rocks along the coast through a vertical range of as much as 100 m above and 150 m below present sea level. Under these conditions some cavernous zones that formed during higher stands of the Pleistocene sea than at present are far above the phreatic zone at present. The cavernous zones that formed in the phreatic zone during low stands of the Pleistocene sea are now far below sea level.

In regions where carbonate rocks were covered by deposits of unconsolidated material, as for example, the blanket sands in the karst region of Puerto Rico, karstification and circulation of groundwater was initiated when surface streams cut through the cover into carbonate rocks. Where the unconsolidated cover on the carbonate rocks adjacent to and parallel to some streams was removed by erosion in the initial stage of karstification, the carbonate rocks became case hardened and resistant to solution and erosion while karstification took place where the limestone was covered by permeable but less soluble deposits. This differential erosion resulted in a ridge of carbonate rocks between the river channel and the area of sinkholes where karstification has lowered the land surface. In some karst areas where the top of the zone of saturation has been lowered as much as 100 m, the solution and collapse of sinkholes lowered the land surface except for the hills (mogotes) which separate sinkholes. Some such hills are more than 100 m high in Puerto Rico and Jamaica, W.I. In Laos, in southeastern Asia, similar hills are reported to be as much as 300 m high. The tops of some of these hills are cone-shaped and others are tower-shaped. Where the cover over the limestone was thin during the initiation of karstification, the carbonate rocks on the top of the tower hills between sinks became case hardened caps which prevented erosion like that on the less resistant cone-shaped hills.

Carbonate aquifers

Although karst carbonate terranes have a large range in permeability and topographic conditions, they include some of the most extensive and productive aquifers in the world. Among the most productive are those which contain buried karst in which the secondary permeability was not destroyed when younger overlying formations were deposited. Carbonate aquifers are the source of water for some of the largest artesian wells and springs. Many of the springs are the source of large surface streams; apparently the Fountain of Vaucluse in southern France as discussed by Professor J. Avias, in his chapter on France, is the largest known spring from a single orifice.

CONCLUSIONS

The water from carbonate aquifers is generally harder than that from other kinds of aquifers and the likelihood of surface pollution is greater in karst areas. However, because of its relatively constant temperature and chemical quality, water from carbonate rocks is not as difficult to treat for human consumption as water from surface streams and lakes. In some coastal karst areas where there was a source of sand during Pleistocene time, many of the sinkholes and solution cavities were filled with relatively permeable sand, which reduce the chances of surface pollution. Artesian water in carbonate rocks far from the recharge area in karst terrane is not subject to surface pollution from sinkholes and other karst features.

The geologic age of highly productive carbonate aquifers ranges from Palaeozoic to Pleistocene. In coastal areas where carbonate aquifers are exposed to sea water the Ghyben–Herzberg principle can be used to estimate the position of the interface between salt water and fresh water. However, where there are dynamic conditions in a very permeable formation and large quantities of fresh water move along the zone of diffusion at this interface and discharge over the salt water, the interface is deeper and nearer the coast than that indicated by the Ghyben–Herzberg principle. Natural vertical solution openings, such as sinkholes and natural wells, which extend from sea level through a relatively dense impervious part of the aquifer to a cavernous zone may serve as "cased wells" through the fresh-water zone into the mixing zone of fresh water with salt water. Such vertical solution channels passing through relatively impervious parts of carbonate rocks can be seen in caves. In coastal areas of carbonate rocks these karst features were formed during a low stand of the Pleistocene sea. Sea level at the present time is high enough for sea water to reach these openings in some places as at Tarpon Springs, Florida, U.S.A. and Cephalonia, Greece, and along the Adriatic coast of Yugoslavia where the differences in head and density between sea water in the vertical solution channel and fresh water in the aquifer control the movement of water from the sea into the aquifer.

Hydrogeology and palaeohydrology

Carbonate terranes show not only a large range in permeability and topographic conditions, but also a large range in practical considerations relating to water supply, pollution from surface water, water-tight reservoirs, and ground stability for building purposes. As stated by H. E. LeGrand (see p. 479) investigations to solve such problems require detailed information on the hydrogeology and palaeohydrology, in which the studies may be divided into three parts as follows:

(1) The depositional period of the carbonate rocks as well as the overlying and underlying rocks. During deposition such significant characteristics as the composition, texture, lithology, primary permeability and the thickness and lateral extent of the deposits are established.

(2) Immediate post-depositional period. Elevation of deposits above sea level may expose the carbonate rocks to meteoric weathering and subaeral erosion, where they

are not overlain by beds that prevent such weathering. In some places all of the carbonate rocks may be removed; in others the upper part may be removed and the permeability of the remaining part may be increased by solution especially if a groundwater circulation system is established. Under some conditions such secondary permeability is not destroyed when younger formations are deposited on the carbonate rocks.

(*3*) *Subsequent pre-Holocene period*. If uninterrupted deposition of long duration produced thick deposits over the carbonate rocks, groundwater circulation in the carbonate rocks may have been sluggish and insignificant, resulting in little or no solution. If deposits over the carbonate rocks were removed by erosion and were elevated during this period so that groundwater could circulate in the rocks, solution openings and increased permeability could develop in some places. During this interval the limey deposits have become compacted and consolidated and in some places perhaps recrystallized. Folding, faulting, and jointing may have occurred.

Geologic structure

In studies of the hydrogeology, geologic structure may be very important, not as a separate entity, but in relation to topography and discharge facilities. Also, geologic structure may have a pronounced effect on the circulation and the direction of movement of water in carbonate rocks. However, there is a tendency for some investigators to assume that water in carbonate rocks always moves downdip notwithstanding the fact that the direction of lateral movement of the water is governed by the recharge–discharge relationship.

In the beginning, downward percolation to the water table and the phreatic zone generally is along joints and fractures, if the permeability of the rocks is small. Lateral movement is generally along joints and fractures, and some lateral movement may be along bedding planes. Where the rocks are approximately horizontal and the hydraulic gradient is approximately parallel to and in the same direction as the dip, water may move along bedding planes and in the direction of dip. However, with steeply dipping beds as on the flanks of a breached anticline, water will not move downdip unless it is a recharge area for an artesian aquifer which discharges at a lower elevation. Where there is no place for discharge downdip the water may move along bedding planes but parallel to the strike of the formations.

Under water-table conditions where carbonate rocks are underlain by shale or other relatively impervious beds which form geologic structure as a gently plunging syncline, the direction of movement of water may be parallel to the direction of the dip on the flanks of the syncline until it reaches the axis of the syncline where it moves at right angles to the dip along the axis of the syncline.

Under artesian conditions, the movement of the water in carbonate rocks is also in accordance with the hydraulic gradient between recharge and discharge areas. In inclined aquifers water may move downdip where: (*1*) the aquifer is overlain by relatively impervious beds; and (2) the relative positions of the recharge and discharge areas are favourable for a hydraulic gradient in the same direction as the dip. Where permeable carbonate artesian aquifers connecting recharge and discharge areas are folded into an anticline and syncline, water may move downdip in one limb of a structure and updip in the adjacent limb.

There is a tendency for some investigators in some areas to overemphasize faults as a controlling factor in the occurrence and movement of water. Faulting has had a significant effect on the formation of some poljes, which are due chiefly to structural movement along faults and to later solution of carbonate rocks. Notwithstanding complex structure and hydrogeology in some countries, such as Yugoslavia, where poljes are relatively numerous at different elevations, the results of hydrogeologic investigations have made it possible to utilize these poljes successfully for water supply and hydroelectric plants, as described by Prof. Milan Herak in his chapter on Yugoslavia. In some regions, such as in the classic Cockpit Country of Jamaica, W.I., as described by H.R. Versey in the chapter on that country, some faults apparently transmitted water laterally from the interior to the coast as karstification was initiated. Vertical movement along some of the faults tilted and changed the altitude of some areas of the carbonate rocks as karstification occurred. The effects of faulting in that region caused some investigators to conclude that faulting is an essential factor in karstification. However, the karst in Puerto Rico, where faulting has not affected the karstification, is similar to that of Jamaica, except that karst in Puerto Rico does not include interior valleys or poljes.

Some of the most complex effects of faulting on the hydrogeology result from movement of large masses of carbonate rocks along overthrust faults as described in some of the above reports.

Faults in water-bearing carbonate rocks may cause serious problems in controlling inflow of water in mining operations where water moves along a fault zone. Also in some places a faults zone may act as a barrier to the movement of water until penetrated in mining operations.

In general, joints and other fractures in carbonate rocks are among the most essential structural features for the initiation of a circulation system and development of secondary permeability.

Methods of investigation

Conventional methods used in investigating the hydrology and hydrogeology of other kinds of geologic formations may be used successfully in studies of carbonate rocks, if the investigators understand the principles of occurrence and movement of water in rocks in which permeability is due chiefly to solution. Studies of the palaeohydrology and conditions of sedimentation are needed in all sedimentary formations especially in carbonate rocks. In water-budget studies in some surface drainage basins underlain by carbonate rocks, the groundwater drainage basin may be connected with adjacent surface basins. Under artesian conditions, of course, the groundwater may be separated from surface drainage by relatively impervious beds.

Dyes, spores, and other tracers including tritium and other radioaction tracers probably have been used more extensively in cavernous carbonate rocks than in other aquifers. Spores and pollen, of course, can be used successfully as water tracers only where the water moves through open channels. Tritium, carbon-14, and other radioelements are being used to obtain the relative age and rate of movement of groundwater.

Geochemical and geophysical studies are used to supplement other conventional methods of investigating the hydrology amd hydrogeology of carbonate rocks. Geochemical studies are especially valuable in investigations of conditions of deposition of the geologic formations, solution and secondary deposition in carbonate rocks after they have been deposited.

Conventional geophysical methods are used with different degrees of success. In general, borehole geophysics is the most successful and gives more information than the other methods. The results of some methods have been disappointing to hydrologists and hydrogeologists and acceptable to others. Some investigators are hopeful that methods in the future will be more successful.

References Index

ABEL, O. and KYRLE, G., 254, *263*
ABRAMI, G. and MASSARI, F., 115, *125*
Abrard, R., 130
AIGROT, M., *183*
Aigrot, M., 130
Albinet, M., 130
ALFIREVIĆ, S., *76*
ALMAGIA, R., 95, *125*
ALMÁSSY, E., *see* SCHMIDT, E. R. et al.
AMPFERER, O., 247, 250, *263*
AMPFERER, O. and HAMMER, W., 247 *263*
ANDRUSOV, D., 298, 300, 302, *323*
ANELLI, F., *125*
ANTONELLI, C., *see* CAPELLO, C. F. et al.
ARANDJELOVIĆ, D. and MILOŠEVIĆ, LJ., 58, *76*
ARCHAMBAULT, G., *183*
ARGENTOV, K. I., *411*
Argentov, K. I., 396
Armand, D. L., 397
Armand, L., 150
ASHMEAD, P., *see* EYRE, J. and ASHMEAD, P.
ATKINSON, T., DREW, D. P. and HIGH, C., 435, 440, *442*
Aub, C. F., 460
AUBERT, D., 165, 181, *183*
AVIAS, J., 181, *183*, *184*
Avias, J., 130, 182
AVIAS, J. and DUBOIS, P., *184*

BABIĆ, Z., *see* MILETIĆ, P. and BABIĆ, Z.
BACK, W., *see* HANSHAW, B. B. et al.
BADINI, G., 116, 117, *125*
BADINI, G. and GECCHELE, G., 105, 111, *125*
BAHUN, S., 57, *76*
BAHUN, S., *see* PAVLIN, B. et al.
BAKIĆ, M., *76*
BALDACCI, O., 103, *125*
BALKOV, V. A., *411*
BALOGH, K., 272, 279, *294*
BALSAN, L., *184*
Balsan, L., 130
BALUK W. and RADWAŃSKI, A., 333, *339*
Barabás, A., 270
Barbot de Mari, N. P., 355
BARKOV, A. S., 356, *411*
BARON, G., CAILLÈRE, S., LAGRANGE, R. and POBEGUIN, T., *184*
BARR JR., T. C., 485, 493, *504*

BARRERE, P., *184*
Barrere, P., 130
BARTEL, K., *76*
BASKOV, K. A. and KORNUTOVA, YE. I., *411*
BASLER, DJ., MALEZ, M. and BRUNNACKER, K., 37, *76*
BATSCHE, H., BAUER, F., BEHRENS, H., BUCHTELA, K., HRIBAR, F., KÄSS, W., KNUTSSON, G., MAIRHOFER, J., MAURIN, V., MOSER, H., NEUMAIER, F., OSTANEK L., RAJNER, V., RAUERT, W., SAGL, H., SCHNITZER, W. A. and ZÖTL, J., 256, *263*
BATURIĆ, J., *76*
BAUČIĆ, I., *76*
BAUČIĆ, I., *see* ROGLIĆ, J. and BAUĆIĆ, I.
BAUER, F., 231, 235, 237, 239, 261, *263*, *264*
Bauer, F., 244, 251
BAUER, F., ZÖTL, J. and MAYR, A., *221*, 241, *264*
BAUER, F., *see* BATSCHE, H. et al.
BAZÝNSKI, J., 338, *339*
BEEDE, J. W., 21, *24*
Begouen, Comte, 161
BEHLILOVIĆ, S., *76*
BEHRENS, H., *see* BATSCHE, H. et al.
BELSER, E., *221*
BENDEFY, L., 282, *294*
Berger, J., 172
BERTALAN, K., 283, *295*
BERTALAN, K., *see* SCHMIDT, E. R. et al.
BERTARELLI, L. V. and BOEGAN, E., 96, *125*
BEŠIĆ, Z., 50, 65, 66, *76*
BEYER, A., TIETZE, E. and PILAR, GJ., 20, *24*, *76*
BIDOVEC, F., *see* MOSETTI, F. et al.
BIELY, A., BYSTRICKÝ, J. and FUSAN, O., 299, 301, *323*
BIESE, W., 213, *221*
BIROT, P., 5, 11, *17*, *184*
Birot, P., 130
BISSANTI, A., 103, *125*
BITTNER, A., *see* LÓCZY, L. et al.
BITTNER, E., *see* MOJSISOVICS, E. et al.
BLEAHU, M. and RUSU, T., 353
BLEAHU, M., DECU, A. and DECU, V., 353, *353*
Blondeau, A., 130
BOCK, H., 226, 252, 254 256, *264*
BOCK, H., LAHNER, G. and GAUNERSDORFER, G., *264*

BOCKH, J., see LÓCZY, L. et al.
BOEGAN, E., 97, 119, *125*
BOEGAN, E., see BERTARELLI, L. V. and
    BOEGAN, E.
BOGAT'KO, N. M., *411*
BÖGLI, A., 15, *17*, 232, *264*, 323, *323*, 427, 435,
    442
BOHINEC, V., 77
BOILLOT, G., 144, *184*
Boillot, G., 130
BOLSENKÖTTER, H., 221
BOL'SHAKOV, N. M., *411*
BOL'SHAKOV, P. M., 386, *411*
Bonnet, A., 130
BONNET, A., DU CAILAR, J., COUDER, J. and
    DUBOIS, P., *184*
BONNET, A., see MARGAT, J. et al.
BONTE, A., 141, 142, *184*
Bonte, A., 130
BORELI, M. and PAVLIN, B., 77
BORZA, K. and MARTÍNY, E. 298, 309, *323*
BOURGIN, A., *184*
Bourgin, A., 130
BOZICEVIĆ, S., 58, 77
Bozicević, S., 35, 59
BOZISEVIĆ, S., see MALEZ, M. and BOZISEVIĆ, S.
BRANDECKER, H., MAURIN, V. and ZÖTL, J.,
    246, 247, *264*
BRAQUE, R., *184*
BRETZ, J. H., 23 24
Breuil Abbé, 161
BRINKMANN, R., 221
BRODAR, S., 35, 77
Brown C. B., 465
BROWN, R. F., 501, *504*
BROWN JR. H. B., see DICKEN, S. N. and
    BROWN JR., H. B.
BRUCKER, R. W. and BURNS, D. P., 500, *504*
BRUNNACKER, K., see BASLER, DJ. et al.
BUCHTELA, K., see BATSCHE, H. et al.
BÜDEL, J., *323*
BUKOWY, S., 332, *339*
BUNINA, M. V., 380, *411*
BURDON, J. D. and PAPAKIS, N., 24
BURKE, A. R., 436, 437, *442*
BURLICA, Č., 77
BURNS, D. P., see BRUCKER, R. W. and BURNS,
    D. P.
BUROV, V. P., *411*
BUROV, V. S., 373, *411*
BUŠALIJA, I., 77
BUTNARIU, E., see NAUM, T. and BUTNARIU, E.
BUTUROVIĆ, A., 77
BYSTRICKÝ, J., 298, *323*
BYSTRICKÝ, J., see BIELY, A. et al.

CADY, R. C., 487, 489, *504*
CAILLÈRE, S., see BARON, G. et al.
CAILLETEAU, P., CARTIER, G. and DREYFUSS,
    M., *184*
Cailleux, A., 142
CAMPOBASSO, V. and OLIVIERI, C., 102, *125*
CAPELLO, C. F., 87, 105, 116, *125*
CAPELLO, C. F., NANGERONI, G., PASA, A.,
    LIPPI BONCAMBI, C., ANTONELLI, C. and
    MALESANI, E., *125*
CAPPA, G., 103, *125*
CARO, R., 181, *184*
Caro, R., 130
CARTER, W. L. and DWERRYHOUSE, A. R.,
    429, 439, *442*
CARTIER, G., see CAILLETEAU, P. et al.
CASTALDI, F., 92, *125*
CASTERET, N., *184*
Casteret, N., 130
CASTIGLIONI, G. B., 98, *125*
CAUBEL, A., see ROUIRE, J. et al.
Caumartin, V., 130
CAVAILLE, A., 145, *184*
Cavaille, A., 130
CHABOT, G., 166, *184*
CHABOT, G., see CHOLLEY, A. and CHABOT, G.
CHERNYSH, L. V., 374, *411*
Chernyshev, F. N., 355
CHEVALLIER, P., 170, *184*
Chevallier, P., 130
CHEVET, B., *184*
CHIKISHEV, A. G., 378, 379, 380, 381, 382, *412*
Chikishev, A. G., 363, 367, 376
CHOLLEY, A., *184*
CHOLLEY, A. and CHABOT, G., 77
CHOPPY, J., *184*
Choppy, J., 130
CHOUTEAU, J., see CORROY, G. et al.
CHURINOV, M. V., 368, *412*
ĆIRIĆ, B., 77
ĆIRIĆ, M., see JURAS, I. and ĆIRIĆ, M.
CIRY, R., 137, 142, 167, *184*
Ciry, R., 130
CLAYTON, K. M., 427, *442*
COLAMONICO, C., 92, 103, *125*
COLIN, J., *184*
COMBES, J.-P., *184*
COOKE, C. W., 477, *504*
COOPER JR., H. H. and KENNER, W. E., 476,
    504
COPPADORO, A., 98, *125*
CORBEL, J., 12, *17*, 146, 148, 149, 170, 172, 181,
    184, *323*, 431, 435, *442*
Corbel, J., 130
CORROY, G., GOUVERNET, C., CHOUTEAU, J.,
    SIVIRINE, A., GILET, R. and PICARD, J., *185*

COTECCHIA, V., 102, 123, *125*, *126*
COTECCHIA, V. and MAGRI, G., 123, *126*
COUDER, J., *see* BONNET, A. et al.
COX, E. T., 3, 4, *17*
CRAMER, H., *221*
CREAC'H, Y., 179, *185*
CRESCENTI, U. and VIGHI, L., 91, *126*
ČUBRILOVIĆ, P., *77*
Cukor, V., 44
CULLINGFORD, C. H. D., 418, 433, *442*
CUNHAC, R., *see* JEZEQUEL, F. and CUNHAC, R.
CUSHMAN, R. V., KRIEGER, R. A. and MCCABE, J. A., 501, *504*
CVIJIĆ, J. 4, 5, 6, 7, 8, 9, 12, 13, 15, *17*, 21, *24*, 77, 312 314, 318. *324*

DABROWSKI, T., 336, 337, *339*
DAINELLI, G., 86, 115, *126*
DALMASSO, E., *185*
DAL PIAZ, G., 108, *126*
D'AMBROSI, C., 98, 119, *126*
DAMIDOT, L., *see* DREYFUSS, M. and DAMIDOT, L.
DANEŠ, J., 6, *17*
DARÁNYI, F., 271, 276, 281, 283, 286, 289, *295*
DA SCHIO, A., TREVISIOL, G. and PERIN, G., 101, *126*
DAVEAU, S., *185*
DAVIES, W. E., 486, *505*
DAVIS, W. M., 4, *17*, 23, 24, 355, *412*
DEBUCHANANNE, G. D. and RICHARDSON, R. M., 487, *505*
DECU, A., *see* BLEAHU, M. et al.
DECU, V., *see* BLEAHU, M. et al.
DE GASPERI, G. B., 98, 115, *126*
DE GASPERI, G. B. and FERUGLIO, G., 98, *126*
DEHEE, R., 135, *185*
De Joly, R., 130, 158
DE LAVAUR, G., *185*
De Lavaur, G., 130
DELL'OCA, S., 124, *126*
DEMANGEON, P., 148, *185*
DEMANGEOT, J., 92, 109, 110, *126*
DE MARCHI L., 99, *126*
DE MARTONNE, E., 172, *185*, 348, *353*
DICKEN, S. N. and BROWN JR., H. B., 501, *505*
DIENER, C., 3. *17*
Di MAIO, M., *see* PASINI G. et al.
DJERKOVIĆ, B., *77*
DJIKIĆ S., *77*
DJURIĆ, V., *77*
DOBROVOL'SKIY, M. N., *412*
Dokuchayev, V. V., 355
DOLGUSHIN, I. YU., *412*
DONINI, L., 115, *126*
DORN, P., *221*

DOSCH, F., 239, *264*
Dosch, F., 241
DOUGHTY, P. S., 420, *442*
DREW, D. P., 439, *442*
DREW, D. P., *see* ATKINSON, T. et al.
DREYFUSS, M., 166, *185*
Dreyfuss, M., 130
DREYFUSS, M. and DAMIDOT, L., *185*
DREYFUSS, M., *see* CAILLETEAU, P. et al.
DROQUE, C., 158, *185*
Droque, C., 130, 182
DROPPA, A., 297, 298, 308, 312, 314, 319, 321, 323, *324*
DUBLYANSKIY, V. N., 368, *412*
DUBOIS, M., *185*
Dubois, M., 130
DUBOIS, P., 154, 156, 157, *185*
Dubois, P., 130
DUBOIS, P. and GRIOSOL, Y., *185*
DUBOIS, P., *see* AVIAS, J. and DUBOIS, P.
DUBOIS, P., *see* BONNET, A. et al.
Du CAILAR, J., *see* BONNET, A. et al.
DUNHAM, K. C. and ROSE, C. C., 431, *442*
DUROZOY, G. and PALOC, H., *185*
DWERRYHOUSE, A. R., *see* CARTER, W. L. and DWERRYHOUSE, A. R.
DYJOR, S., 328, *339*
DZENS-LITOVSKIY, A. I., 356, 375, *412*
DŽULYŃSKI, S., HENKIEL, A., KLIMEK, K. and POKORNY, J., 334, *339*

ECKERT, M., *221*, 259, *264*
Elhai, H., 130
EMBER, K., *see* SCHMIDT, E. R. et al.
ENDRISS, G., *221*
EPIFANOV, M. I., *see* GOLTS, S. I. and EPIFANOV, M. I.
ERHARDT, GY., *see* SCHMIDT, E. R. et al.
ERIKSSON, E., *see* MOSETTI, F. et al.
EYRE, J. and ASHMEAD, P., 430, *442*

FABIANI, R., 101, *126*
FABRY, J., *see* ROUIRE, J. et al.
FEARNSIDES, W. G., 422, *442*
FEDOROVICH, B. A., 386, *412*
FELS, E., 249, *264*
FENELON, P., *185*
Fenelon, P., 130
FERENC, K., *see* SCHMIDT, E. R. et al.
FERSMAN, A. YE., 397, *412*
FERUGLIO, E., 98, 108, *126*
FERUGLIO, G., *see* DE GASPERI, G. B. and FERUGLIO, G.
FIEDLER, A. G. and NYE, S. S., 502, *505*
Filhol, H., 161
FINK, M. H., 251, *264*

FLIS, J., 336, *339*
FLÜGEL, H., 252, *264*
FÖLDVÁRY, A., 268, *295*
Földváry-Vogl, A., 268
FORD, D. C., 435, *442*
FORD, T. D., 438, *442*
Ford, T. D., 432
FORKASIEWICZ, J. and PALOC, H., *186*
FORTI, F., 98, *126*
FORTI, F. and TOMMASINI, T., 98, *126*
FOURNET, J., 2, *17*
FOURNIER, E., 166, *186*
FRANCOIS, S., *see* GLANGEAUD, L. et al.
FRANIĆ, D., 77
Frey, A., 434
FRICKE, W., *221*
FRIESE, H., *221*
FRITZ, F., *see* PAVLIN, B. et al.
FUGANTI, A., *see* VENZO, G. A. and FUGANTI, A.
FUKAREK, P., 77
FÜLÖP, J., 271, *295*
FUREDDU, A. and MAXIA, C., 93, 112, *126*
FUSAN, O., *see* BIELY, A. et al.

GALLE–CAVALLON, H. and PALOC, H., *186*
GALLOCHER, P., *186*
GAMS, I., 40, 46, 47, 52, 77, 78, 418, *442*
GAREVSKI, R., 37, *78*
GARWOOD, E. J., 422, *442*
GARWOOD, E. J. and GOODYEAR, E., 420, *442*
GAUNERSDORFER, G., *see* BOCK, H. et al.
GAVELA, B., 37, *78*
GAVRILOV, A. M., *412*
GAVRILOVIĆ, D., *78*
GECCHELE, G., *see* BADINI, G. and GECCHELE, G.
GELLERT, J., 205, *221*
GEORGE, T. N., 419, *442*
GERASIMOV, I. P., 387, *412*
GERECKE, F. and NEUNHÖFER, H., *222*
GEYER, O. F. and GWINNER, M. P., *222*
GÈZE, B., 141, 145, 181, 183, *186*
Gèze, B., 130
GÈZE, B. and POBEGUIN, T., 181, *186*
GIGNOUX, M., 172, *186*
GIGOU, R. and MONNIN, J., *186*
GILET, R., *see* CORROY, G. et al.
GILEWSKA, S., 330, *339*
GJURAŠIN, K., 44, 68, *78*
GLANGEAUD, L., PÉZARD, R., FRANÇOIS, S., PERRENOUD, M.-J. and TOITOT, M., *186*
GLAZEK, J., 330, 333, 335, *339*
GLUSHKO, V. V., *412*
GMELIN, I. G., 399, *412*
GODEK, I., *78*
GOLTS, S. I. and EPIFANOV, M. I., 387, *412*

GOODWIN JR., M. H., *see* HENDRICKS, E. L. and GOODWIN JR., M. H.
GOODYEAR, E., *see* GARWOOD, E. J. and GOODYEAR, E.
Gorbunov, A. P., 397
GORBUNOVA, K. A., *see* MAKSIMOVICH, G. A. and GORBUNOVA, K. A.
GORDEYEV, D. I., 356, *412*
GORJANOVIĆ–KRAMBERGER, D., 37, *78*
GÓRZYŃSKI, Z., 332, *339*
GOSPODARIČ, R., 37, 46, 47, *78*
GOSSELET, J., 142, *186*
GÖTZINGER, G., 239, *264*
GOUVERNET, C., *see* CORROY, G. et al.
GRÄBE, H., *222*
GRADZIŃSKI, R., 334, *339*
GRADZIŃSKI, R. and WÓJCIK, Z., 330, *339*
GRADZIŃSKI, R. and WRÓBLEWSKI, T., 335, *339*
GRAHMANN, R., *222*
GREEN, G. W. and WELCH, F. B. A., 423, *442*
GRESSEL, W. and PICHLER, H., 257, *264*
GRIOSOL, Y., *see* DUBOIS, P. and GRIOSOL, Y.
GROOM, G. E. and WILLIAMS, V. H., 436, *442*
GRUBER, TH., 2, *17*
GRUBIĆ, A., *78*
GRUND, A., 4, 5, 6, 8, 9, *17*, 21, 23, *24*, *78*, 226, *264*
GUŠIĆ, B., *78*
GVOZDETSKIY, N. A., 312, *324*, 356, 368, 369, 370, 371, 372, 393, 395, 396, 397, 398, *412*, *413*
GWINNER, M. P., *see* GEYER, O. F. and GWINNER, M. P.

HAASE, H., 201, *222*
HABE, F., 77
HABIČ, P., *78*
HACQUET, B., 2, *17*
HADŽI, J., *78*
HAEFKE, F., *222*
HALAVÁTS, GY., *see* LÓCZY, L. et al.
HAMMER, W., *see* AMPFERER, O. and HAMMER, W.
HAMRLA, M., *78*
HANSHAW, B. B., BACK, W. and RUBIN, M., 479, *505*
HENDRICKS, E. L. and GOODWIN JR., M. H., 477, *505*
HENKEL, L., *222*
HENKJEL, A., *see* DŽULYŃSKI, S. et al.
Henry, J.-P., 130
HERAK, M., *78*
HERRICK S. M. and LEGRAND, H. E., 477, *505*
HERRMANN, A., *222*

REFERENCES INDEX 523

HIGH, C., see ATKINSON, T. et al.
HODOSCEK, K., see MOSETTI, F. et al.
HOMOLA, V., 298, *324*
HORNINGER, G., 260, *264*
HOSE, H. R. and VERSEY, H. R., 447, *466*
HOSKINS, H. A., see PRICE, P. H. et al.
HRANILOVIĆ, H., *78*
HRIBAR, F., see BATSCHE, H. et al.
HUBER, F., *222*
HUDSON, R. G. S., 420, 426, *442*
HUNDT, R., *222*

IL'INA, S. M., see IVANOV, B. N. and IL'INA, S. M.
IVANOV, A. A. and LEVITSKIY, YU. F., 375, *413*
IVANOV, B. N. and IL'INA, S. M., 368 *413*
Ivanov, D. L., 355

JANÁČEK, J., 297, *324*
JANÁČIK, P., 314, *324*
JANJIĆ, M., *78*
JASKÓ, S., 288, *295*
JEANNEL, R., *186*
JEANNEL, R. and RACOVITZA, E. G., 341, *353*
JEDLICKA, F., 258, 264
JELAVIĆ, A., 75, *78*
JENKO, F., *78, 79*
JEVREMOVIĆ, M., 45, *79*
JEZEQUEL, F. and CUNHAC, R., *186*
JOHNSTON JR., W. D., 494, *505*
JOVANOVIĆ, P. S., 8, *17, 79*
JULIAN, M., *186*
JURAS, I. and ĆIRIĆ, M., *79*
JURKOVIĆ, I. and SAKAČ, K., *79*
JUX, U., 199, *222*

KAJMAKOVIĆ, R. and PETROVIĆ, B., *79*
KAMMERER, F., *222*
Karakash, N. I., 355
KARZENKOV, G. I., see PECHERKIN, I. A. and KARZENKOV, G. I.
KÄSS, W., see BATSCHE, H. et al.
KATZER, F., 5, 9, *17*, 21, 24, *79*, 226, *264*
KAYSER, K., 10, *17, 324*
KEIL, K., *222*
KEILHACK, K., 24, *222*
KEMÉNY, A., 297, 318, *324*
KENDALL, P. F. and WROOT, H. E., 429, *442*
KENNER, W. E., see COOPER JR., H. H. and KENNER, W. E.
KENT, P. E., 432, *442*
KERNER, F., *79*
KESSLER, H., *222*, 284, 285, 286, 289, *295*
KEVO, R., *79*
KHOROSHIKH, P. P., *413*

KIKNADZE, T. Z., 372, *413*
King, R. J., 432
KIRILOV, M. V., *413*
KIŠPATIĆ, M., *79*
KISS, J., 268, *295*
KLEYNER, YU. M., 387, *413*
KLIMAZEWSKI, M., 334, *339*
KLIMEK, K., see DŽULYŃSKI, S. et al.
KNEBEL, W., 5, *18*
KNEŽEVIĆ, B., 45, *79*
KNUTSSON, G., see BATSCHE, H. et al.
KOBER, L., *79*
KOCH, F.,*79*
KODRIČ, M., *79*
KOEHNE, W., *222*
KOISAR, B., 335, *339*
KÓKAY, J., 278
KOLODYAZHNAYA, A. A., 371, *413*
KOLOSKOV, P. I., 386, *413*
KOLOTIL'SHCHIKOV, V. K., 408, *413*
KOMATINA, M., *79*
KONČEK, M., 310, *324*
KORENEVSKIY, S. M., 374, 375, *413*
KORNUTOVA, YE. I., see BASKOV, K. A. and KORNUTOVA, YE. I.
KOROTKEVICH, G. V., 373, *413*
KOROZHUYEV, S. S., *413*
KOSACK, H.-P., *222*
KOSMAC, J., *79*
KOSTAREV, V. P., see MAKSIMOVICH, G. A. and KOSTAREV, V. P.
KOTLICKA, G., 336, *339*
KOVAČEVIĆ, P., *79*
KOWALSKI, K., 333, *339*
Kozlowski, A., 338
KRANZ, W., *222*
KRASOŃ, J. and WÓJCIK, Z., 337, 338, *340*
KRAUS, F., 3, *18*
KREBS, N., *79*
Krebs, N., 5
KRIEGER, R. A., see CUSHMAN, R. V. et al.
KROMBELIN, K., see LEHMANN, H. et al.
KRUBER, A. A., 355, 368, *413*
KRULC, Z., 58, *79*
KSIAZKIEWICZ, M., SAMSONOWICZ, J. and RÜHLE, E., 327, *339*
KUBÍNY, D., 297, 319, *324*
KUDELIN, B. I., 405, *413*, 414
KULLMAN, E., 298, *324*
KUNAVER, P., *79*
KUNSKÝ, J., 314, *324*
KUŠČER, 45, 68, *79*
Kutek, J., 338
Kuzmenko, A. S., 355
KUZNETSOV, I. G., 371, *414*
KUZNETSOV, YU. YA., 387, *414*

KYRLE, G., 256, *264*
Kyrle, G., 226
KYRLE, G., *see* ABEL, O. and KYRLE, G.

LACKÖ, D., *see* LÓCZY, L. et al.
LAENG, G., 101, *126*
LAGRANGE, R., *see* BARON, G. et al.
LAHNER, G., *80*
LAHNER, G., *see* BOCK, H. et al.
LÁNG, G., *see* SCHMIDT, E. R. et al.
LÁNG, S., 297, *324*
LAPTEV, F. F., 356, *414*
LAURA, R., 146, *186*
Laures, M., 130
LAURES, M. and PALOC, H., *186*
LAZZARI, A., 92, *126*
LECHNER, A., 248, 249, 250, *264*
Lefeld, J., 338
LEGRAND, H. E., *see* HERRICK, S. M. and LEGRAND, H. E.
LEGRAND, H. E., *see* STRINGFIELD, V. T. and LEGRAND, H. E.
LEHMANN, H., 10, *18*, 98, 109, *127, 324*
LEHMANN H. and SUNARTADIRDJA, M., *324*
LEHMANN, H., KROMBELIN, K. and LOTSCHERRT, W., 11, *18*
LEHMANN, O., 9, 10, 18, 22, *24*, 44, 69, *80, 222*, 226, 256, *264*
LENČO, V., 308, *324*
Lepekhin, I. I., 355
LERICHE, M., *186*
LEROUX, E., RICOUR, J. and WATERLOT, G., *186*
LETOURNEUR, J., *186*
LEVEN, YA. A., 397, *414*
LICHTENBERGER, E., 258, *264*
LICHTENECKER, N., 229, 239, *264*
LIGASACCHI, A. and RONDINA, G., 106, *127*
LILIENBERG, D. A., 371, *414*
LINDNER, H., 222
LIPPI BONCAMBI, C., 110, *127*
LIPPI BONCAMBI, C., *see* CAPELLO, C. F. et al.
LIS, J. and WÓJCIK, Z., 332, *339*
LISZKOWSKI, J. 333, 338, *339*
LÓCZY, L., BOCKH, J., BITTNER, A., HALAVÁTS, GY., LACKÖ, D., PAPP, K., VADÁSZ, E. and VITALIS, J., 269, *295*
LOHMAN, S. W., 489, *505*
LORENZI, A., 115, *127*
LOTSCHERRT, W., *see* LEHMANN, H. et al.
LOUČEK, D., 298, 323, *324*
Louis, H., *324*
LUBURIĆ, P., *see* PAPEŠ, J. et al.
LUKNIŠ, M., 297, 316 318, *324*
LUKOVIĆ, M., *80*
LUPPOV, N. P., 387, *414*

LUTAUD, L., 142, *186*
LUTOVAC, M., *80*
LYELL, Ch., 2, *18*
LYKOSHIN, A. G., 357, *414*

MADEYSKA–NIKLEWSKA, T., 334, *339*
MAGAŠ, N., 44, *80*
MAGDALENIĆ, A., 43, 65, 68, *80*
MAGDALENIĆ, A., *see* PAVLIN, B. et al.
Magniez, G., 130
MAGRI, G., *see* COTECCHIA, V. and MAGRI, G.
MAHEL, M., 298, *324*
MAHEU, J., *186*
MAIRHOFER, J., *see* BATSCHE, H. et al.
MAKHAYEV, V. N., *414*
MAKSIMOVICH, G. A., 356, 357, 409, *414*
MAKSIMOVICH, G. A. and GORBUNOVA, K. A., 313, *324*, 380, *414*
MAKSIMOVICH, G. A. and KOSTAREV, V. P., *414*
MALESANI, E., *see* CAPELLO, C. F. et al.
MALEZ, M., 35, 37, *80*
MALEZ, M., *see* BASLER, DJ. et al.
MALEZ, M., *see* RACZ, Z. et al.
MALEZ, M. and BOŽIČEVIĆ, S., *80*
MANAKOVIK, D., *80*
MANFREDINI, M., 122, *127*
MARCACCINI, P., 92, 103, 111, *127*
MARGAT, J., MOLINARD, L., PALOC, H. and BONNET, A., *186*
MARIĆ, L., *80*
MARINELLI, O., 93, 98, 115, 116, 117, *127*
MARKOWICZ–LOHINOWICZ, M., 333, 336, 337, *340*
Markowicz–Lohinowicz, M., 338
MARRES, P., *186*
MARTEL, E. A., 5, *18*, 21, *24*, *80*, 142. 150, 169, 172, 175, 177, *186*, *187*, 372, *414*
Martel, E. A., 150, 158, 178
MARTIN, L., 130
MARTINIS, B., 91, 95, 96, 102, 113, 114, *127*
MARTÍNY, E., *see* BORZA, K. and MARTÍNY, E.
MARUASHVILI, L. I. and TINTILOZOV, Z. K., 372, *414*
MARUSSI, A., 98, *127*
MARUSZCZAK, H., 328, 333, 335, *340*
MÄRZ, CH., *264*
MASLOV, V. P., *414*
MASSARI, F., *see* ABRAMI, G. and MASSARI, F.
MATONIČKIN, I. and PAVLETIĆ, Z., *80*
MATONIČKIN, I., *see* PAVLETIĆ, Z. and MATONIČKIN, I.
MAUCCI, W., 91, 97, 98, *127*
MAUCHA, L., 291, *295*
MAURIN, V., 254, 256, *264*, 265
MAURIN, V. and ZÖTL, J., 241, 256, *265*

MAURIN, V., see BATSCHE, H. et al.
MAURIN, V., see BRANDECKER, H. et al.
MAXIA, C., see FUREDDU, A. and MAXIA, C.
MAYR, A., see BAUER, F. et al.
MAZÚR, E., 298, 307, 314, 316, 323, *324*
MCCABE, J. A., see CUSHMAN, R. V. et al.
MCCUE, J. B., see PRICE, P. H. et al.
MEAD, D.G., see SMITH, D.I. and MEAD, D.G.
MEDWENITSCH, W., see SIKOSEK, B. and MEDWENITSCH, W.
MEGNIEN, C., 135, 137, 138, 141, 142, *187*
Megnien, C., 130
MEINZER, O. E., 23, *24*
MELIK, A., *80*
MENGAUD, L., *187*
MENNESSIER, G., 168, *187*
Mennessier, G., 130
MICHOVSKA, J., 312, 314, *324*
MIČIAN, Ľ., 309, *324*
MIJATOVIĆ, B., *80*
MIKULEC, S., 65, *80*
MIKULEC, S. and TRUMIĆ, A., 51, 52, *80*
MILETIĆ, P. and BABIĆ, Z., *80*
MILIĆ, Č. S., *80*
MILIKHIKER, SH. G., 380, *414*
MILOJEVIĆ, B., *80*
MILOJEVIĆ, N., 52, *81*
MILOJEVIĆ, S., *81*
MILOŠEVIĆ, LJ., *81*
MILOŠEVIĆ, LJ., see ARANDJELOVIĆ, D. and MILOSEVIĆ, LJ.
MILOVANOVIĆ, B., *81*
MIROSHNIKOV, L. D., *414*
MOISLEY, H. A., 427, *442*
MOJSISOVICS, E., *18*
MOJSISOVICS, E., TIETZE, E. and BITTNER, E., 3, *18*
MOLINARD, L., see MARGAT, J. et al.
MOLITVIN, P. V., 409, *414*
MONNIN, J., see GIGOU, R. and MONNIN, J.
MORELLI, C., 118, *127*
MORETTI, T., see ORTOLANI, M. and MORETTI, T.
MOSER, H., see BATSCHE, H. et al.
MOSETTI, F., 118, 119, 120, *127*
MOSETTI, F., ERIKSSON, E., BIDOVEC, F., HODOSCEK, K. and OSTANEK, L., 118, *127*
MOTTS, W. S., 502, *505*
MOWSZOWICZ, J. and OLASZEK, R., 336, 337, *340*
MÜCKENHAUSEN, E., 222
MUGNIER, C., *187*
Mugnier, C., 130
MURKO, V., *81*
MURKO, V., see POTOCIĆ, Z. and MURKO, V.
Muxart, R., 130

MUXART, R. and STCHOUZKOY, T., *187*

NANGERONI, G., 86, 95, 108, 109, 112, 121, *127*
NANGERONI, G., see CAPELLO, C. F. et al.
NAPIAS, M., *187*
NAUM, T. and BUTNARIU, E., 351, *353*
NAUMANN, C. FR., 1, *18*
NÉMETH, L., see SCHMIDT, E. R. et al.
NEUMAIER, F., see BATSCHE, H. et al.
NEUNHÖFER, H., see GERECKE, F. and NEUNHÖFER, H.
NICOD, J., 178, 179, 181, *187*
Nicod, J., 130
NOINSKIY, M. E., 356, *415*
Nonweiller, E., 65
NOSAN, A., *81*
NOSZKY, J., 271, *295*
NOVAK. D., *81*
NOVAK, G., 37, *81*
NYE, S. S., see FIEDLER, A. G. and NYE, S. S.

OLASZEK, R., see MOWSZOWICZ, J. and OLASZEK, R.
OLIVIERI, C., see CAMPOBASSO, V. and OLIVIERI, C.
OLLI, YE. I., 396, *415*
ORAVECZ, J., 267, *295*
ORTOLANI, M. and MORETTI, T., 92, 109, 110, *127*
OSOKIN, I. M., 399, *415*
OSOLÉ, F., 35, *81*
OSTANEK, L., see BATSCHE, H. et al.
OSTANEK, L., see MOSETTI, F. et al.
OWEN, D. D., 3, *18*
OZONAI, GY., see SCHMIDT, E. R. et al.

PAL, A. M., 222
PALOC, H., 157, 181, *187*
Paloc, H., 130
PALOC, H., see DUROZOY, G. and PALOC, H.
PALOC, H., see FORKASIEWICZ, J. and PALOC, H.
PALOC, H., see GALLE-CAVALLON, H. and PALOC, H.
PALOC, H., see LAURES, M. and PALOC, H.
PALOC, H., see MARGAT, J. et al.
PANOŠ, V., 313, *324*
PANOŠ, V. and ŠTELCL, O., *324*
PAPAKIS, N., see BURDON, J. D. and PAPAKIS, N.
Papeš, J., 49
PAPEŠ, J. and SRDIĆ, R., *81*
PAPEŠ, J., LUBURIĆ, P., SLISKOVIĆ, T. and RAIĆ, V., *81*
PAPIU, V. C., POPESCU, A. and SERAFIMOVICI, V., 341, *353*
PAPP, F., 291, *295*
PAPP, K., see LÓCZY, L. et al.

PARDE, M., *187*
PASA, A., 99, 107. *127*
PASA, A., *see* CAPELLO, C. F. et al.
PASINI, G., RIBALDONE, G. and DI MAIO, M., 100, *127*
Patrulius, D., 348
PAVLETIĆ, Z. and MATONIĆKIN, I., *81*
PAVLETIĆ, Z., *see* MATONIĆKIN, I. and PAVLETIĆ, Z.
PAVLIN, B., 64, *81*
PAVLIN, B. and MLADINEO, L., 65, *81*
PAVLIN, B., BAHUN, S. and FRITZ, F., *81*
PAVLIN, B., MAGDALENIĆ, A. and ZLATOVIĆ, D., *81*
PAVLIN, B., *see* BORELI, M. and PAVLIN, B.
PAVLOVIĆ, P., *see* RADOVANOVIĆ, M. and PAVLOVIĆ, P.
PECHERKIN, I. A., *415*
PECHERKIN, I. A. and KARZENKOV, G. I., 380, *415*
PENCK, A., 1, 4, 5, 7, 8, 13, *18, 222, 265*
Penck, A., 226
PERIĆ, J., 81
PERIN, G., *see* DA SCHIO, A. et al.
PERRENOUD, M.-J., *see* GLANGEAUD, L. et al.
PETKOVIĆ, K., *81*
PETREQUIN, P., *187*
PETRIK, M., 45, *81*
PETROVIĆ, B. and PREVELIĆ, B., *81*
PETROVIĆ, B., *see* KAJMAKOVIĆ, R. and PETROVIĆ, B.
PETROVIĆ, J. B., *81*
PÉZARD, R., *see* GLANGEAUD, L. et al.
PFEIFFER, D., *222*
PICARD, J., *see* CORROY, G. et al.
PICHLER, H., *see* GRESSEL, W. and PICHLER, H.
PIETZSCH, K., *223*
PIGOTT, C. D., 425, 433, *442*
PILAR, GJ., *see* BEYER, A. et al.
PINCHEMEL, P., 142, *187*
PITTER, T. M., *415*
PITTY, A. F., 439, *442*
Pitty, A. F., 433
PLENIČAR, M., 46, 47, *81*
PLOTNIKOV, N. I., 386
POBEGUIN, T., 181, *187*
Pobeguin, T., 130
POBEGUIN, T., *see* GÈZE, B. and POBEGUIN, T.
POBEGUIN, T., *see* BARON, G. et al.
POBORSKI, J., 336, 338, *340*
POKORNY, J., 334, *340*
POKORNY, J., *see* DŽULYŃSKI, S. et al.
POLJAK, J., 16, *18, 81, 82*
POLŠAK, A., *81*
POMEROL, C., 141, 143, *187*
Pomerol, C., 130

Pommier, C., 130
POPESCU, A., *see* PAPIU, V. C. et al.
POPOV, I. V., 356, *415*
POSLAVSKAYA, O. YU., 386, *415*
POTOČIĆ, Z. and MURKO, V., *82*
POWELL, R. L., 501, *505*
PRACCHI, R., 91, 101, 106, *127, 128*
PRESTWICH, J., 3, *18*
PREVELIĆ, B., *see* PETROVIĆ, B. and PREVELIĆ, B.
PRICE, P. H., MCCUE, J. B. and HOSKINS, H. A., 489, *505*
PROSKURYAKOV, P. S., *415*
PRUVOST, P., 135, *187*
PULINA, M., 333, 334 *340*
Pulina, M., 334

RACOVITZA, E. G., *see* JEANNEL, R. and RACOVITZA, E. G.
RACZ, Z., ŠILJAK, M. and MALEZ, M., *82*
RAD'KO, N. I., *415*
RADOVANOVIĆ, M., *82*
RADOVANOVIĆ, M. and PAVLOVIĆ, P., *82*
Radwański, A., 338
RADWAŃSKI, A., *see* BALUK, W. and RADWAŃSKI, A.
RAIĆ, V., *see* PAPEŠ, J. et al.
RAJNER, V., *see* BATSCHE, H. et al.
RAKOVEC, I., *82*
RALJEVIĆ, B., 14, *18*
RAUERT, W., *see* BATSCHE, H. et al.
RAVASZ, C., *see* SCHMIDT, E. R. et al.
REDENŠEK, V., *82*
REEVES, F., 489, *505*
Remezov, S., 355
REMY, P. A., *82*
RÉMY-BERZENCOVITCH, E., 258, *265*
RENAULT, P., *187*
Renault, P., 130, 176
RENAULT. P. and SAUMADE, P., *187*
REŠTAROVIĆ, S., 62, *82*
REUTER, F. and REUTER, R., *223*
REUTER, R., *see* REUTER, F. and REUTER, R.
RHOADES, R. F. and SINACORI, M. N., 23. *24*
RIBALDONE, G., *see* PASINI, G. et al.
RIBIS, R., *187*
RICHARDSON, R. M., *see* DEBUCHANANNE, G. D. and RICHARDSON, R. M.
Richter, E., 5
RICHTER, E., 9, *18*
RICHTER, M., *223*
RICHTER-BERNBURG, G., *223*
RIDJANOVIĆ, J., *82*
RIEDL, H., 260, *265*
RICOUR, J., 142, *187*
RICOUR, J., *see* LEROUX, E. et al.
ROBINSON, E., 447, *466*

RODIONOV, N. V., 356, 368, 387, *415*
ROGLIĆ, J., 10, 12, 13, 15, 16, *18*, *82*, *325*
ROGLIĆ, J. and BAUČIĆ, I., *82*
ROMANYUK, A. F., *415*
RONDINA, G., *see* LIGASACCHI, A. and RONDINA, G.
ROQUES, H., 181, *188*
Roques, H., 130
ROSE, C. C., *see* DUNHAM, K. C. and ROSE, C. C.
Rosenmüller, J. C., 118
ROTH, Z., 297, 318, *325*
ROUIRE, J., *188*
Rouire, J., 130
ROUIRE, J., CAUBEL, A. and FABRY, J., *188*
ROUSSET, C., *188*
Rousset, C., 130
ROVERETO, G., 92, 105, *128*
ROZEN, M. F., 401, *415*
RÓŻYCKI, S. Z., 333, 334, *340*
RUBIN, M., *see* HANSHAW, B. B. et al.
RUDNICKI, 333, 335, 336, 337, *340*
Rudnicki, J., 338
RÜHLE, E., *see* KSIAZKIEWICZ, M. et al.
RUSU, T., *see* BLEAHU, M. and RUSU, T.
RUTTE, E., *223*
Rychkov, P. P., 355
RŽEHAK, V., *82*

SACHANBIŃSKI, M., 336, *340*
SADOV, V. P., *415*
SAGL, H., *see* BATSCHE, H. et al.
SAIBENE, C., 91, 106, 112, *128*
SAKAČ, K., *82*
SAKAČ, K., *see* JURKOVIĆ, I. and SAKAČ, K.
SALONENKO, V. P., 404, *415*
SALVAYRE, H., 135, 150, *188*
Salvayre, H., 130
SAMSONOWICZ, J., 333, *340*
SAMSONOWICZ, J., *see* KSIAZKIEWICZ, M. et al.
SAUMADE, P., *see* RENAULT, P. and SAUMADE, P.
SAVARENSKIY, F. P., 358, *415*
Savarenskiy, F. P., 356
SAVNIK, R., *82*
SAWICKI, L., 5, 9, *18*, 297, 313, 316, 318, *325*
SAWKINS, J. Q., 3, *18*
Šćavničar, B., 28
SCHACHTSCHABEL, P., *see* SCHEFFER, F. and SCHACHTSCHABEL, P.
SCHAUBERGER, O. and TRIMMEL, H., 226, *265*
SCHEFFER, F. and SCHACHTSCHABEL, P., *223*
SCHEUERMANN, R., *223*
SCHLUMBERGER, R., *188*
SCHMIDL, A., 3, *18*
SCHMIDT, E. R., 277, 278, *295*
Schmidt, E. R., 283

SCHMIDT, E. R., ALMÁSSY, E., BERTALAN, K., EMBER, K., ERHARDT, GY., FERENC, K., LÁNG, G., NÉMETH, L., OZONAI, GY. and RAVASZ, C., 274, 290, 292, *295*
SCHMIDT–THOMÉ, P., 258, 259, *265*
SCHNITZER, W. A., *see* BATSCHE, H. et al.
SCHOELLER, H., 181, *188*
Schoeller, H., 130
SCHRÉTER, Z., 273, *295*
SCOTT, J., 495, *505*
SEGRE, A. G., 92, 108, 109, 110, 121, *128*
SEKYRA, J., 298, 308, 314, 321, 323, *325*, 398, *415*
SENCU, V., 351, *353*
SENEŠ, J., 297, *325*
SERAFIMOVICI, V., *see* PAPIU, V. C. et al.
ŠERKO, A., 40, 82
SERONIE–VIVIEN, M. R., 146, *188*
Seronie–Vivien, M. R., 130
SERVICIO GEOLOGICO D'ITALIA, 102, 117, *128*
SESTINI, A., 109, *128*
Severgin, V. M., 355
SHCHERBAKOV, A. V., 359, *415*
SHCHERBAKOVA, YE. N., *416*
SHCHUKIN, I. S., 398, *416*
SHTUKENBERG, A. A., 355
SIFFRE, A., 130
SIFFRE, M., *188*
Siffre, M., 130
SIKOŠEK B. and MEDWENITSCH, W., *82*
ŠILJAK, M., *see* RACZ, Z. et al.
Simpson, F., 338
SINACORI, M. N., *see* RHOADES, R. F. and SINACORI, M. N.
SIVIRINE, A., *see* CORROY, G. et al.
SKUODIS, V. P., 361, *416*
SKORTSOV, T. G., *416*
SLAVIN, V. I., 374, *416*
SLIŠKOVIĆ, T., *see* PAPEŠ, J. et al.
SMITH, D. I. and MEAD, D. G., 434, *443*
SMITH, P. M., 499, *505*
SMOLÍKOVÁ, L., 298, 309, *325*
SOKOLOV, D. S., 356, 357, 368, *416*
SOKOLOV, N. I., 356, 372, *416*
SOMMARUGA, C., 112, *128*
SPENGLER, E., 226, *265*
SPIROVSKI, J., *82*
SPÖCKER, R. G., *223*, 259, *265*
SRDIĆ, R., *see* PAPEŠ, J. and SRDIĆ, R.
SREBENOVIĆ, D., *83*
Stchouzkoy, T., 130
STCHOUZKOY, T., *see* MUXART, R. and STCHOUZKOY, T.
ŠTELCL, O., *see* PANOŠ, V. and ŠTELCL, O.
STEMBERGER, M., *83*
STEPANOVIĆ, B., *83*

STEPINAC, A., 61, *83*
STILLE, H., *223*
Stille, H., 276
STINY, J., 258 *265*
STOLBERG. F., *223*
STRINGFIELD, V. T., 474, 477, 478, 479, *505*
STRINGFIELD, V. T. and LEGRAND, H. E., 475, 477, *505*
STUPISHIN, A. V., 356, *416*
SULIMSKI, A., 333, *340*
SUNARTADIRDJA, M., *see* LEHMANN, H. and SUNARTADIRDJA, M.
Svetina, G., 118
SWEETING, G. S. and SWEETING, M. M., 426, *443*
SWEETING, M. M., *83*, 421, 425, 427, 430, *443*, 460, 462, 466
SWEETING, M. M., *see* SWEETING, G. S. and SWEETING. M. M.
SWINNERTON, A. C., 23, *24*
SZABÓNÉ–DRUBINA, M., 277, *295*
SZAFER, W., 333, *340*
SZALAI, T., 276, 277, 281, *295*

TARÁBEK, K., 309, *325*
TARKINGTON, T. W., VARNEDOE JR., W. W. and VEATCH, J D., 493 *505*
TARLO, L. B., 332, *340*
TATARINOV, K. A., 367, *416*
Tatishchev, V. N., 355
TELEGDI–ROTH, K., 271, *295*
TERZAGHI, K., 6, 8, 9, 10, *18*, *83*, 226, *264*
THEOBALD, N., *188*
Theobald, N., 130
THOMAS, T. M., 436, *443*
TIETZE, E., *see* BEYER, A. et al.
TIETZE, E., *see* MOJSISOVICS, E. et al.
TIKHOMIROV, N. K., *416*
Tillius, A., 118
TINTILOZOV, Z. K., *see* MARUASHVILI, L. I. and TINTILOZOV, Z. K.
TOITOT, M., *see* GLANGEAUD, L. et al.
TOMMASINI, T., *see* FORTI, F. and TOMMASINI, T.
TONIOLO, A. R., 113, *128*
TREIBS, W., *83*
TRENER, G. B., 108, 115, *128*
TREVISIOL, G., *see* DA SCHIO, A. et al.
TRICART, J., 137, 142, *188*
Tricart, J., 130
TRIMMEL, 235, 259, *265*
TRIMMEL, H., *see* SCHAUBERGER, O. and TRIMMEL, H.
TROMBE, F., *188*
Trombe, F., 130
TUCÁN, F., 75, *83*

TURINA, L., *83*
TYCZYŃSKA, M., 333, 340

UFFICIO IDROGRAFICO MAGISTRATO ACQUE VENEZIA, 92, 96, 113, 114, *128*
UGRENOVIĆ, A., *83*

VADÁSZ, E., 268, 270, 280, *295*
VADÁSZ, E., *see* LÓCZY, L. et al.
VALVASOR, J. W., 2, *18*
VANCON, J.–P., 166, *188*
VANDEL, A., *188*
Vandel, A., 130, 181
VANDENBERGHE, A., *188*
VARDABASSO, S., 88, 90, 112, *128*
VARNEDOE JR., W. W., *see* TARKINGTON, T. W. et al.
VARSONOF'YEVA, V. A., 356, *416*
VAUGHAN, A., 423, *443*
VEATCH, J. D., *see* TARKINGTON, T. W. et al.
VENZO, G. A. and FUGANTI, A., 97, *128*
Venzo, S., 92, 113
VERDEIL, P., 161, *188*
Verdeil P., 130
VERSEY, H. R., 447, 448, *466*
VERSEY, H. R., *see* HOSE, H. R. and VERSEY, H. R.
VIDOVIĆ, M., *83*
VIGHI, L., *see* CRESCENTI, U. and VIGHI, L.
VÍLA, G., *188*
VILIMONOVIĆ, J., *83*
VIRLET, J., 2, *18*
VITALIS, J., *see* LÓCZY, L. et al.
VITÁSEK, F., 297, *325*
VLAHOVIĆ, V., 60, 66, *83*
VON BUBNOFF, S., *223*
VON KNEBEL, W., *223*
VORMAIR, F., 253, *265*
VUKOVIĆ, S., 35, *83*
Vulić, Z., 55

WAGER, L. R., 420, *443*
WAGNER, G., 258, *265*
WAHL, J.–B., *188*
WARING, G. A., 489, *505*
WARWICK, G. T., 419, 423, 433, *443*
Waterlot, G., 130
WATERLOT, G., *see* LEROUX, E. et al.
Watham, A., 430
WATSON, R. A., 497, *505*
WEBER, H., 214, *223*
WEIDENBACH, F., 219, *223*
WEIN, GY., 274, 280, *295*
WELCH. F. B. A., *see* GREEN, G. W. and WELCH, F. B. A.
WEYDERT, P., *188*

Weydert, P., 130
WHITE, CH. A., *18*
WIEFEL, J., *223*
WILLIAMS, V. H., see GROOM, G. E. and WILLIAMS, V. H.
WINKLER-HERMADEN, A., 252, *265*
WÓJCIK, Z., 335, *340*
WÓJCIK, Z., see GRADZIŃSKI, R. and WÓJCIK, Z.
WÓJCIK, Z., see KRASOŃ, J. and WÓJCIK, Z.
WÓJCIK, Z., see LIS, J. and WÓJCIK, Z.
WORM, G., *265*
WRABER, M., *83*
WRIGHT, R. M., 448, *466*
WRÓBLEWSKJ, T., see GRADZIŃSKI, R. and WRÓBLEWSKI, T.
WROOT, H. E., see KENDALL, P. F. and WROOT, H. E.

YAKUSHOVA, A. F., 356, *416*
YORKE, C., 432, *443*

ZANS, V. A., 445, 452, 463, *466*
ZAYTSEV, I. K., 356, 368, *416*
ZLATOVIĆ, D., see PAVLIN, B. et al.
ZOGOVIĆ, D., *83*
ZÖTL, J., 231, 233, 241, *265*
ZÖTL, J., see MAURIN, V. and ZÖTL, J.
ZÖTL, J., see BATSCHE, H. et al.
ZÖTL, J., see BAUER, F. et al.
ZÖTL, J., see BRANDECKER, H. et al.
ZOTTER, H., 494, *505*
ZUBČEVIĆ, D., *83*

Index

Aach Spring, 208
Abakansk Range, 403
Abkhaziya Range, 372
Abogy–Dzhe Cavern, 395
Abrskil Cavern, 372
Abruzzi, 87, 90, 122
Absorbent valleys, 143
Abtashi Range, 396
Acquafredda Cave, 116
Acquasanta Cave, 95, 125
Addura Cave, 95, 112
Adirondacks, 483
Adriatic coast, 71, 512
— insular and coastal region, 42
— Sea, 2, 20, 42, 46–48, 103, 118, 123
Aeration, zone of, 513
Aggressive water, 13
Aggtelek, 318
— karst, 267, 273, 274, 279, 280, 284, 286, 290, 291, 294
Agly River, 161
Agriculture, 75, 76, 260
Aisne, 143
Akburinsk karst, 398
Aksaya Basin, 396
Aksu River, 401
Alabama, 72, 472–474, 481, 484, 486, 490, 492, 493
Alaska, 12, 469
Alatyrsk–Gor'kovsk uplift, 364
Alaya, 395, 398, 399, 408
— Range, 397, 398
Alaysk Range, 399
Alb Plateau, 207, 208
Alburni Mountains, 110, 121
Aldan River, 387
Aldansk Shield, 393, 394
Aleysk karst region, 401
Alimini Lakes, 123
Alkalinity, 72
Allegheny Plateau, 469
— type, 486, 493
Allgäuer Alpen, 225, 230
Almagià, 95
Almaj Mountains, 342, 343
Alpe Campigne, 106
— di Siusi, 108
— Flavona, 108

Alpine chains, 136
— Dinaric mountains, 309
— folded zone, 358
— folding, 344, 345
— karst, 52, 167, 229–251, 309, 313, 314, 318, 319
— movements, 279
— orogeny, 29, 30, 510
— phases, 305, 345
— Range, 85, 115
— tectogenesis, 88, 90
— tectonic movements, 399
— type structure, 299
— units, 29
Alps, 1, 2, 27, 29, 52, 86, 87, 91, 93, 98, 105, 107, 116, 129, 133, 168, 170, 189, 211, 231, 234, 257, 258, 260, 263, 269, 309, 335, 511
Altay, 402
— karst, 401, 402
Alum Pot-hole, 428
Amankutan Grotto, 398
Amir–Temir Cavern, 398
Amur River, 407
Anabarsk Massif, 393
Anakopiysk chasm, 372
Andiysk Range, 371
Andossi Mountain, 106
Andrews Well, 440
Angara–Katangsk uplift, 390
— karst, 393
— Lena marginal trough, 392
— River, 389, 390, 392, 393
Antillean Island Arc, 445
Antro del Corchia, 95, 111, 124
Anuysk–Chuysk karst, 401
Aosta Valley, 116
Apennines, 85, 89, 92, 103, 105, 108–111, 115, 116, 121, 122
Appalachian cave type, 482, 483
— karst, 472, 495
— Mountains, 467, 469, 471, 481, 495
— plateaus, 482, 490–494
Apuanian Alps, 87, 92, 108, 111, 121
Apulia, 85, 87–89, 91–93, 101, 102, 122–124
Apuseni Mountains, 342, 344, 347
Aquitanian Basin, 136, 144–146, 160
Aradus, 19
Aral Sea, 387

Aravan karst, 398
— River, 397
Archi Cave, 95
Arco Cave, 95
Ardèche River, 154, 158
Ardennes, 133, 143
Arene Candide caves, 95
Ariège River, 161
Arize River, 161, 162
Arizona, 471, 473
Arkansas, 468, 472, 503
Armand aven, 150
Arma Pollera, 95
Armorican Massif, 133
Armyansk Highland, 368
Arno River, 108, 111
Arrangorena River, 164
Artois, 142
Ashiyakhta River, 401
Asia, central, 385, 386, 395–399.
Asiago, 98
Assyntian orogenesis, 192, 345
Aterno Valley, 122
Atlantic and Gulf Coastal Plain, 473–481
— Coast, 481
— Coastal Plain, 467, 471
Atlas, 511
Attian phase, 31
Aude River, 135, 161
Aurignac, 162
— Cave, 182
Aurunci Mountains, 110
Ausoni Mountains, 110
Austria, 225–265, 408
Austrian Catalogue of Caves, 250
— orogeny, 277
Austro-sub-Hercynian movements, 276, 277, 279
Aven Armand, 182
— de la Soeur near Sauve, 182
— de Marzal, 158
— d'Orgnac, 182, 129, 158
— du Mont Marcou, 129
Avène, Source d', 135
Aven of Castor, 176
— of the Nouguière, 178
Avens of Caldaire, 176
Axbridge, 440
Ay River, 380, 409
Azerbaydzhan, 371, 372
Azykhsk Cavern, 372

Babiseysk Cavern, 404
Bacteriological picture, 72
Baden-Württemberg, 190

Baikal folded zone, 405
— orogenic belt, 358
Bakony, 267, 270, 277, 278
— Mountains, 277, 282–284, 288–290
Balagansk Cavern, 392
Balakhoninsk Cavern, 364
Balaton Lake, 267, 268
Balck Lake, 50
Balenská Cave, 320
Balma di Rio Martino, 95
Baltic Shield, 359
Banat Mountains, 345, 346, 353
Baračeve pećine, 36
Baragal River, 401
Barbecue Bottom, 454, 455
Bare karst, 136
Barnet Well, 440
Bas Dauphiné, 168
Basin karst, 313, 320
Basis of corrosion, 6
Basque Country, 162
Basses Pyrenees, 164
Basso Carso, 95
— —, anticline 88
Battisti Cave, 95
Baubashat Massif, 397
Bavarian Forest, 205
Bay of Bakar, 55
Baysuntau Mountain, 399
Bear Cave, 318
Begovac Lake, 55
"Bei den Flüssen", 248, 250
Belanské Tatry, 298, 323
Belaya River, 366, 369, 370, 380, 381
— Stena Cave, 374
— Tissa River, 374
Belle Foux de Nans, 177
Belluno pre-Alps, 108
Beljanička pećina, 36
Belorusskaya Arch, 360, 361
Bel'skiy Trough, 378
Belyy Spoy Range, 380
Berda River, 402
Bergamo pre-Alps, 116
Bergisches Land, 198
Berici Hills, 98, 100, 101, 125
Bermuda, 465
Betalov Spodmol, 35, 36
Betharram Grotto, 162, 182
Bétoires, 142, 143
Bezdonnoye Lake, 381
Biambarska pećina, 36
Bichel'ya River, 389
Bidouze River, 164
Biely Váh, 304
Bifurto Abyss, 111

SUBJECT INDEX 533

Bigonoda Cave, 95
Bihor, 347, 348
— Mountains, 343, 345
Biochemical oxygen demand, 72
Biokovo Mountain, 32
Biraya River, 407
Birsk River, 366
Biryusa River, 389, 390, 404
Biscayne aquifer, 468
Biševo, 36
Black Forest, 205, 206
Black Hills, 473
Black River, 447
Black Sea, 46, 47, 50, 352, 372, 407
Blinovka River, 409
Bloška Planota, 47
Blue Mountains, 449
Blue Ridge, 481, 483, 484, 495
Böcc della Tuera Cave, 124
Bogovina pećina, 35, 36
Bogue and Elim springs, 463
Bohemian upland, 297
Bois de Monié, 154
— de Pailolives, 182
— de Paris, 154
Bojana River, 45, 46
Bologna Apennines, 115
Bol'shaya Usa River, 378
Bolshebnyy Labyrinth, 405
Bol'shoy Baritovoy Cavern, 397
Bol'shoy Kavkaz, 368
Borodinsk Cavern, 404
Boron'yevsk–Ternovsk region, 375
Borovnica, 47
Bosna River, 51
Bosnia, 3, 5, 27, 35, 48, 49, 61
Bossea Cave, 92, 95, 105, 121
Bossico Plateau, 106
Bouillidoux, 145
Bourbouillous, 166
Bourne River, 174
Bournillon subterranean river, 175
Bowle, 433
Bowl-shaped dolines, 212
Brač, 44
Bradano Valley, 103
Bramabiau Grotto, 150
Brembo Valley, 107
Brie, 137
Brina Caves, 37
Brudoux River, 172
Bründl Spring, 257
Buca del Diavolo, 95
Bucegi Massif, 344, 345
— Mountains, 343, 347
Buchkogel, 257

Buco Cattivo, 95
— del Corno, 95
— del Frate, 95, 124
— del Fuso Caves, 107
— della Bondaccia, 95
— della Volpe, 95, 107, 121, 124
— dell'Orso, 121, 124
— del Piombo, 95, 107
Bue Marino, 124
— — Cave, 95, 112
Bükk, 267
— Mountains, 273, 274, 276, 279, 280, 282, 286, 290
Bukovac pećina, 35, 36
Bulge Cave, 203
Bull Cave, 495
Bulukha River, 401
Buna, 70
Buranco Rampium, 95
Burgundy, 142, 164, 166
Busa del Geson, 99
— del Sciason, 95
Bus Coalghes, 95
— de la Bela, 95
— de la Lum, 95
— de la Lum abyss, 98
— de la Rana, 95, 99
— de la Scondurava, 95
— del Diaol, 95
— del Piatte, 95
— del Quai, 95
— di Remeron, 95
— di Tacoi, 95
Bussento caves, 95
— River, 121
Buško Blato, 14, 65
Buyg River, 399
Bybrook Spring, 458
Byrrang Mountains, 395

Čachtice karst, 320
Cades Cove, 495
Čaića ponor, 59
Calabria, 87, 108, 111
Cala delle Ossa, 95
Calanques, 177
Caledonian, 395
California, 473
Calvana Mountains, 111
Cambridge Valley, 463
Camera Cave, 95
Camonica Valley, 116
Campania, 90, 109
Campi, 109
Campobasso, 102

Campo Imperatore Depression, 110
— Lasca and Antrodoco, 109
Campu e Susu, 112
Canadian River, 472
— Rock, 439
Canale di Brenta, 99
Canterno Lake, 109
Caoügno de los Gouffres, 161
Căpăţîna Mountains, 343
Cape of Izenty–Aral, 387
— Palinuro, 111
— Santa Maria di Leuca, 104
Capivento, 104
Cappa, 103
Carbonate aquifers, 514, 515
Caribbean area, 12
Carlsbad Caverns, 467, 502
Carpathian karst, 309, 373
— nappes, 279, 280
Carpathians, 276, 297–300, 302, 303, 305, 310, 312–314, 318, 323, 342, 344, 345, 347, 348, 351, 352, 373–375, 511
Carpatho–Balkanian Belt, 29, 52
Carso, 2, 507
— proprio, 95
Cased wells, 515
Caspian Sea, 372
Castelcivita Cave, 95, 110, 121
Castel del Monte, 109, 110
Castellana caves, 95, 104, 124
Castel Tesino caves, 95, 108
Caucasus, 368–372, 410, 411
Caudano caves, 95, 105
Caumont caves, 143
Causse d'Aumelas, 154, 158
— de Blandas, 146, 150
— de la Selle, 154
— de l'Hortus, 154
— de Marvejols, 148
— de Mende, 148
— de Pompignan, 154
— de Rodez, 148
— de Sauveterre, 146, 149, 150
— de Severac, 146
— de Viols-le-Fort, 154, 158
— du Larzac, 146, 150, 154
— Méjean, 146, 149, 150
— Noir, 146, 148
Causses, 136, 141, 143, 154
— Majeurs, 144, 146, 149
— Mineurs, 144
Cavallone Cave, 92, 95, 110
Cave River, 462
Caverna Grande delle Streghe, 105
Caverns, 23, 35, 36, 510
Caves, 23, 35, 36, 234, 503, 510

— of Plitvice Lakes, 36
Cave Valley, 462
Cavottes Grotto, 166
Cayman Trench, 445
Ceahlău Massif, 343
— Mountains, 347
Cedar Sink, 497, 500
Cefn caves, 436
Celebes, 11
Central Alps, 91, 105, 116, 231, 234, 260
— Asia, 385, 386, 395–399
— Carpathian karst, 373
— European karst, 313, 314
— Styrian karst, 252, 253
— Virginia, 487
Cephalonia, 515
Ceremošnja, 36
Cerkniščica, 48
Cerkniško polje, 2, 46, 47, 48, 63
Cerovačke pećine, 35, 36
Československý kras, 298
Cetina River, 46, 50, 62, 64, 65
Cevennes, 135, 153, 154
Cèze gorge, 154
Chakyl-Kalyan Mountains, 398
Chalk karst, 367
Champigny, 143
Charysh River, 401
Cheddar Cave, 435
— Gorge, 434
— Lake Springs, 440
— Second Feeder, 440
Chemical composition, 291, 292
— constituents, 290
Cheposh River, 401
Cherek River Basin, 371
— — Valley, 371
Chernaya Cheremosh River, 374
— Tissa River, 374
Chiese, 107
China, 11
Chil'Ustunsk Cavern, 397
Chirchik River, 396
Chivchinsk Mountains, 373
Chiya River, 401
Chloride concentration, 72
Choč Mountains, 308, 310, 320
— nappe, 300–306
Chon-Kyzylsu River, 396
Chouruns, 169
Choux fleurs, 181
Chusovaya River, 379, 381
Čierny Váh, 303, 304
Cimmerian movements, 194, 276, 279
— orogeny, 277, 281
— phase, 30

SUBJECT INDEX 535

Ciscarpathian karst, 373
— Trough, 366, 373
Ciscaucasus, 368, 372
Cisuralian Trough, 359
Cisural Trough, 376
Ciucash Massif, 343
Clamouse Grotto, 158
Clarendon Block, 445, 447–449
Classification, 507, 511, 512
Clear Spring, 464
Climate, 131–132, 148, 154–156, 313, 333, 349, 405, 502, 508, 509
Climatic factors, 310
Clints, 429
Clos del Fayoum, 178
Cloups, 144
Clues, 178
Coalesced dolines, 491
Cockpit, 3
— Country, 6, 453, 460, 463, 464, 469, 517
— karst, 460, 461
Codru-Moma Mountains, 343, 345
Coefficient of absorption, 408
Coefficients of correction, 285
Coefficient of underground flow, 410
Col del Pass Abyss, 95
Collepardo Cave, 95
Colombi Cave, 95
Colt Park, 419
Commission des Phénomènes Karstiques du Comité National de Géographie, 130
— de Spéléologie, 130
— de Spéléologie du Club-Alpin Francais, 129
— of Karst Phenomena, 11
Como Lake, 107, 120
Complete karst, 314
Conclusions, 507–518
Consumption of potassium permanganate or biochromate, 72
Corno d'Aquiglio, 99, 100
Correction percentage of precipitation, 285
Corrosion, 507, 510
Corrosional intensity, 40
— peneplain, 6
— plains, 10, 11, 40
Costalta Cave, 95
Cottian Alps, 105
Country of lapiez, 169
Courry la Cocalière, 158
Covered karst, 9, 309, 313, 121, 512
Coves, 491
Covoli di Costovizza, 125
Covolo della Guerra, 95
Cracow Cave, 338
Crawford Upland, 501
Cres, 44

Creuse River, 166
Creuses, 142
Crimea, 355, 367, 368, 398, 406, 407, 408
Cristalli Abyss, 95
Crna Gora, 65
Črni Kal, 35, 36
Crno jezero, 50
Croatia, 6, 26, 27, 35, 58, 61
Croatian karst, 20
— Littoral, 62
Crvena stijena, 36
Cuba, 11, 445
Cuesta landscape, 205
Cunardo caves, 95, 121
Cutaneous karst, 167
Cyclic evolution of relief, 9
Czarna Cave, 330, 335
Czechoslovakia, 297

Dabarsko polje, 49
Dachstein, 225, 229, 231, 235, 236, 237, 238, 241, 243–245, 258
Dalmatia, 4, 48, 49, 61
Danube, 2
Dan Yr Ogof, 436
Dargilan Grotto, 150
David Cave, 144
Deep karst, 9, 15, 219, 512
Degraded cockpit karst, 462
Dell'Oca, 124
Demänovské Cave, 303, 304
Demänovské jaskyne, 320
Demoiselles Grotto, 158, 182
Denuded karst, 509
Derbyshire, 419, 420, 425, 432, 433, 439, 441
Devero Valley, 105
Devon, 438
Diavolo Cave, 95, 124
Dimbau Mountain, 347
Dinaric karst, 1–3, 5, 13, 15, 16, 22, 25, 33, 46, 55, 58, 59, 62, 67, 74, 75, 507
— Mountains, 25, 27, 29–31, 52, 54, 60
— region, 71, 72, 225
— units, 29
Dinarides, 86, 88, 309
Dikobraz Cavern, 397
Diois karst, 167, 174, 175
Dissolution caves, 141–143
Divljakovačka pećina, 36
Divnaya Cavern, 404
Div'ya Cavern, 379
Djulin ponor, 34
Djurkovina, 36
Dnestr River, 366
Dnieper–Donets Basin, 367

Dnieper River, 362
Doberdò Plateau, 97
Dobra River, 34, 40
Dobrogea, 342–345, 349, 352, 353
Dobšina Ice Cave, 318
Doline, 6, 507
— caves, 435
Dolines, 16, 33, 237, 435, 503, 509
— bowl-shaped, 212
Dôme du Coutach, 154, 158
Domica caves, 318
Don Figuerero Mountains, 449
Dorog, mines of, 289
Dossi upper Cave, 95
Doubs River, 166
Dourbie gorge, 150
— River, 146, 150
Doux, 145
Drachenhöhle, 253
Dragone Cave, 95
Dragon's Cave, 253, 254
Drauzug, 257
Dragovskiy Menchul Mountain, 374
Dretulja River, 55
Drina River, 62
Dripping Springs, 497
— — escarpment, 496
Drugeon River, 166
Druzhba Cavern, 374
Dry Harbour Mountains, 460, 464
Dry valleys, 35, 142, 509
Dry valley systems, 423
Duanvale, 463
Dubačka pećina, 36
Duboki do, 35
Dunántuli Középhegység, 267
Dunn's River, 458
Durmitor Mountains, 27
Dwarf holes, 213
Dye tests, 59, 176, 178
Dzhungarsk Alatau Mountain system, 395

Ease Gill system, 430
Eastern Alps, 91, 107, 116
Eastern Mountain Mass, Jamaica, 445
Easton, 440
Eastwater Swallet, 440
Eaux Chaudes Grotto, 162
Eaux d'extravasement, 150
Economic karst problems, 260
Ecrêtement de crue, 183
Edwards Limestone aquifer, 468
— Plateau, 468, 473, 503
Egau Spring, 208
Eggegebirge, 208, 209, 210

Eifel, 190, 198, 208
Eisgrube, 255, 256
Eisriesenwelt Cave, 236
El Fontanon Cave, 124
Elm Lake, 242
Elwy Valley, 436
Embuts, 177
Emposieux, 166
Engineering geology, problems of, 220
England, 418, 420, 436
Enns River, 229
Enrico Revel Abyss, 111
Epicontinental karst areas, 511
Erbezzo-Bosco Chiesanuova, 99
Erdfalls, 239
Ernici region, 110
— Mountains, 109
Erosion concretion mechanisms, 181
Estonia, 360
Estonian Archipelago, 360
— Plateau, 360
Etagement, 145
ETG climate, 310
— European karst, central, 313, 314
Evaporation, 286, 336
Evaporite karst, 373
Evapotranspiration, 285
Extravasating water, 150
"Eyes", 46, 52

Fagarash Mountains, 343
Farneto Cave, 95, 116
Fate Cavern, 95
Fatničko polje, 49
Fédération Française de Spéléologie, 129
Fella River, 116
Fell Beck, 430
Feltre pre-Alps, 108
Fergana Range, 399
Fergansk Range, 397
Fern Cave, 490
Fichtelgebirge, 192, 205
Fidelya Grotto 404, 405
Fisht–Oshten–Lagonaki Massiv, 369, 370
Fiumelaffe Cave. 95, 120, 124
Fleischhacker Cave, 254
Fleurs de gypse, 181
Flint Ridge Cave, 499, 500
— River Valley, 476, 477
Florida, 467–469, 471–474, 477–481, 515
Floridan aquifer, 468
Flow, ratio of, 287
Fluviokarst, 13, 315, 512
Fontabelle Lagoon, 464
Fontana Buia Cave, 111

SUBJECT INDEX 537

Font de Sorps, 178
Foresta Umbra area, 103
Forestry, 74, 75, 260
Fossa di Pranu Piastru cave, 124
Fossil dolines, 212
— karst, 140
Fountain of Vaucluse, 514
Foux de Dardenne, 177
France, 129–188, 408, 511, 514
Franconia, 190, 205, 206
Franconian Alps, 190, 215–219
— Forest, 205
Franzei Cave, 95
Frasassi Cave, 95
Frauen Cave, 254
Frauenmauer–Langstein–Tropfstein Cave, 236
Free carbon dioxide, 72
Friulian pre-Alps, 89, 93, 98, 115
Fumane Valley, 99
Funés valleys, 108
Funnel-shaped dolines, 212
Fužine, 39

Gabauer Spring, 251, 252
Gacka River, 6, 55, 62
Gacko polje, 6, 10
Gailtaler Alpen, 225, 257
Garagai of Gaspard de Besse, 177
Gardiole, 158
Gardons gorge, 154
Gargano, 88, 89, 91, 93, 101, 103, 123
— Plateau, 103
Garonne River, 135, 161
Garrigues, 153
Garza River, 107
Gassenlandschaft, 239
Gatačko polje, 48, 49
G. B. Trener Cave, 95
Geldloch, 235
Gelo Cave, 95
Gemeride nappes, 303
Gemerides, 299, 300–302, 304–306
Genesis of karst features, 507, 508, 510
Genovese Cave, 95
Geochemical studies, 518
Geologic setting, 474–476
— structure, 516
Geophysical methods, 56–58, 518
Georgia, 372, 468, 471, 472, 474, 476–478, 481
Geosynclinal karst areas, 510, 511
Gerecse, 278
— Mountains, 277
German cuesta landscape, 205, 206
— hardness, 216
Germany, 189–223, 508

Ghyben–Herzberg principle, 123, 469, 515
Gill Hole, 430
Gineppa Spring, 461
Gissarsk Range, 399
Giurgeu Mountains, 343
Giusti Cave, 95, 125
Goatchurch Cave, 435
Gojak, 59
Golema peštera, 36
Golgo Abyss, 112
Golubac, 53
Golubarnik Cave, 58
Gombasek Cave, 318
Gordale Scar, 427
Gorge du Verdon, 182
— l'Ardèche du Tarn, 182
Gorjanci, 276
Gorski kotar, 39, 54, 61, 62
Gottesacker Plateau, 211, 259
Gouffre Berger, 129
— de la Pierre Saint-Martin, 129
— de Padirac, 182
— des Corbeaux, 161
— Jean Nouveau, 129
— Bouchet, 129
Goule de Foussoubie, 129
Gours, 145, 181
Gračanica River, 65
Grančarevo, 65
Grand Dome, 144
Grande Chartreuse, 170
— — Massif, 169
Grandioznyy Grotto, 405
Grand Roc Grotto, 182
Grands Causses, 144, 148, 153
Gran Sasso d'Italia, 109
Gran Sasso–Maiella Mountains, 110
— — Massif, 110
Grapčeva pećina, 36, 37
Grassy Cove, 492
Grave, 104
Gravine, 104
Grazer Bergland, 252, 254, 257
Great arch, 175
Great Britain, 417–443
Great Plains, 467
Great Plan de Canjuers, 178
Great Smokies, 481, 483
Great Smoky Mountains, 494
Great Valley, 483–487, 495
Greben Cave, 374
Greece, 512
Greenbrier county, 491
Green Mountains, 483
Green River, 497
Grikes, 429

Grosshöhlen, 234
Grotta Azzurra of Capri, 95, 111, 124
— — of Palinuro, 95
— di Smeraldo, 111
— Gigante, 95, 97, 124, 125
— Grande di Ciolo, 95
— Maggiore of Pugnetto, 105
— Nuova di Villanova, 95
— Vecchia di Villanova, 95
Grottone di Biddiriscottai, 95
Groundwater circulation, 23
— development, 67–68
Guadalupe Mountains, 502
Guattari Cave, 95
Guglielmo Cave, 95, 107
Guiers River, 170
Gulf Coastal Plain, 467, 468, 470
— of Carpiane, 177
— of Finland, 408
— of Mexico, 471, 478
Gumbeysk–Suunduksk karst, 384
Gypsum–anhydrite karst, 389
— karst, 115, 212, 215, 351, 360, 361, 364, 366, 367, 387, 388, 390, 391, 394, 399, 486

Hagengebirge, 236
Hăghimash Massif, 343
Haiti, 445
Halbkarst, 8, 9, 13
Hammerbach, 255, 257
— Spring, 254, 256
Hamshire Valley, 460
Hardness, 72, 216, 284, 290–292, 336
Harz foreland, 201
— Mountains, 190, 192, 199, 200, 202–204, 209, 210, 213, 218
Haute Garonne, 161, 162
Haute Savoie, 167
Hawaii, 466
Hellenic Mountains, 511
Hellenides, 309
Helvetian nappe system, 258
Hemmer Spring, 254
Hérault, 144, 182
— gorge, 154
— River, 153, 154, 160
— valley, 156, 157
Herbert Hills Creek, 488
Hercegovina, 3–5, 35, 48, 50, 61, 62, 65
Hercynian folds, 385, 395
— massifs, 133
— orogeny, 30
— peneplain, 135, 136
— phase, sub-, 277, 299
Hermanec Cave, 320

Herm Grotto, 161
Hessian Depression, 190, 205
High Alpine karst, 167, 169
Highimash Massif, 346
High karst region, 46
Highland of Paderborn, 208
— rim, 472, 504
High Pyrenean karst, 162
High Tatra Mountains, 300, 308, 327, 329, 330, 335
Hinterau Valley, 250
Hochifen area, 258
— Massif, 258
Hochkönig, 225, 229, 244, 246
Hochschwab, 225, 229, 236, 261
Hochstradner Niveau, 253, 254
Hoher Ifen, 225
Höllengebirge, 229
Hölloch, 259
Höll Spring, 251
Hollybrook, 440
Holokarst, 8, 509
Holy Cross Mountains, 327–329, 334, 336–338
Honexhurst Borehole, 440
Horse Cave, 497
Horvat Cave, 58
Hron River, 309
— Valley, 308, 321
Hruštovača, 36
Huda luknja, 36
Hudson–Champlain Lowland, 495
— River, 481, 484
Hums, 31, 39, 509, 510
Hungary, 267–295, 318
Hunsrück, 198
Hušnjakovo, 36, 37
Hutton Roof Crag, 431
Hvar, 43, 67
Hydraulic gradient, 517
— permeability, 512
Hydro-electric power plants, 261
Hydrogeological concepts, 19–24
— conditions, 352, 438, 439
— consequences, 336, 337
— problems, 180
Hydrogeology, 140, 477–480, 512–518
Hydrographic zones, 7
Hydrography, 313
Hydrologic conditions, 284, 493
— regime, 488
Hydrology, 451, 512–518
Hydrorelief, 286, 289

Ice caves, 169
Idaho, 473

Idrian thrust fault, 63
Igues, 144
Ile de France, 143
Ilim River, 393
Imeretiya Range, 372
Imperfect karst, 313, 314, 318, 319
Ina River, 402
— — Basin, 403
Indiana, 3, 467, 471, 472, 495, 496, 501
Indo-China, 11
Indoga River, 400
Inek Magarasa Cavern, 372
Inferniglio Cave, 95
Infiltration, ratio of, 285
Infinito Cave, 95
Ingual River, 359
Inner Carpathians, 298–300, 302, 303, 305, 306, 308
Inn River, 229
Interior Lowlands, 471, 495
— Low Plateau, 467
— valleys, 462, 463
Intermediate karst, 314, 319
International Geographical Union, 11
Interrupted caves, 435
Intrusion of saline water, 69
Ionian Sea, 123
Irena River, 366
Irkutsk oblast, 390
— piedmont trough, 390, 391
— River, 390
Irregular salt leaching, 204
Irrigation, 76
Iseo Lake, 107
Iskanderdar'ye River, 398
Isonzo River, 119
Ister, 2
Isthme Durancien, 131
Istria, 35, 90
Istrian Plain, 40
Italian Karst, 85, 86, 88, 91, 95, 96, 101
Italy, 85–128, 408
Ivdel' revealed karst cavities, 382

Jama, 15, 507
— at Raspor, 35
— Dubinka, 35
Jamaica, 3, 11, 439, 445, 469, 514, 517
Jamas, 33, 509
Jasovská Cave, 318
Jaur, source du, 135
Java, 10, 11
Jazben, 35
Jelšansko polje, 43, 67
Joint caves, 212

Jonte Gorge, 150
— River, 146, 150
Joppa Ridge, 500
Jordan, 511
Julian Alps, 27, 52
Jura, 129, 136
— Mountains, 2, 164
Jux, 199

Kaibab, 473
Kal'ya River, 382
Kalymsk Massif, 405
Kama River, 365
Kamenica, 16
Kamenka River, 409
Kamnik Alps, 52
Kan-i-Gut Cavern, 398
Kansas, 471
Kapova Cavern, 382
"Kar", 2
Karabakhsk Range, 372
Kara-Chumysh River, 402
Karatau Range, 386, 396
Karatyube Massif, 398
Karawanken, 27, 225, 257, 258
Karlyuksk Cavern, 399
Karnische Alpen, 225, 257
Karren, 33, 485, 488, 509
Karst barré, 432, 508, 511
— basins, 108
— cove, 491
Karstgassen, 231
Karst horsts, 320
Karstification, 283, 507, 513, 514
— phases, 160, 161, 168
Karst massifs, 109
— of Austria, 225–265
— of Czechoslovakia, 297–325
— of France, 129–188
— of Germany, 189–223
— of Gorizia, 95
— of Great Britain, 417–443
— of Hungary, 267–295
— of Italy, 85–128
— of Jamaica, 445–463
— of Monfalcone, 88, 90, 95, 97
— of Poland, 327–340
— of Rumania, 341–353
— of the basins, 314
— of the Bohemian Massif, 313
— of the horsts, 314
— of the monoclinal ridges, 313, 314, 319
— of the United States, 467–505
— of the U.S.S.R., 355–419
— of Trieste, 2, 3, 88, 95, 97, 118

Karst of Yugoslavia, 25–83
Karstology, 149
Karst plains, 510
— plateaus, 349
Karsts barrés, 156
— cutanés, 167
— nus, 136
Karst soil, 75
— uplands, 459
Karwendel Mountains, 225, 229, 247–249
Katerloch, 254
Kattaopa Cavern, 398
Katuna River, 402
Katunsk karst, 401
Kazakhstan, 358
Kegelkarst, 10, 16, 211, 460
Kel'-Ketchkhen sink, 371
Kentucky, 3, 467, 468, 471, 472, 483, 490, 495, 496, 498, 501
— -Indiana area, 495
— karst, 500
Khantaysk medial massif, 408
— — Rybninsk uplift, 388
Kharonorsk brown-coal region, 400
Khudugunsk Cavern, 392
Kinderlinsk Cavern, 402
Kirenga River, 392
Kitoy River, 390
Kiya River, 403
— — basin, 403
Kizelovsk Cavern, 380
Kizel River, 379
Klammbach spring, 250
Klammkalk zone, 225
Klikchinsk Range, 399
Kluftkarren, 232, 236, 237, 249, 254
Kokkiy River, 396
Kolvo-Vishersk karst, 379
— — region, 379
Komarnica River, 50
Kopet-Dag, 358
Koppenbrüller Cave, 236, 243
Korana River, 40, 41
Korčula, 44
Kos'va River, 379
Kotliči, 52, 438
Kotuy–Nuku Range, 400
Koyvo–Serebryansk region, 381
Kraljica pećina, 36
Kras, 2, 25, 507
Krasniye caverns, 368
Krest Cavern, 403
Kristal'naya Cavern, 367
Križna jama, 35, 36
Križná nappe, 300–306
Krka River in Dalmatia, 40, 45, 46

— — in Dolenjsko, 40
Krš, 25, 507
Kubinsk Cavern, 404
Kućaj–Beljanica belt, 53
Kudepsta River, 372
Kugitangtau Range, 399
Kulaya River, 406
Kuloy River, 363
Kuloysk Plateau, 407, 408
Kungurs Cavern, 355
Kupa River, 40, 54, 62
Kupica, 54
Kupreško polje, 48
Kurutkan River, 401
Kuznetsk Alatau Range, 403
— — Alatausk karst province, 403
Kyl'su Canyon, 396
Kyrktau Plateau, 396, 398
Kyr-Kyr River, 396
Kyzylkum Desert, 386, 387

Labinian phase, 30
Labouiche River, 162
Lacca del Roccolino, 95
Lacus Timavi, 118
Lagonaki Plateau, 370
La Grand Foux, 177
Lake Apopka, 477
— Baikal, 387, 392, 407
— Balaton, 276, 278, 288
— Beloye, 363
— Champlain, 481, 484
— Ertso, 372
— Lacha, 363
— Maggiore, 87
— of Canterno, 124
— of Doberdò, 95
— of Matese, 124
— Onezhskoye, 363
— Tarpon, 469
— Teletskoye, 402
— Yelm, 359
La Madeleine Cave, 181
Laman Abyss, 95
Lame, 104
Lamprechtsofen cave, 236
Lanches, 166
Land use, 440–442
Languedoc, 129, 131, 136, 154, 182, 183
Languedocian coast, 131
— karst, 153, 181
Lanzo Valley, 105
Laos, 514
Lapies, 236, 237, 506
Laramian movements, 278, 348

— orogeny, 277
— phase, 30
Laramide system, 449
Lascaux Grotto, 182
Lathkill River, 433
Latium, 108, 110
— Campania zones, 87
Latvian saddle, 360, 361
Laubram Batch, 440
Laugerie Basse, 182
Laurel Creek Cave, 486
Laurin Spring, 255, 256
Lausitz, 192
Lavarone Plateau, 107
La Vettica Abyss, 95
Lavognes, 150
Laysu River, 399
Leaching caves, 212
Lebanon, 3, 511
Lechkhumi Range, 372
Lechtaler Alpen, 225, 230
Lecias, 162
Le Coppe at Campo Imperatore, 110
Ledena pećina, 36
Ledenica, 36
Ledro Valley, 115
Lena River, 387, 392, 407, 408
Leningrad oblast, 360
Lepini, 110
— - Aurunci Mountains, 110
Lepontine Alps, 105
Les Abysses, 183
Les Barronies, 174
Lesina Lake, 123
Lespugne grottos, 162
Lessini area, 122
— Mountains, 92, 99
— Plateau, 98
Lete in Campnia, 124
L'Evêque Fountain, 178
Lexington–Nashville plains, 472
— plains, 504
Lez River, 156, 158, 182
Ličanka, 61, 62
Lička Jesenica River, 55
Lienzer Dolomiten, 225, 257, 258
Lika, 26, 35, 57
— River, 40, 45, 58, 62
Limestone karst, 211, 215
Limy plateau, 345, 346
— ridges, 346
Lippe Spring, 209
Lipska pećina, 36
Liptov Basin, 308
Lirou River, 156
Lithostratigraphic aspects, 267, 327, 341–344

Lithostratigraphy, 130, 136, 137, 148, 153, 160, 164, 167
Lithuania, 361
Little Carpathians, 298, 308, 310, 320
— Sinking Creek, 497
Livanjsko polje, 14, 64, 65
Livenza River, 120
Ljubljanica River, 40, 62
Ljubljansko barje, 47
Lluidas Vale, 465
Logaško polje, 47
Lokvarka, 62
Lokvarska pećina, 36
Lombardian pre-Alps, 89, 93, 98, 106, 121
Lombardy, 106, 115, 120
Lombrive Cavern, 161
Longwood Swallet, 440
Lorraine, 137, 143
Lost River, 501
Loško polje, 47
Lot River, 146
Lotru Mountains, 343
Loue River, 166
Lows, 432
Low Tatras, 297, 298, 300, 303, 308
Luberon karst, 167
Lublin upland, 335
Lu Marmuru Cave, 124
Lumini Plain, 107
Luppa swallow hole, 95
Lurbach, 255–257
Lur Grotto, 254, 255, 256
Lyakan Cavern, 397
"Lyustra", 404

Macedonia, 35, 37
Maddalena Mountains, 106, 107
Madonie Mountains, 112
Madonna di Praia a Mare Cave, 95
Magian River, 397
Maiella Massif, 110
Main Block, Jamaica, 445, 449
Mainz Basin, 190
Makarovec, 36, 37
Malá Fatra, 298, 300, 308, 309, 320, 323
Mala Karlovica, 35
Malaya Chuya River, 392
— Ugol'ka River, 374
— Usa River, 378
Malham Cove, 427
— District, 431
— Tarn area, 430
Malyy Kavkaz, 368
— Khingan Range, 407
Mammoth Cave, 467, 472, 497–501

Mammoth Cave National Park, 501
— — Plateau 495–497, 499–501
Mammut Cave, 231, 235, 236, 243–245
Manaccore Cave, 95
Mangyshlak–Ustyurtsk karst, 386, 387
Manifold-Dove, 433
Manin Valley, 322
Manor Farm Swallet, 440
Mansfelder Schlotten, 202
Mansfeld Trough, 202
Maona Cave, 95
Maramuresh, 344
— Mountains, 343
Maratea Cave, 95
Marches, 87, 116, 122
— region, 108
Marginal corrosion, 10, 12, 13, 15, 16
— denudation, 6
Marguareis Massif, 105
— Mountain, 105
Maritime Alps, 105, 179
— pre-Alps, 105
Martha Brae River, 463, 464
Maryland, 484, 487
Massa Marittima, 111
Massif Central, 133, 146, 167
— de la Fage, 154
— de Logrian, 154
— du Dic St-Loup, 154
— du Thaurac, 154
Matese Group, 110
— Mountain, 110
Matlock Valley, 433
Mauria Pass, 116
Mecsek, 267
— Mountains, 275, 280–284, 291, 292, 294
Mediterranean geosyncline, 26
— karst, 9, 136, 310, 313
— Sea, 511
— types of karst, 7
Medvednica, 276
Medvezh'ya Cavern, 379
— Gora, 359
— pećina, 36
Megreliya Range, 372
Mehedintzi Mountains, 343, 347
— Plateau, 343, 353
Méjean causse, 150
Mellte River, 436
Mendip area, 419
— Hills, 424, 433
Mendips, 423, 424, 425, 433, 438–440
Meraviglia Cave, 95
Merokarst, 8, 509
Metaliferi Mountains, 343, 344, 346, 347
Methods of investigation, 517, 518

Mexico, 11
— Cave, 463
Mezen River, 363
Mezzogiorno Cave, 95
Middle-European karst, 9
Middle Polish uplands, 329, 332, 333, 336
Mietusia Cave, 335
Mill Hole, 497, 498
Minervois, 160
Minnonica Well, 95
Minor Carpathians, 302, 309
Mirna River, 45, 46
Mitchell Plain, 501
Missouri, 468, 472, 503
Mixed corrosion, 15
Mlynka Cavern, 367
Mocho Mountains, 460
Močiljska pećina, 36
Modulus of underground flow, 405, 410
Moesian Shield, 344
Mogotes, 11, 511
Mokrička jama, 35, 36
Moldavian Plateau, 344, 345
Molochnyy Kamen Cave, 374
Monard Grotto, 177
Moncenisio karst area, 115
— Pass, 116
Monfalcone karst, 119
Monginevro–Ambin area, 116
Mongolo–Okhotsk folded region, 405
Monrupino Uvala, 97
Montagne de Berg, 154
— de la Gardiole, 154
— de la Sellette, 154
— de la Sérane, 154
— Noire, 133, 135, 160
Montana, 471, 473
Monta Alben, 106
— Albo, 112
— Argentario, 111
— Baldo Group, 108
— Campo, 107
— — dei Fiori, 121
— Cancervo, 106
— Catria 87, 108
— Cavallo, 120
— Cetona, 111
— Cronio Cave, 125
— Cucco Cave, 95
— Donato hills, 116
Montego Bay, 447, 449
Monte Grande, 101
— Montea, 87, 108
— Morello, 111
— Mottolone, 101
Montenegrinean phase, 30

SUBJECT INDEX

Montenegro, 35, 36, 50, 60, 65
Monte Palosso, 107
— Part, 98
— Pastello, 99
— Pellegrino caves, 112
— Pernice, 107
— Pisano, 111
— Sacro, 103
— San Gottardo, 101
— San Martino, 106
— Sarri, 112
— Sei Busi, 97
— Spigno, 103
— Tambura, 111
— Tesoro, 99
— Tomba, 99
— Tre Crocette Cave, 95
Montello Hill, 91, 114, 115
Monterosso Rise, 95
Mont Luberon, 174
Montpellieran karst, 157, 158
Montpellier-le-Vieux, 182
Mont Ventoux, 174
Morača River, 62
Moravian karst, 2, 297
Morecambe Bay district, 419, 420, 422, 424, 425, 431, 436
Mornova Zijalka, 35, 36
Morocco, 408
Morphological aspects, 313
— concepts, 1–18
— consequences, 333–336
— features, 282, 284, 341–351, 425–436
Morvan, 137
Moscow Depression, 362
— tectonic basin, 361–364
Mosors, 509
Mount Mitchell, 481
— Tesla, 347
Moustiers caves, 181
Mouth River, 458, 465
Mrežnica River, 40
Mrózna Cave, 338
Msta River, 362
Muches, 143
Münsterland Basin, 208, 209
Murán karst, 309
— Plateau, 303, 305, 306, 308, 314
— Platform, 302
Muránska planina, 314
Murge, 88, 91, 93, 104
— of Salento, 90
— plateaus, 101, 103
— region, 89, 103, 104
Murg River, 206
Mur Valley, 256

Nadezhd Grotto, 405
Naryn River, 396, 397
Narynsk Range, 396
Nashville, 504
— Basin, 468
— – Lexington plains, 468
Nassau Mountains, 463
— Valley, 462
Natural shafts, 135
Neretva River, 40, 45, 46, 62, 70
Nettuno, 93
— Cave, 95, 124
Nevada, 473
Nevesinjsko polje, 49
New England, 481, 483, 484
New Jersey, 484
New Mexico, 467, 471, 472, 501
New York, 472, 483, 484
Niaux Cavern, 161
Niedźwiedzia Cave, 330, 334
Nikšićko polje, 60, 65, 66
Nive River, 162
Nixsee Basin, 202
Nizhnaya Tunguska River, 388
Nizke Tatry, 303, 309, 310, 320, 323
— — Mountains, 301
Njivice, 35, 36
Non Valley, 107
Norcia Basin, 122
Nördliche Kalkalpen, 225
— —, see Northern Limestone Alps
Normandy, 143
North Alpine karst, 229–251
North Carolina, 471, 473, 474, 479, 481
Northeast Range, 272–274, 279, 280, 282–284, 290, 291, 294
Northern karst, 9
Northern Limestone Alps, 225–230, 234, 235, 258, 259
Northern types of karst, 7
North Sudetic Basin, 329, 336–338
Nouveau Jean aven, 176
Novakova pećina, 36
Novoselitsk–Voditsk karst, 375

Obodska pećina, 36
Obrh, 48
Ofanto Valley, 103
Ogliastra plateaus, 112
Ogulinska Dobra River, 59
Ohio, 483
Ojcowski National Park, 338
"Oka", 46, 52
Oklahoma, 468, 500
Oksko–Tsninsk Arch, 364

Oleniy Island, 359
Oliero Caves, 99, 124
Omon River, 400
Onego-Severo-Dvinsk plateau, 407, 408
— — — watershed, 409
Onezhsk Peninsula, 363
Opreno Cave, 125
Optimisticheskaya Cavern, 367
Oregon, 470
Ore Mountains, 312
Orgaleysk Cavern, 392
Oriente Deept, 445, 449
Oshskikh Mountains, 397
"Os'minog", 404
Ottuk River, 397
Oubliettes de Gargas, 162
Outer Carpathians, 298, 299
Ozark Highlands, 471
Ozark plateaus, 468, 503
Ozarks, 472
Ozernaya Cavern, 367
Ozernyy Grotto, 405

Paderborn Plateau, 209
Pader Spring, 209
Pădurea Craiului Mountains, 343, 345, 348, 353
Pag, 43, 68
Palaeohydrology, 513, 515
Palaeokarst, 141, 436
"Paläotraun", 245
Palazzese di Polignano Cave, 95
Palosso Mountains, 106, 107
Pamir–Alaya Mountain, 397
Pamirs, 358, 395
Pandivera upland, 360
Pannonian Basin, 2, 29, 31
Pantena Valley, 99
Paris Basin, 129, 135–138, 142–144, 153, 164, 167
Parmelan Plateau, 169
Parska Golobina, 35, 36
Pashiysk Cavern, 380
Pashiysk–Chusovsk district, 380
— karst, 379
Pàstena Cave, 95, 110
Pau Cave, 162
— River, 162
Pay-Khoy karst, 381
—— Range, 376, 378
Pays des Charentes, 145
— d'Othe, 143
Pech-Merle-Marcenac caves, 145
Pechora River, 379
Pecos River, 501, 502

Pecos River Valley, 467, 471, 472, 501
Pécina below Jerino brdo, 36
— nearLičko Lešće, 36
— Velike Paklenice, 36
Pederobba Cave, 95
Peloritani Mountains, 112
Pelse Plateau, 108
Peniny, 322
— Mountains, 321
Pennines, 425, 433
Percées, 161
Pennsylvania, 471, 481, 484, 487, 490, 495, 501
Pennyroyal, 495, 497, 501
— Plain, 472, 496, 497, 500
Perched karst, 149
Perciata Cave, 95
Peredovyy Range, 370
Pergelisol, 137
Peri-Adriatic overthrust, 89
Périgord, 145
Pershani Mountains, 343
Pertosa Cave, 95, 121, 124
Peruča, 48, 64
Peštera, Bela voda, 36
— Dona Duha, 36
Petnička pećina, 36
Petrefakten Cave, 231, 243
Pfalzian phase, 30
Phases of karstification, 137–140, 165, 210, 281, 282, 324, 325, 330–333, 349, 450
Phreatic zone, 513
Piaggia Bella, 105
Pian del Lago, 111, 124
Piani, 103, 109
— Calascio, 109
— Castelluccio, 109
— Chiano, 109
— di Artavaggio, 106
— di Bobbio, 106
— di Colfiorito, 124
— of Carseolani, 109
— of Castelluccio, 110
— of Colfiorito, 109
— of Cornino, 109
— of Rocca di Cambio, 109
— of Santa Scolastica, 109
— Prosciuta, 109
— Tagno, 109
Piano Canale, 103
— Ceresaldi, 103
— del Dragone, 121
— di Arcinnazzo, 109
— Moltigno, 110
— of Castelluccio, 122
— San Martino, 103
— San Vito, 103

SUBJECT INDEX 545

Piatra Craiului Massif, 343, 346
— — Mountains, 350
— Mare, 347
— — Massif, 343
Piave River, 116
Picardy, 142, 143
Piedmont, 492
Pieniny Klippen Balt, 329
Piercys Mill Cave Spring, 493
Pietrasecca swallow-hole, 95
Pietra Selvaggia Abyss, 95
Pieve di Cadore, 116
Piezometric map, 475
Pinega River, 363
Pisana Stina, 36
Pišurka, 36
Pits, 33, 488, 509
Pitted dolines, 16
Pivka Basin, 46, 47
— sinking river, 46
Plain of Gacka, 40
— of Lika, 40
— of northern Dalmatia, 40
— of St. Maurice de Navacelle, 150
— of the Carnac, 150
— of Una and Korana, 40
— of Zadvarje, 40
Plains, 5, 40
Plan d'Aups, 177
— de Bauduen-Majastre, 178
— de Canjuers, 179
— de Caussols, 178
Plans du Verdon, 179
Planinsko polje, 46, 63
Plateau de Lussan, 154
— des Gras, 154
— d'Uzès, 154
— karst, 313, 314
— of Asiago, 99
Plateaus of Bernadia, 98
Platé Desert, 169
Plattenlandschaft, 231
Pli de Montpellier, 154, 158
Plitvica River, 41
— Lakes, 41
Pliva River, 51
Plumets, 181
Poches de dissolution, 141–143
Pock-marked karst, 28
Podkamennaya Tunguska River, 389, 407
Podpećka pećina, 36
Podvelebitski Canal, 44
Poiana Rusca Mountains, 341, 343
Pointed funnel-shaped dolines, 212
Pokrovka River, 409
Poland, 327–340

Poljak Cave, 58
Polje, 16, 47, 507, 510
Polje of Štikada, 55
Poljes, 5, 31, 37, 40, 509
Pollaccia Rise, 121
Pollino Massif, 111
Pollution, 72, 73, 501
Polyarnyy Range, 376
Polygenic karst, 181
Ponornice, 46
Ponors, 15, 35, 507, 509, 510
Pont d'Arc, 158
Ponte di Veia northern Cave, 95
Po Plain, 89
Popoli springs, 122
Popovo polje, 35, 49, 50
Portel Grotto, 161
Port Mion resurgence, 183
Port Royal Mountains, 445, 449
Postavarul Massif, 343
Postglacial Thermal Maximum, 237, 260
Post-Hercynian peneplain, 135
Postojna, 37
— Cave, 38
Postojnska jama, 35, 36, 37, 46
Pot holes, 430
Potočka Zijalka, 35, 36
Potosi Waterboil, 464
Pots, 172
Pozzatina doline, 103
Pozzolo-Bocca d'Orno, 101
Pranjčevići, 64
Pre-Alpine karst, 170
— — shields, 344, 347, 349
Pre-Alps, 93, 136, 167, 174, 177, 178, 228–230, 232, 251
Predjama, 35, 46
Predjamski grad, 36
Predkarpatskiy Trough, 366
Predkavkaz'ye, 368
Predural'skiy Trough, 359
Prehistoric question, 181
Prekarst, 140
Prekonoška pećina, 36
Pre-Laramide System, 449
Preluca Massif, 343
Pre-Senonian orogenic movements, 299
Prezid, 47
Prisayan–Baikal extension, 392
— — Kansk region, 389
Prisayansk region, 390
Prodanova pećina, 36
Propastva, 36
Provence, 131, 136, 168, 174, 178, 182, 183
— karst, 167, 176, 181
— Pre-Alps, 177

Prut River, 366
Psekha River Basin, 369
"Pseudokarst", 1
Psou River, 372
Puerto Rico, 11, 469, 514, 517
Pugnetto Cave, 95
Puisards, 143
Puits, 145
Puli, 104
Punta degli Stretti, 111
Purifications, 73
Putignano Cave, 95
P'yana River, 406
Pyarga River Basin, 409
Pyrenean Chain, 136
— phase, 31
— ranges, 161
Pyrenees 129, 160, 163

Queen of Spains Valley, 463, 464
Quellungshöhlen, 202, 203, 213
Quercy, 144
Quinzano Cave, 125

Racha Range, 372
Rachinsk Range, 372
Radavačka pećina, 36
Radobolja, 61
Raetikon, 225, 230
Ragas, 177
Raj Cave, 330, 335, 338
Rakhov Massif, 373
Rampsit, 440
Ranconi-Ceré area, 99
Rarău Massif, 343
Rascino Lake, 109
Rascles, 177, 178
Raša River, 45
Raška pećina, 36
Ratio of flow, 287
— of infiltration, 285
Ravanička pećina, 36
Rax, 225
Raxalpe, 229, 239, 241
Raxlandschaft, 229, 239
Raymond Gaché abyss, 105
Reculées, 149, 150
Reforestation, 75
Reg d'Argent, 182
Regular salt leaching, 214
Relative barriers, 67
Réseau de Courry-la Cocalière, 129
— de la Dent de Grolles, 129
— Trombe, 129

Réseaux cutanés, 137
Resurgent caves, 435
Retezat Mountains, 343
Revel Abyss, 95
Reyt River, 367
Rhenisch Schiefergebirge, 190, 192, 198, 199, 209, 210, 215–217
Rhodanian movement, 279
— phase, 31, 278
— tectonic movements, 316
Rhône River, 153, 182, 183
Ridge karsts, 350
Riseneis Cave, 231, 243, 244
Riff-Tell, 511
Rinnenkarren, 238
Rio Martino Cave, 105
Risovača, 36, 37
Riverhead Spring, 455
Rječina River, 45, 46, 62, 63
Roaring River, 458, 463
Rocca di Mezzo piano, 109
"Rock-milk", 405
Rocky Mountains, 471
Rodna Mountains, 343
Rodney Stoke, 440
Romanelli Cave, 95
Romualdo Cave, 35, 36
Roswell artesian basin, 468
Rovereto, 92
Rumania, 341–353
Rumanian Plain, 341, 345
Russia, 355
Russian Platform, 358–367
— Shield, 345
Rybinsk Depression, 389
Saalach River, 229
— — Valley, 244
Saarland, 190, 208
Sabini Mountains, 92
Sabrat cavern, 161
Saint-Pons, 135
Saint Sol Lacave Cave, 144
Sakhalin Island, 405
Saksagana River, 359
Salair, 408
— karst region, 402
Salento area, 123
— Peninsula, 91, 123
Saline karst, 213
Salt-diapir structure, 375
— karst, 351, 388, 394
— water encroachment, 44, 45
Salzach River, 229, 260
Samogradska pećina, 36
San Bernardino Maggiore Cave, 124
Šandalja, 35, 36

SUBJECT INDEX

San Canziano caves, 118, 119
San Giovanni Cave, 95, 124
— — d'Antro Cave, 95
— — di Domusnovas Cave, 112
— — di Duino, springs of, 119
— — di Su Anzu Cave, 124
— — Rotondo polje, 103
San Gottardo Cave, 101
San Marco in Lamis polje, 103
San Pellegrino Pass, 116
Santa Cruz Mountains, 449
Santa Fe River, 479
Sant'Egidio polje, 103
San Teodoro Cave, 95
Santo Stefano, 109
Santullo Well, 95, 110
San Zeno di Montagna, 107
Sapins d'argile, 181
Saratov-Ul'yanovsk Trough, 367
Sarayna River, 382
Sardinia, 85, 88, 90, 91, 93, 105, 112, 124
Sary-Chat River, 397
Sarydzhaz River, 397
Sarykamyshsk Basin, 387
Saturation with oxygen, 71
Sault karst, 161
Sava River, 46
Savian phase, 31, 299
Saxonian orogeny, 209
— Schiefergebirge, 200
Saxony, 190, 192
Sayan, 404
— Altay fold-mountain region, 407, 408
Schichtfugenkarren, 237
Schichttreppenkarst, 232, 427
Schlagerboden, 225
— area, 251
Schmelzbach, 255
Schneckenloch, 259
Schneealpe, 225, 229, 239–241
Schneeberg, 225, 229, 239, 241
— Alpen, 239, 258, 261
Schwarzer See, 242
Scialets, 170
Scientific interest, 180
Scotland, 436, 437
Scuttari Lake, 10, 46
Sebesh Mountains, 343, 345
Secondary permeability, 513, 516
Selvino Plateau, 106
Semriach-Peggau karst, 253
Sens region, 141
Serbia, 4, 27, 37, 52
Serbo-Macedonian Belt, 29
Serezha River, 406
Serle near Brescia, 98

Serre of Salento, 88–90, 101, 104
Sesia-Toce, 105
Seven River, 455
Severnaya Dvina River, 363
Shaft caves, 101
Shake holes, 430
Shakhdagsk Range, 372
Shallow karst, 9, 21, 218, 512
Shelterbelts, 75
Shezhimsk cave-shaft, 379
Shim lake, 363
Shurobsaya Basin, 399
Shuul'gan River, 382
Siberia, 358, 382, 401–405
Siberian Platform, 358–367, 376, 387–395, 405, 407
Sibillini Mountains, 110, 121, 122
Sicani, 112
Sicily, 85, 88, 90, 93, 105, 112, 115, 116, 125
Siegerland, 198
Sierra de los Organos, 11
Sikhote-Alin' folded region, 405
Silesia–Cracow upland, 327, 328, 332, 333, 337, 338
Silica karst plateau, 316, 317, 319
Silver Springs, 467
Sim River, 380
Simsk Trough, 378, 381
Sinjsko polje, 48
Sink, 440
Siphonbach, 256
Šipun, 36
Skadarsko jezero, 10, 46
Skalistyy Range, 370, 371
Skifsk Platform, 368
Skocyan caves, 118
Škocijanska jama, 36
Skrytyy Grotto, 405
Slatina karst, 320
Slovakia, 279–325
Slovakian karst, 5, 297, 303, 306, 308, 310, 315–319
— Ore Mountains, 306, 309, 315
Slovenia, 2, 35, 47, 61, 63
Slovenský kras, 298, 314, 316
— Raj, 297, 306, 308, 314
Small Plan de Canjuers, 178
Smeraldo Cave, 95
Smocza Cave, 338
Śniezna Cave, 330, 335
Soča River, 40
Sogno Cave, 95
Soils, 216, 492
Solotvinsk karst, 375
Solution process, 507, 510
Solutré Cave, 182

Sommaruga, 112
Sorgue River, 176
— Spring, 171
Sorrento Peninsula, 92
Soucis, 145
South Carolina, 472, 474, 479, 481
South Dakota, 473
Southern Alps, 86, 87, 89, 91, 98, 257, 258
Specific resistivity, 284
Špehovka, 35, 36
Spelaion Carso, 130
Speleogic Federation of Vaucluse, 176
Spelunca, 130
Spipola Cave, 95, 116, 124
Spiš, 322
— Gemer Ore Mountains, 305, 306
— nappe, 302
Spluga della Preta, 95
— — — Abyss, 100
Spluga Pass, 116
Sporadic karst, 313, 314
Sporožna, 36
Spring Bayou, 466
Springhead, 440
Sprofondi, 110
Sredneyuryuzansk karst, 381
St. Andrews Well Main Rising, 440
St. Ann River, 458
Stanovoy Range, 387
Stara Planina, 53
Stare Gliny Cave, 332
Staševica, 39
Stavropolsk upland, 372
St. Catherine River, 458
St. Cuthberts Swallet, 440
Steep-walled dolines, 212
Steinernes Meer, 225, 229, 244, 246
St. Elizabeth River, 458
Steppenheide, 211
Steyrer See, 241, 242
Štikadsko polje, 57
St. Lawrence Lowland, 495
— — Valley, 483
Stoanhaus pit, 99
Stolovaya River, 389
Stopića pećina, 36
Stratená Mountains, 303, 305
— upland, 302
Strášov Mountains, 303, 308, 309, 320
— nappes, 305, 306
— upland, 302
Strega Cave, 95
Strona Valley, 105
Structural-lithological character of karst, 313
— types of karst, 346
St. Thomas in the Vale, 458

Stura di Demonte, 116
Styrian karst, central, 252, 253
— movements, 276, 279
— phase, 31, 278, 299
Su Anzu Cave, 95, 112
Su Bentu Cave, 95, 112
Subterranean streams, traces of, 494
Sudetes, 328, 334
Sugana Valley, 108
Sugomaksk Cavern, 384
— Miassk karst, 384
Su Gorgovone Cave, 124
Sukhaya Sheleksna Basin, 409
Su Marmuri Cave, 95, 124
Sumgan Cavern, 382, 383
Surprise Pit, 493
Suwannee River, 478
Svetlaya Cavern, 382
Svilaja Mountains, 64
Swabia, 205, 206
Swabian Alps, 190, 215–217, 219
Swabian-Franconian Alps, 206, 207
Swago Creek, 494
Swallet caves, 435
Swallets, 434
Swallow holes, 16, 35
Swildon's Hole, 435, 440
Switzerland, 435
Syi River basins, 403
Sylva River, 366

Taborska jama, 36
Taconic Mountains, 483, 484
Tada River, 403
Tagilo-Verkhnechusovsk district, 384
Tagil River, 384
Tana che urla, 95, 111
Tana dell'Uomo Selvatico, 95, 111
Tana del Re Tiberio, 95
Tange gorge, 397
Tantal Cave, 236
Tanzerloch, 99
Tarana Basin, 399
Tarbagatay Range, 395
Tarn gorge, 150
— River, 146
Tarpon Springs, 469, 515
Taseyeva River Valley, 389
Tassare Cave, 95
Tatra National Park, 338
Tatras, 297, 298, 300, 303, 308, 320, 327–330, 333, 335–338
Tatrides, 299, 300, 303–305
Taunus, 198
Tavaran Lungo, 113

SUBJECT INDEX 549

Tavolara island, 112
Taymyr, 358
—— folded zone, 395
Techensk–Uvel'sk karst, 384
Tectonic setting, 276, 328–330, 344–349
Tekelik River canyon, 396
Temperature of waters, 71
Tennengebirge, 225, 229, 236
Tennessee, 471, 472, 481, 483–485, 487, 490–492, 494, 504
Tennessee Valley Authority dams, 487
Termenevka River, 409
Terminology, 507, 508
Terrace of San Giovanni Rotondo, 103
Tersk–Alatau Range, 396, 397
Tesha River, 406
Tethys, 26
Têt River, 135
Teutoburg Forest, 208, 209, 210
Texas, 468, 471, 473, 501, 503
Thuringia, 190
Thuringian Basin, 190, 200, 201, 202, 214
— Forest, 192, 200, 202, 205
— Schiefergebirge, 200
Tiber River, 108, 111
Timan, 407
— Ridge, 364, 365
Timavo resurgences, 124
— River, 98, 118–120
Tisovec karst, 298
Tissa Basin, 374, 375
Titova pećina near Drvar, 36
— — on Vis, 36
Toce Valley, 116
Toddeitto Cave, 95
Toirano Cave, 124
Tomsk–Kolyvansk karst province, 403
Tom's Spring, 464
Toros Mountains, 511
Torrione di Vallesinella Cave, 95, 108
Totes Gebirge, 225, 229, 237, 238, 241, 242, 258
Tourism, 182
Trans-Baikal region, 399, 400
Trans-Carpathian karst province, 374
Transdanubian Central Mountains, 267–269, 271, 276, 277, 279–284, 288–291, 294
Trascau Massif, 346
— Mountains, 342, 343, 350
Transylvanian Basin, 344
Trebiciano Cave, 97, 118
Trebišnjica River, 40, 50, 62, 65
Tre Cantoni Cave, 95
Trelawny, 457, 463, 464
Trémusa Cave, 95
Trento, 107
Treppo Carnico, 116

Treviso Plain, 113
Tribeč, 300
Trittkarren, 237
Troglav, 64
Trois Frères Cave, 161
Trompia Valley, 107
Tropical cone karst, 318
— karst, 310, 318, 398, 431
— Tertiary karst, 431
Trout Cave, 486
Tserik-Kyel Lake, 371
Tserkovnaya Mountain, 403
Tubića pećina, 36
Tumba e Nurai Abyss, 95
Tunguska Basin, 393
Turansk Lowland, 385
— Platform, 358, 385–387
Tura River, 384
Turbidity, 72, 73
Turga River, 400
Turgaysk Trough, 376
Turkestan Range, 398
Turňa karst, 308
Turnhole Bend, 497
Turukhansk karst, 388
Tuscan area, 87
Tuscany, 108, 116, 125
Tuvinsk Trough, 405
Two-cycles theory, 352
Tyan'Shan, 395, 396, 397, 399
Typological division of the karst, 312–323
Tyrolean Alps, 229
Typ River, 396

Ufa River, 366, 406
Ufimsk, 408
— Arch, 365
— Serginsk karst, 379
— Plateau, 356, 406, 407
Ugam River, 396
Ugol'ka Basin, 374
Ultratatrides, 299, 301, 305
Ultraveporides, 299, 305
Uluir River Basin, 409
Umbria, 108, 122, 124
Umbrian zone, 87
Una River, 62, 379
Underground flow, 405–411
— — coefficient, 410
Un'insk Cavern, 379
— –Podcheremsk karst, 379
United States, 467–505
Untersberg area, 229
Upper Cimmerian movements, 209
Upper Rhine Graben, 195

Upper Silesian Coal Basin, 328, 329
Uppony Block, 267, 280
Urals, 355, 376, 406–408, 511
— orogenic belt 358, 376–385
Ustyrtsk Plateau, 387
Ustyurt Cavity, 386
Us'va River, 379
Ušačka pećina, 36
Utah, 473
Uvala, 16, 507
Uvalas, 37, 40, 509
Uzornaya Cavern, 400

Vagran River, 382
Váh Valley, 322
Val Cavallina, 106
— Cellina caves, 95
— del Gesso, 122
— d'Era, 116
Valence, 153
Val Fredda, 99
Vallachian movements, 348
Valle Cupa, 110
Vallées absorbantes, 143
Valle Sella, 116
Vallestra Cave, 95
Valley and Ridge section, 481, 483, 487–490
"Vallone" of Doberdò, 95
Val Maira, 116
— Seriana, 107
Varano lake, 123
Varaždinske Toplice caves, 35, 36
Varese, 106
Vaucluse, 129
— fountain, 158, 171
— karst, 167
— Plateau, 174–176
— spring, 70, 175
Vaymuga River basin, 406
Važec karst, 321
Velebit, 32
Velichest-Vennaya shaft, 372
Velika and Mala Kapela, 55
— Karlovica, 35
— pećina, 36, 37
Velikaya River Basin, 361
—— Ugol'ka River, 374
Veliki Rumin, 59, 60
Velino Mountains, 109, 110
Vel'ka Fatra, 302, 303, 309, 320
—— Mountains, 300, 323
Vene Cave, 95
Venetian pre-Alps, 89, 93, 98, 115
Ventoux, 167
Veporides, 299–302, 305, 306

Verband Österreichischer Höhlenforscher, 226
Vercors, 129, 163, 170, 171, 175
— mass, 167
Vereinigung für Hydrogeologische Forschungen in Graz, 226
Verkhnebel'sk district, 382
Verkhnesimsk karst, 381
Verkhneural'sk karst, 384
Verkhnevishersk karst, 379
Verkhoyano Chukotsk region, 405
Verkinjica, 36
Vermont, 472
Vernaison River, 174
Verona Hills, 99
Verteba Cavern, 367
Vértes, 289
— Gerecse, 277
— Mountains, 284
Veternica, 36, 37
Vetrena Dupka, 36
Vette Plateau, 108
Vidourle River, 156
— Valley, 156
Vienna Basin, 229
Vienna Speleological Institute, 226
Viganti Cave, 95
Vilenica, 35, 36
Villacher Alpe, 225, 257, 258
Villanova caves, 98
Villány Block, 280
— Mountains, 267, 282, 291, 292
Vil'va River, 379
Vilyusk Basin, 393
Vindija, 35, 36
Vipacco River, 97, 119
Vipava River, 46
Virenque River, 157
Virghish Valley Basin, 343
Virginia, 484, 487, 495
Vis gorge, 154
— River, 150, 157, 158
— Valley, 156, 157
Vitebsk rapids, 361
Vitezić pećina, 36
Vitims Plateau, 400
Vitunj, 55
Vjetrenica, 35, 36
Vocontian region, 168
— Trough, 174
Void volume, 287
Volga Basin, 407
— -Ural Arch, 365, 366
Volkhov River, 360
Volturno Valley, 108
Vore, 104
Vorondovskiy Cavern, 372

SUBJECT INDEX

Vosges, 133
Vostechnaya Khosta River, 372
Voutes mouillantes, 180
Vrbas River, 62
Vrbovsko–Jesenica anticline, 56
Vrelo Cave, 39
Vrlovka, 36
Vrnjača, 36
Vrulja at Donja Brela, 44
Vruljas, 44
Vrutak, 44
Vulcan Mountains, 343, 353
Vyatsk swell, 365
Vym' River Basin, 365
Vysoká, 300
Vysoká nappes, 300
Vysoké Tatry, 300

Wagwater Belt, 445
Wales, 419, 424, 425, 435–437
Warta River, 337
Waschberg, 225
Washington, 470
Water-absorbing capacity, 286
— level gradient, 476
— supply, 182, 261, 439, 440
— table, 513
Weidenbach, 219
Werra Basin, 214
Westbury Main, 440
Western Alps, 86, 87, 89, 91, 92, 105, 115
Westerwald, 198
West Virginia, 483, 485, 486, 488–491, 493–495
Wettersteingebirge, 229
White Sea, 362
Whitings Neck, 485
Wieliczka salt mine, 336
Windbreaks, 75
Windows, 483
Windsor River, 463
Wooded karst, 9
Wookey Hole, 434, 435, 440

Würm spring, 250
Wyoming, 473

Yangel'ka River, 384
Yayvo–Kos'vinsk karst, 379
Yemtsa River, 406, 409
Yenisei Ridge, 407, 408
— River, 387, 404
Yettykyz Cavern, 398
York Dam Spring, 461
Yorkshire, 419–421, 425–428, 431, 438, 441
Y.S. River, 465
Ystradfellte district, 437
Yucatan, 11
Yugoslavia, 25–83, 129, 276, 442, 467, 517
— inner region, 52
Yuryuzan River, 380
Yvillers, 141

Zabaykal'ye, 399
Zagorska Mrežnica River, 55, 57
Zangezursk Range, 372
Zaozernyy Grotto, 405
Zapadnaya Dvina River, 361, 362
Zay River, 365
Zechstein depression, 202
Zeravshan River, 397
Zeravshansk Range, 396, 398
Zeta River, 62, 66
Zeysk-Bureinsk Platform, 358
Zheltaya River, 359
Žiar Mountain, 309
Zimna Cave, 335
Zinzulusa Cave, 93, 95, 104, 124
Zlotska pećina, 36
Zocca d'As Cave, 95
Zrmanja River, 40, 45, 46, 62
Žrnovnica, 44
Žumberak, 276
Zvolen Basin, 308, 320, 321